Progress in Mathematics

SADLIER-OXFORD

Rose Anita McDonnell

Catherine D. LeTourneau

Anne Veronica Burrows

Anne Brigid Gallagher

Francis H. Murphy

M. Winifred Kelly

with

Dr. Elinor R. Ford

Series Consultants

Tim Mason
Math Specialist
Palm Beach County School District
West Palm Beach, FL

Margaret Mary Bell, S.H.C.J., Ph.D.
Director, Teacher Certification
Rosemont College
Rosemont, PA

Dennis W. Nelson, Ed.D.
Director of Basic Skills
Mesa Public Schools
Mesa, AZ

Sadlier-Oxford
A Division of William H. Sadlier, Inc.

The publisher wishes to thank the following teachers and administrators, who read portions of the program prior to publication, for their comments and suggestions.

Mrs. Maria Bono
Whitestone, NY

Mrs. Jennifer Fife
Yardley, PA

Ms. Donna Violi
Melbourne, FL

Sr. Lynn Roebert
Covina, CA

Ms. Anna Cano-Amato
Brooklyn, NY

Mr. Galen Chappelle
Los Angeles, CA

Mrs. Ana M. Rodriguez
Miami, FL

Sr. Ruthanne Gypalo
East Rockaway, NY

Sr. Anita O'Dwyer
North Arlington, NJ

Mrs. Madonna Atwood
Creve Coeur, MO

Mrs. Marlene Kitrosser
Bronx, NY

Acknowledgments

Every good faith effort has been made to locate the owners of copyrighted material to arrange permission to reprint selections. In several cases this has proved impossible. The publisher will be pleased to consider necessary adjustments in future printings.

Thanks to the following for permission to reprint the copyrighted materials listed below.

"A Dividend Opinion" (text only) by John Ciardi from MUMMY TOOK COOKING LESSONS by John Ciardi. Text copyrighted © 1990 by Judith C. Ciardi. Reprinted by permission of Houghton Mifflin Co. All rights reserved.

"Compass" (text only) by Georgia Heard.

Excerpt from *Counting on Frank* (text only), written and illustrated by Rod Clement, published by Gareth Stevens, Inc., Milwaukee, Wisconsin.

"Joan Benoit: 1984 U.S. Olympic Marathon Gold Medalist" (text only by Rina Ferrarelli. Copyright Rina Ferrarelli. First appeared in MORE GOLDEN APPLES (Papier Mache Press, 1986).

"Lunch Time" (text only) by Lee Bennett Hopkins in FOOD FIGHT, POETS JOIN THE FIGHT AGAINST HUNGER WITH POEMS TO FAVORITE FOODS by Micael J. Rosen, copyright © 1996 by Lee Bennett Hopkins, reprinted by permission of Harcourt Brace & Company.

Excerpt from MATH CURSE (text only) by Jon Scieszka. Copyright © 1995 by Jon Scieszka. Used by permission of Viking Penguin, a division of Penguin Putnam Inc.

Excerpt from MATH FOR SMARTY PANTS (text only) by Marilyn Burns. Published by Little, Brown and Company.

Excertp from "The Old Math. One." (text only) by Arnold Adoff from CHOCOLATE DREAMS by Arnold Adoff. Copyright © 1989 by Arnold Adoff. By permission of Lothrop, Lee and Shepard Books, a division of William Morrow & Company, Inc.

"Praise Song for a Drummer" (text only) translated by Mary Smith.

"Sky" (text only) copyright © 1997 by Lee Bennett Hopkins. Now appears in MARVELOUS MATH published by Simon & Schuster. Reprinted by permission of Curtis Brown, Ltd.

"Some Opposites" (text only) by Richard Wilbur. "Opposites 8, 12, 18, 22, and 39" from OPPOSITES: POEMS AND DRAWINGS, copyright © 1973 by Richard Wilbur, reprinted by permission of Harcourt Brace & Company.

"Symmetry" (text only) by Du Fu from MAPLES IN THE MIST:Children's Poems of the Tang Dynasty. Translated by Minforng Ho. Copyright © 1996 by Minfong Ho. By permission of Lothrop, Lee & Shepard Books, a division of William Morrow & Company, Inc.

"To Build A House" (text only) by Lillian M. Fisher. Used by permission of the author, who controls all rights.

Excerpt "Where is math in dinnertime?" (text only) reprinted with the permission of Simon & Schuster Books for Young Readers, an imprint of Simon & Schuster Children's Publishing Division, from MATH IN THE BATH by Sara Atherlay. Text copyright © 1995 Sara Atherlay.

Anastasia Suen, Literature Consultant

Photo Credits

Math manipulatives supplied by ETA.
Diane J. Ali: 142, 173.
Cate Photography: 7, 8, 82, 204, 237, 316, 380.
Corbis: 418.
Leo deWys: 112, 143.
Neal Farris: 38–39, 456.
Kathy Ferguson: 320.
Digital Stock: 492.
FPG/Telegraph Colour Library: 56; Elizabeth Simpson: 228.
The Image Bank/Elle Schuster: 50;

Luis Veiga: 78–19; Francesco Ruggeri: 98; M. Tcherevkoff: 105, 190; Gary Gay: 52; Steven Hunt: 150.
The Image Works/B. Daemmrich: 32; Larry Kolvoord: 364.
Ken Karp: 117.
Greg Lord: 103.
Clay Patrick McBride: 23, 37, 63, 93, 121, 138, 149, 161, 162, 201, 225, 257, 260, 282, 295, 296, 331, 332–333, 334, 375, 411, 424, 441, 446, 473.

NASA: 178.
Richard & Amy Hutchings: 12, 24, 46, 74, 128–129, 66.
Parish Kohanim: 411 background.
Richard Levine: 87.
The Stock Market/Chris Hamilton: 39; Michal Heron: 100; Myron J. Dorf: 125; Jose L. Palaez: 214, 336; Kunio Owaki: 266; Mugshots: 474.
Tony Stone Images/Phillipp Englehorn: 100; Robert Stahl: 37, 140; Hulton Getty: 68; Cameron Hewet: 48–49; Tom Bean: 81; Bob

Krist: 103, 450; Peter Dean: 104; Robert Frerk: 108; Lori Adamski Peek: 124; Janet Gill: 126; Mary Kate Denny: 134; Trevor Mein: 298; Bertrand Rieger: 352–353; Michael Orton: 375; Kim Heacox: 442; David Young-Wolff: 428–429; Ian Shaw: 453; Mark Lewis: 478; Christopher Bissell: 242; Hans Strands: 258; Anthony Blake: 382; Stewart Cohen: 484; Stephen Frink: 490; Uniphoto: 262; Daemmrich: 272.

Illustrators

Diane Ali
Bob Berry
Robert Burger

Pasquale Fusco
Adam Gordon
Batelman Illustration

Dave Jonason
Robin Kachantones
Bea Leute

Blaine Martin
Wendy Pierson
Fernando Rangel

Contents

✱ **Algebraic Reasoning**

✶ Algebraic Reasoning

* Algebraic Reasoning

✳ **Algebraic Reasoning**

CHAPTER 9
Geometry

＊ Algebraic Reasoning

CHAPTER 10
Measurement

* **Algebraic Reasoning**

Progress in Mathematics

Whether you realize it or not, you see and use mathematics every day!

The lessons and activities in this textbook will help you enjoy mathematics *and* become a better mathematician.

This year you will build on the mathematical skills you already know, as you explore *new* ideas. Working in groups, you will solve problems using many different strategies. You will also have opportunities to make up your own problems and to keep a log of what you discover about math in your own personal Journal. You will compute with fractions, decimals, and integers. You will learn more about algebra, geometry, measurement, probability, statistics, proportion, percent, and percent applications.

You will become a *Technowiz* by completing the computer and calculator lessons and activities. These not only will teach you valuable skills but will also expose you to how technology is used in different situations by many people.

Throughout the year, you can use the Skills Update section at the beginning of this book to sharpen and review any skills you need to brush up on.

We hope that as you work through this program you will become aware of how mathematics really is a *big* part of your life.

An Introduction to Skills Update

Progress in Mathematics includes a "handbook" of essential skills, Skills Update, at the beginning of the text. These one-page lessons review skills you learned in previous years. It is important for you to know this content so that you can succeed in math this year.

If you need to review a concept in Skills Update, your teacher can work with you, using ideas from the Teacher Edition. You can practice the skill using manipulatives, which will help you understand the concept better.

Your class may choose to do these one-page lessons at the beginning of the year so that you and your teacher can assess your understanding of these previously learned skills. Or you may choose to use Skills Update as a handbook throughout the year. Many lessons in your textbook refer to a particular page in the Skills Update. This means you can use that Skills Update lesson at the beginning of your math class as a warm-up activity. You may even want to practice those skills at home.

If you need more practice than what is provided on the Skills Update page, you can use exercises in the *Skills Update Practice Book*. It has an abundance of exercises for each lesson.

Place Value

The value of a digit (0, 1, 2, . . . , 9) in a number depends on its position, or **place**, in the number. Each place is 10 times the value of the next place to its right. Each **period** contains 3 digits.

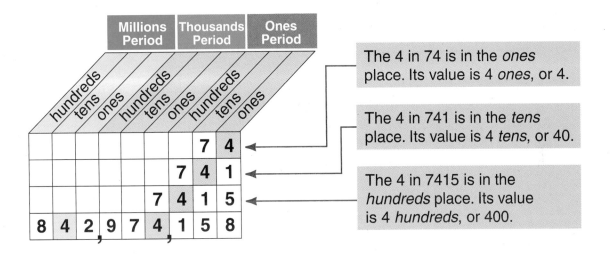

The 4 in 74 is in the *ones* place. Its value is 4 *ones*, or 4.

The 4 in 741 is in the *tens* place. Its value is 4 *tens*, or 40.

The 4 in 7415 is in the *hundreds* place. Its value is 4 *hundreds*, or 400.

In 842,974,158 each 4 has a different value.

4 = 4 ten millions, or 40,000,000

4 = 4 thousands, or 4000 or 4,000 ← Four-digit numbers may be written with or without a comma.

Name the period of the underlined digits.

1. <u>943</u>,862
2. <u>802</u>,400,253
3. 603,411,<u>218</u>

4. 740,<u>200</u>,335
5. <u>41</u>,252,231
6. <u>9</u>,527,000

Write the place of the underlined digit. Then write its value.

7. 7<u>3</u>
8. 1<u>2</u>3
9. 3<u>4</u>25

10. 753,<u>6</u>41
11. 6,<u>4</u>23,728
12. 3<u>6</u>,250

13. <u>8</u>31,429,678
14. 64<u>8</u>,279,240
15. <u>2</u>4,983,402

Numeration II

Comparing and Ordering

Compare the number of acres on Assateague Island with the number of acres on Cumberland Island.

National Seashores	Acreage
Assateague Island (MD-VA)	39,630
Fire Island (NY)	19,578
Padre Island (TX)	130,434
Cumberland Island (GA)	36,415

39,630
36,415 $3 = 3$

39,630
36,415 $9 > 6$

39,630 > 36,415   or   36,415 < 39,630

▶ **To order whole numbers:**

- Align the digits by place value.

- Compare the digits in each place, starting with the greatest place.

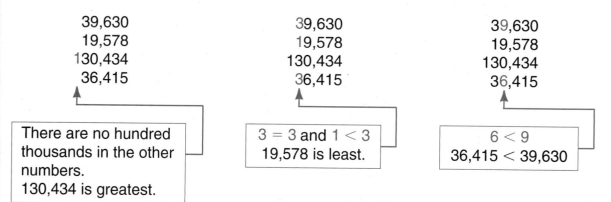

39,630	39,630	39,630
19,578	19,578	19,578
130,434	130,434	130,434
36,415	36,415	36,415

There are no hundred thousands in the other numbers.
130,434 is greatest.

$3 = 3$ and $1 < 3$
19,578 is least.

$6 < 9$
36,415 < 39,630

In order from greatest to least, the numbers are:
130,434; 39,630; 36,415; 19,578.

Write in order from greatest to least.

1. 9996; 999; 10,000; 9997

2. 32,423; 38,972; 36,401; 31,276

3. 8472; 8927; 8352; 8179

4. 600,201; 599,763; 601,501; 600,202

5. 9878; 1,072,000; 99,869

6. 896,387; 812,000; 99,785

Rounding Whole Numbers

The population of Midway is 83,524. Since populations change frequently, a **rounded number** may be used instead of the exact number.

▶ **To round a number to a given place:**

- Find the place you are rounding to.

- Look at the digit to its right.
  If the digit is *less than 5*, round *down*.
  If the digit is *5 or more*, round *up*.

Round 83,524 to the nearest ten.

| 83,524 ↓ 83,520 | The digit to the right is 4. 4 < 5 Round down to 83,520. |

83,524

83,520 ———●——— 83,530

Round 83,524 to the nearest hundred.

| 83,524 ↓ 83,500 | The digit to the right is 2. 2 < 5 Round down to 83,500. |

83,524

83,500 ——●———— 83,600

Round 83,524 to the nearest thousand.

| 83,524 ↓ 84,000 | The digit to the right is 5. 5 = 5 Round up to 84,000. |

83,524

83,000 ————●—— 84,000

Round each to the nearest ten, hundred, and thousand.
Use a number line to help you.

1. 6709　　　**2.** 1256　　　**3.** 7893　　　**4.** 5649　　　**5.** 42,314

6. 11,987　　**7.** 49,678　　**8.** 76,432　　**9.** 148,786　　**10.** 940,067

Numeration IV

Factors and Multiples

▶ **Factors** are numbers that are multiplied to find a product.

$$8 \times 3 = 24$$
factors

$$4 \times 2 \times 3 = 24$$
factors

To find all the factors of a number, use multiplication sentences.

Find all the factors of 20.

$$5 \times 4 = 20$$
$$10 \times 2 = 20$$
$$20 \times 1 = 20$$

Factors of 20: 1, 2, 4, 5, 10, 20

▶ The **multiples** of a number are the products of that number and any whole number.

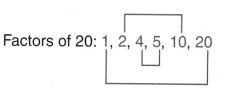

| $3 \times 0 = 0$ | $3 \times 4 = 12$ | $3 \times 8 = 24$ |

Multiples of 3: 0, 3, 6, 9, 12, 15, 18, 21, 24, . . .

Multiples of 6: 0, 6, 12, 18, 24, 30, 36, 42, 48, . . .

Multiples of 8: 0, 8, 16, 24, 32, 40, 48, 56, 64, . . .

List all the factors of each number.

1. 12 2. 18 3. 22 4. 35 5. 7

6. 13 7. 9 8. 60 9. 72 10. 100

11. 90 12. 57 13. 73 14. 61 15. 108

List the first ten nonzero multiples of each number.

16. 4 17. 5 18. 6 19. 7 20. 8 21. 9

22. 10 23. 11 24. 12 25. 15 26. 20 27. 25

4

Divisibility: 2, 5, 10

A number is **divisible** by another number when you divide
and the remainder is zero.

$14 \div 2 = 7$ 14 *is* divisible by 2.
$15 \div 2 = 7$ R1   15 *is not* divisible by 2.

▶ **Divisibility by 2**
   A number is divisible by 2 if its ones digit is divisible by 2.

   80, 32, 294, 856, and 1908 are divisible by 2.

   > All even numbers are divisible by 2.

▶ **Divisibility by 5**
   A number is divisible by 5 if its ones digit is 0 or 5.

   60, 225, 400, 1240, and 125,605 are divisible by 5.

▶ **Divisibility by 10**
   A number is divisible by 10 if its ones digit is 0.

   40, 280, 500, 2070, and 46,790 are divisible by 10.

Which numbers are divisible by 2? by 5? by 10?

1. 300	**2.** 270	**3.** 315	**4.** 455
5. 7875	**6.** 9730	**7.** 6660	**8.** 6584
9. 22,892	**10.** 16,710	**11.** 195,000	**12.** 360,000

True or false? If false, give an example that shows it is false.

13. All numbers divisible by 5 are divisible by 10.

14. All numbers divisible by 10 are divisible by 2 and by 5.

15. All even numbers are divisible by 10.

16. All numbers divisible by 25 are divisible by 5.

Numeration VI

Decimals to Hundredths

The value of a digit in a decimal depends on its position, or **place**, in the decimal. Each place is 10 times the value of the next place to its right.

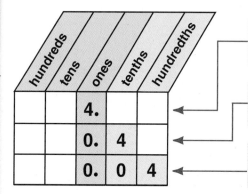

The 4 is in the *ones* place.
Its value is 4 *ones*, or 4.

The 4 in 0.4 is in the *tenths* place.
Its value is 4 *tenths*, or 0.4.

The 4 in 0.04 is in the *hundredths* place.
Its value is 4 *hundredths*, or 0.04.

▶ **To read a decimal less than 1:**

- Start at the decimal point.

- Read the number as a whole number. Then say the name of the place.

0.92

hundredths

Read: ninety-two hundredths

Study this example.

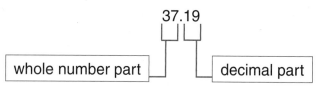

37.19

whole number part   decimal part

Read: thirty-seven *and* nineteen hundredths

Read each decimal. Then write the place of the underlined digit and its value.

1. 0.<u>8</u>
2. 0.0<u>2</u>
3. 0.1<u>3</u>
4. <u>5</u>.6

5. 7.<u>1</u>
6. 0.<u>4</u>5
7. 9.6<u>3</u>
8. 10.<u>0</u>1

9. 42.7<u>8</u>
10. <u>2</u>6.9
11. <u>3</u>00.09
12. 1<u>5</u>6.8

Adding Whole Numbers

Add: 8164 + 4676 = __?__

Estimate: 8000 + 5000 = 13,000

▶ **To add whole numbers:**

- Align the numbers.

- Add, starting with the ones. Regroup when necessary.

Add the ones. Regroup.	Add the tens. Regroup.	Add the hundreds.	Add the thousands. Regroup.
1 8 1 6 4 + 4 6 7 6 0	1 1 8 1 6 4 + 4 6 7 6 4 0	1 1 8 1 6 4 + 4 6 7 6 8 4 0	1 1 8 1 6 4 + 4 6 7 6 1 2,8 4 0
10 ones = 1 ten 0 ones	14 tens = 1 hundred 4 tens		12 thousands = 1 ten thousand 2 thousands

The sum is 12,840.

12,840 is close to the estimate of 13,000.

Estimate. Then add and check.

1. 536 + 143	**2.** 749 + 250	**3.** 1578 + 6421	**4.** 3664 + 6323	**5.** 1768 + 63

6. 2369 + 48	**7.** 17,243 + 13,963	**8.** 12,872 + 46,931	**9.** 567,892 + 132,104	**10.** 467,893 + 413,540

Align and estimate. Then add.

11. 5751 + 756   **12.** 63,689 + 2468   **13.** 63,620 + 76,237

Operations II

Subtracting Whole Numbers

Subtract: 4816 − 1932 = __?__

Estimate: 5000 − 2000 = 3000

▶ **To subtract whole numbers:**

- Align the numbers.

- Subtract, starting with the ones. Regroup when necessary.

Subtract the ones.	More tens needed. Regroup. Subtract.	More hundreds needed. Regroup. Subtract.	Subtract the thousands.
4 8 1 6 − 1 9 3 2 4	7 11 4 8 $\cancel{1}$ 6 − 1 9 3 2 8 4	17 3 $\cancel{7}$ 11 $\cancel{4}$ $\cancel{8}$ $\cancel{1}$ 6 − 1 9 3 2 8 8 4	17 3 $\cancel{7}$ 11 $\cancel{4}$ $\cancel{8}$ $\cancel{1}$ 6 − 1 9 3 2 2 8 8 4

8 hundreds 1 ten = 7 hundreds 11 tens

4 thousands 7 hundreds = 3 thousands 17 hundreds

The difference is 2884.

2884 is close to the estimate of 3000.

Estimate. Then subtract and check.

1. 489 − 366	**2.** 818 − 405	**3.** 5969 − 1867	**4.** 5461 − 3320	**5.** 3598 − 679
6. 6244 − 29	**7.** 36,243 − 13,963	**8.** 74,112 − 53,921	**9.** 456,781 − 179,660	**10.** 587,893 − 498,721

Align and estimate. Then subtract.

11. 6862 − 867   **12.** 42,578 − 1579   **13.** 74,731 − 65,126

8

Adding Decimals to Hundredths

Add: 0.44 + 0.3 + 0.85 = __?__

Estimate: 0.4 + 0.3 + 0.9 = 1.6

▶ **To add decimals:**

- Line up the decimal points.
- Add as with whole numbers.
- Write the decimal point in the sum.

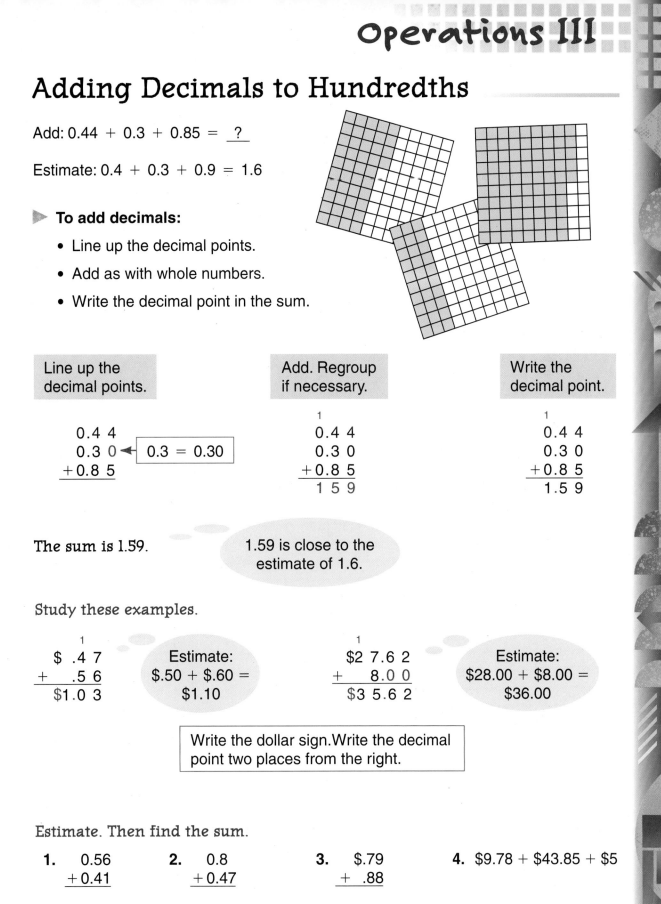

Line up the decimal points.	Add. Regroup if necessary.	Write the decimal point.
0.4 4 0.3 0 ◄ 0.3 = 0.30 +0.8 5	¹ 0.4 4 0.3 0 +0.8 5 1 5 9	¹ 0.4 4 0.3 0 +0.8 5 1.5 9

The sum is 1.59.

1.59 is close to the estimate of 1.6.

Study these examples.

$$\begin{array}{r} {\scriptstyle 1} \\ \$\ .4\ 7 \\ +\quad .5\ 6 \\ \hline \$1.0\ 3 \end{array}$$

Estimate:
$.50 + $.60 = $1.10

$$\begin{array}{r} {\scriptstyle 1} \\ \$2\ 7.6\ 2 \\ +\quad\ 8.0\ 0 \\ \hline \$3\ 5.6\ 2 \end{array}$$

Estimate:
$28.00 + $8.00 = $36.00

Write the dollar sign. Write the decimal point two places from the right.

Estimate. Then find the sum.

1. 0.56
      +0.41

2. 0.8
      +0.47

3. $.79
      + .88

4. $9.78 + $43.85 + $5

Operations IV

Subtracting Decimals to Hundredths

Subtract: 0.7 − 0.46 = ___?___

Estimate: 0.7 − 0.5 = 0.2

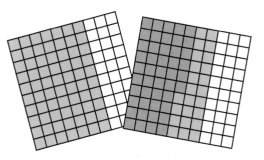

▶ **To subtract decimals:**

- Line up the decimal points.
- Subtract as with whole numbers.
- Write the decimal point in the difference.

Line up the decimal points.	Subtract. Regroup if necessary.	Write the decimal point.
$\begin{array}{r} 0.7\,0 \\ -0.4\,6 \end{array}$ ◀ 0.7 = 0.70	$\begin{array}{r} ^{6\ \ 10} \\ 0.\not7\,\not0 \\ -0.4\,6 \\ \hline 0\,2\,4 \end{array}$	$\begin{array}{r} ^{6\ \ 10} \\ 0.\not7\,\not0 \\ -0.4\,6 \\ \hline 0.2\,4 \end{array}$

The difference is 0.24.

0.24 is close to the estimate of 0.2.

Study these examples.

$\begin{array}{r} ^{7\ \ 13} \\ \$.\not8\,\not3 \\ -\ \ .4\,9 \\ \hline \$.3\,4 \end{array}$

Estimate:
$.80 − $.50 =
$.30

$\begin{array}{r} ^{7\ \ 16} \\ \$5\,\not8.\not6\,4 \\ -\ \ \ \ 5.8\,3 \\ \hline \$5\,2.8\,1 \end{array}$

Estimate:
$59.00 − $6.00 =
$53.00

Write the dollar sign. Write the decimal point two places from the right.

Estimate. Then find the difference.

	1.	2.	3.	4.	5.
	0.74	0.81	$.95	$14.97	0.8 − 0.29
	− 0.39	− 0.6	− .59	− 10.49	

Multiplying Whole Numbers: One Digit

Multiply: $7 \times 27 = \underline{\ ?\ }$

Estimate 7×27.
Round 27 to its greatest place.

$$7 \times 27$$
$$\downarrow \quad\ \downarrow$$
$$7 \times 30 = 210$$

▶ **To multiply by a one-digit number:**

- Multiply the ones. Regroup if necessary.

- Multiply the tens. Add and regroup if necessary.

- Continue to multiply, add, and regroup in the remaining places.

Multiply the ones by 7. Then regroup.	Multiply the tens by 7. Then add the 4 tens.

$$\begin{array}{r} \overset{4}{} \\ 2\ 7 \\ \times\ \ 7 \\ \hline 9 \end{array}$$

$$\begin{array}{r} \overset{4}{} \\ 2\ 7 \\ \times\ \ \ 7 \\ \hline 1\ 8\ 9 \end{array}$$

The product is 189.

189 is close to the estimate of 210.

Study these examples.

$$\begin{array}{r} \overset{3\ \ 2}{} \\ 5\ 6\ 4 \\ \times\ \ \ \ \ 5 \\ \hline 2\ 8\ 2\ 0 \end{array}$$

Estimate:
$5 \times 600 = 3000$

$$\begin{array}{r} \overset{5\ \ 3}{} \\ \$\ 7.8\ 5 \\ \times\ \ \ \ \ \ \ 6 \\ \hline \$\ 4\ 7.1\ 0 \end{array}$$

Estimate:
$6 \times \$8 = \$48$

Write the dollar sign. Write the decimal point two places from the right.

Estimate. Then find the product.

1. $\begin{array}{r} 55 \\ \times\ 6 \\ \hline \end{array}$

2. $\begin{array}{r} 95 \\ \times\ 4 \\ \hline \end{array}$

3. $\begin{array}{r} 613 \\ \times\ \ 9 \\ \hline \end{array}$

4. $8 \times \$.72$

5. $7 \times \$8.64$

Operations VI

Multiplying Whole Numbers: Two Digits

Multiply: $32 \times 46 = \underline{\ ?\ }$

Estimate: $30 \times 50 = 1500$

▶ **To multiply by a two-digit number:**

- Multiply by the ones.
- Multiply by the tens.
- Add the partial products.

Multiply by 2 ones.	Multiply by 3 tens.	Add the partial products.

$$
\begin{array}{r}
4\ 6 \\
\times 3\ 2 \\
\hline
9\ 2
\end{array}
\leftarrow 2 \times 46
$$

$$
\begin{array}{r}
4\ 6 \\
\times 3\ 2 \\
\hline
9\ 2 \\
1\ 3\ 8\ 0
\end{array}
\leftarrow 30 \times 46
$$

You may omit this 0.

$$
\begin{array}{r}
4\ 6 \\
\times 3\ 2 \\
\hline
9\ 2 \\
+\ 1\ 3\ 8\ 0 \\
\hline
1\ 4\ 7\ 2
\end{array}
$$

partial products

The product is 1472.

1472 is close to the estimate of 1500.

Study these examples.

$$
\begin{array}{r}
6\ 4\ 2 \\
\times\ \ \ 4\ 1 \\
\hline
6\ 4\ 2 \\
2\ 5\ 6\ 8 \\
\hline
2\ 6,3\ 2\ 2
\end{array}
$$

Estimate:
$40 \times 600 = 24{,}000$

$$
\begin{array}{r}
\$3.7\ 5 \\
\times\ \ \ \ 6\ 2 \\
\hline
7\ 5\ 0 \\
2\ 2\ 5\ 0 \\
\hline
\$2\ 3\ 2.5\ 0
\end{array}
$$

Estimate:
$60 \times \$4 = \$240$

Write the dollar sign. Write the decimal point two places from the right.

Estimate. Then find the product.

1. $\begin{array}{r} 67 \\ \times 34 \\ \hline \end{array}$ **2.** $\begin{array}{r} 56 \\ \times 78 \\ \hline \end{array}$ **3.** $\begin{array}{r} 329 \\ \times\ \ 43 \\ \hline \end{array}$ **4.** $64 \times \$.36$ **5.** $92 \times \$7.68$

Trial Quotients

Divide: 2183 ÷ 46 = __?__

Follow these steps to divide:

- *Decide* where to begin the quotient.

$$4\ 6\overline{)2\ 1\ 8\ 3} \qquad 46 < 218$$

> The quotient begins in the tens place.

- *Estimate.*

Think: ④6$\overline{)②①8\ 3}$ ⟶ 4 × __?__ = 21 ⟶ Try 5.

- *Divide.*

- *Multiply:* 5 × 46 = 230

$$\begin{array}{r} 5 \\ 4\ 6\overline{)2\ 1\ 8\ 3} \\ -2\ 3\ 0 \end{array}$$

> The digit used in the quotient is too large.

- *Subtract* and *compare* remainder with divisor.

$$\begin{array}{r} 4 \\ 4\ 6\overline{)2\ 1\ 8\ 3} \\ -1\ 8\ 4 \\ \hline 3\ 4 \end{array}$$

Try 4.

34 < 46

- *Bring down* the next digit from the dividend and repeat the steps.

$$\begin{array}{r} 4\ 7 = 47\ \ R21 \\ 4\ 6\overline{)2\ 1\ 8\ 3} \\ -1\ 8\ 4\downarrow \\ \hline 3\ 4\ 3 \\ -3\ 2\ 2 \\ \hline 2\ 1 \end{array}$$

21 < 46

- *Check.*

46 × 47 = 2162   2162 + 21 = 2183

Estimate to find the missing digit in the quotient.
Complete the division.

$$\begin{array}{r} 8? \\ \textbf{1.}\ 49\overline{)4018} \\ -392 \\ \hline 98 \end{array}$$

$$\begin{array}{r} 7? \\ \textbf{2.}\ 67\overline{)5226} \\ -469 \\ \hline 536 \end{array}$$

$$\begin{array}{r} 3? \\ \textbf{3.}\ 65\overline{)2573} \\ -195 \\ \hline 623 \end{array}$$

$$\begin{array}{r} 4? \\ \textbf{4.}\ 27\overline{)1234} \\ -108 \\ \hline 154 \end{array}$$

13

Dividing Whole Numbers

Divide: 4782 ÷ 83 = __?__

Estimate: 4800 ÷ 80 = 60

$4782 \div 83$

Decide where to begin the quotient.

$8\,3\overline{)4\ 7\ 8\ 2}$ 83 > 47

$8\,3\overline{)4\ 7\ 8\ 2}$ 83 < 478

The quotient begins in the tens place.

Divide the tens.	Divide the ones.	Check.

Divide the tens.

```
        5
8 3)4 7 8 2
  - 4 1 5
      6 3
```

Divide the ones.

```
      5 7 R 5 1
8 3)4 7 8 2
  - 4 1 5 ↓
      6 3 2
    - 5 8 1
        5 1
```

Check.

```
      5 7
    × 8 3
    1 7 1
    4 5 6
    4 7 3 1
  +     5 1
    4 7 8 2
```

The quotient is 57 R51.

57 R51 is close to the estimate of 60.

Study these examples.

```
    $.2 8
3)$.8 4
  - 6
    2 4
  - 2 4
```

Estimate: $.90 ÷ 3 = $.30

```
      $.1 7
23)$ 3.9 1
  - 2 3
    1 6 1
  - 1 6 1
```

Estimate: $4.00 ÷ 20 = $.20

Estimate. Then find the quotient.

1. $24\overline{)522}$ **2.** $45\overline{)3268}$ **3.** $79\overline{)5576}$ **4.** $65\overline{)\$9.10}$

5. $38\overline{)1589}$ **6.** $17\overline{)1634}$ **7.** $59\overline{)4267}$ **8.** $19\overline{)\$18.24}$

Adding Fractions: Like Denominators

▶ **To add fractions with *like* denominators:**

- Add the numerators.

$$\frac{2}{8} + \frac{4}{8} = \frac{2+4}{8}$$

- Write the result over the common denominator.

$$= \frac{6}{8}$$

- Express the sum in simplest form.

$$\frac{6}{8} = \frac{6 \div 2}{8 \div 2} = \frac{3}{4}$$ ◄  Divide $\frac{6}{8}$ by $\frac{2}{2}$.

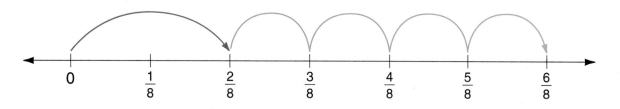

Study these examples.

$$
\begin{array}{r}
\frac{1}{12} \\
+\frac{7}{12} \\
\hline
\frac{8}{12} = \frac{8 \div 4}{12 \div 4} = \frac{2}{3}
\end{array}
$$

$$
\begin{array}{r}
\frac{7}{8} \\
+\frac{3}{8} \\
\hline
\frac{10}{8} = 1\frac{2}{8} = 1\frac{1}{4}
\end{array}
$$ ◄ Rename $\frac{10}{8}$ as a mixed number.

Add the fractions. Write each answer in simplest form.
You may use models or a number line to check your work.

1. $\begin{array}{r}\frac{3}{5}\\+\frac{1}{5}\\\hline\end{array}$

2. $\begin{array}{r}\frac{2}{3}\\+\frac{1}{3}\\\hline\end{array}$

3. $\begin{array}{r}\frac{1}{8}\\+\frac{3}{8}\\\hline\end{array}$

4. $\begin{array}{r}\frac{3}{4}\\+\frac{3}{4}\\\hline\end{array}$

5. $\begin{array}{r}\frac{5}{9}\\+\frac{1}{9}\\\hline\end{array}$

6. $\frac{8}{9} + \frac{1}{9}$

7. $\frac{3}{10} + \frac{3}{10}$

8. $\frac{5}{6} + \frac{5}{6}$

9. $\frac{11}{12} + \frac{7}{12}$

10. $\frac{4}{9} + \frac{7}{9}$

11. $\frac{8}{10} + \frac{7}{10}$

12. $\frac{4}{6} + \frac{5}{6}$

13. $\frac{11}{12} + \frac{11}{12}$

14. $\frac{12}{25} + \frac{8}{25}$

15. $\frac{8}{15} + \frac{2}{15}$

16. $\frac{9}{24} + \frac{7}{24}$

17. $\frac{5}{16} + \frac{13}{16}$

15

Fractions II

Subtracting Fractions: Like Denominators

▶ **To subtract fractions with *like* denominators:**

- Subtract the numerators.

- Write the result over the common denominator.

$$\begin{array}{r} \frac{4}{8} \\ -\frac{2}{8} \\ \hline \frac{2}{8} \end{array}$$

- Express the difference in simplest form.

$$\frac{2}{8} = \frac{2 \div 2}{8 \div 2} = \frac{1}{4}$$ ◀ Divide $\frac{2}{8}$ by $\frac{2}{2}$.

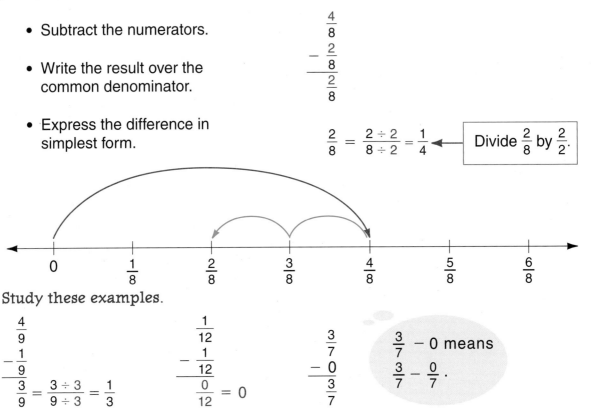

Study these examples.

$$\begin{array}{r} \frac{4}{9} \\ -\frac{1}{9} \\ \hline \frac{3}{9} \end{array} = \frac{3 \div 3}{9 \div 3} = \frac{1}{3}$$

$$\begin{array}{r} \frac{1}{12} \\ -\frac{1}{12} \\ \hline \frac{0}{12} \end{array} = 0$$

$$\begin{array}{r} \frac{3}{7} \\ -0 \\ \hline \frac{3}{7} \end{array}$$

$\frac{3}{7} - 0$ means $\frac{3}{7} - \frac{0}{7}$.

Subtract the fractions. Write each answer in simplest form.
You may use models or a number line to check your work.

1. $\begin{array}{r} \frac{2}{3} \\ -\frac{1}{3} \\ \hline \end{array}$

2. $\begin{array}{r} \frac{9}{10} \\ -\frac{1}{10} \\ \hline \end{array}$

3. $\begin{array}{r} \frac{2}{9} \\ -\frac{2}{9} \\ \hline \end{array}$

4. $\begin{array}{r} \frac{4}{5} \\ -\frac{1}{5} \\ \hline \end{array}$

5. $\begin{array}{r} \frac{7}{12} \\ -\frac{5}{12} \\ \hline \end{array}$

6. $\frac{5}{8} - \frac{1}{8}$

7. $\frac{8}{11} - \frac{8}{11}$

8. $\frac{8}{9} - \frac{2}{9}$

9. $\frac{11}{12} - \frac{1}{12}$

10. $\frac{7}{10} - \frac{3}{10}$

11. $\frac{4}{10} - 0$

12. $\frac{6}{6} - \frac{4}{6}$

13. $\frac{10}{10} - \frac{6}{10}$

14. $\frac{11}{24} - \frac{2}{24}$

15. $\frac{23}{28} - \frac{5}{28}$

16. $\frac{12}{12} - 0$

17. $\frac{12}{12} - \frac{12}{12}$

16

Statistics and Graphs I

Understanding Graphs

A **bar graph** is used to compare information (data). A **line graph** is used to show changes in data over time.

Solve. Use the bar graph and the line graph at the right.

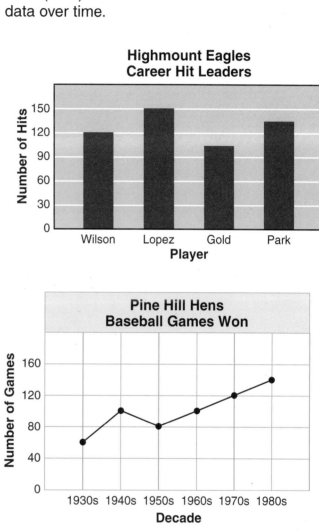

1. What information is given on the horizontal axis of each graph?

2. Which graph is used to compare four items of data?

3. Which graph shows changes in data over time?

4. Which graph shows data that were recorded at regular times?

5. On the bar graph, how many hits does each interval on the vertical scale represent?

6. On the line graph, how many games won does each interval on the vertical scale represent?

7. Which of Highmount's career hit leaders had the greatest number of hits? Who had 15 fewer hits than Wilson? 30 more hits than Wilson?

8. During which decade did Pine Hill win the most games? What trend, if any, does the line graph show for games won by the Pine Hill Hens?

Statistics and Graphs II

Making Pictographs

You can make a **pictograph** to display the data in the table.

▶ **To make a pictograph:**

- List each category of music.

- Choose a symbol or picture to use to represent a number of CDs sold. Examine your data. Select a convenient value for the symbol.

  Let ◎ = 50 CDs.

- Draw the symbols to represent the data. Round data to help you do this. For example:

  $$391 \rightarrow 400 \qquad 247 \rightarrow 250$$

- Write a **key** to show the value of the symbol used.

- Give your graph a title.

CDs Sold at Al's Audio Outlet	
Category	Number Sold
Rock	391
Classical	151
Folk	77
R&B	247
Jazz	126
World Music	169

CDs Sold at Al's Audio Outlet	
Rock	◎ ◎ ◎ ◎ ◎ ◎ ◎ ◎
Classical	◎ ◎ ◎
Folk	◎ (
R&B	◎ ◎ ◎ ◎ ◎
Jazz	◎ ◎ (
World Music	◎ ◎ ◎ (
Key: Each ◎ = 50 CDs.	

Solve. Use the pictograph above.

1. What does (represent?

2. How many symbols were used for R&B? for Jazz?

3. About how many more CDs are needed so that Jazz and R&B would have the same number?

4. Which categories had between 100 CDs and 200 CDs?

5. About how many CDs were sold in all? How can you use multiplication to help you answer?

6. Describe what this pictograph would look like if each symbol represented 10 CDs. What would it look like if each represented 100 CDs?

Statistics and Graphs III

Making Bar Graphs

Henry displayed the data at the right in a **horizontal bar graph**.

▶ **To make a horizontal bar graph:**

- Draw horizontal and vertical axes on grid paper.

- Use the data from the table to choose an appropriate scale. (The data range from 26 to 82. Choose intervals of 10.)

- Draw and label the scale along the horizontal axis. Start at 0 and label equal intervals.

- Label the vertical axis. List the name of each dinosaur.

- Draw horizontal bars to represent each length. Make the bars of equal width.

- Write a title for your graph.

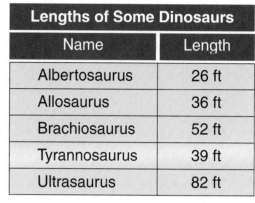

Lengths of Some Dinosaurs	
Name	Length
Albertosaurus	26 ft
Allosaurus	36 ft
Brachiosaurus	52 ft
Tyrannosaurus	39 ft
Ultrasaurus	82 ft

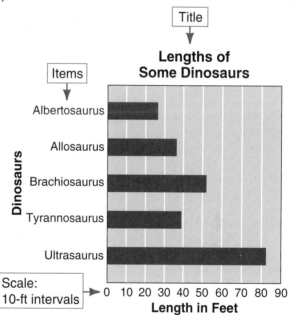

▶ To make a **vertical bar graph**, place the scale along the vertical axis and the items along the horizontal axis.

Solve. Use the bar graph above.

1. What data are along the vertical axis? the horizontal axis?

2. Which dinosaurs have lengths between 30 and 40 feet?

3. How would using 5-foot intervals change the appearance of the graph? 20-foot intervals?

Geometry I

Geometric Figures

Some simple geometric figures:

G and H are points.

- **point**—an exact position or location

- **line**—a set of points that extends forever (infinitely) in opposite directions

$\overleftrightarrow{GH}$ is a line.

- **line segmen**t—part of a line that has two endpoints

$\overline{GH}$ is a line segment.

- **plane**—a flat surface that extends forever (infinitely) in all directions

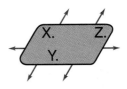

Plane *XYZ* contains the points *X*, *Y*, and *Z*.

▶ A **ray** is part of a line with one endpoint. An **angle** is formed by two rays with a common endpoint. The common endpoint is called the **vertex** of the angle. The letter naming the vertex is always in the middle.

Point *E* is the vertex of ∠*E*.

Ray *EC* ($\overrightarrow{EC}$) has endpoint *E*.

Angle *CEB* (∠*CEB*) is formed by $\overrightarrow{EC}$ and $\overrightarrow{EB}$.

Identify each figure. Then name it using symbols.

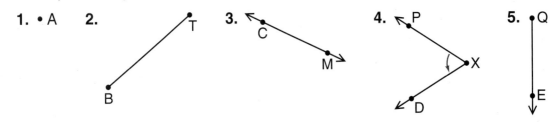

1. •A   2.   3.   4.   5. •Q

20

Lines: Intersecting and Parallel

Lines in the same plane either **intersect** (meet at a point) or are **parallel** (never meet).

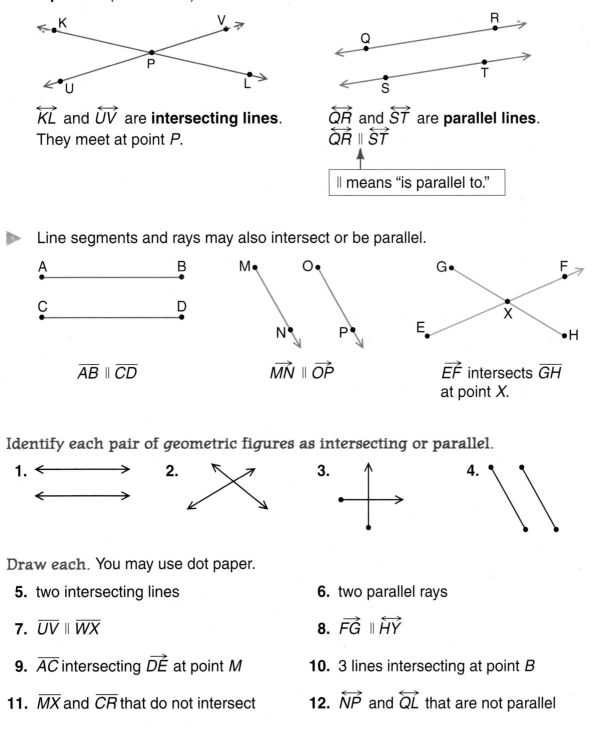

$\overleftrightarrow{KL}$ and $\overleftrightarrow{UV}$ are **intersecting lines**. They meet at point P.

$\overleftrightarrow{QR}$ and $\overleftrightarrow{ST}$ are **parallel lines**.
$\overleftrightarrow{QR} \parallel \overleftrightarrow{ST}$

‖ means "is parallel to."

▶ Line segments and rays may also intersect or be parallel.

$\overline{AB} \parallel \overline{CD}$

$\overrightarrow{MN} \parallel \overrightarrow{OP}$

$\overrightarrow{EF}$ intersects $\overline{GH}$ at point X.

Identify each pair of *geometric figures* as intersecting or parallel.

1.

2.

3.

4.

Draw each. You may use dot paper.

5. two intersecting lines

6. two parallel rays

7. $\overleftrightarrow{UV} \parallel \overline{WX}$

8. $\overrightarrow{FG} \parallel \overleftrightarrow{HY}$

9. $\overline{AC}$ intersecting $\overrightarrow{DE}$ at point M

10. 3 lines intersecting at point B

11. $\overrightarrow{MX}$ and $\overline{CR}$ that do not intersect

12. $\overrightarrow{NP}$ and $\overleftrightarrow{QL}$ that are not parallel

Geometry III

Polygons

A **polygon** is a closed plane figure formed by line segments. The line segments are the **sides** of the polygon.

The point where any two sides of a polygon meet is called a **vertex** (plural: vertices) of the polygon.

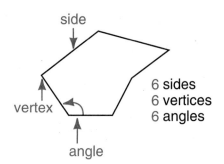

side

vertex

angle

6 sides
6 vertices
6 angles

> The sides of a polygon do not cross each other.

Polygon	Number of Sides	Number of Vertices	Examples
Triangle *tri* means 3	3	3	
Quadrilateral *quad* means 4	4	4	
Pentagon *penta* means 5	5	5	
Hexagon *hexa* means 6	6	6	
Octagon *octa* means 8	8	8	

Decide if each figure is a polygon. Write Yes or No. Then name the polygons.

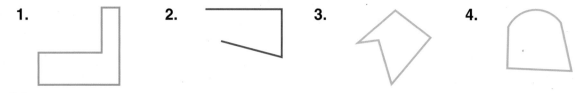

1.   **2.**   **3.**   **4.**

Metric Units of Length

The most commonly used units of **length** in the **metric system** of measurement are given below.

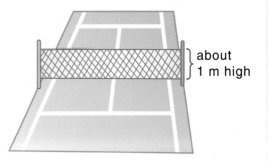

about 1 m high

meter (m)—standard unit of length in the metric system

The height of a net on an actual tennis court is about 1 m.

millimeter (mm)

An actual dime is about 1 mm thick.

centimeter (cm)

The width of a finger is about 1 cm.

decimeter (dm)

A cassette tape is about 1 dm long.

kilometer (km)

Complete. Use mm, cm, dm, m, or km.

1. The length of a pair of scissors is about 20 _?_ .

2. The width of your hand is about 85 _?_ .

3. The swimming freestyle distance is 1500 _?_ .

4. The width of an electric wire is about 1 _?_ .

It takes about 15 minutes to walk 1 km.

Measurement II

Metric Units of Capacity

The most commonly used units of **capacity** in the **metric system** of measurement are given below.

There are about 20 drops of water in 1 mL.

- **liter (L)**—standard unit of capacity in the metric system

- **milliliter (mL)**—used to measure very small capacities

It takes 1000 milliliters to make one liter.

Multiply to rename liters as milliliters:

25 L = (25 × 1000) mL = 25 000 mL

Divide to rename milliliters as liters:

4000 mL = (4000 ÷ 1000) L = 4 L

Which metric unit of capacity is better to measure each?
Write m*L* or *L*.

1. milk jug   **2.** bottle of eyedrops   **3.** glass of juice

4. water cooler   **5.** tablespoon   **6.** gas tank of car

Complete. Multiply or divide to rename the units.

7. 9 L = _?_ mL   **8.** 10 000 mL = _?_ L   **9.** 72 L = _?_ mL

Metric Units of Mass

The most commonly used units of **mass** in the **metric system** of measurement are given below.

The mass of an actual paper clip is about 1 g.

- **gram (g)**—standard unit of mass in the metric system

- **kilogram (kg)**—used to measure large masses

▶ It takes 1000 grams to make one kilogram.

Multiply to rename kilograms as grams:

$72 \text{ kg} = (72 \times 1000) \text{ g} = 72\,000 \text{ g}$

The mass of an actual baseball bat is about 1 kg.

Divide to rename grams as kilograms:

$9000 \text{ g} = (9000 \div 1000) \text{ kg} = 9 \text{ kg}$

Which metric unit of mass is better to measure each?
Write *g* or *kg*.

1. safety pin

2. banana

3. dime

4. a dozen bananas

5. personal computer

6. pencil

Complete. Multiply or divide to rename the units.

7. 23 kg = __?__ g

8. 3000 g = __?__ kg

9. 8 kg = __?__ g

10. 50 000 g = __?__ kg

11. 50 kg = __?__ g

12. 12 000 g = __?__ kg

25

Measurement IV

Customary Units of Length

The **customary units of length** are
the *inch*, *foot*, *yard*, and *mile*.

1 foot (ft) = 12 inches (in.)
1 yard (yd) = 36 in. = 3 ft
1 mile (mi) = 5280 ft = 1760 yd

about 1 in. long

The length of an actual
shoe box is about 1 ft.

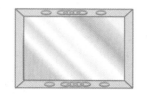

The length of an actual
mirror is about 1 yd.

A person walks a distance of
about 1 mile in 20 minutes.

Multiply to rename larger
units as smaller units.

Divide to rename smaller
units as larger units.

5 mi = __?__ yd   1 mi = 1760 yd

5 mi = (5 × 1760) yd

= 8800 yd

816 in. = __?__ ft   12 in. = 1 ft

816 in. = (816 ÷ 12) ft

= 68 ft

Write the letter of the most reasonable estimate.

1. length of a pen **a.** 6 ft **b.** 6 in. **c.** 6 yd

2. height of a table **a.** $2\frac{1}{2}$ ft **b.** $2\frac{1}{2}$ mi **c.** $2\frac{1}{2}$ in.

3. distance between two cities **a.** 225 mi **b.** 225 yd **c.** 225 ft

Multiply or divide to rename the units. Use a calculator to help you.

4. 8 ft = __?__ in. 5. 25 yd = __?__ ft 6. 252 in. = __?__ yd

7. $2\frac{1}{2}$ mi = __?__ yd 8. 126 in. = __?__ yd 9. 26,400 ft = __?__ mi

Customary Units: Capacity and Weight

The **customary units of capacity** are the *fluid ounce*, *cup*, *pint*, *quart*, and *gallon*.

1 cup (c) = 8 fluid ounces (fl oz)
1 pint (pt) = 2 c = 16 fl oz
1 quart (qt) − 2 pt − 1 c
1 gallon (gal) = 4 qt = 8 pt

The **customary units of weight** are the *ounce*, *pound*, and *ton*.

1 pound (lb) = 16 ounces (oz)
1 ton (T) = 2000 lb

1 pt	1 qt	1 gal	about 1 lb	about 2 T

▶ Multiply and divide to rename units of measurement.

Multiply to rename larger units as smaller units.	Divide to rename smaller units as larger units.

18 gal = __?__ qt   1 gal = 4 qt

18 gal = (18 × 4) qt

$\qquad$ = 72 qt

56 oz = __?__ lb   16 oz = 1 lb

56 oz = (56 ÷ 16) lb

$\qquad = 3\frac{1}{2}$ lb

Write the letter of the most reasonable estimate.

1. capacity of a can of soup   **a.** 2 pt   **b.** 2 c   **c.** 2 gal

2. weight of a tennis ball   **a.** 2 lb   **b.** 2 T   **c.** 2 oz

3. capacity of a large bowl   **a.** 4 qt   **b.** 4 fl oz   **c.** 4 c

Multiply or divide to rename the units. Use a calculator to help you.

4. 6 pt = __?__ qt   **5.** 22 gal = __?__ pt   **6.** 144 oz = __?__ lb

7. $10\frac{1}{2}$ c = __?__ fl oz   **8.** 5000 lb = __?__ T   **9.** 5000 T = __?__ lb

27

Measurement VI

Perimeter and Area Formulas

Formulas can be used to find the perimeter and area of rectangles.

Perimeter of Rectangle	**Area of Rectangle**
$P = (2 \times \ell) + (2 \times w)$	$A = \ell \times w$
ℓ = length, w = width	ℓ = length, w = width

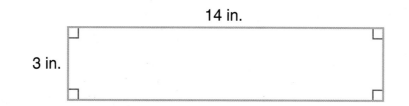

14 in.

3 in.

$P = (2 \times \ell) + (2 \times w)$	$A = \ell \times w$
$P = (2 \times 14 \text{ in.}) + (2 \times 3 \text{ in.})$	$A = 14 \text{ in.} \times 3 \text{ in.}$
$P = 28 \text{ in.} + 6 \text{ in.}$	$A = 42 \text{ sq in.}$
$P = 34 \text{ in.}$	(square inches)
The distance around the rectangle is 34 in.	The surface covered is 42 sq in.

Find the perimeter of each rectangle. Use the perimeter formula.

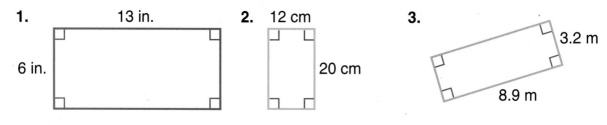

1.   13 in.   6 in.

2.   12 cm   20 cm

3.   3.2 m   8.9 m

Find the area of each rectangle. Use the area formula.

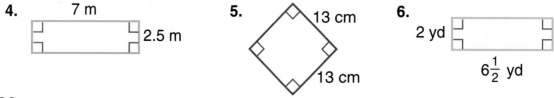

4.   7 m   2.5 m

5.   13 cm   13 cm

6.   2 yd $6\frac{1}{2}$ yd

Key Sequences

Calculator: Parentheses Keys

You can use the parentheses keys on a calculator to compute the value of some mathematical expressions.

Compute: $9 − (3 × $2.68). Use a calculator.

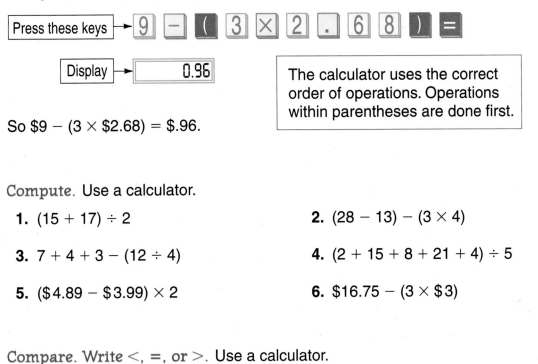

Press these keys ➝ 9 $−$ $($ 3 $×$ 2 $.$ 6 8 $)$ $=$

Display ➝ `0.96`

The calculator uses the correct order of operations. Operations within parentheses are done first.

So $9 − (3 × $2.68) = $.96.

Compute. Use a calculator.

1. (15 + 17) ÷ 2

2. (28 − 13) − (3 × 4)

3. 7 + 4 + 3 − (12 ÷ 4)

4. (2 + 15 + 8 + 21 + 4) ÷ 5

5. ($4.89 − $3.99) × 2

6. $16.75 − (3 × $3)

Compare. Write <, =, or >. Use a calculator.

7. 14 − (12 − 6) __?__ 24 + 2 − 18

8. $1.25 + $.75 − $1.80 __?__ $.50 − ($4.89 − $4.51)

9. 2367 − (468 + 367) __?__ 2491 − 962 + 146

10. 4699 − (8674 − 3993 − 2055) __?__ (1078 − 289 + 1091) + 220

11. $44 − (5 × $6.66) __?__ $45 − (6 × $5.55)

12. 1000 × (5596 − 96) __?__ (1000 × 5596) − 96

13. (42,000 ÷ 3) × 100 __?__ 42,000 ÷ (3 × 100)

Introduction to Problem Solving

Dear Student,

Problem solvers are super sleuths. We invite you to become a super sleuth by using these *five steps* when solving problems.

1 IMAGINE
Create a mental picture.

2 NAME
List the facts and the questions.

3 THINK
Choose and outline a plan.

4 COMPUTE
Work the plan.

5 CHECK
Test that the solution is reasonable.

Sleuths use clues to find a solution to a problem. When working together to solve a problem, you may choose to use one or more of these *strategies* as clues:

USE THESE STRATEGIES:
Missing/Extra Information
Guess and Test
Interpret the Remainder
Use a Graph
Write a Number Sentence
Use Simpler Numbers

USE THESE STRATEGIES:
Hidden Information
More Than One Solution
Logic/Analogies
Use a Diagram
Find a Pattern
Multi-Step Problem

USE THESE STRATEGIES:
Working Backwards
Organized List
Use Drawings/Formulas
Combining Strategies
Write an Equation
Make a Table

1 ▶ IMAGINE

Create a mental picture.

As you read a problem, create a picture in your mind. Make believe you are there in the problem. This will help you think about:
- what facts you will need;
- what the problem is asking;
- how you will solve the problem.

After reading the problem, it might be helpful to sketch the picture you imagined so that you can refer to it.

2 ▶ NAME

List the facts and the questions.

Name or list all the facts given in the problem. Be aware of *extra* information not needed to solve the problem. Look for *hidden* information to help solve the problem. Identify the question or questions the problem is asking.

3 ▶ THINK

Choose and outline a plan.

Think about how to solve the problem by:
- looking at the picture you drew;
- thinking about what you did when you solved similar problems;
- choosing a strategy or strategies for solving the problem.

4 ▶ COMPUTE

Work the plan.

Work with the listed facts and the strategy to find the solution. Sometimes a problem will require you to add, subtract, multiply, or divide. Multi-step problems require more than one choice of operation or strategy. It is good to *estimate* the answer before you compute.

5 ▶ CHECK

Test that the solution is reasonable.

Ask yourself:
- "Have I answered the question?"
- "Is the answer reasonable?"

Check the answer by comparing it to your estimate. It the answer is not reasonable, check your computation. You may use a calculator.

Problem Solving

Strategy: Guess and Test

Problem: Last summer Jane earned $75.50 mowing lawns. From these earnings, she saved $2.50 more than she spent. How much money did Jane save?

1 IMAGINE Picture yourself in the problem.

2 NAME

Facts: Jane saves $2.50 more than she spends.
Jane earned $75.50.

Question: How much money did she save?

3 THINK Since Jane made $75.50, choose a reasonable guess for the amount of money spent, such as $30.00. Make a table and compute the amount saved. Find the total to test your guess.

4 COMPUTE

Spent	$30.00	$33.00	$36.00	$39.00
Saved	$32.50	$35.50	$38.50	$41.50
Total	$62.50	$68.50	$74.50	$80.50
Test	too low	too low	too low	too high

So the amount spent is between $36.00 and $39.00.
Try $37.00.

Spent	$37.00	$36.50
Saved	$39.50	$39.00
Total	$76.50	$75.50
Check	too high	correct

Jane saved $39.00.

5 CHECK Subtract the amount saved from the amount earned to see if $36.50 was spent.

$75.50 − $39.00 = $36.50   Enter: 75.5 − 39 = 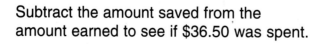 36.5

The answer checks.

32

Strategy: Hidden Information

Problem: In a typical week, a chicken farmer collects about 1164 eggs each day. If all of the eggs are sent to the market, how many dozen eggs are sent each week?

1 IMAGINE Visualize packaging eggs in cartons.

2 NAME

Fact: Daily 1164 eggs are collected.

Question: How many dozen eggs are collected in 1 week?

3 THINK Is there hidden information in the problem? Yes, there are two hidden facts.

7 days = 1 week
12 eggs = one dozen

First, to find how many eggs are collected in one week, multiply:

7 × 1164 = ?
days   eggs per day   eggs in one week

Then, to find how many dozen eggs are sent to the market each week, divide:

eggs collected ÷ 12 = number of dozens
in one week   eggs   sent to the market

4 COMPUTE

$$
\begin{array}{r}
{\scriptstyle 1\ 4\ 2}\\
1\ 1\ 6\ 4\\
\times\qquad 7\\
\hline
8\ 1\ 4\ 8
\end{array}
$$

eggs collected each week

$$
\begin{array}{r}
6\ 7\ 9\\
12\overline{)8\ 1\ 4\ 8}\\
-7\ 2\\
\hline
9\ 4\\
-8\ 4\\
\hline
1\ 0\ 8\\
-1\ 0\ 8\\
\hline
\end{array}
$$

dozen eggs sent to the market

Each week 679 dozen eggs are sent to the market.

5 CHECK Check your computations by using opposite operations.

8148 ÷ 7 $\overset{?}{=}$ 1164 Yes.   12 × 679 $\overset{?}{=}$ 8148 Yes.

Problem Solving

Strategy: Two-Step Problem

Problem: The science class planted 40 seeds. The students display the number of seeds that sprout each day on this graph. How many seeds have not sprouted by May 10?

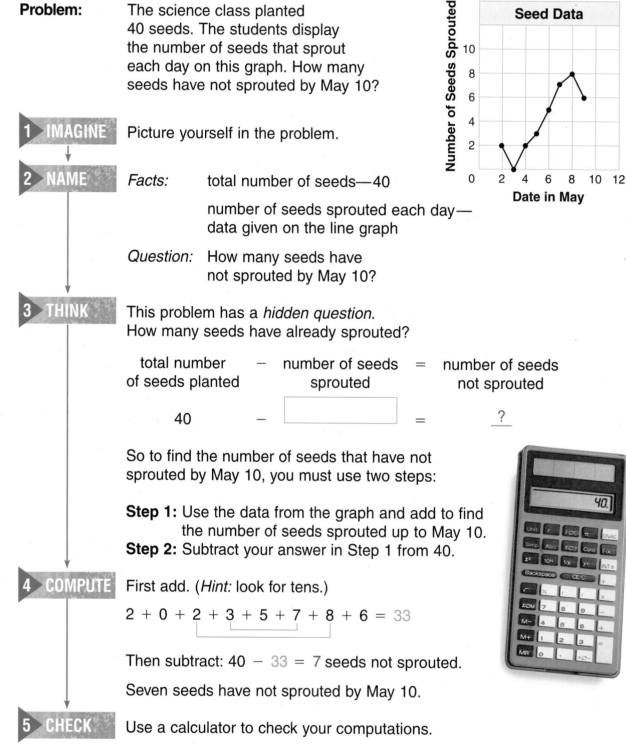

1 IMAGINE Picture yourself in the problem.

2 NAME

Facts: total number of seeds—40

number of seeds sprouted each day— data given on the line graph

Question: How many seeds have not sprouted by May 10?

3 THINK This problem has a *hidden question.* How many seeds have already sprouted?

total number of seeds planted	−	number of seeds sprouted	=	number of seeds not sprouted
40	−		=	?

So to find the number of seeds that have not sprouted by May 10, you must use two steps:

Step 1: Use the data from the graph and add to find the number of seeds sprouted up to May 10.
Step 2: Subtract your answer in Step 1 from 40.

4 COMPUTE First add. (*Hint:* look for tens.)

$2 + 0 + 2 + 3 + 5 + 7 + 8 + 6 = 33$

Then subtract: $40 - 33 = 7$ seeds not sprouted.

Seven seeds have not sprouted by May 10.

5 CHECK Use a calculator to check your computations.

Does $2 + 0 + 2 + 3 + 5 + 7 + 8 + 6 + 7 = 40$? Yes.

Strategy: Write a Number Sentence

Problem: A survey showed that a group of sixth graders owns 56 pets. Twelve of the pets are dogs, thirteen are cats, and the rest are birds. How many of the pets are birds?

Pet Survey

Animal	Tally	Total
Dog	~~HHT~~ ~~HHT~~ II	12
Cat	~~HHT~~ ~~HHT~~ III	13
Bird	?	?
		56

1 ▶ IMAGINE Picture yourself taking a survey.

2 ▶ NAME

Facts: 56 pets
12 dogs
13 cats
The rest are birds.

Question: How many pets are birds?

3 ▶ THINK Use the information to write a number sentence. Write words first.

Number of birds	+	Number of dogs and cats	=	Total number of pets

Use ? or *n* to represent number of birds.

$? + \underline{12 + 13} = 56$

$? + 25 = 56$

addition equation or number sentence

You can solve the number sentence by the Guess and Test strategy or by using a related number sentence.

4 ▶ COMPUTE

$? + 25 = 56$

Try 30. $30 + 25 = 55$

Try 31. $31 + 25 = 56$

There are 31 birds.

$? + 25 = 56$ and

$? = 56 - 25$

$? = 31$

5 ▶ CHECK Substitute 31 for ? in the number sentence to test whether the number sentence is true.

$? + 25 = 56$
$31 + 25 = 56$ The answer checks.

35

Problem Solving

Applications

Choose a strategy from the list or use another strategy you know to solve each problem.

1. Blanca has collected 59 boxes of paper clips. The paper clips in each box make a chain about 312 in. long. Does Blanca have enough clips to make a mile-long chain? (*Hint:* 1 mi = 63,360 in.)

2. Newgate School makes a chain with 12,250 paper clips and rubber bands. The chain uses four times more paper clips than rubber bands. How many paper clips does the chain use? how many rubber bands?

3. Each rubber band in the Newgate chain is 5 cm long. How many rubber bands are in a length of chain that measures 1695 cm?

4. A team of 18 students collects paper clips. The team collects an average of 375 paper clips per student. About how many paper clips did the entire team collect?

5. Cathy and Bill spent $8.89 on rubber bands. Each box cost $1.27, and Cathy bought 3 more boxes than Bill. How many boxes of rubber bands did each student buy?

USE THESE STRATEGIES:
Write a Number Sentence
Guess and Test
Hidden Information
Two-Step Problem

Use the table for problems 6–8.

6. Susan buys 27 boxes of medium paper clips. How much does she spend?

7. Which is more expensive: 8 boxes of large paper clips or 11 boxes of small paper clips?

8. Mr. Steel buys 20 boxes of small paper clips and 14 boxes of medium paper clips. Which purchase costs less?

Paper Clip Cost			
Qty.	Small (500/box)	Medium (225/box)	Large (115/box)
1–9	$1.75	$2.25	$2.15
10–19	$1.50	$2.00	$1.90
20–39	$1.40	$1.85	$1.75
40–	$1.20	$1.65	$1.55

Numeration, Addition, and Subtraction

1

TO BUILD A HOUSE

Here on this plot
Our house will rise
Against the hill
Beneath blue skies

Ruler and tape
Measure the size
Of windows and cupboards
The floors inside

We add, subtract,
Multiply, divide
To build closets and stairs
The porch outside

Without numbers and measure
Would our house ever rise
Against the hill
Beneath blue skies?

Lillian M. Fisher

In this chapter you will:

Explore one trillion
Use properties, estimation, and rounding
Use inverse operations
Use the STORE and RECALL keys
Solve problems that have missing or
   extra information

Critical Thinking/Finding Together

Find Robert's house number if it is the
seventh number in this sequence: 4119
4008 4037 3926

37

1-1 How Much Is a Trillion?

Discover Together

Materials Needed: calculator, paper, pencil, place-value chart

The students at I.S. 218, a middle school in New York City, along with 29 other schools in the area, have an annual project of collecting pennies to buy clothes for homeless people and to help stock food kitchens for needy people.

The project has been successful beyond most people's imagination. Here is some data from their collection in a recent school year:

- about 2055 sacks of pennies containing $50 each were collected

- pennies collected weighed about $4\frac{1}{2}$ tons

- total collection amounted to a whopping $102,738.67

1. Use a place-value chart like the one below. Assume that the collection amounted to $100,000. Write 100,000 in your chart.

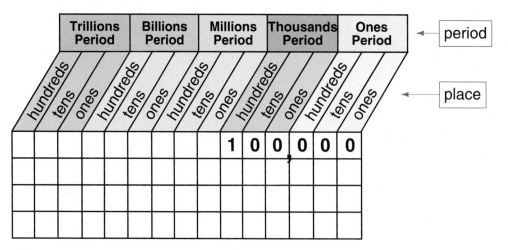

Trillions Period			Billions Period			Millions Period			Thousands Period			Ones Period		
hundreds	tens	ones	hundreds	tens	ones	hundreds	tens	ones	hundreds	tens	ones	hundreds	tens	ones
								1	0	0	0	0	0	

Each place in the chart is 10 times the next place to its right.

2. The hundred millions place is 10 × 10 × 10 times the hundred thousands place. 100,000,000 = <u>1000</u> times 100,000. Write 100,000,000 (one hundred million) in your place-value chart.

3. The hundred billions place is 10 × 10 × 10 times the hundred millions place.
   Complete: 100,000,000,000 = <u>?</u> times 100,000,000.
   Write 100,000,000,000 (one hundred billion) in your chart.

4. The hundred trillions place is <u>?</u> × <u>?</u> × <u>?</u> times the hundred billions place?
   Complete: 100,000,000,000,000 = <u>?</u> times 100,000,000,000.
   Write 100,000,000,000,000 (one hundred trillion) in your chart.

The students at the 30 schools collected about $100,000.

5. How many pennies are there in 100,000 dollars?
   Complete: 100,000 × 100 = <u>?</u> pennies.

Communicate

6. If many schools work together to collect 100 billion pennies, how many dollars would they have? Discuss with other groups how to find the number of dollars. Then write a number sentence. Do the same for 100 trillion pennies.

   Discuss ✓

7. The students at these 30 schools collected about 10 million pennies. About how many such schools would it take, at the same rate of collection, to have 1 billion pennies? 1 trillion pennies?

PROJECT

8. Plan a 1–2 month class project to collect pennies for a worthy cause.

   • Decide how much money (dollars) the class wants to collect. Then decide how many pennies must be collected to reach the class goal.

   • Make a bar graph to show the number of pennies the class actually collects each week.

   • At the end of the project, determine about how many weeks, at the same rate, it would take your class to collect 1 trillion pennies.

1-2 Expanded Form

The value of each digit in a number is shown
by writing the number in **expanded form**.

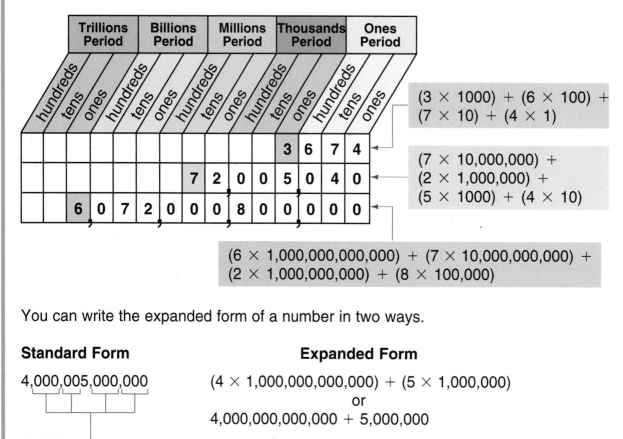

Trillions Period			Billions Period			Millions Period			Thousands Period			Ones Period		
hundreds	tens	ones	hundreds	tens	ones	hundreds	tens	ones	hundreds	tens	ones	hundreds	tens	ones
												3	6	7	4
							7	2	0	0	5	0	4	0
		6	0	7	2	0	0	0	8	0	0	0	0	0

$(3 \times 1000) + (6 \times 100) + (7 \times 10) + (4 \times 1)$

$(7 \times 10,000,000) + (2 \times 1,000,000) + (5 \times 1000) + (4 \times 10)$

$(6 \times 1,000,000,000,000) + (7 \times 10,000,000,000) + (2 \times 1,000,000,000) + (8 \times 100,000)$

You can write the expanded form of a number in two ways.

Standard Form

4,000,005,000,000

Expanded Form

$(4 \times 1,000,000,000,000) + (5 \times 1,000,000)$

or

$4,000,000,000,000 + 5,000,000$

Omit place values of 0 when you write in expanded form.

Copy and complete.

1. $506,039 = (\underline{?} \times 100,000) + (\underline{?} \times 1000) + (\underline{?} \times 10) + (\underline{?} \times 1)$

2. $3,000,750 = (\underline{?} \times 1,000,000) + (\underline{?} \times 100) + (\underline{?} \times 10)$

3. $16,009,000,000 = (1 \times \underline{?}) + (6 \times \underline{?}) + (9 \times \underline{?})$

4. $9,000,005,000,000 = (\underline{?} \times 1,000,000,000,000) + (\underline{?} \times 1,000,000)$

5. $38,500,000,000,000 = (3 \times \underline{?}) + (8 \times \underline{?}) + (5 \times \underline{?})$

Write in expanded form in two ways.

6. 16,527

7. 41,293

8. 359,048

9. 702,468

10. 7,042,102

11. 21,407,098

12. 3,400,000,000

13. 17,800,000,000

14. 800,000,000,000

15. 75,000,000,000

16. $55,000,000

17. $55,000,000,000

18. $9,500,000,000,000

19. $18,000,000,000,000

20. 4,050,009,000,000

21. 1,500,400,000,000

Write the numbers in each chart in expanded form.

	Popular Vote Total for Winning Presidential Candidates	
	Year	Votes (nearest thousand)
22.	1976	40,829,000
23.	1980	43,899,000
24.	1984	54,282,000
25.	1988	48,881,000
26.	1992	44,908,000

	Television Facts	
	Households with:	Number (nearest thousand)
27.	At least 1 TV	93,100,000
28.	Only 1 TV	33,516,000
29.	2 or more TVs	59,584,000
30.	Color TVs	91,238,000
31.	Black and White TVs	1,862,000

 CRITICAL THINKING

Write whether the statement is true about (a) all of the numbers, (b) some of the numbers, or (c) none of the numbers.

16,043,011
6,057,022
6,610,043,000

32. My trillions digit is 6.

33. My ten-millions digit is 1.

34. My millions digit is 3 less than my thousands digit and the same as my tens digit.

1-3 Rounding to Greatest Place

For a social studies assignment, Zaida needs to find the size of each of the four largest oceans.

She uses her computer to get the information given in the table.

Ocean	Number of Square Miles
Pacific	64,186,300
Atlantic	33,420,000
Indian	28,350,500
Arctic	5,105,700

Zaida is considering using **rounded numbers** to report her findings. She wants to round each number in the table to its greatest place.

▶ **To round a number to its greatest place:**

- Find the digit in the greatest place.

- Look at the digit to its right.
  If the digit is *less than 5*, round *down*.
  If the digit is *5 or more*, round *up*.

5,105,700   28,350,500

The greatest place is millions.   The greatest place is ten millions.

5,105,700 | 1 < 5   28,350,500 | 8 > 5
↓ | Round down.   ↓ | Round up.
5,000,000   30,000,000

Replace all digits to the right of the greatest place with zeros.

64,186,300   rounds to   60,000,000
33,420,000   rounds to   30,000,000
28,350,500   rounds to   30,000,000
05,105,700   rounds to   5,000,000

The numbers are **ordered** from greatest to least.
6 > 3   3 > 2   2 > 0

Write the place to which the number was rounded.

1. 6,504 to 7,000

2. 14,540 to 10,000

3. 4,973,622 to 5,000,000

4. 824,900 to 800,000

5. 75,029,998 to 80,000,000

Round each number in the table to its greatest place.

6.

Ocean	Average Depth (feet)
Pacific	12,925
Atlantic	11,730
Indian	12,598
Arctic	3,407

7.

Continent	Number of Square Miles
Europe	3,800,000
Asia	17,200,000
Africa	11,700,000
Australia	3,071,000

Write the numbers in the table from least to greatest.

8.

National Monuments	Acreage
Congaree Swamp	22,200
Effigy Mounds	1,481
Death Valley	2,067,628
White Sands	143,733
Dinosaur	210,844

9.

Planet	Average Distance from the Sun (miles)
Venus	67,240,000
Earth	92,900,000
Neptune	2,796,460,000
Mars	141,710,000
Pluto	3,666,000,000

PROBLEM SOLVING

10. Tell your teacher or another adult whether Zaida should use the numbers in the table at the top of page 42 or the rounded numbers at the bottom of the page to report the sizes of the oceans. To what place would you recommend that she round the numbers?

Communicate ✓

CHALLENGE

Use each of the digits 2, 3, 4, 6, 7, 8, 9 once in each number described.

11. Write the greatest number.

12. Write the least number.

13. Write the greatest number that is less than 6,000,000.

14. Write the least number that is more than 8,000,000.

1-4 Properties of Addition

When you add numbers, the statement of the addition is called an **addition sentence** or **addition equation**.

addition sentence

$$\underline{13 + 4 + 2} = \underline{19}$$

addends · · · · sum

The following properties of addition are true for any numbers a, b, and c.

- **Commutative Property of Addition**
  Changing the *order* of the addends does not change the sum.

  $a + b = b + a$ · · · · · · · · · · $16 + 44 = 44 + 16$
  $60 = 60$

  Think: "order."

- **Associative Property of Addition**
  Changing the *grouping* of the addends does not change the sum.

  $(a + b) + c = a + (b + c)$ · · · · $(1 + 4) + 7 = 1 + (4 + 7)$
  $5 \quad + 7 = 1 + \quad 11$
  $12 = 12$

  Think: "grouping."

- **Identity Property of Addition**
  The sum of zero and a number is that number.

  $a + 0 = a \qquad 0 + a = a \qquad 89 + 0 = 89 \qquad 0 + 32 = 32$

  Think: "same."

Name the property of addition used.

1. $18 + 53 = 53 + 18$

2. $(7 + 8) + 2 = 7 + (8 + 2)$

3. $90 + 0 = 90$

4. $78 + 142 = 142 + 78$

5. $0 + 142 = 142$

6. $62 + (14 + 12) = (62 + 14) + 12$

Complete. Name the properties of addition used.

7. $(7 + 2) + 3 = \underline{\ ?\ } + (2 + 3)$

8. $18 + \underline{\ ?\ } = 18$

9. $98 + 17 = \underline{\ ?\ } + 98$

10. $(81 + 27) + 73 = 81 + (27 + \underline{\ ?\ })$

11. $n + 0 = \underline{\ ?\ }$

12. $0 + c = \underline{\ ?\ }$

13. $b + a = \underline{\ ?\ } + b$

14. $\underline{\ ?\ } + c = c + b$

15. $(a + b) + \underline{\ ?\ } = a + (b + c)$

16. $\underline{\ ?\ } + b = b + 0$

Using Properties to Compute

The properties of addition may help you add numbers more easily.

$$8 + 139 + 42 + 61$$
$$8 + 42 + 139 + 61 \longleftarrow \text{Commutative Property}$$
$$(8 + 42) + (139 + 61) \longleftarrow \text{Associative Property}$$
$$50 + 200$$
$$250$$

Explain how to use the properties of addition to make these computations easier. Then compute.

Communicate

17. $(32 + 155) + 45$

18. $36 + (64 + 81)$

19. $19 + 14 + 16$

20. $202 + 65 + 32 + 68$

21. $91 + 47 + 9 + 53$

MENTAL MATH

Compute mentally.

22. $22 + 57 + 8$

23. $8 + 129 + 2$

24. $25 + 48 + 25$

25. $150 + 25 + 25$

26. $4500 + 2832 + 500$

27. $1981 + 6200 + 800$

28. $1500 + 7600 + 2400$

29. $5900 + 2500 + 4100$

1-5 Estimating Sums and Differences

▶ You can use front-end estimation or rounding to estimate sums. 47,362 + 38,651 is about ？ .

Front-end estimation

- Add the front digits (digits in the greatest place).
- Write zeros for the other digits.
- Adjust the estimate with the back digits.

$$\begin{array}{r} 47{,}362 \\ +\,38{,}651 \\ \hline \end{array}$$
about　70,000

$$\left.\begin{array}{r} 47{,}362 \\ +\,38{,}651 \end{array}\right\}\ \begin{array}{l}\text{about}\\ 15{,}000\end{array}$$

Rough estimate: 70,000　　　Adjusted estimate: 70,000 + 15,000 = 85,000

Estimation by rounding

- Round each number to the greatest place of the lesser number.
- Add the rounded numbers.

$$\begin{array}{r} 47{,}362 \longrightarrow\ \ 50{,}000 \\ +\,38{,}651 \longrightarrow\ +40{,}000 \\ \hline \text{about}\ \ \ 90{,}000 \end{array}$$

Both 85,000 and 90,000 are reasonable estimates.

▶ Use the same two methods to estimate differences.

Front-end estimation	**Estimation by rounding**
• Subtract the front digits.	• Round to the greatest place of the lesser number.
• Write zeros for the other digits.	• Subtract the rounded numbers.

$$\begin{array}{r} 38{,}642 \longrightarrow\ \ 38{,}642 \\ -\ 7{,}780 \longrightarrow\ -07{,}780 \\ \hline \text{about}\ \ 30{,}000 \end{array}$$

$$\begin{array}{r} 38{,}642 \longrightarrow\ \ 39{,}000 \\ -\ 7{,}780 \longrightarrow\ -\ 8{,}000 \\ \hline \text{about}\ \ 31{,}000 \end{array}$$

Both 30,000 and 31,000 are reasonable estimates.

Estimate the sum or difference in two ways.

1. 232
   + 298

2. 415
   + 326

3. 526
   − 437

4. 982
   − 629

5. 2526
   + 2844

6. 7625
   + 3475

7. 3275
   − 1098

8. 8298
   − 4721

9. 47,280
   + 31,560

10. 20,987
    + 59,676

11. 360,493 + 528,761

12. 813,604 − 394,678

13. 189,362 − 54,364

Estimating Money Sums and Differences

To estimate money sums and differences, round to the greatest place of the lesser number; then add or subtract the rounded numbers.

$421.89 ⟶ $400.00
+ 379.50 ⟶ + 400.00
about $800.00

$76.85 ⟶ $77.00
− 5.47 ⟶ − 5.00
about $72.00

Estimate the sum or difference.

14. $7.65
    + 1.42

15. $8.09
    − 4.95

16. $97.38
    − 45.86

17. $48.98
    + 22.10

18. $887.56
    − 259.60

19. $687.98
    + 550.09

20. $87.52
    + 7.52

21. $455.95
    − 6.87

22. $947.60
    − 25.89

23. $7,259,886 + $4,950,000

24. $9,499,867 − $459,860

PROBLEM SOLVING

25. Which stadium holds about 20,000 more people than the Astrodome? about 10,000 more?

26. Which stadium holds about 15,000 fewer people than the Silverdome? about 10,000 fewer?

27. Suppose a company manufactures seats for all of these stadiums. About how many seats will the company make altogether?

Stadium	Capacity
Astrodome	60,502
Kingdome	64,984
Superdome	69,065
Silverdome	80,500

1-6 Adding Larger Numbers

The table shows an estimate of the number of travelers to the United States from nine different locations in a recent year.

How many travelers came from these three countries: Japan, Germany, and Venezuela?

Travelers to U.S.	
Location	**Number**
Japan	3,301,000
U.K.	3,075,000
Germany	1,962,000
France	821,000
Italy	590,000
Brazil	576,000
Australia	502,000
Venezuela	453,000
South Korea	427,000

To find the total number of travelers, add: $3,301,000 + 1,962,000 + 453,000 = $ _?_

First estimate the sum. Round to the greatest place of the least number.

Round to hundred thousands.

$$
\begin{array}{r}
3,301,000 \longrightarrow 3,300,000 \\
1,962,000 \longrightarrow 2,000,000 \\
+\quad 453,000 \longrightarrow +\quad 500,000 \\
\hline
\text{about} \quad 5,800,000
\end{array}
$$

Then add.

$$
\begin{array}{r}
\overset{1\ 1}{} \\
3,301,000 \\
1,962,000 \\
+\quad 453,000 \\
\hline
5,716,000
\end{array}
$$

The total number of travelers was 5,716,000.

Check.

Enter: 3,301,000 [+] 1,962,000 [+]

453,000 [=] [5716000.]

5,716,000 is close to the estimate of 5,800,000.

Estimate. Then find the total number of travelers.

1. U.K., France, South Korea

2. Germany, Italy, Brazil

Find the sum. Use a calculator to check your answers.

3.	**4.**	**5.**	**6.**
3,465,892	1,096,784	8,723,306	524,876
+ 2,396,087	+ 7,854,209	+ 946,782	+ 7,502,782

Align the numbers and then add.

7. 36,789 + 42,197 + 61,432

8. 18,432 + 106,714 + 75,941

9. $12.96 + $42.76 + $21.50

10. $36.72 + $464.97 + $642.30

11. 5,924,687 + 6,872,309

12. 8,511,076 + 294,755

13. 12,366,055 + 56,744,906

14. 9,376,000 + 12,982,000

PROBLEM SOLVING
Use the table on page 48.

15. How many travelers came from France, Italy, and Brazil combined?

16. How many travelers came from Japan, U.K., and Germany combined?

17. How many travelers came from the nine locations?

MENTAL MATH

Add. Explain any shortcuts that you use.

Communicate ✓

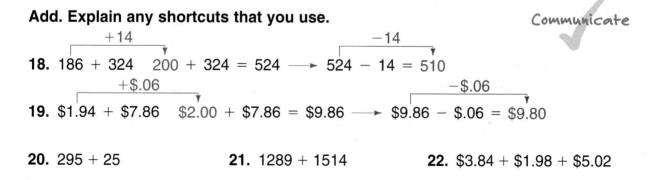

18. 186 + 324   200 + 324 = 524 $\longrightarrow$ 524 − 14 = 510   (+14 / −14)

19. $1.94 + $7.86 $2.00 + $7.86 = $9.86 $\longrightarrow$ $9.86 − $.06 = $9.80   (+$.06 / −$.06)

20. 295 + 25

21. 1289 + 1514

22. $3.84 + $1.98 + $5.02

1-7 Zeros in Subtraction

Chen needs 2000 copies of the program for the 6th grade pageant. The copy machine stopped working after it made 967 copies. How many more copies are needed?

To find the number of copies needed, subtract: 2000 − 967 = _?_

Estimate: 2000 − 1000 = 1000

▶ **To subtract when the minuend has zeros:**

- Regroup as many times as necessary before starting to subtract.

- Subtract.

More hundreds, tens, and ones are needed. Regroup thousands.	Regroup hundreds.	Regroup tens.	Subtract.
$\begin{array}{r} 1\ \ 10 \\ 2\!\!\!\not0\ 0\ 0 \\ -\ \ \ 9\ 6\ 7 \\ \hline \end{array}$	$\begin{array}{r} 9 \\ 1\ \not10\ 10 \\ 2\!\!\!\not0\ \not0\ 0 \\ -\ \ \ 9\ 6\ 7 \\ \hline \end{array}$	$\begin{array}{r} 9\ \ 9 \\ 1\ \not10\ \not10\ 10 \\ 2\!\!\!\not0\ \not0\ \not0 \\ -\ \ \ 9\ 6\ 7 \\ \hline \end{array}$	$\begin{array}{r} 9\ \ 9 \\ 1\ \not10\ \not10\ 10 \\ 2\!\!\!\not0\ \not0\ \not0 \\ -\ \ \ 9\ 6\ 7 \\ \hline 1\ 0\ 3\ 3 \end{array}$

- Check. Enter: 2000 ⊟ 967 ▭ [1033.]

Chen needs 1033 more copies.

Study this example.

$$\begin{array}{r} 8,309,000 \\ -\ \ \ 777,625 \\ \hline \end{array} \longrightarrow \begin{array}{r} 12\ \ \ \ \ \ \ \ \ 9\ \ 9 \\ 7\ \not18\ 10\ 8\ \not10\ \not10\ 10 \\ 8,\!\not3\ \not0\ \not9,\!\not0\ \not0\ \not0 \\ -\ \ \ 7\ 7\ 7,\!6\ 2\ 5 \\ \hline 7,\!5\ 3\ 1,\!3\ 7\ 5 \end{array}$$

Estimate:
8,300,000 − 800,000
= 7,500,000

Find the difference. Use a calculator to check your answers.

1. 3000 − 1739	**2.** 4000 − 1176	**3.** 30,000 − 14,698	**4.** 60,000 − 56,240

Align the numbers and then subtract.

5. 40,300 − 17,415

6. 20,205 − 4289

7. $500.82 − $112.91

8. $610.07 − $48.89

9. 320,000 − 103,400

10. 475,000 − 49,090

11. 6,864,000 − 2,086,045

12. 1,618,000 − 308,015

PROBLEM SOLVING

Find the difference between the projected income and the actual income for each week. Use the graph at the right.

13. Week 1 actual income: $49,642

14. Week 2 actual income: $37,928

15. Week 3 actual income: $76,450

16. Week 4 actual income: $70,400

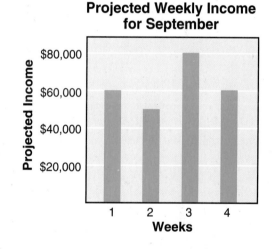

Projected Weekly Income for September

SHARE YOUR THINKING

Communicate ✓

Predict the number of digits in the answer for each exercise. Compare your predictions with those of a classmate.

17. 12,596,424 + 69,085,209

18. 98,423,086 + 5,652,050

19. 124,008,565 − 5,476,293

20. 67,542,312 − 58,623,098

21. Check your predictions on a calculator or by paper and pencil.

22. Can you use estimation to make your predictions? Explain.

Inverse Operations

Addition and subtraction are **inverse operations**.
You can use related addition and subtraction
sentences to help solve problems.

▶ Angie saved some money. She spent $12.52
of it on a new bicycle tire. She now has $9.21.
How much money did Angie have saved?

$$\underline{\ ?\ } - \$12.52 = \$9.21$$

Use the inverse operation $$\$9.21 + \$12.52 = \underline{\ ?\ }$$ ◄┤ related
to write a related sentence.   addition
                                                          sentence

Add to solve. $\$9.21 + \$12.52 = \$21.73$ *Algebra* ✓

Angie had saved $21.73.

▶ Justin bought a cassette at the record store. He
also bought a CD for $12.60. The total bill was
$21.10. What was the cost of the cassette?

$$\underline{\ ?\ } + \$12.60 = \$21.10$$

Use the inverse operation $$\underline{\ ?\ } = \$21.10 - \$12.60$$ ◄┤ related
to write a related sentence.   subtraction
                                                          sentence

Subtract to solve. $\$21.10 - \$12.60 = \$8.50$

The cost of the cassette was $8.50.

Write four related sentences for these numbers.

1. 13, 4, 17   **2.** 28, 31, 59   **3.** $1.79, $2.56, $4.35

**In your Math Journal, solve the problem and then
explain how you solved it.**   *Math
                                                       Journal* ✓

4. The sum is 1746. One addend is 389. What is the other addend?

Match the number sentence to the problem.
Then use a related sentence to solve.

5. David has 26 trees to plant. He has dug
holes for 16 trees. How many holes are
still to be dug?

6. Coretta had 321 baseball cards. She sold some
cards to Tanya. Coretta now has 90 cards. How
many cards did she sell to Tanya?

7. Yori studied 90 minutes for a test.
This was 40 minutes more than Ramon
studied. How many minutes did Ramon study?

a. $321 - \underline{\ ?\ } = 90$

b. $90 = \underline{\ ?\ } + 40$

c. $\underline{\ ?\ } + 16 = 26$

Related Sentences with Variables

You can use a variable (letter) to stand for an unknown
number in a number sentence.

number sentence: $\underline{\ ?\ } + 15 = 25$ $\underline{\ ?\ } - 9 = 36$

$n + 15 = 25$ $c - 9 = 36$

related number sentence: $n = 25 - 15$ $c = 36 + 9$

$n = 10$ $c = 45$

Use a related sentence to find the missing number.

8. $8 + \underline{\ ?\ } = 12$

9. $36 - \underline{\ ?\ } = 9$

10. $\underline{\ ?\ } + 110 = 367$

11. $\underline{\ ?\ } - 368 = 682$

12. $60 = \underline{\ ?\ } - 15$

13. $92 = 47 + \underline{\ ?\ }$

14. $n - 40 = 56$

15. $a + 12 = 37$

16. $b + 52 = 96$

17. $y - 61 = 44$

18. $21 = z + 9$

19. $18 = m - 10$

20. $x - \$1.37 = \$3.47$

21. $r + \$2.96 = \$10.00$

22. $\$14.85 + w = \$90.00$

TECHNOLOGY

Store and Recall Keys

You can use a calculator to find the standard form of a
number written in expanded form.

▶ Use a calculator to find the standard form of:
   $(7 \times 10{,}000) + (2 \times 1000) + (5 \times 100) + (4 \times 10) + (0 \times 1)$.

Press these keys

→ ON/AC ← Clears the calculator's memory.

$(7 \times 10{,}000)$ STO (2×1000) 2nd SUM (5×100)

Stores the value displayed
in memory.

This key combination adds the displayed
value to the value stored in memory.

2nd SUM (4×10) 2nd SUM (0×1) 2nd SUM RCL

Display ⟶ $\boxed{72540.}$ Recall key displays the value stored in memory.

The standard form of
$(7 \times 10{,}000) + (2 \times 1000) + (5 \times 100) + (4 \times 10) + (0 \times 1)$ is 72,540.

**Enter the following key sequences into your calculator.
Write each number in standard form.**

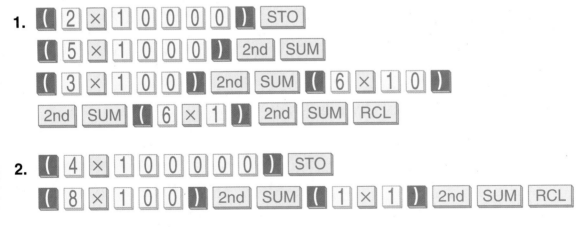

1. (2 × 1 0 0 0 0) STO
   (5 × 1 0 0 0) 2nd SUM
   (3 × 1 0 0) 2nd SUM (6 × 1 0)
   2nd SUM (6 × 1) 2nd SUM RCL

2. (4 × 1 0 0 0 0 0) STO
   (8 × 1 0 0) 2nd SUM (1 × 1) 2nd SUM RCL

Use the STO , 2nd , SUM , and RCL keys
to find each number in standard form.

3. $(9 \times 100{,}000) + (6 \times 10{,}000) + (3 \times 1000) +$
$(2 \times 10) + (5 \times 1)$

4. $(7 \times 1{,}000{,}000) + (4 \times 100{,}000) + (1 \times 10{,}000) +$
$(4 \times 1000) + (0 \times 100) + (2 \times 10) + (2 \times 1)$

5. $(5 \times 1{,}000{,}000) + (7 \times 100{,}000) +$
$(2 \times 10{,}000) + (3 \times 1{,}000)$

6. $(7 \times 1{,}000{,}000) + (1 \times 1{,}000) + (9 \times 1)$

7. $(8 \times 10{,}000) + (6 \times 1{,}000) + (2 \times 10)$

8. $(5 \times 10{,}000) + (5 \times 1)$

9. $(3 \times 100{,}000) + (2 \times 1{,}000)$

SHARE YOUR THINKING

Communicate ✓

10. Write the number 500,042,000.
Using a single operation, change the 4 to a zero without
changing any other digits. Write this new number.
Explain the strategy you used to eliminate the 4.

11. Write the number 1,435,906,200.
Using a single operation, change the 3 to a zero without
changing the other digits. Write this new number.
Explain the strategy you used to eliminate the 3.

12. Write the number 1,023,456,789.
Using a single operation, change the 6 to a zero. You may
or may not change the other digits. Write this new number.
Can you eliminate the 6 another way? Explain your reasoning.

1-10 Problem Solving: Missing/Extra Information

Problem: On Friday, 52,345 tickets were sold for the baseball game. On Saturday, 58,468 tickets were sold. The stadium holds 60,000 people. How many tickets were sold on both days?

1 IMAGINE Picture yourself in the problem.

2 NAME *Facts:* Friday—52,345 tickets sold
Saturday—58,468 tickets sold
The stadium holds 60,000 people.

Question: How many tickets were sold on both days?

3 THINK Look at the facts.
What information is needed?
   You need to know that 52,345 tickets were sold
   Friday and 58,468 tickets were sold Saturday.

What information is not needed?
   You *do not* need to know that the stadium
   holds 60,000 people.

To find out how many tickets were sold, add:
   $52,345 + 58,468 = $ _?_ number of tickets sold

4 COMPUTE First estimate.

$50,000 + 60,000 = 110,000$

$$
\begin{array}{r}
\overset{1}{}\overset{1}{}\overset{1}{} \\
5\,2{,}3\,4\,5 \\
+\ \ 5\,8{,}4\,6\,8 \\
\hline
1\,1\,0{,}8\,1\,3
\end{array}
$$
   Enter: 52,345 $+$ 58,468 $=$ | 110813. |

On both days 110,813 tickets were sold.

5 CHECK Use a calculator and change the order of the addends.

$58,468 + 52,345 = 110,813$

The answer checks.

Tell the information *not* needed or supply the missing information. Then solve.

1. Burrows Farm uses 450 acres for corn. The remaining acres are used for potatoes. How many acres are used for potatoes?

Burrows Farm

Corn 450 acres	Potatoes ?

? acres

IMAGINE   Draw a diagram of crops to be grown.

NAME   *Facts:*   450 acres for corn
                remaining acres for potatoes

Question: How many acres are used for potatoes?

THINK   Do you have enough information to solve the problem? No, the number of acres on Burrows farm is missing. Make up the number of acres: 720 acres.

As stated, the problem cannot be solved. There is missing information.

To find how many acres are used for potatoes, subtract:
720 − 450 = _?_ number of acres for potatoes

➤ **COMPUTE** ➤ **CHECK**

2. Bernice has a collection of 90 stamps from Europe, Asia, and Africa. Sixty stamps are from Europe. How many stamps does she have from Africa?

3. On Monday, Ginger packed 426 boxes, Tuesday 573, and Wednesday 685. She worked with 3 other people. How many boxes did she pack in 3 days?

4. Frank paid $40 for a pair of shoes, $78 for a jacket, and $6.75 for a pair of socks. He chose not to buy a $32.95 shirt. After purchasing these articles he had $20.50 left. How much money did Frank have at first?

5. Carol read 120 pages of a book on Saturday. She read more of the book on Sunday. On Monday she read the same number of pages that she read on Sunday. She has 76 pages left to read. How many pages did she read on Sunday and Monday?

6. Marco earned $240 baby-sitting. He worked 7 weeks for a total of 80 hours. How much did Marco earn per hour?

1-11 Problem-Solving Applications

Solve each problem and explain the method you used.

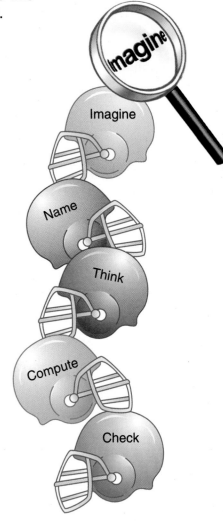

Imagine

Name

Think

Compute

Check

1. A survey found that forty million, twenty thousand households watched football games on Saturday November 15. Write this number in standard form. Then write it in expanded form.

2. On Saturday 2,959,016 viewers watched the Sox play on TV. On Sunday, 2,959,061 viewers watched them play. On which day did more viewers watch the Sox on TV?

3. About 460,000 more households watched football on Saturday November 22 than on the 15th. How many households watched football on November 22? (*Hint:* See problem 1.)

4. About 93,100,000 U.S. households own at least one television. About 33,516,000 homes own only one television. About how many homes own more than one television?

5. In 1970 there were 7085 fewer cable TV systems than in 1990. There were 9575 cable TV systems in 1990. How many were there in 1970?

6. In 1980 there were 17,671,490 cable TV subscribers. By 1990, that number had increased to 54,871,330. To the nearest thousand, how many more cable subscribers were there in 1990 than in 1980?

7. On September 8, 15 million households watched a popular prime-time TV show. On September 15, 250 thousand more households watched the show. Then on September 22, the number of households watching that show dropped by 400 thousand. How many households watched the show on September 22?

8. An advertiser bought a prime-time commercial spot on a local channel for $10,800 and a late-night spot for $3209. She also bought a Saturday morning spot. If the total cost was $19,059, how much did the advertiser spend on the Saturday morning spot?

Use a strategy from the list or another strategy you know to solve each problem.

9. Ms. Vaccaro's class surveyed 1500 sixth graders about television viewing habits. They surveyed twice as many girls as boys. How many boys did they survey?

10. Of the students surveyed, 621 had cable TV. How many of those surveyed did not?

11. The average sixth grader watches 34,320 minutes of television each year. Abigail watched 28,540 minutes of television last year and 29,790 minutes two years ago. How does her viewing time last year compare to the time of the average sixth grader?

12. If 962 of the students in the survey in problem 9 watch situation comedies, how many students in the survey do not watch situation comedies?

13. Students rated two dozen programs *excellent* and 5 *poor*. How many fewer *poor* than *excellent* ratings were given?

USE THESE STRATEGIES:
Write a Number Sentence
Hidden Information
Missing Information
Extra Information
Guess and Test

Use the bar graph for problems 14–16.

14. About how many minutes did the average viewer watch television on weeknights?

15. About how many minutes did the average viewer watch on Saturday?

16. About how many more people watch daytime television during the week than on Saturday?

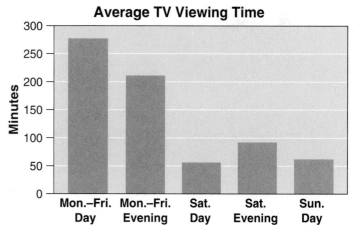

Average TV Viewing Time

MAKE UP YOUR OWN

17. Make up an extra information problem using the bar graph data. Have a classmate solve it.

Write in expanded form in two ways.   *(See pp. 38–41.)*

1. 3,009,000

2. 8,001,009,000

3. 64,000,000,000

Round each number to its greatest place.   *(See pp. 42–43.)*

4. 239,872

5. 1,535,682

6. 14,763,000

Write in order from greatest to least.   *(See pp. 42–43.)*

7. 36,000,000,000; 49,000,000; 30,050,000,000

Complete. Use the properties of addition.   *(See pp. 44–45.)*

8. $123 + 82 = \underline{\ ?\ } + 123$

9. $d + 0 = \underline{\ ?\ }$

10. $(75 + 208) + 92 = 75 + (\underline{\ ?\ } + 92)$

Estimate the sum or difference.   *(See pp. 46–47.)*

11.   155,989
        $+\ \ \ 45,086$

12.   463,123
        $-\ 129,876$

13. $455.00
        $+\ \ \ \ \ 89.69$

Compute. Watch for $+$ and $-$ signs.   *(See pp. 48–51.)*

14.   7,198,629
        $+\ \ \ \ 815,222$

15.   15,032,786
        $+\ 29,988,532$

16. $199.99
        $+\ \ \ 287.56$

17.   600,000
        $-\ 320,900$

18.   750,000
        $-\ \ 69,852$

19.   9,056,300
        $-\ 3,593,500$

20. $82,396,000 + 9,746,000$

21. $5,050,000 - 2,099,000$

Use a related sentence to find the missing number.   *(See pp. 52–53.)*

22. $y - 52 = 96$

23. $\$17.96 + m = \$50.00$

PROBLEM SOLVING   *(See pp. 56–59.)*

24. Argentina has an area of 1,068,296 square miles. India has an area of 1,296,338 square miles. How much greater is the area of India?

25. Ms. Yee makes a down payment of $18,250 on a $162,500 house and borrows the rest. Her annual income is $200,000. How much does she borrow?

(See *Still More Practice,* p. 507.)

ROMAN NUMERALS

The ancient Romans used the symbols given below
to form numerals.

Symbol	I	V	X	L	C	D	M
Value	1	5	10	50	100	500	1000

To find the value of a Roman numeral:

Add

- if the symbol is repeated.

  XX = 10 + 10 = 20   CC = 100 + 100 = 200

- if a symbol for a greater number comes before
  that of a lesser number.

  DC = 500 + 100 = 600   MD = 1000 + 500 = 1500

Subtract

- if a symbol for a lesser number comes before
  that of a greater number.

  XC = 100 − 10 = 90   CM = 1000 − 100 = 900

Sometimes you must both add and subtract.

  MCMIV = 1000 + (1000 − 100) + (5 − 1) = 1904

A bar over a Roman numeral *multiplies* its value by 1000.

  $\overline{X}$ = 1000 × 10 = 10,000 $\overline{D}$ = 1000 × 500 = 500,000

Write each as a standard numeral.

1. CL   **2.** XL   **3.** MM   **4.** CDIX

5. $\overline{D}$CIX   **6.** $\overline{MCMXCV}$ **7.** $\overline{XDIX}$ **8.** $\overline{MCLV}$

Write the Roman numeral for each.

9. 127   **10.** 1914   **11.** 4300   **12.** 6320

13. 61,612   **14.** 15,000   **15.** 214,000   **16.** 948,000

Performance Assessment

Use the digits 2, 4, 6, 7, 8, 9. Write each number described below and explain your choices.

1. greatest number

2. greatest number less than 600,000

3. least number

4. least number more than 800,000

Write in expanded form.

5. 42,000,000

6. 5,008,000,000

7. 16,500,000,000,000

Round each number to its greatest place.

8. 74,239

9. 369,059

10. 2,596,000

Write in order from least to greatest.

11. 3,432,000,000; 42,976,000; 10,000,000,000

Complete. Use the properties of addition.

12. $74 + \underline{\ ?\ } = 89 + 74$

13. $(26 + 8) + 12 = 26 + (\underline{\ ?\ } + 12)$

Estimate and then compute. Watch for $+$ and $-$ signs.

14. $4,203,873$
$+\,2,956,729$

15. $\$529.87$
$+\ \ \ 391.50$

16. $70,000 - 32,000$

Use a related sentence to find the missing number.

17. $n + 45 = 92$

18. $a - 22 = 68$

PROBLEM SOLVING   *Use a strategy or strategies you have learned.*

19. In a recent year, 1,962,000 travelers came to the United States from Germany and 576,000 from Brazil. How many travelers is that altogether?

20. Albertsville's population increased from 192,875 to 243,620. By how many people did the population increase?

Multiplication and Division

The Old Math. One.

If a train leaves Union Station,
in Chicago, at eight in the morning
carrying three thousand dozen gross of dark
almond bark and travels the average speed
of fifty-seven miles per hour for one
day, then c o l l i d e s with a train
that left San Francisco one day
earlier full of fifteen hundred dozen
bite-sized chocolate puppies,
how many days will the residents
of Left Foothills, Colorado, have to
spend in the high school gym
while the National Guard, the
Environmental Protection Agency,
and the local sheriff's department
remove the worst bite-sized bark bits
(or the worst bark-sized bite bits)
and return the area to its former
h a b i t a b l e   c o n d i t i o n ?

Arnold Adoff

In this chapter you will:

Use properties and estimate products
  and quotients
Discover patterns in multiplication
  and division
Learn about exponents, short division,
  and divisibility
Compute using the order of operations
Use the y^x calculator key
Solve problems by interpreting the
  remainder

Critical Thinking/Finding Together

Our product is less than 1125 and our
quotient is about $1\frac{1}{10}$. What two 2-digit
numbers are we?

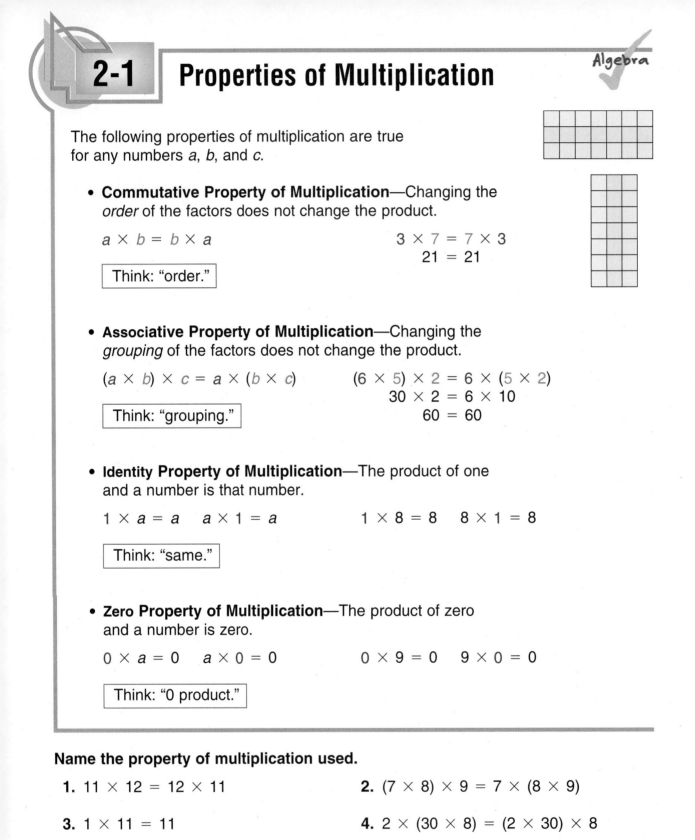

2-1 Properties of Multiplication

The following properties of multiplication are true for any numbers *a*, *b*, and *c*.

- **Commutative Property of Multiplication**—Changing the *order* of the factors does not change the product.

  $a \times b = b \times a$

  Think: "order."

  $3 \times 7 = 7 \times 3$
  $21 = 21$

- **Associative Property of Multiplication**—Changing the *grouping* of the factors does not change the product.

  $(a \times b) \times c = a \times (b \times c)$

  Think: "grouping."

  $(6 \times 5) \times 2 = 6 \times (5 \times 2)$
  $30 \times 2 = 6 \times 10$
  $60 = 60$

- **Identity Property of Multiplication**—The product of one and a number is that number.

  $1 \times a = a \quad a \times 1 = a$

  Think: "same."

  $1 \times 8 = 8 \quad 8 \times 1 = 8$

- **Zero Property of Multiplication**—The product of zero and a number is zero.

  $0 \times a = 0 \quad a \times 0 = 0$

  Think: "0 product."

  $0 \times 9 = 0 \quad 9 \times 0 = 0$

Name the property of multiplication used.

1. $11 \times 12 = 12 \times 11$

2. $(7 \times 8) \times 9 = 7 \times (8 \times 9)$

3. $1 \times 11 = 11$

4. $2 \times (30 \times 8) = (2 \times 30) \times 8$

5. $25 \times 0 = 0$

6. $9 \times 1 = 9$

7. $1 \times 1 = 1$

8. $0 \times 1 = 0$

Distributive Property

Distributive Property of Multiplication over Addition—When the same number is *distributed* as a factor across two addends, the product does not change.

$$a \times (b + c) = (a \times b) + (a \times c)$$

$$8 \times (5 + 9) = (8 \times 5) + (8 \times 9)$$
$$8 \times \quad 14 \quad = \quad 40 \quad + \quad 72$$
$$112 = 112$$

Think: "same factor across addends."

Complete. Use the properties of multiplication.

9. $8 \times 19 = 19 \times \underline{\ ?\ }$ **10.** $26 \times \underline{\ ?\ } = 26$ **11.** $7 \times \underline{\ ?\ } = 4 \times 7$

12. $18 \times \underline{\ ?\ } = 0$ **13.** $11 \times 0 = \underline{\ ?\ } \times 11$ **14.** $\underline{\ ?\ } \times 6 = 6$

15. $5 \times (1 \times 2) = (5 \times \underline{\ ?\ }) \times 2$ **16.** $\underline{\ ?\ } \times (9 \times 3) = (4 \times 9) \times 3$

17. $7 \times (3 + 1) = (\underline{\ ?\ } \times 3) + (\underline{\ ?\ } \times 1)$ **18.** $\underline{\ ?\ } \times (\underline{\ ?\ } + \underline{\ ?\ }) = (2 \times 5) + (2 \times 4)$

19. $6 \times c = \underline{\ ?\ } \times 6$ **20.** $1 \times n = \underline{\ ?\ }$

21. $5 \times (4 \times m) = (5 \times 4) \times \underline{\ ?\ }$ **22.** $t \times (6 + 7) = (t \times \underline{\ ?\ }) + (t \times \underline{\ ?\ })$

Explain how to use the properties of multiplication to make these computations easier. Then compute.

Communicate ✓

23. $5 \times 9 \times 2$ **24.** $8 \times 9 \times 0$ **25.** $(7 \times 2) \times 50$

26. $(7 \times 5) \times 2$ **27.** $9 \times 2 \times 9$ **28.** $4 \times 6 \times 25$

MENTAL MATH

Compute mentally.

29. $3 \times 23 = (3 \times 20) + (3 \times 3) = \underline{\ ?\ }$ **30.** $6 \times 41 = \underline{\ ?\ }$

31. $52 \times 5 = (50 \times 5) + (2 \times 5) = \underline{\ ?\ }$ **32.** $35 \times 4 = \underline{\ ?\ }$

33. $122 \times 4 = (100 \times 4) + (20 \times 4) + (2 \times 4) = \underline{\ ?\ }$ **34.** $5 \times 108 = \underline{\ ?\ }$

2-2 Special Patterns

Discover Together

Materials Needed: calculator, paper, pencil

Use a calculator to compute the products and quotients in exercise 1. Write each number sentence on your paper.

1.

Column A	Column B
$1 \times 34 = \underline{\ ?\ }$	$34{,}000 \div 1 = \underline{\ ?\ }$
$10 \times 34 = \underline{\ ?\ }$	$34{,}000 \div 10 = \underline{\ ?\ }$
$100 \times 34 = \underline{\ ?\ }$	$34{,}000 \div 100 = \underline{\ ?\ }$
$1000 \times 34 = \underline{\ ?\ }$	$34{,}000 \div 1000 = \underline{\ ?\ }$
$34 \times 2 = \underline{\ ?\ }$	$68{,}000 \div 2 = \underline{\ ?\ }$
$34 \times 20 = \underline{\ ?\ }$	$68{,}000 \div 20 = \underline{\ ?\ }$
$34 \times 200 = \underline{\ ?\ }$	$68{,}000 \div 200 = \underline{\ ?\ }$
$34 \times 2000 = \underline{\ ?\ }$	$68{,}000 \div 2000 = \underline{\ ?\ }$

> 20, 200, and 2000 are multiples of 10.

2. What patterns do you notice in your number sentences in Column A?

3. Will the total number of zeros in the factors *always* be the same as the number of zeros in the product? Test your answer by finding these products: 34×5, 34×50, 34×500, and 34×5000. Explain what happened.

4. How many zeros will there be in the product for each of these examples?

100×56 1000×389 72×5000

5. In your Math Journal, summarize the two steps you use to multiply a whole number by 10, 100, 1000 or by their multiples. Be sure to include:

Math Journal

- how the nonzero digits are multiplied (1×34, 34×5, and so on); and

- how you decide the number of zeros to write in the product.

6. What patterns do you notice in your number sentences in Column B?

7. For each number sentence in Column B, write the number of zeros in the dividend and in the divisor. Then write the number of zeros in your answer (the quotient).

8. In the example $34{,}000 \div 100$,
   $34{,}000$ has $\underline{3}$ zeros.
   $\qquad 100$ has $\underline{2}$ zeros.
   $\underline{3} - \underline{2} = \underline{1}$ The quotient has $\underline{1}$ zero.

9. In the example $68{,}000 \div 20$,
   $68{,}000$ has $\underline{3}$ zeros.
   $\qquad 20$ has $\underline{1}$ zero.
   $\underline{3} - \underline{1} = \underline{2}$ The quotient has $\underline{\ ?\ }$ zeros.

10. Copy and complete the steps below for dividing a whole number by 10, 100, 1000 or their multiples (no remainders).

    Step 1: Use basic facts to divide. ($34 \div 1$, $68 \div 2$, and so on)

    Step 2: Count the number of $\underline{\ ?\ }$ in the dividend and in the divisor.

    Step 3: Subtract to find the number of zeros in the $\underline{\ ?\ }$.

Communicate

11. How can you quickly find the product: 72×3000?

Discuss

12. How can you quickly find the quotient: $44{,}000 \div 200$?

13. Compute: 60×2800, 400×500, 7000×2100. Write a rule in your Math Journal for finding products when both factors contain zeros.

Math Journal

14. Compute: $3000 \div 50$, $10{,}000 \div 200$, $200{,}000 \div 400$. Explain in your Math Journal why you must be especially careful in determining the number of zeros in the quotient.

2-3 Estimating and Finding Products

In ancient times, the land an ox could plow in a day was called an "acre." Today, an acre is defined as 4840 square yards. How many square yards could an ox plow in one year (365 days)?

Estimate by **rounding each factor to its greatest place**.
Then multiply.

> ≈ means approximately equal to

Estimate: 365 × 4840 ≈ 400 × 5000
             About 2,000,000 square yards

▶ **To multiply whole numbers:**

- Multiply by the ones, then by the tens, then by the hundreds, and so on.

- Add the partial products.

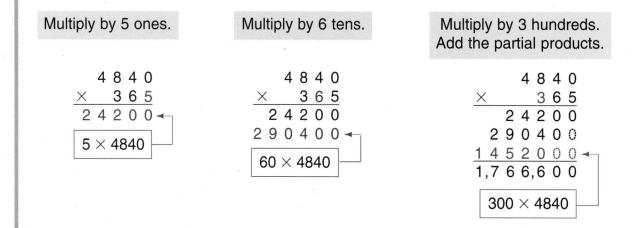

Multiply by 5 ones.	Multiply by 6 tens.	Multiply by 3 hundreds. Add the partial products.

```
      4 8 4 0
  ×     3 6 5
    2 4 2 0 0  ◄
  ┌──────────┐
  │ 5 × 4840 │
  └──────────┘
```

```
      4 8 4 0
  ×     3 6 5
    2 4 2 0 0
  2 9 0 4 0 0  ◄
    ┌───────────┐
    │ 60 × 4840 │
    └───────────┘
```

```
        4 8 4 0
  ×       3 6 5
      2 4 2 0 0
    2 9 0 4 0 0
  1 4 5 2 0 0 0  ◄
  1,7 6 6,6 0 0
      ┌────────────┐
      │ 300 × 4840 │
      └────────────┘
```

So an ox can plow 1,766,600 square yards in one year.

> 1,766,660 is close to the estimate of 2,000,000.

Study this example.

622 × $659.38 ≈ 600 × $700   About $420,000

Estimate and then find each product.

1. 335
   × 129

2. 824
   × 617

3. 925
   × 376

4. 5847
   × 219

5. 7932
   × 324

6. $44.25
   × 142

7. $53.38
   × 319

8. $847.69
   × 293

9. $795.20
   × 498

10. 932 × 898

11. 813 × 678

12. 741 × 2361

13. 612 × 1472

14. 835 × $45.72

15. 746 × $69.58

Estimate to compare. Write <, =, or >. Use a calculator to check.

16. 679 × 325 _?_ 679 × 425

17. 7976 × 853 _?_ 7976 × 753

18. 225 × 1125 _?_ 425 × 1300

19. 9651 × 438 _?_ 438 × 9651

20. 3425 × 1225 _?_ 4425 × 225

21. 1110 × 2222 _?_ 2222 × 1110

PROBLEM SOLVING

22. Each month Express Package Delivery must deliver
    50,000 packages to cover its expenses. Last month
    each of its 20 employees delivered more than
    124 packages each day. If there were 21 working days
    last month, did the company cover its expenses? Explain.

Communicate ✓

CONNECTIONS: HISTORY

23. When the Northwest Territory of the United States was to be
    surveyed, the U.S. Congress determined the number of
    acres in each township of this new land.

    (a) Look up the Articles of Confederation and find the
        number of acres in each township.

    (b) How many acres would there be in 120 such townships?

2-4 Zeros in Multiplication

When multiplying by a number with a zero, you may save a step by omitting a partial product.

Multiply: 620 × 372 = _?_

Estimate: 600 × 400 = 240,000

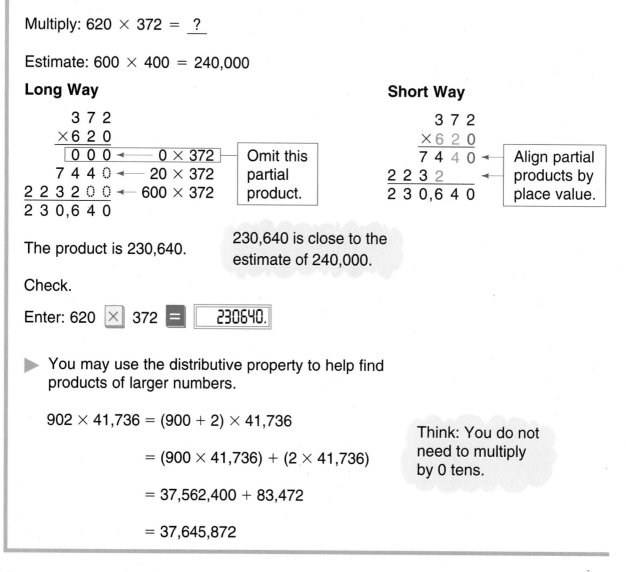

Long Way

```
      3 7 2
    × 6 2 0
    0 0 0  ←—— 0 × 372      Omit this
    7 4 4 0 ←—— 20 × 372     partial
  2 2 3 2 0 0 ←— 600 × 372   product.
  2 3 0,6 4 0
```

The product is 230,640.

Short Way

```
      3 7 2
    × 6 2 0
    7 4 4 0  ←—   Align partial
  2 2 3 2         products by
  2 3 0,6 4 0     place value.
```

230,640 is close to the estimate of 240,000.

Check.

Enter: 620 ☒ 372 ▤ | 230640. |

▶ You may use the distributive property to help find products of larger numbers.

$$902 \times 41{,}736 = (900 + 2) \times 41{,}736$$

$$= (900 \times 41{,}736) + (2 \times 41{,}736)$$

$$= 37{,}562{,}400 + 83{,}472$$

$$= 37{,}645{,}872$$

Think: You do not need to multiply by 0 tens.

Complete the estimate shown at the right of each exercise. Then multiply.

1.
```
  237      200
× 380    × 400
```

2.
```
  593      600
× 506    × 500
```

3.
```
 $8.17    $8.00
× 609    × 600
```

4.
```
 $3.85    $4.00
× 806    × 800
```

5.
```
 2365     2000
× 580    × 600
```

6.
```
 6549     7000
× 302    × 300
```

70

Estimate and then find each product.

7. 403 × 585

8. 209 × 791

9. 601 × 482

10. 830 × 793

11. 740 × 5565

12. 310 × $18.93

13. 240 × $35.48

14. 902 × 6071

15. 4003 × 4203

Use the distributive property and a calculator to compute.

16. 506 × 831

17. 780 × 342

18. 470 × 1211

19. 209 × 4921

20. 640 × 39,215

21. 640 × 390,215

22. In exercise 21, were you able to multiply 600 by 390,215 on your calculator? Explain how you can use the distributive property, calculator, and mental math to compute the product, without using paper and pencil calculation.

Computation
Method

PROBLEM SOLVING

23. There are 375 audience tickets available for each taping of the Win It All game show. If 204 shows are taped each year, how many tickets are there in all?

24. The producers of Win It All hand out 150 contestant applications for each show. If there are 204 shows, how many applications do they hand out?

 SHARE YOUR THINKING

25. Write in your Math Journal three things you needed to know before you studied today's math lesson. What was easy for you in this lesson? What was difficult for you?

Math
Journal

2-5 Exponents

Another way to write 100 is 10 × 10 or 10^2.

10^2 ← **exponent**

10^2 — **base**

Read: "ten squared" or
"ten to the second power."

▶ An **exponent** tells how many times to use the base as a factor.

10^2 means 10 is used as a factor 2 times: 10 × 10.

10^3 means 10 is used as a factor 3 times: 10 × 10 × 10.

Read: "ten cubed" or
"ten to the third power."

10 as a factor
3 times

▶ Powers of 10 are used to show place value.

$10^1 = 10$
$10^2 = 10 × 10 = 100$
$10^3 = 10 × 10 × 10 = 1000$
$10^4 = 10 × 10 × 10 × 10 = 10,000$
$10^5 = 10 × 10 × 10 × 10 × 10 = 100,000$

Any number to the first power
is that number: $10^1 = 10$.

The exponent tells the number of zeros.

Exponents may be used to write standard numerals in expanded form.

8467 = (8 × 1000) + (4 × 100) + (6 × 10) + (7 × 1)

= (8 × 10^3) + (4 × 10^2) + (6 × 10^1) + (7 × 1)

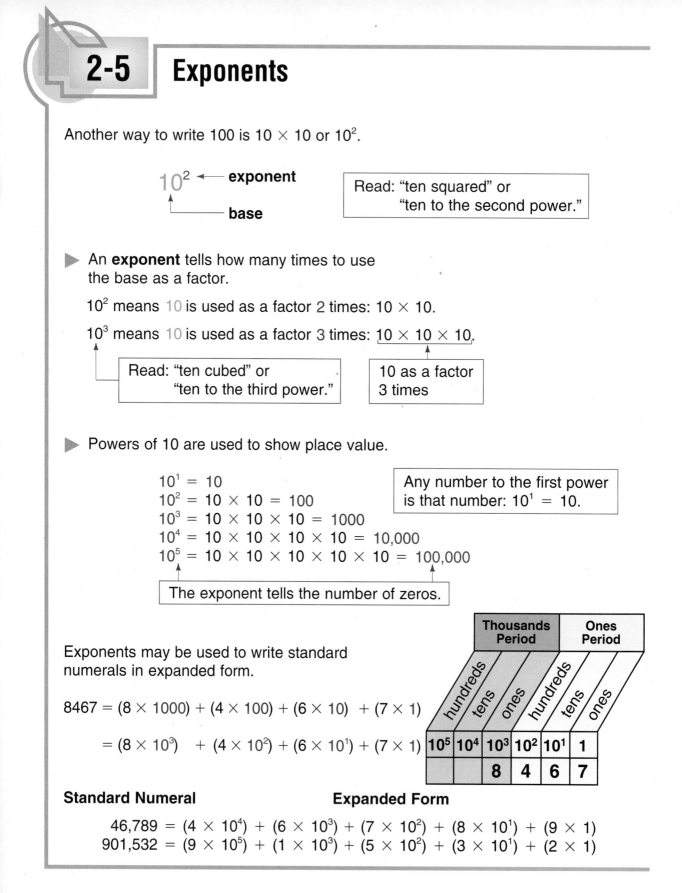

Thousands Period			Ones Period		
hundreds	tens	ones	hundreds	tens	ones
10^5	10^4	10^3	10^2	10^1	1
		8	4	6	7

Standard Numeral   **Expanded Form**

46,789 = (4 × 10^4) + (6 × 10^3) + (7 × 10^2) + (8 × 10^1) + (9 × 1)
901,532 = (9 × 10^5) + (1 × 10^3) + (5 × 10^2) + (3 × 10^1) + (2 × 1)

Write the standard numeral.

1. $(4 \times 10^3) + (3 \times 10^2) + (5 \times 10^1) + (6 \times 1)$ 2. $(8 \times 10^5) + (2 \times 10^4)$

3. $(9 \times 10^4) + (7 \times 10^3) + (2 \times 10^2) + (3 \times 10^1)$

4. $(3 \times 10^2) + (4 \times 10^1) + (3 \times 1)$

5. $(9 \times 10^6) + (6 \times 10^4) + (2 \times 10^2) + (3 \times 10^1) + (2 \times 1)$

Write in expanded form using exponents.

6. 872   7. 463   8. 5798   9. 4267

10. 14,906   11. 92,800   12. 602,081   13. 842,006

Bases Other than Ten

Exponents are used with other numbers.

2^4 means 2 as a factor 4 times, or $2 \times 2 \times 2 \times 2 = 16$.

$3^5 = 3 \times 3 \times 3 \times 3 \times 3 = 243$ $4^3 = 4 \times 4 \times 4 = 64$

Write in exponent form.

14. $8 \times 8 \times 8$ 15. 16×16 16. $2 \times 2 \times 2 \times 2 \times 2$

17. $5 \times 5 \times 5 \times 5$ 18. $25 \times 25 \times 25$ 19. 100×100

Write the standard numeral.

20. 5^2 21. 7^2 22. 12^3 23. 10^4 24. 24^2

25. 9^3 26. 1^4 27. 0^7 28. 2^1 29. 16^1

CHALLENGE

Write each using a base number and an exponent.

30. 25   31. 16   32. 36   33. 49   34. 27

35. 8   36. 1000   37. 10,000   38. 81   39. 64

2-6 Short Division and Divisibility

Leilani's sister saved the same amount of money each month for 9 months for a vacation. The vacation cost $1908. How much did she save each month?

To find the amount she saved each month, divide: $1908 ÷ 9 = __?__

Estimate: $2000 ÷ 10 = $200

When dividing by a one-digit divisor, you can use **short division**.

$$\begin{array}{r} \$\ 2 \\ 9\overline{)\$\ 1\ 9^{1}0\ 8} \end{array}$$

| 2 × 9 = 18 |
| 19 − 18 = 1 |

$$\begin{array}{r} \$\ 2\ 1 \\ 9\overline{)\$\ 1\ 9^{1}0^{1}8} \end{array}$$

| 1 × 9 = 9 |
| 10 − 9 = 1 |

$$\begin{array}{r} \$\ 2\ 1\ 2 \\ 9\overline{)\$\ 1\ 9^{1}0^{1}8} \end{array}$$

| 2 × 9 = 18 |
| 18 − 18 = 0 |

So Leilani's sister saved $212 each month.

Leilani remembered the divisibility rules in the chart so she knew that $1908 ÷ 9 would divide evenly, with zero remainder.

$1908

$$1 + 9 + 0 + 8 = 18 \qquad 18 ÷ 9 = 2$$

| **Number is divisible by:** |
| • 3 if the sum of its digits is divisible by 3. |
| • 9 if the sum of its digits is divisible by 9. |

> Think: $1908 is divisible by 9 since 18 is divisible by 9.

Study these examples. Use divisibility rules to predict which examples will have remainders.

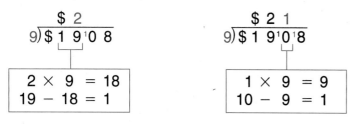

$$\begin{array}{r} 5\ 6 \\ 3\overline{)16^{1}8} \end{array}$$

$$\begin{array}{r} 1\ 3\ 5 \\ 5\overline{)6^{1}7^{2}9^{4}} \end{array} \text{R4 or } 135\frac{4}{5}$$

$$\begin{array}{r} 6\ 6\ 8\ 3 \\ 9\overline{)6\ 0,^{6}1^{7}4^{2}7} \end{array}$$

$$\begin{array}{r} 2\ 1,8\ 6\ 4 \\ 2\overline{)4\ 3,^{1}7^{1}2\ 8} \end{array}$$

Divide using short division.

1. $3\overline{)81}$ **2.** $6\overline{)84}$ **3.** $5\overline{)490}$ **4.** $7\overline{)315}$

5. $588 \div 4$ **6.** $912 \div 8$ **7.** $2496 \div 2$ **8.** $6975 \div 3$

Find each quotient by short division. Use R to write remainders. Check on a calculator by multiplying the divisor and the quotient and then adding the remainder.

9. $4\overline{)137}$ **10.** $9\overline{)836}$ **11.** $5\overline{)1398}$ **12.** $7\overline{)1805}$

13. $815 \div 5$ **14.** $644 \div 4$ **15.** $36{,}570 \div 7$ **16.** $19{,}580 \div 6$

Complete. Name the divisor. Use divisibility rules to help you.

17. $?\overline{)5782}$ 2891
18. $?\overline{)5898}$ 1966
19. $?\overline{)67{,}404}$ $7\ 489$ R3
20. $?\overline{)63{,}327}$ $7\ 915$ R7

PROBLEM SOLVING

21. Brandi has 368 bottles to put into 8-bottle cartons. How many cartons does she need?

22. Air Ways shipped 2079 radios. The radios were packed 9 to a box. How many boxes of radios were shipped?

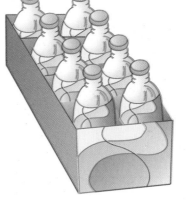

 **FINDING TOGETHER**

Think of the divisibility rules you know.
Write a divisibility rule for 6. (*Hint:* Remember $2 \times 3 = 6$.)

Use your rule to tell which numbers are divisible by 6.

23. 84 **24.** 120 **25.** 432 **26.** 455 **27.** 6291 **28.** 8496

29. Find five numbers between 100 and 200 that are divisible by 3, 6, and 9.

2-7 Estimating Quotients

The entire Beaufort School District is going by bus to the science fair. There are 1895 students and teachers in the district. The buses the district is renting hold 48 people each. About how many buses are needed?

To find about how many buses are needed, estimate: $1895 \div 48$.

▶ **To estimate the quotient of two numbers:**

- Write compatible numbers for the dividend and the divisor.

- Divide.

$$\underline{1895} \div \underline{48} \longrightarrow 2000 \div 50 = 40$$

about 2000   about 50

> Two numbers are **compatible numbers** when one number divides the other evenly.

2000 and 50 are compatible numbers. They are easy to divide mentally.

About 40 buses are needed.

Study these examples.

$$893\overline{)27{,}056} \longrightarrow 900\overline{)27{,}000} \;\;\overset{30}{}$$

$$893\overline{)27{,}056} \approx 30$$

> Both dividend and divisor are changed to get compatible numbers.

"is approximately equal to"

$$420\overline{)\$415{,}786} \longrightarrow 420\overline{)\$420{,}000} \;\;\overset{\$1000}{}$$

$$420\overline{)\$415{,}786} \approx \$1000$$

> Only the dividend is changed to get compatible numbers.

Estimate the quotient.

1. 2164 ÷ 43

2. 5838 ÷ 28

3. 7842 ÷ 37

4. 3984 ÷ 19

5. 82,461 ÷ 41

6. $51,206 ÷ 53

7. 13,642 ÷ 206

8. 85,136 ÷ 409

9. $485,725 ÷ 520

10. 672,385 ÷ 710

11. 879,500 ÷ 425

12. $972,360 ÷ 325

Choose the best estimate.

13. $32\overline{)2940}$ ≈ ?   **a.** 1   **b.** 10   **c.** 100   **d.** 1000

14. $19\overline{)6248}$ ≈ ?   **a.** 3   **b.** 30   **c.** 300   **d.** 3000

15. $210\overline{)380,493}$ ≈ ?   **a.** 2   **b.** 20   **c.** 200   **d.** 2000

16. $389\overline{)792,432}$ ≈ ?   **a.** 2   **b.** 20   **c.** 200   **d.** 2000

PROBLEM SOLVING

17. A truck driver drove 5845 miles in 19 days. Did he average more than 250 miles a day?

18. Sheila's company mails 3580 advertising flyers in 25 days. Do the mailings average more than 200 flyers per day?

19. The mileage on Michael's new car is 686 miles. The mileage on his sister's car is 45,650. About how many times greater is the mileage on her car?

SKILLS TO REMEMBER

Estimate. Then multiply.

20. 54 × 426

21. 76 × 549

22. 65 × 5305

23. 48 × 4017

24. 630 × 4454

25. 801 × 7182

26. 420 × $17.82

27. 350 × $24.37

Update your skills. See pages 13–14.

2-8 Zeros in Division

The mass of the largest uncut diamond ever found is 62 120 centigrams. How many carats is this? (1 carat = 20 centigrams)

To find the number of carats, divide: 62 120 ÷ 20 = ?

Estimate: 60 000 ÷ 20 = 3000

Decide where to begin the quotient.	$20\overline{)62120}$ 20 > 6
	$20\overline{)62120}$ 20 < 62

The quotient begins in the thousands place.

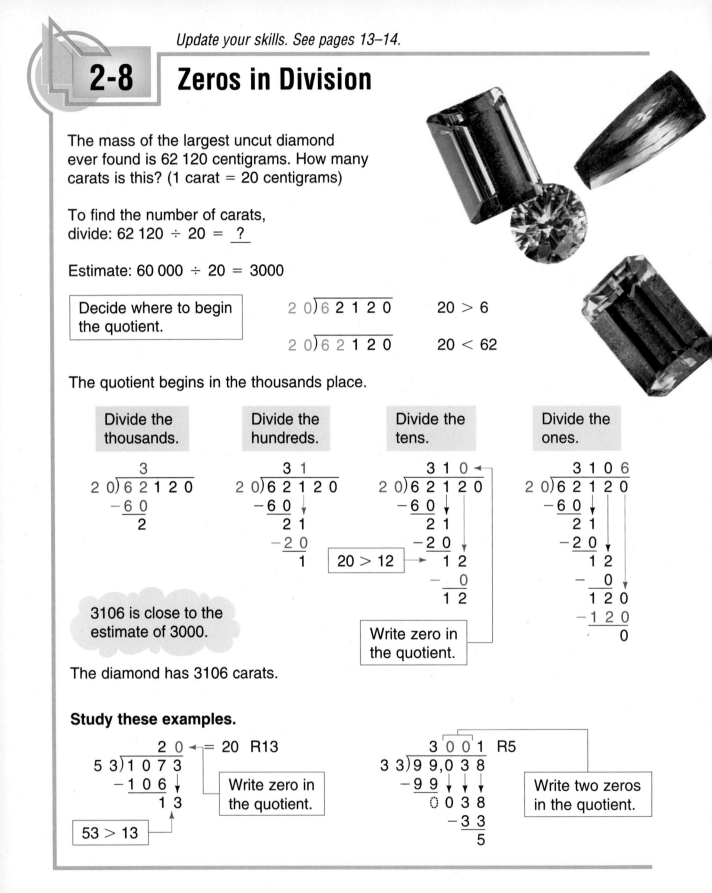

Divide the thousands.	Divide the hundreds.	Divide the tens.	Divide the ones.

$$\begin{array}{r} 3 \\ 20\overline{)62120} \\ \underline{-60} \\ 2 \end{array}$$

$$\begin{array}{r} 31 \\ 20\overline{)62120} \\ \underline{-60}\downarrow \\ 21 \\ \underline{-20} \\ 1 \end{array}$$

$$\begin{array}{r} 310 \\ 20\overline{)62120} \\ \underline{-60}\downarrow \\ 21 \\ \underline{-20}\downarrow \\ 12 \\ \underline{-\ 0} \\ 12 \end{array}$$

20 > 12

$$\begin{array}{r} 3106 \\ 20\overline{)62120} \\ \underline{-60}\downarrow \\ 21 \\ \underline{-20}\downarrow \\ 12 \\ \underline{-\ 0}\downarrow \\ 120 \\ \underline{-120} \\ 0 \end{array}$$

3106 is close to the estimate of 3000.

Write zero in the quotient.

The diamond has 3106 carats.

Study these examples.

$$\begin{array}{r} 20 \\ 53\overline{)1073} \\ \underline{-106}\downarrow \\ 13 \end{array} = 20 \ R13$$

53 > 13

Write zero in the quotient.

$$\begin{array}{r} 3001 \ R5 \\ 33\overline{)99,038} \\ \underline{-99}\downarrow\downarrow\downarrow \\ 0\ 0\ 3\ 8 \\ \underline{-33} \\ 5 \end{array}$$

Write two zeros in the quotient.

78

Estimate and then find each quotient.

1. $30\overline{)126}$ **2.** $38\overline{)230}$ **3.** $25\overline{)266}$ **4.** $44\overline{)882}$

5. $15\overline{)1634}$ **6.** $12\overline{)1312}$ **7.** $40\overline{)2060}$ **8.** $50\overline{)3602}$

9. $32\overline{)9843}$ **10.** $41\overline{)8547}$ **11.** $84\overline{)25,297}$ **12.** $66\overline{)39,967}$

13. $72\overline{)72,072}$ **14.** $43\overline{)86,129}$ **15.** $82\overline{)24,637}$ **16.** $91\overline{)36,408}$

Find the value of _n_. You may use a calculator to help you.

17. $n = 28,671 \div 57$ **18.** $n = 14,558 \div 29$ **19.** $504,144 \div 36 = n$

20. $696,024 \div 24 = n$ **21.** $n = 186 \times 400,472$ **22.** $n = 223 \times 681,009$

Use the table to find the number of carats in each gem. (1 carat = 20 centigrams)

Masses of Largest Gems	
Gem	**Mass (in centigrams)**
Cut diamond	10 600
Ruby	170 000
Emerald (single crystal)	140 500
Sapphire (carved)	46 040
Opal	527 000
Black opal (rough)	40 400

23. Cut diamond   **24.** Ruby   **25.** Emerald

26. Sapphire   **27.** Opal   **28.** Black opal

CRITICAL THINKING

Write number sentences using any two of these numbers: 133, 1, 0, 133, 4056. Tell whether these statements are (a) always true; (b) sometimes true; or (c) never true.

29. The product is zero.   **30.** The quotient is zero.

31. The sum is zero.   **32.** The product is greater than 4000.

33. The difference is zero.   **34.** The quotient is greater than 30.

2-9 Finding Quotients

Alaska's shoreline is how many times the length of Connecticut's shoreline?

To find how many times the length, divide:

33,904 ÷ 618 = __?__

Estimate: 30,000 ÷ 600 = 50

Miles of Shoreline	
Alaska	.33,904
Florida.	.8,426
Maine.	.3,478
Connecticut.	.618
Alabama.	.607
New Hampshire.	.131
Pennsylvania.	.89

Divide the tens.	Divide the ones.	Check.

```
       5              54 R532            6 1 8
6 1 8)3 3,9 0 4   6 1 8)3 3,9 0 4      ×    5 4
     -3 0 9 0         -3 0 9 0          2 4 7 2
      3 0 0            3 0 0 4          3 0 9 0
                     -2 4 7 2          3 3,3 7 2
                        5 3 2        +     5 3 2
                                      3 3,9 0 4
```

Alaska's shoreline is about 55 times the length of Connecticut's shoreline.

54 R532 is close to the estimate of 50.

Study this example of dividing a money amount by a whole number.

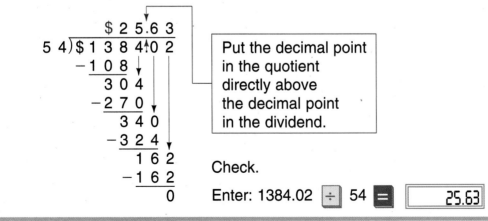

```
        $ 2 5.6 3
5 4)$ 1 3 8 4.0 2
   -1 0 8
     3 0 4
    -2 7 0
       3 4 0
      -3 2 4
         1 6 2
        -1 6 2
             0
```

Put the decimal point in the quotient directly above the decimal point in the dividend.

Check.

Enter: 1384.02 ÷ 54 = [25.63]

80

Estimate and then find each quotient.

1. $45\overline{)568}$
2. $62\overline{)928}$
3. $52\overline{)6638}$
4. $34\overline{)5777}$

5. $36{,}389 \div 82$
6. $30{,}139 \div 93$
7. $2710 \div 759$
8. $2350 \div 315$

9. $423\overline{)98{,}982}$
10. $514\overline{)88{,}408}$
11. $480\overline{)\$5784.00}$
12. $350\overline{)\$7374.50}$

How many times greater is Alaska's shoreline than the shoreline of each of these states? Use the chart on page 80.

13. Pennsylvania
14. New Hampshire
15. Alabama

16. Maine's shoreline is how many times that of New Hampshire?

17. Florida's shoreline is how many times that of Connecticut's and Alabama's shorelines combined?

PROBLEM SOLVING

18. Arizona's land area is 113,642 square miles and its water area is 364 square miles. How many times greater is the land area than the water area?

19. Kansas' land area is 81,823 square miles and its water area is 459 square miles. How many times greater is the land area than the water area?

  **CALCULATOR ACTIVITY**

20. On planet NO-LEAP each year has exactly 365 days. EU-2's father is 14,977 days old. How many days ago was his birthday?

Think: $14{,}977 \div 365 = \underline{?}$ years $\underline{?}$ days

[1] [4] [9] [7] [7] [INT÷] [3] [6] [5] [=]   Q 41   R 12

The birthday was 12 days ago.

21. How many days ago was the birthday?

   a. Mom: 14,418 days old   **b.** Grandpa: 30,441 days old

2-10 Order of Operations

Frances wants to determine the number of apples she is carrying in her pickup truck. Two large crates contain 215 apples each. Three other crates contain 150 apples each.

Frances computes as follows:
Number of apples $= 2 \times 215 + 3 \times 150$
$= 430 + 450$
$= 880$

When a mathematical expression contains more than one operation, mathematicians have agreed to follow these **order of operations** rules:

- First *multiply* or *divide*.
  Work from left to right.

$$6 + \underbrace{4 \times 2}$$

- Then *add* or *subtract*.
  Work from left to right.

$$6 + \quad 8 \quad = 14$$

▶ When there are parentheses, do the operations within the parentheses first. Then follow the order of operations.

Compute: $(42 \div 7) \times 3 - 4 \div 2 + 1$

$$(42 \div 7) \times 3 - 4 \div 2 + 1$$ Do operations within parentheses.

$$6 \quad \times 3 - 4 \div 2 + 1$$ Multiply. Divide.

$$18 \quad - \quad 2 \quad + 1$$ Subtract.

$$16 \quad\quad + 1$$ Add.

$$17$$

Tell which operation is to be done first. Then compute.

1. $3 \times 9 + 8$

2. $16 \div 4 + 2$

3. $15 - 6 \div 3$

4. $7 + 11 \times 3$

5. $21 - 9 + 11$

6. $27 \div 9 \times 3$

7. $(14 \div 2) + 6$

8. $2 \times (15 - 3)$

9. $64 \div (8 \times 8)$

Compute. Watch the order of operations.

10. $4 \times 8 \times 3 - 2$

11. $18 \div 6 \div 3 - 1$

12. $9 + 3 \times 2 + 4$

13. $12 - 3 \times 1 + 2$

14. $(40 \div 4) + 5 - 3 + (6 \times 4)$

15. $5 + (34 - 2) \div 8 + (17 + 2)$

16. $10 \times 3 + (48 \div 6) \times 4$

17. $(50 \div 10) \times 2 + 6 \times 6$

18. $7 + 3 - (5 \div 5 \times 2)$

19. $(19 + 1 \times 7 - 6) \div 5$

Choose the correct mathematical expression to solve each problem. Then solve.

20. Catherine works after school at a job for which she is paid $25 a day. She makes tips of $15 a day. How much money does she receive in 5 days?

 a. $5 \times 25 + 15$ b. $5 \times 25 + 5 \times 15$ c. $25 + 15 \times 5$

21. Leon packs 30 business envelopes in each of 25 boxes and 30 regular envelopes in each of 20 boxes. How many envelopes does he pack?

 a. $30 \times 25 + 20$ b. $(30 \times 25) + 20$ c. $30 \times 25 + 30 \times 20$

CHALLENGE

Compute. Watch the order of operations.

Do powers first. Then do the other operations.

22. 4×10^3 $4 \times 10^3 = 4 \times 10 \times 10 \times 10 = 4000$

23. $10^2 \times 3$

24. 6×10^4

25. $10^3 \div 5$

26. $10^6 \div 10^2$

TECHNOLOGY

Use the Calculator Key $\boxed{y^x}$

You can use a calculator to find the standard numeral for 5^4.

▶ Here are two key sequences that can be used to find the standard numeral for 5^4.

exponent

5^4 means 5 is used as a factor 4 times.

base

1. Press these keys → $\boxed{5}$ $\boxed{\times}$ $\boxed{5}$ $\boxed{\times}$ $\boxed{5}$ $\boxed{\times}$ $\boxed{5}$ $\boxed{=}$ ⟶ $\boxed{625.}$ ← Display

or

2. $\boxed{5}$ $\boxed{y^x}$ $\boxed{4}$ $\boxed{=}$ ⟶ $\boxed{625.}$ ← Display

base — exponent

> exponent
>
> base ⟶ y^x ⟶ raises the number entered, 5, to the power named by the next number entered, 4.

▶ Use the $\boxed{y^x}$ to find the product of $4^2 \times 4^3 = \underline{\ ?\ }$.

Press these keys → $\boxed{4}$ $\boxed{y^x}$ $\boxed{2}$ $\boxed{\times}$ $\boxed{4}$ $\boxed{y^x}$ $\boxed{3}$ $\boxed{=}$ ⟶ $\boxed{1024.}$

So $4^2 \times 4^3 = 1024$.

Display

▶ Use the $\boxed{y^x}$ key to find the quotient of $10^6 \div 10^2 = \underline{\ ?\ }$.

Press these keys → $\boxed{1}$ $\boxed{0}$ $\boxed{y^x}$ $\boxed{6}$ $\boxed{\div}$ $\boxed{1}$ $\boxed{0}$ $\boxed{y^x}$ $\boxed{2}$ $\boxed{=}$ ⟶ $\boxed{10000.}$

So $10^6 \div 10^2 = 10,000$.

Display

Use the $\boxed{y^x}$ key to find each standard numeral.

1. 10^2　　**2.** 3^2　　**3.** 10^0　　**4.** 6^2　　**5.** 2^3　　**6.** 3^4

7. 9^3　　**8.** 1^9　　**9.** 0^7　　**10.** 5^5　　**11.** 8^3　　**12.** 6^4

84

Find the product or quotient. Watch for × and ÷.

13. $3^2 \times 3^3 =$?

14. $4^5 \times 4^0 =$?

15. $6^2 \div 6^1 =$?

16. $10^5 \div 10^3 =$?

17. $10^3 \times 10^4 =$?

18. $10^8 \div 10^3 =$?

19. $8^2 \times 8^5 =$?

20. $9^4 \times 9^2 =$?

21. $10^{10} \div 10^7 =$?

Compare. Write <, =, or >.

22. $2^2 \times 2^3$? 2^5

23. $4^8 \times 4^1$? 4^9

24. $7^0 \times 7^5$? 7^5

25. $8^4 \times 8^3$? 8^7

26. $5^3 \times 5^5$? 5^8

27. $3^4 \times 3^4$? 3^8

28. $9^4 \div 9^1$? 9^3

29. $10^5 \div 10^3$? 10^2

30. $6^9 \div 6^4$? 6^5

31. $4^5 \div 4^3$? 4^2

32. $11^6 \div 11^2$? 11^4

33. $5^{10} \div 5^7$? 5^3

34. Study the exponents in exercises 22–33.
What do you discover?

Write the missing standard numeral using exponents.

35. ? $\times 6^4 = 6^9$

36. ? $\div 10^9 = 10^3$

37. $5^7 \times$? $= 5^{10}$

38. $10^8 \div$? $= 10^0$

39. ? $\div 10^4 = 10^2$

40. $7^3 \times$? $= 7^{11}$

41. $9^0 \times$? $= 9^8$

42. $10^9 \div$? $= 10^8$

43. $12^5 \times$? $= 12^{14}$

Compare. Write <, =, or >.

44. $8^4 \times 8^1$? $10^5 \div 10^2$

45. $3^4 \times 3^1$? $10^7 \div 10^5$

46. $4^2 \times 4^2$? $10^{10} \div 10^8$

47. $9^0 \times 9^2$? $10^7 \div 10^5$

48. $10^0 \div 10^0$? $8^2 \times 8^0$

49. $10^6 \div 10^3$? $5^1 \times 5^3$

PROBLEM SOLVING

50. Find a pattern in the powers
of 3. Predict the last digit of
the standard numeral for 3^{12}.

51. How would you find the
standard numeral for 8^5
if $8^4 = 4096$?

2-12 | Problem Solving: Interpret the Remainder

Problem: To celebrate Somerville's 200th anniversary 2000 people are invited to a formal dinner. If 12 people are seated at each table, how many tables will be needed?

1 IMAGINE Picture tables that seat 12 people.

2 NAME *Facts:* 1 table—12 people
total—2000 people

Question: How many tables will be needed for 2000 people?

3 THINK Since one table holds 12 people, divide 2000 people by 12 people.

$2000 \div 12 = \underline{\ ?\ }$

4 COMPUTE

```
        1 6 6  R8
1 2)2 0 0 0
   −1 2 ↓
      8 0
     −7 2 ↓
         8 0
        −7 2
           8
```

> There is a remainder of 8 people so 1 more table is needed.

Somerville needs 167 tables to seat 2000 people.

5 CHECK Multiply and add to check your answer.

```
      1 6 6
    ×   1 2
      3 3 2
    1 6 6
    1 9 9 2
  +       8
    2 0 0 0
```

> Check by multiplying the quotient and the divisor. Then add the remainder.

The answer checks.

Solve.

1. The city has planned a 15 km walk for hunger. The goal is to raise $25,000. If the pledge is $1.00 per km, how many people will need to walk 15 km to reach or go beyond the goal?

IMAGINE   Put yourself in the problem.

NAME   *Facts:*   walk for hunger—15 km
                          pledge per km—$1.00
                          goal—$25,000

             Question: How many people will need to walk 15 km
                             to reach or go beyond the goal?

THINK   Each person who walks 15 km will raise $15.

             To find how many people will need to raise $15
             in order to reach the goal of $25,000, divide:
             $25,000 ÷ $15 = ?

             *Think: What will a
             remainder mean?*

COMPUTE ➤ **CHECK**

2. Each touring van will accommodate 22 people. If one group has 170 people, how many touring vans will the group need?

3. Two hundred twenty-five dignitaries are invited to a parade. There are three reviewing stands that each seat 70 people. How many extra chairs will be needed to seat all the dignitaries?

4. Each float for the parade is to be decorated with 1025 carnations. There are 16 floats. If carnations come 500 to a box, how many boxes will be needed?

5. Festival organizers plan to have 170 fireworks set off at night. The show will last $\frac{1}{2}$ hour excluding the finale. If the show calls for the same number of fireworks to be set off each minute, how many should go off each minute? How many fireworks will there be for the finale?

6. Local vendors plan to sell hot dogs during the festival. Their goal is to sell 5000 hot dogs. If hot dogs are packed 48 to a box, how many boxes should the vendors order?

Problem-Solving Applications

Solve each problem and explain the method you used.

1. Rachel's craft group is building a collection of model ships. Rachel cuts 5-in. masts out of balsa wood dowels. How many masts can she cut from a 72-in. dowel?

2. Can Eric equally divide an 87-in. dowel into 3-in. lengths?

3. Ramon cuts 100 miniature wooden planks to build decks for 11 ships. What is the greatest equal number of planks he can use for each deck?

4. A ball of thin twine makes convincing ropes for the model ships. Inez cuts 15 equal pieces from one 375-in. ball of twine. How long is each piece of twine?

5. Ramon counts the wooden planks he uses for one ship. He discovers that the number of planks is divisible by 2, 3, 5, 9, and 10. What is the least number of planks he could have used?

6. Models built by four craft groups will be exhibited together. Each display case will hold 6 model ships. There will be 117 model ships in the exhibit. How many display cases will be needed?

7. The sails in one model are 128 mm long. How many sails can Kevin cut from a strip of cloth 1200 mm long?

8. The sails in a smaller model are 109 mm long. Talia cuts as many sails as she can from a 2000-mm strip of cloth. Is there enough cloth left over to make a flag that is 50 mm long?

9. The craft group's collection of model ships has a mass of 8064 g. The mass of each ship is 448 g. How many ships are in the collection?

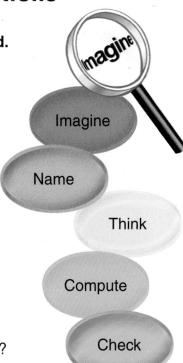

Imagine

Name

Think

Compute

Check

Solve. Use a strategy from the list or another strategy you know to do each problem.

USE THESE STRATEGIES:
Interpret the Remainder
Write a Number Sentence
Guess and Test
Extra Information
Use a Graph
Hidden Information

10. A craft group of 10 people spent $1496.88 on materials for 22 model ships. If the cost of materials was the same for each, what did it cost to build one model ship?

11. The History Museum displays a variety of models. A model fort is built entirely from miniature bricks. Each brick has a mass of 115 g. The model fort has a mass of 143 175 g. How many bricks are used in this model?

12. Another historical model shows a Civil War battlefield. There are twice as many Confederate soldiers as Union soldiers in the model. There are 639 soldiers in the display. How many Confederate soldiers are there?

13. A model of the *Monitor* requires 2350 bolts. The bolts are produced in sets of 15. How many sets of bolts must be ordered to make this model?

14. Many museum visitors enjoy the model of the *Spirit of St. Louis* built entirely from toothpicks. The model has a mass of 19 124 g. Each toothpick has a mass of 4 g. How many toothpicks were used in the model?

15. The model builder of the *Spirit of St. Louis* took 45 seconds to place each toothpick. How many minutes did it take her to build the entire model?

Use the bar graph for problems 16 and 17.

16. How many more kits for model aircraft were sold than for model buildings?

17. Model car kits cost $8.95 each. How much money was spent on model car kits?

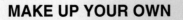

MAKE UP YOUR OWN

18. Write a problem modeled on problem 13 above. Have a classmate solve it.

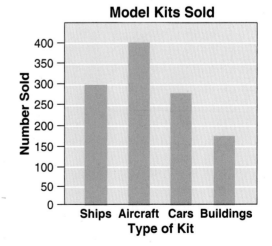

Model Kits Sold

Number Sold (y-axis: 0, 50, 100, 150, 200, 250, 300, 350, 400)

Type of Kit: Ships, Aircraft, Cars, Buildings

Complete. Identify the property of multiplication used. *(See pp. 64–65.)*

1. $6 \times (5 \times 9) = ($ _?_ $\times 5) \times 9$

2. $12 \times (8 + 4) = (12 \times 8) + (12 \times$ _?_ $)$

3. $c \times 9 = 9 \times$ _?_

4. $(7 \times 5) \times$ _?_ $= 7 \times (5 \times 4)$

Multiply or divide as indicated. Look for patterns. *(See pp. 66–67.)*

5.
10×9
100×9
1000×9

6.
10×47
100×47
1000×47

7. $138,000 \div 30$
$138,000 \div 300$
$138,000 \div 3000$

Estimate and then find each product. *(See pp. 68–71.)*

8.
446
$\times\ 127$

9.
752
$\times\ \ \ 683$

10.
$\$54.79$
$\times\ \ \ 246$

11.
5062
$\times\ \ 805$

12.
8009
$\times 3206$

Write the standard numeral. *(See pp. 72–73.)*

13. 10^3

14. 10^6

15. 2^5

16. 23^2

17. 30^3

Which numbers are divisible by 3? by 6? by 9? *(See pp. 74–75.)*

18. 155

19. 2850

20. 1278

21. 21,669

22. 834,102

Estimate and then find each quotient. *(See pp. 76–81.)*

23. $53\overline{)769}$

24. $35\overline{)3579}$

25. $432\overline{)9510}$

26. $389\overline{)\$2789.13}$

Compute. Watch the order of operations. *(See pp. 82–83.)*

27. $10 \times 4 + 3 - 5$

28. $(72 \div 12) \times 4 + 5 \times 5$

PROBLEM SOLVING *(See pp. 86–89.)*

29. Walking at 3 miles per hour, Lida burns about 5 Calories per minute. At that rate, how many Calories are burned in 5 hours?

30. The sixth and seventh grades have 362 students taking buses for a field trip. Each bus holds 46 people. What is the fewest number of buses needed?

(See *Still More Practice,* pp. 507–508.)

NAPIER'S BONES

John Napier (1550–1617) invented a simple "calculator" called **Napier's Bones** or **Napier's Rods**. This calculator can be used to multiply.

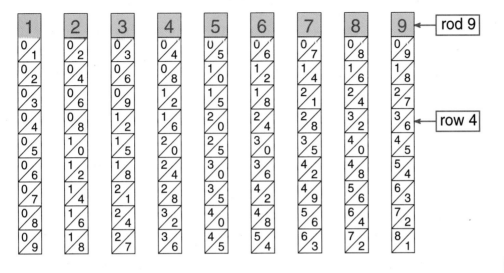

To multiply 93 × 4:

- Place rods 9 and 3 side by side.
- Go down 4 rows.
- Add along the diagonals.

So 93 × 4 = 372.

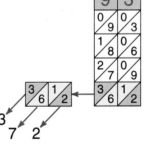

To multiply 753 × 24:

- Find rods 7, 5, and 3.
- Line up the boxes in row 2 with the boxes in row 4.
- Add diagonally. Regroup to the next diagonal if necessary.

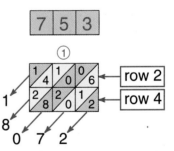

753 × 24 = 18,072

Use Napier's Bones to find each product.

1. 56 × 4   **2.** 89 × 3   **3.** 647 × 8   **4.** 372 × 6

5. 259 × 16   **6.** 822 × 15   **7.** 246 × 23   **8.** 3476 × 58

Check Your Mastery

Tell how many digits there are in the product or the quotient. Explain your answer.

1. 3-digit number × 3-digit number

2. 4-digit number × 3-digit number

3. 5-digit number ÷ 2-digit number

4. 6-digit number ÷ 3-digit number

Estimate and then find each product.

5.
$$\begin{array}{r} 5300 \\ \times\ 200 \\ \hline \end{array}$$

6.
$$\begin{array}{r} 863 \\ \times 794 \\ \hline \end{array}$$

7.
$$\begin{array}{r} \$72.19 \\ \times\ \ \ \ 350 \\ \hline \end{array}$$

8.
$$\begin{array}{r} 4057 \\ \times\ \ 908 \\ \hline \end{array}$$

Write the standard numeral.

9. 6^2

10. 4^3

11. 10^1

12. 10^6

Estimate and then find each quotient.

13. $47\overline{)865}$

14. $62\overline{)6389}$

15. $321\overline{)8760}$

16. $534\overline{)\$4763.28}$

Compute. Watch the order of operations.

17. $40 - 6 \div 3 - 2 + 7$

18. $19 + 1 + (3 + 5 \times 2)$

PROBLEM SOLVING *Use a strategy or strategies you have learned.*

19. A restaurant estimates that it sells 575 dinners per day. At that rate, how many dinners will it sell in a year?

20. Rhode Island has an area of 1545 square miles and Texas has an area of 268,601 square miles. How many times greater is the area of Texas than that of Rhode Island?

21. Paisley's Videos had 1206 videotapes in stock. Selling them in packages of 6 tapes each, 185 packages were sold. The extra packages of tapes were given away as door prizes. How many packages were given away? How many tapes were given away?

Decimals: Addition and Subtraction

3

PRAISE SONG FOR A DRUMMER

The drum drums health,
The drum drums wealth,
He takes his wife six hundred thousand cowries.
The drum drums health,
The drum drums wealth,
He takes his son six hundred thousand cowries,
The drum drums health,
The drum drums wealth.

Mary Smith, translator

In this chapter you will:

Read, write, round, compare, and order decimals
Estimate, add, and subtract decimals
Solve problems using simpler numbers

Critical Thinking/Finding Together

Research how the cowrie has been used as money in Africa and elsewhere. If one cowrie = $.95, find the value of 15 cowries.

3-1 Decimals

To read a decimal, you need to know the **place** of each digit. The place-value chart below will help you.

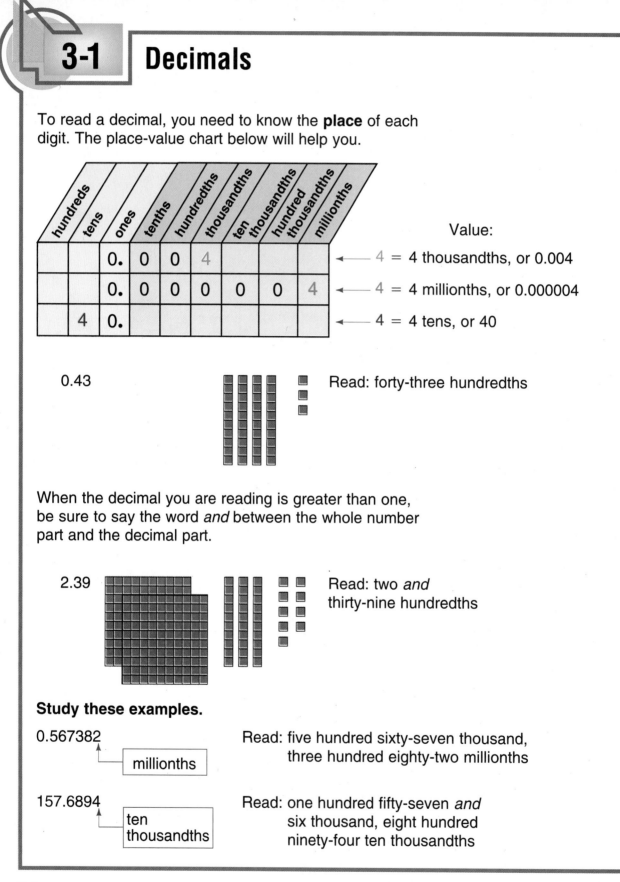

Value:

4 = 4 thousandths, or 0.004

4 = 4 millionths, or 0.000004

4 = 4 tens, or 40

0.43   Read: forty-three hundredths

When the decimal you are reading is greater than one, be sure to say the word *and* between the whole number part and the decimal part.

2.39   Read: two *and* thirty-nine hundredths

Study these examples.

0.567382
└── millionths

Read: five hundred sixty-seven thousand, three hundred eighty-two millionths

157.6894
└── ten thousandths

Read: one hundred fifty-seven *and* six thousand, eight hundred ninety-four ten thousandths

Write the place of the underlined digit. Then write its value.

1. 0.0<u>5</u> **2.** 0.00<u>8</u> **3.** 0.0000<u>6</u> **4.** 0.000<u>1</u>

5. 0.42<u>9</u>7 **6.** 0.8<u>1</u>523 **7.** 1.9876<u>2</u>3 **8.** 7.01432<u>5</u>

9. 1<u>6</u>.1876 **10.** 187.927<u>4</u>3 **11.** 550.1976<u>0</u>8

Use 346.789125. Name the digit in the given place.

12. thousandths **13.** tenths **14.** ones

15. tens **16.** millionths **17.** hundredths

18. hundreds **19.** ten thousandths **20.** hundred thousandths

Read each decimal. Then write the word name.

21. 0.004 **22.** 0.0006 **23.** 0.000079 **24.** 0.00062

25. 79.8 **26.** 8.408 **27.** 200.02 **28.** 0.314

29. 5.042019 **30.** 12.00678 **31.** 1.568970

In which decimal does the digit 4 have the greater value?
By how many times?

32. a. 2.473 **33. a.** 35.674 **34. a.** 0.576843

    b. 9.846   **b.** 8.37984   **b.** 126.97426

 CRITICAL THINKING

Write in your Math Journal whether the statement
is true about (a) all of the numbers; (b) some
of the numbers; or (c) none of the numbers.

16.07965
123.00938
3.789340

35. My millionths digit is 5. **36.** My thousandths digit is 9.

37. My ten thousandths digit is 3 more than my millionths
digit and the same as my ones digit.

3-2 Decimals and Expanded Form

The value of each digit in a decimal is shown
by writing the decimal in **expanded form**.

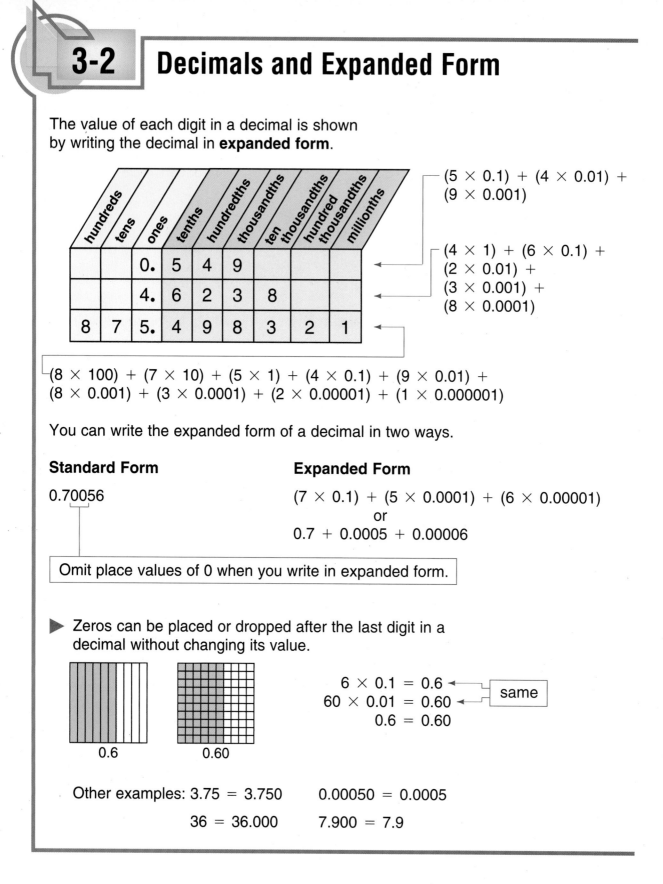

hundreds	tens	ones	tenths	hundredths	thousandths	ten thousandths	hundred thousandths	millionths
		0.	5	4	9			
		4.	6	2	3	8		
8	7	5.	4	9	8	3	2	1

(5 × 0.1) + (4 × 0.01) +
(9 × 0.001)

(4 × 1) + (6 × 0.1) +
(2 × 0.01) +
(3 × 0.001) +
(8 × 0.0001)

(8 × 100) + (7 × 10) + (5 × 1) + (4 × 0.1) + (9 × 0.01) +
(8 × 0.001) + (3 × 0.0001) + (2 × 0.00001) + (1 × 0.000001)

You can write the expanded form of a decimal in two ways.

Standard Form

0.70056

Expanded Form

(7 × 0.1) + (5 × 0.0001) + (6 × 0.00001)
or
0.7 + 0.0005 + 0.00006

Omit place values of 0 when you write in expanded form.

▶ Zeros can be placed or dropped after the last digit in a
decimal without changing its value.

6 × 0.1 = 0.6
60 × 0.01 = 0.60   same
0.6 = 0.60

0.6   0.60

Other examples: 3.75 = 3.750   0.00050 = 0.0005

36 = 36.000   7.900 = 7.9

Complete.

1. $0.075 = (\underline{\ ?\ } \times 0.01) + (\underline{\ ?\ } \times 0.001)$

2. $0.00040 = (\underline{\ ?\ } \times 0.0001)$

3. $4.0008 = (\underline{\ ?\ } \times 1) + (\underline{\ ?\ } \times 0.0001)$

4. $0.000009 = (\underline{\ ?\ } \times 0.000001)$

5. $0.005496 = (5 \times \underline{\ ?\ }) + (4 \times \underline{\ ?\ }) + (9 \times \underline{\ ?\ }) + (6 \times \underline{\ ?\ })$

6. $38.000502 - (3 \times \underline{\ ?\ }) + (8 \times \underline{\ ?\ }) + (5 \times \underline{\ ?\ }) + (2 \times \underline{\ ?\ })$

Write in expanded form in two ways.

7. 0.0023

8. 0.000590

9. 5.0138

10. 0.00098

11. 0.045678

12. 78.5009

13. 600.000006

14. 8000.079

15. 30,509.09

Short Word Names for Decimals

Standard Form: 0.000301
Word Name: three hundred one
             millionths
Short Word Name: 301 millionths

Standard Form: 14.0065
Word Name: fourteen and
            sixty-five ten thousandths
Short Word Name: 14 and 65
            ten thousandths

Write the decimal in standard form.

16. 4 thousandths

17. 32 hundred thousandths

18. 8 and 48 hundredths

19. 8000 and 8 thousandths

20. 900 and 7 millionths

21. 907 millionths

22. 330 hundred thousandths

23. 57,820 hundred thousandths

SKILLS TO REMEMBER

Round to the greatest place.

24. 88

25. 63

26. 45

27. 34

28. 924

29. 655

30. 8332

31. 4569

Rounding Decimals

The actual diameter of a water pipe is 1.9229 in. You can use a **rounded number** to express this measurement.

To round a decimal to a given place:

- Find the place you are rounding to.

- Look at the digit to its right.
  If the digit is *less than 5*, round *down*.
  If the digit is *5 or more*, round *up*.

- Drop all the digits to the right of the place you are rounding to.

Round 1.9229 to the nearest thousandth.

1.9229 ↓ 1.923	The digit to the right of the thousandths place is 9. 9 > 5 Round up to 1.923. Do not write a zero.

Round 1.9229 to the nearest tenth.

1.9229 ↓ 1.9	The digit to the right of the tenths place is 2. 2 < 5 Round down to 1.9. Drop all the digits to the right.

Study these examples.

Round to the greatest nonzero place.

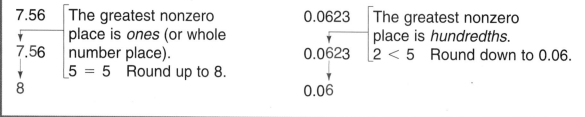

7.56 ↓ 7.56 ↓ 8	The greatest nonzero place is *ones* (or whole number place). 5 = 5 Round up to 8.	0.0623 ↓ 0.0623 ↓ 0.06	The greatest nonzero place is *hundredths*. 2 < 5 Round down to 0.06.

Round each number to the underlined place.

1. 0.8̲2
2. 0.7̲9
3. 0.29̲3
4. 0.14̲8
5. 0.416̲9
6. 0.834̲4
7. 8̲.3
8. 2̲.4
9. 38̲.62
10. 75̲.203
11. 5.5̲31
12. 9.8̲59
13. 15.13̲32
14. 2.14̲55
15. 14.063̲01
16. 23.987̲65
17. 0.9̲873
18. 12.9̲56
19. 0.399̲6
20. 6.789̲0

Round each number to the greatest nonzero place.

21. 0.37
22. 0.69
23. 2.814
24. 7.099
25. 0.4296
26. 0.5535
27. 0.078
28. 0.064
29. 68.347
30. 47.8243
31. 609.8432
32. 2046.73

Round to the nearest cent.

33. $4.368 $4.368   8 > 5; Round up: $4.37
34. $5.472
35. $35.476
36. $12.525
37. $.463
38. $.085

Place a decimal point in each numeral so that the sentence seems reasonable. Then round the decimal to the nearest tenth.

39. Conrad rode his racing bicycle 1575 miles in an hour.

40. The record in the marathon was 26693 hours.

41. Maria's science grades averaged 894 for the month.

PROBLEM SOLVING

42. A measurement of one inch is equal to 0.0254 meter. Round this decimal to the nearest hundredth and thousandth.

43. The price of gasoline at three service stations was $1.129, $.959, and $1.299. Round each price to the nearest cent.

3-4 Compare and Order Decimals

To compare two decimals, start at the left and compare digits in the same places. Start with the greatest place, then the next greatest place, and so on.

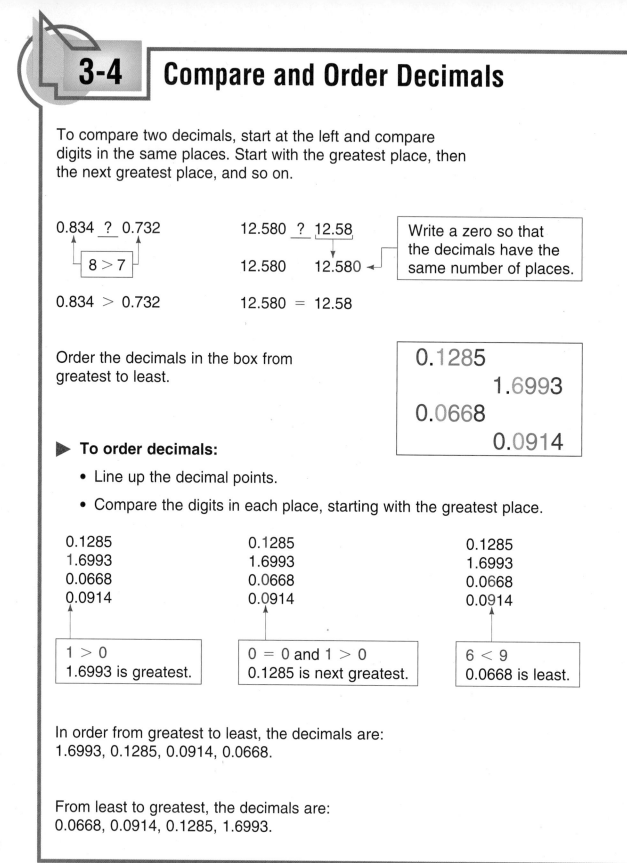

0.834 ? 0.732

8 > 7

0.834 > 0.732

12.580 ? 12.58

12.580   12.580

Write a zero so that the decimals have the same number of places.

12.580 = 12.58

Order the decimals in the box from greatest to least.

> 0.1285
> 1.6993
> 0.0668
> 0.0914

▶ **To order decimals:**

• Line up the decimal points.

• Compare the digits in each place, starting with the greatest place.

0.1285	0.1285	0.1285
1.6993	1.6993	1.6993
0.0668	0.0668	0.0668
0.0914	0.0914	0.0914

1 > 0
1.6993 is greatest.

0 = 0 and 1 > 0
0.1285 is next greatest.

6 < 9
0.0668 is least.

In order from greatest to least, the decimals are:
1.6993, 0.1285, 0.0914, 0.0668.

From least to greatest, the decimals are:
0.0668, 0.0914, 0.1285, 1.6993.

Compare. Write <, =, or >.

1. 0.46 ? 0.39

2. 0.709 ? 0.921

3. 0.06 ? 0.60

4. 9.8 ? 9.80

5. 0.509 ? 0.510

6. 0.623 ? 0.627

7. 0.4286 ? 0.4190

8. 0.5691 ? 0.5690

9. 0.53 ? 0.6

10. 0.8 ? 0.78

11. 7.610 ? 7.61

12. 7.3 ? 7.31

13. 2.34 ? 2.513

14. 91.42 ? 90.425

15. 0.059 ? 0.59

Write in order from greatest to least.

16. 0.75, 0.39, 0.2, 0.35

17. 0.484, 0.495, 0.523, 0.54

18. 8.63, 8.6, 8.65, 7.99

19. 9.21, 9.0, 9.2, 9.06

20. 0.5478, 0.546, 0.5462, 0.5593

21. 8.134, 8.215, 8.2152, 8.2052

Write in order from least to greatest.

22. 2.7054, 0.9832, 1.2396, 0.9276

23. 2.7993, 0.0803, 0.0779, 0.2396

24. 0.1211, 0.12, 0.121, 0.0911

25. 0.052387, 0.52386, 0.05023, 0.0523

Order the decimals in each table from greatest to least.

26.

Batting Averages	
Ira	0.278
Henry	0.302
Sam	0.099
Steve	1.000
Mario	0.525

27.

Masses of Five Objects (kilograms)	
A	0.206
B	2.7564
C	0.2
D	0.8384
E	2.76

CHALLENGE

Computation Method

Solve. Use mental math or a calculator to help you.

28. I am a decimal. I am more than 10 times greater than 0.029. I am between 0.2 and 0.3. Who am I?

3-5 Estimating Decimal Sums and Differences

▶ You can use front-end estimation or rounding to estimate decimal sums.
0.82 + 0.29 + 0.36 is about __?__

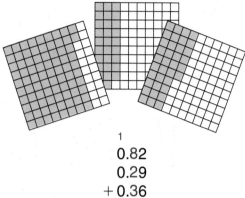

Front-end estimation

- Add the *nonzero* front digits.

- Drop the other digits in the decimal part of the number.

$$\begin{array}{r} 1 \\ 0.82 \\ 0.29 \\ +\,0.36 \\ \hline \text{about } 1.3 \end{array}$$

Estimation by rounding

- Round each decimal to the greatest *nonzero* place of the least number.

- Add the rounded numbers.

$$\begin{array}{rcl} & & 1 \\ 0.82 & \rightarrow & 0.8 \\ 0.29 & \rightarrow & 0.3 \\ +\,0.36 & \rightarrow & +\,0.4 \\ \hline & & \text{about } 1.5 \end{array}$$

Both 1.3 and 1.5 are reasonable estimates of the sum.

▶ Use the same two methods to estimate differences of decimals.

Front-end estimation

- Subtract the *nonzero* front digits.

- Write zeros for the other digits in the whole-number part of the number.

$$\begin{array}{r} 93.36 \\ -\,45.09 \\ \hline \text{about } 50 \end{array}$$

Estimation by rounding

- Round to the greatest *nonzero* place of the lesser number.

- Subtract the rounded numbers.

$$\begin{array}{rcl} 93.36 & \rightarrow & 90 \\ -\,45.09 & \rightarrow & -\,50 \\ \hline & & \text{about } 40 \end{array}$$

Both 40 and 50 are reasonable estimates of the difference.

Estimate the sum or difference. Use front-end estimation.

1. 31.6
   + 18.1

2. 68.7
   − 63.9

3. 7.5
   − 2.9

4. 9.1
   − 3.6

5. 0.87
   − 0.54

6. 0.74
   − 0.15

7. 23.89
   + 69.47

8. 44.59
   + 39.07

9. 0.66
   0.7
   + 0.19

10. 0.84
    0.59
    + 0.8

Estimate the sum or difference by rounding.

11. 7.9
    + 4.3

12. 8.45
    + 7.09

13. 0.82
    + 0.74

14. 0.456
    + 0.819

15. 47.59
    − 12.32

16. 88.51
    − 43.09

17. 49.34
    − 5.72

18. 92.76
    − 9.44

19. 5.41
    0.89
    + 4.3

20. 23.08
    9.5
    + 4.72

Estimate by first rounding each amount to the nearest dollar.

21. $78.34
    − 25.29

22. $156.39
    + 45.48

23. $89.96
    + 12.59

24. $702.66
    − 55.45

PROBLEM SOLVING

25. Kathleen has run 6.75 km in the long-distance race. About how much farther must she run if the race distance is 10 km?

26. During a tour of Europe, Alfredo flew 112.5 km, 41.8 km, and 109.5 km. Estimate the total distance that Alfredo traveled?

27. Juan earned $15.63, $8.95, and $19.82 over a 3-day period. About how much is that?

28. About how much greater is the difference 325.87 − 42.76 than the sum 109.53 + 59.87? Explain the method of estimation you used to get your answer.

Communicate

3-6 Adding Decimals

Mr. Kopald's physical science class is studying electricity. The class has just learned that electric power is measured in kilowatts.

Three of Mr. Kopald's students have conducted an experiment and obtained the data in the table. They must now find the total number of kilowatts used by the three appliances.

Electric Power Usage	
Appliance	**Kilowatts (kw)**
Microwave oven	1.45
Clothes dryer	4
Clock	0.003

Each student computed the total in a different way. Study their computations and decide which student has the correct answer.

Charles

```
  1.4 5
  4
+ .0 0 3
  1.8 5 3
```

Answer: 1.853 kw

Anetta

```
      1
  1.4 5
  .0 0 3
+     4
  1.5 2
```

Answer: 1.52 kw

Luz

```
  1.4 5
  4
+ 0.0 0 3
  5.4 5 3
```

Answer: 5.453 kw

Mr. Kopald reminded the students of these two important facts about adding decimals:

- Decimal points in the numbers must be lined up—one underneath the other—before you add.

- In a whole number, the decimal point is understood to be "on the right of the whole number." So 4 = 4.000. You may add as many zeros as needed.

So Luz had the correct total number of kilowatts.

Estimate. Then find the sum.

1. $\begin{array}{r} 7 \\ +\,8.56 \\ \hline \end{array}$

2. $\begin{array}{r} 8.4 \\ +\,0.93 \\ \hline \end{array}$

3. $\begin{array}{r} 2.57 \\ +\,3.5 \\ \hline \end{array}$

4. $\begin{array}{r} \$5.95 \\ +\,6.98 \\ \hline \end{array}$

5. $\begin{array}{r} \$4.09 \\ +\,3.99 \\ \hline \end{array}$

6. $\begin{array}{r} 0.429 \\ +\,0.316 \\ \hline \end{array}$

7. $\begin{array}{r} 0.519 \\ +\,0.623 \\ \hline \end{array}$

8. $\begin{array}{r} 0.479 \\ +\,0.085 \\ \hline \end{array}$

9. $\begin{array}{r} 0.078 \\ +\,0.789 \\ \hline \end{array}$

10. $\begin{array}{r} 0.008 \\ +\,0.009 \\ \hline \end{array}$

Align and estimate. Then add.

11. 3.9 + 4.96

12. 5.39 + 1.8

13. 0.26 + 0.4 + 0.15

14. 0.8 + 0.75 + 0.6

15. 6 + 9.42

16. 3 + 3.36

17. 0.554 + 0.723 + 0.08

18. 0.352 + 0.07 + 0.425

19. $9.78 + $43.85 + $5

20. $8 + $19.53 + $4.70

PROBLEM SOLVING

21. The amount of electric power used by three appliances is as follows: hair dryer: 1 kw; dishwasher: 2.3 kw; and toaster: 0.7 kw. What is the total number of kilowatts used?

22. Find the total number of kilowatts used by the three appliances on page 104 and the three appliances in problem 21.

 SHARE YOUR THINKING   Communicate

23. Write a letter to your parent(s) or another adult to tell them what you have just learned about adding decimals. Be sure to talk about lining up the decimal points and when extra zeros must be used.

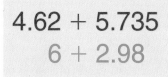

4.62 + 5.735
6 + 2.98

More Adding Decimals

A recent survey shows the amount of liquid that the average person consumes per year: 43.7 gal of soda, 37.3 gal of water, 27.3 gal of coffee, 21.1 gal of milk, and 8.1 gal of juice. What is the total number of gallons of liquid consumed?

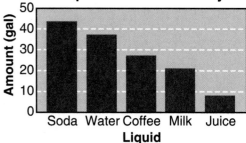

Liquid Consumed Yearly

To find the total number of gallons, add: $43.7 + 37.3 + 27.3 + 21.1 + 8.1 = $ _?_

Estimate: $44 + 37 + 27 + 21 + 8 = 137$

　　　　　　　About 137 gal

Add.

```
    2 1
    4 3.7  ── Line up the decimal points.
    3 7.3
    2 7.3
    2 1.1
  +   8.1
  1 3 7.5  ── Write a decimal point in the sum.
```

The total number of gallons of liquid consumed is 137.5 gal.

137.5 is close to the estimate of 137.

Study this example.

$52.9045 + 63 + 0.7386 + 5.92 = $ _?_

```
   12   11
   52.9045
   63.0000  ── Place a decimal point after 63
    0.7386     and write 4 zeros.
 +  5.9200  ── Write 2 zeros.
  122.5631
```

Check.

Enter: 52.9045 $+$ 63 $+$ 0.7386 $+$ 5.92 $=$ | 122.5631 |

Estimate. Then find the sum. Use a calculator to check.

1.	0.324	**2.**	0.836	**3.**	9.647	**4.**	7.392	**5.**	12
	0.935		0.143		8.932		0.499		8.67
	+ 0.872		+ 0.248		+ 7.456		+ 9.037		+ 9.0013

6.	13.6	**7.**	2.3142	**8.**	0.82	**9.**	10.3564	**10.**	3.2106
	23		0.0859		5.2369		4.2837		0.0008
	+ 1.259		+ 0.6		+ 0.0815		3.0045		0.4426
							+ 9.2304		+ 21.7454

Align and estimate. Then add. Use a calculator to check.

11. 6.8703 + 0.3425 + 4.39 + 0.252   **12.** 0.786 + 0.4908 + 4.75 + 9.8676

13. 9 + 88.431 + 9.0047 + 3.2021   **14.** 12 + 168.9 + 3.0063 + 29.0753

15. $9.10 + $4.83 + $.63 + $10   **16.** $19.53 + $4.48 + $3 + $.70

Choose the correct addends for each sum. Use estimation to help you.
Explain in your Math Journal the method you used for each exercise.

Math Journal ✓

	Sum	Addends			
17.	6.0108	0.6	4.321	2.1408	3.27
18.	1.4868	0.814	0.143	0.6293	0.7145
19.	1.3861	1.2314	0.005	0.1497	1.147
20.	0.011	0.0009	0.009	0.0201	0.0011

PROBLEM SOLVING

21. One week a pilot flew a plane the following distances: 247.6 mi, 80.5 mi, 536.8 mi, 198.2 mi, and 360 mi. What was the total number of miles flown?

22. A businesswoman has $1123.56 in her checking account. She makes the following deposits: $23.82, $507.88, $595, $678.20. How much is in her account now?

23. The odometer on Anna's car showed | 2 | 2 | 4 | 5 | 6 | . | 8 |. She drove 234.7 mi. What did the odometer show then?

24. One bar of gold weighs 9.3145 lb. Another bar weighs 12.9045 lb. What is their combined weight?

3-8 | Subtracting Decimals

The Panama Canal is 81.6 km long and the Suez Canal is 175.5 km long. How much longer than the Panama Canal is the Suez Canal?

To find how much longer, subtract: 175.5 − 81.6 = ?

Estimate: 180 − 80 = 100
        about 100 km longer

Subtract.

Line up the decimal points.	Subtract. Regroup if necessary.	Write the decimal point.
1 7 5.5 − 8 1.6	0 17 4 15 1̸ 7̸ 5̸.5̸ − 8 1.6 9 3 9	0 17 4 15 1̸ 7̸ 5̸.5̸ − 8 1.6 9 3.9

So the Suez Canal is 93.9 km longer than the Panama Canal.

93.9 is close to the estimate of 100.

Study these examples.

9 − 5.25 = ?

Estimate: 9 − 5 = 4

```
    9
  8 10 10
  9̸.0̸ 0̸  ◄—  9 = 9.00
 −5.2 5
  3.7 5
```

0.8 − 0.314 = ?

Estimate: 0.8 − 0.3 = 0.5

```
    9
  7 10 10
  0̸.8̸ 0̸ 0  ◄—  Write 2 zeros.
 −0.3 1 4
  0.4 8 6
```

Check.

Enter: 9 ⊟ 5.25 ⊟ | 3.75 |

Check.

Enter: 0.8 ⊟ 0.314 ⊟ | 0.486 |

Estimate. Then find the difference.

1. 0.586
   − 0.492

2. 0.739
   − 0.548

3. 0.587
   − 0.49

4. 0.893
   − 0.2

5. 0.6
   − 0.49

6. 0.47
   − 0.295

7. 2.5
   − 0.205

8. 1.2
   − 0.189

9. 222.9
   − 15.8

10. 50.7
   − 5.47

Align and estimate. Then subtract.

11. 0.91 − 0.745

12. 0.915 − 0.74

13. 16 − 1.55

14. 8 − 2.047

15. 53.9 − 6

16. $24.98 − $3

17. $28.95 − $.99

18. $147.28 − $65.79

19. $100.28 − $.59

Find the change. Use estimation to help you.

20. Purchase: $9.63
    Give clerk: $10.03

21. Purchase: $7.57
    Give clerk: $8.02

22. Purchase: $6.79
    Give clerk: $10.04

 CHOOSE A COMPUTATION METHOD

Sometimes you can use mental math
to compute quickly. At other times,
you may need to use paper and pencil.

17 + 0.896

Paulo

I can add 17.000 to
0.896 mentally.
17 + 0 = 17 and
17 + 0.896 = 17.896

4.3 − 2.967

Eiko

I must add zeros to
4.3 and then subtract.
I will use paper and pencil.

23. Write two examples of addition of decimals in your Math Journal—
    one that is easy to compute mentally and one that requires paper and
    pencil. Then do similarly with subtraction of decimals.

Math Journal ✓

3-9 More Subtracting Decimals

The table below shows the densities of some common substances. Find the difference between the density of gold and the density of wood.

To find the difference, subtract: 19.3 − 0.85 = ?

Estimate: 19.3 − 0.9 = 18.4

Subtract.

$$\begin{array}{r} \overset{12}{\cancel{}} \\ \overset{8\ \ \cancel{18}\ 10}{\cancel{}} \\ 1\ \cancel{9}.\cancel{3}\ \cancel{0} \\ -\quad 0.8\ 5 \\ \hline 1\ 8.4\ 5 \end{array}$$

Substance	Density
Air	0.0013
Gasoline	0.7
Wood (oak)	0.85
Water (liquid)	1.0
Steel	7.8
Gold	19.3

The difference is 18.45.

18.45 is close to the estimate of 18.4.

Study these examples.

69 − 8.245 = ?

Estimate: 69 − 8 = 61

$$\begin{array}{r} \overset{9\ \ 9}{} \\ \overset{8\ \cancel{10}\ \cancel{10}\ 10}{} \\ 6\ \cancel{9}.\cancel{0}\ \cancel{0}\ \cancel{0} \\ -\quad 8.2\ 4\ 5 \\ \hline 6\ 0.7\ 5\ 5 \end{array}$$

0.067 − 0.0095

Estimate: 0.07 − 0.01 = 0.06

$$\begin{array}{r} \overset{16}{} \\ \overset{5\ \cancel{17}\ 10}{} \\ 0.0\ \cancel{6}\ \cancel{7}\ \cancel{0} \\ -0.0\ 0\ 9\ 5 \\ \hline 0.0\ 5\ 7\ 5 \end{array}$$

Estimate. Then find the difference. Use a calculator to check.

1.	0.8904 − 0.4134	2.	0.6655 − 0.3265	3.	8.7 − 5.923	4.	15.4 − 2.633	5.	$56.83 − 29

6.	$45.09 − 39	7.	0.52 − 0.0816	8.	0.4 − 0.0725	9.	923 − 18.05	10.	220 − 9.9

110

Compare. Write <, =, or >.

11. $29.6814 - 3.3141$ _?_ 26.341 **12.** $8.6114 - 4.31$ _?_ 4.6114

13. $8 - 4.1259$ _?_ $8.1259 - 4$ **14.** $2.491 - 0.4103$ _?_ $4.491 - 2.4103$

15. $3.245 - 1.9831$ _?_ $1.341 - 0.267$ **16.** $0.9 - 0.816$ _?_ $3.241 - 3.239$

Align and estimate. Then subtract. Use a calculator to check.

17. $0.6548 - 0.529$ **18.** $0.653 - 0.2436$ **19.** $80 - 1.552$

20. $17 - 2.495$ **21.** $8.724 - 6.0006$ **22.** $9.7862 - 3.00007$

23. $809.246 - 9.246$ **24.** $176.235 - 39.235$ **25.** $300.9 - 7.9362$

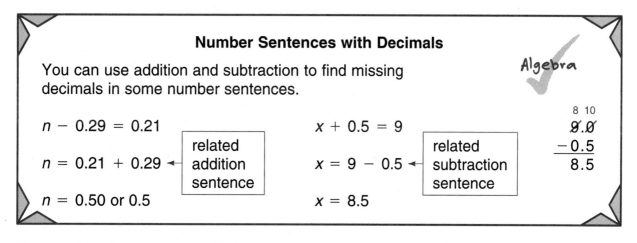

Number Sentences with Decimals

You can use addition and subtraction to find missing decimals in some number sentences.

Algebra

$n - 0.29 = 0.21$

$n = 0.21 + 0.29$ ← related addition sentence

$n = 0.50$ or 0.5

$x + 0.5 = 9$

$x = 9 - 0.5$ ← related subtraction sentence

$x = 8.5$

$$\begin{array}{r} \overset{8\ 10}{\cancel{9}.\cancel{0}} \\ -0.5 \\ \hline 8.5 \end{array}$$

Use a related sentence to find the missing decimal or whole number.

26. $n - 0.2 = 0.8$ **27.** $b + 0.3 = 0.9$ **28.** $m + 0.3 = 0.54$ **29.** $c - 0.5 = 0.49$

30. $z - 0.1 = 2$ **31.** $y + 0.1 = 2$ **32.** $f + 8.6 = 25.6$ **33.** $g - 5.01 = 0.99$

 **CRITICAL THINKING**

34. A student presses the wrong button on the calculator and adds 235.7 instead of subtracting it. The incorrect answer is 817.2. What is the correct answer? Explain how you got your answer.

Communicate

111

3-10 | Problem Solving: Use Simpler Numbers

Problem: A scientist conducted a series of experiments with a liter of polluted water. At the beginning of the experiment, the pollutants reached a volume of 13.17 cubic centimeters (cm^3). Over the next two weeks there was a decrease of 2.08 cm^3 and then an increase of 1.19 cm^3. What was the volume of pollutants after two weeks?

1 IMAGINE

Draw and label a diagram of the problem.

2 NAME

Facts: 13.17 cm^3 at the beginning
2.08 cm^3 decrease
1.19 cm^3 increase

Question: What was the volume of pollutants after two weeks?

2.08 cm^3 decrease   1.19 cm^3 increase   13.17 cm^3

3 THINK

Substitute simpler numbers:
Use: 13 for 13.17 cm^3, 2 for 2.08 cm^3, 1 for 1.19 cm^3.

Reread the problem using these simpler numbers.
Start with 13 cm^3.
Subtract the decrease of 2 cm^3.
Then add the increase of 1 cm^3.

$$13 \text{ cm}^3 \text{ (start)} - 2 \text{ cm}^3 \text{ (decrease)} = 11 \text{ cm}^3$$
$$11 \text{ cm}^3 + 1 \text{ cm}^3 \text{ (increase)} = 12 \text{ cm}^3$$

Now solve the problem using the actual numbers.

4 COMPUTE

$$\begin{array}{r} 13.17 \text{ cm}^3 \\ - 2.08 \text{ cm}^3 \\ \hline 11.09 \text{ cm}^3 \end{array} \qquad \begin{array}{r} 11.09 \text{ cm}^3 \\ + 1.19 \text{ cm}^3 \\ \hline 12.28 \text{ cm}^3 \end{array}$$

After two weeks the volume of pollutants was 12.28 cm^3.

5 CHECK

Work backwards from the answer. Subtract 1.19 cm^3 and then add 2.08 cm^3 to find the original amount.

$$(12.28 \text{ cm}^3 - 1.19 \text{ cm}^3) + 2.08 \text{ cm}^3 = 13.17 \text{ cm}^3$$

The answer checks.

Solve. Use simpler numbers to help you decide what to do.

1. Eva had $60.14 in her bank account on September 1. She then withdrew $12, $18.50, and $14.25 on three separate occasions. She also deposited a check for $76.14. How much is in her account now?

IMAGINE Picture yourself in the problem.

NAME
Facts: in her account—$60.14
withdrew—$12, $18.50, $14.25
deposited—$76.14

Question: How much is in her account now?

THINK Substitute simpler numbers:
Use $60 for $60.14, $10 for $12, $20 for $18.50,
$10 for $14.25, and $80 for $76.14.

Reread and solve the problem using these simpler numbers.
$10 + $20 + $10 = $40 amount withdrawn
$60 − $40 = $20 balance
$20 + $80 = $100 in her present account

Now solve the problem using the actual numbers.

COMPUTE ⟶ **CHECK**

2. A manufacturer makes a certain machine part that measures 26.4 cm in length. A part will pass inspection if it is no more than 0.04 cm shorter than 26.4 cm or no more than 0.04 cm longer than 26.4 cm. What is the shortest measure that can pass inspection?

3. Andy owed Lynn $35.50. He paid back $20.75 but borrowed $5 more. Then he borrowed $8.50. When he was paid, he gave her $25. How much money does Andy still owe Lynn?

4. Ryan earned $122.75 baby-sitting. Vinnie earned $37.15 more than that. Sharon earned $70.95 less than Vinnie. How much money did Sharon earn?

5. Craig is on the school track team. He practices seven days a week. On each of the first five days he runs 4 km. On the next day he runs 6.1 km, and on the last day he runs 3.4 km. How far does Craig run in one week?

3-11 | Problem-Solving Applications

Solve each problem and explain the method you used.

1. Moki and Meg set up a model railroad. The engine of the train is 10.205 cm long. Write 10.205 in expanded form.

2. The caboose is 9.826 cm long. How long is it to the nearest hundredth of a centimeter? to the nearest tenth of a centimeter?

3. Fred connects three freight cars. The red car is 12.64 cm long, the blue car is 12.4 cm long, and the steel-colored car is 12.6 cm long. Write the lengths in order from longest to shortest?

4. Marva puts together three sections of railroad track that are 20.5 cm, 22 cm, and 9.75 cm long. How long is the section of track that Marva creates?

5. A coal car is 3.87 cm tall. A refrigerator car is 5.02 cm tall. Which car is taller? by how much?

6. Bud buys two miniature buildings for his train set. A railroad station costs $15.95 and a gas station costs $1.19 less. How much does the gas station cost?

7. Loretta buys a set of miniature trees that costs $8.59. How much change does she receive from a $20 bill?

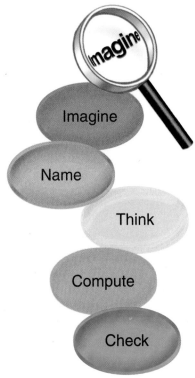

Imagine

Name

Think

Compute

Check

Use the table for problems 8–10.

8. Jackson bought a freight car and a coal car at the sale. How much money did he save?

9. On which type of car do buyers save the most money during this sale?

10. Lea buys a caboose, a passenger car, and an engine on sale. Does she spend more than $26.00?

Item	Original Price	Sale Price
Engine	$11.95	$9.50
Freight Car	$ 8.95	$7.69
Passenger Car	$ 9.50	$8.55
Coal Car	$ 7.75	$6.99
Caboose	$10.29	$8.09

Use a strategy from the list or another strategy you know to solve each problem.

USE THESE STRATEGIES:
Use Simpler Numbers
More Than One Solution
Missing Information
Extra Information
Use a Diagram
Guess and Test

11. A train set has 48 pieces of track. What is the longest track you can build with this set?

12. There are two bridges in the set. One is 17.875 cm long. The other is 5.42 cm longer. How long is the second bridge?

13. Pam's train can travel one fourth of the track in 25.48 seconds. If the train continues at the same rate of speed, how long will it take to travel the entire track?

14. Daryll spends $6.08 for three signs for the train set. None of the signs costs the same amount, but each sign costs more than $2. What is the price of each sign?

15. One train travels at a rate of 55.5 cm per second and the other travels at a rate of 54.9 cm per second. Which train will cover a 250-cm track more quickly?

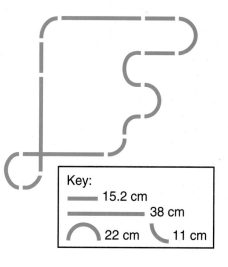

16. Mai builds a mountain for her train set. She uses 2.4 yd of green cloth as grass, 1.75 yd of white cloth as snow, and 0.8 yd of brown cloth as dirt roads. Does Mai use more than 4.5 yd of cloth?

Use the diagram for problems 17–20.

17. What is the total length, in centimeters, of all of the half-circles?

18. What is the entire length of this train track?

19. Is more of the track straight or curved? How much more?

MAKE UP YOUR OWN

20. Draw your own train track pattern using the lengths of track shown above. Is your track longer or shorter than the track above? by how much?

Key:
———— 15.2 cm
——————— 38 cm
⌒ 22 cm   ⌐ 11 cm

Chapter Review and Practice

Read each decimal. Then write the word name. *(See pp. 94–95.)*

1. 0.0008   **2.** 0.000064   **3.** 7.509   **4.** 6.053028

Write in expanded form. *(See pp. 96–97.)*

5. 0.0057   **6.** 0.06795   **7.** 20.600907

Compare. Write <, =, or >. *(See pp. 100–101.)*

8. 0.08 _?_ 0.8   **9.** 0.4287 _?_ 0.429   **10.** 8.720 _?_ 8.72

Estimate. Then add or subtract. *(See pp. 102–111.)*

11.	**12.**	**13.**	**14.**
0.529	7.48	0.009	36
+ 0.427	+ 9	+ 0.008	9.78
			+ 9.0025

15. $8 + 77.321 + 8.0058 + 4.3032$ **16.** $8.20 + $5.94 + $.74 + 21

17.	**18.**	**19.**	**20.**
0.959	0.785	0.4	500
− 0.369	− 0.4	− 0.209	− 8.6

21. $0.826 - 0.69$ **22.** $7.93 - 5.0986$ **23.** $400.8 - 6.9473$

Use a related sentence to find the missing decimal. *(See pp. 110–111.)*

24. $n + 0.2 = 0.65$ **25.** $b - 0.04 = 0.6$ **26.** $y + 0.5 = 5$

PROBLEM SOLVING
(See pp. 104–105, 108–109.)

27. Eileen wrote checks for $26.74, $9.95, and $39.82 over a 3-day period. Estimate the total value of the checks.

28. Pierce bought a cassette for $7.89. He gave the clerk $10.04. How much change did he receive?

29. One bar of gold has a mass of 7.5267 kg. Another bar has a mass of 11.8023 kg. What is their combined mass?

30. The sum of 0.4576 and another number is 12. Find the other number.

(See *Still More Practice,* p. 508.)

LOGIC: OPEN AND CLOSED STATEMENTS

In logic, a **statement** is a sentence that states a
fact. A statement is true or false, but not both.

▶ A **closed statement** can be judged true or false.

All animals have wings.	False
A triangle has exactly 3 sides.	True
Ten is an odd number.	False

▶ An **open statement** contains an unknown.

All squares have exactly ? sides.   ? and x are unknowns.
$10 - x = 5$

If you replace the unknown, the statement becomes
closed. It can then be judged true or false.

All squares have exactly 3 sides. False $10 - 5 = 5$ True

**Tell whether each statement is closed or open.
If the statement is closed, write *True* or *False*.**

1. Alabama is a continent.

2. A horse has 4 legs.

3. $7 - a = 5$

4. $15 \times 5 = 155$

5. Twenty-one is an even number.

6. Six ten thousandths = 0.006

7. $0.45 +$? > 9

8. $16 - 0.75 = 15.25$

**Find a number or numbers that make each open
statement true.**

9. A pentagon has exactly ? sides.

10. 0.456 rounded to nearest tenth is ? .

11. $30 \times f = 270$

12. $48 \div m = 12$

13. $2 \times$? $\times 2 = 16$

14. $0.52 + 0.6 + 3 =$?

15. $0.24 + 0.34 >$?

117

Performance Assessment

Use the figure at the right.

1. Place a decimal in each circle so that the sum on each side of the figure equals the number in the center. Use these decimals! 1.006, 0.4257, 0.4028, 0.9831, 0.0526, 0.6329.

2. Explain the strategy or strategies you used.

Write the place of the underlined digit. Then write its value.

3. 0.67<u>4</u>8

4. 6.1987<u>2</u>

5. 23.42065<u>9</u>

Write in order from greatest to least.

6. 0.373, 0.384, 0.412, 0.43

7. 7.52, 7.5, 7.54, 6.88

Estimate. Then add or subtract.

8. $\begin{array}{r} 0.74 \\ +0.6 \\ \hline \end{array}$

9. $\begin{array}{r} 9 \\ +7.42 \\ \hline \end{array}$

10. 0.37 + 0.5 + 0.26

11. $\begin{array}{r} 8.536 \\ 7.821 \\ +6.345 \\ \hline \end{array}$

12. $\begin{array}{r} 24.7 \\ 34 \\ +\ 2.369 \\ \hline \end{array}$

13. $28.64 + $5.59 + $4 + $.80

14. $\begin{array}{r} 0.848 \\ -0.659 \\ \hline \end{array}$

15. $\begin{array}{r} 0.7 \\ -0.58 \\ \hline \end{array}$

16. 18 − 1.25

17. $\begin{array}{r} 16.5 \\ -\ 3.744 \\ \hline \end{array}$

18. $\begin{array}{r} \$76.55 \\ -\ 59 \\ \hline \end{array}$

19. 0.5437 − 0.428

PROBLEM SOLVING   *Use a strategy or strategies you have learned.*

20. What is the sum of eighty-six thousandths and ninety-five hundredths?

21. Dallas receives an average of 34.55 in. of rain and 2.3 in. of snow each year. How much more rain than snow is that?

Cumulative Review I

Choose the best answer.

1. Choose the standard form of the number.

   four trillion, two hundred eight thousand

   a. 4,000,208
   b. 4,028,000
   c. 4,000,208,000
   d. 4,000,000,208,000

2. Choose the missing addend.

   297,494 + ? = 306,297

   a. 8793
   b. 8803
   c. 603,781
   d. 603,791

3. Estimate.

   816,000
   245,000
   + 97,527

   a. 10,100
   b. 101,000
   c. 900,000
   d. 1,100,000

4. 219,827
    4,000,019
   + 387,055

   a. 4,596,901
   b. 4,606,891
   c. 46,069,901
   d. not given

5. Subtract.

   30,880,200 − 2,665,050

   a. 28,215,150
   b. 28,225,150
   c. 33,542,250
   d. not given

6. 548 × $53.76

   a. $5,268.48
   b. $29,454.48
   c. $29,460.48
   d. not given

7. 2200
   × 300

   a. 660
   b. 6600
   c. 66,000
   d. not given

8. 300,000 ÷ 6000

   a. 5
   b. 50
   c. 500
   d. not given

9. Estimate.

   694,719 ÷ 489

   a. 1400
   b. 2000
   c. 3500
   d. 6000

10. 428)‾9280

   a. 21
   b. 21 R292
   c. 21 R302
   d. not given

11. Find the value.

   24 ÷ 6 + (3 × 5)

   a. 1 R3
   b. 12
   c. 19
   d. 35

12. Round to the greatest nonzero place.

   0.4723

   a. 0.4
   b. 0.47
   c. 0.5
   d. 1.0

13. 22.6
     27
    + 2.53

   a. 51
   b. 52.13
   c. 74.9
   d. not given

14. Subtract.

   14 − 2.45

   a. 11.55
   b. 12.55
   c. 16.45
   d. not given

Ongoing Assessment I

For Your Portfolio

**Solve each problem. Explain the steps and the strategy
or strategies you used for each. Then choose one from
problems 1–4 for your Portfolio.**

1. A roll of ribbon is 250 in. long.
   How much longer should the roll
   be so that Ed can cut an exact
   number of 15 in. streamers, with
   no ribbon left over?

2. Posio School has 1045 graduating
   sixth graders. The school district
   has 35 buses available to take the
   students to the graduation ceremony.
   How many buses are needed?

**Choose the number sentence that can be used to solve
the problem. Then solve it.**

3. Arnetta has flown 7785 miles. She
   needs a total of 10,000 miles in order
   to get an upgrade to first class. How
   many more miles does Arnetta have
   to fly to get the upgrade?

   a. $7785 + y = 10{,}000$
   b. $y + 10{,}000 = 7785$
   c. $y - 7785 = 10{,}000$
   d. $7785 - y = 10{,}000$

4. Rachel and Samantha are salespeople
   at Larry's Audio Outlet. Rachel
   sold $2487.56 worth of merchandise
   this week. Samantha sold $203.88
   more than Rachel. What is the
   total value of the merchandise sold?

   a. $\$2487.56 - \$203.88 = t$
   b. $\$2487.56 + \$203.88 = t$
   c. $\$2487.56 + (\$2487.56 + \$203.88) = t$
   d. $(\$2487.56 + \$203.88) + \$203.88 = t$

Tell about it.

5. Explain how the strategy Interpret the Remainder can
   be used to help you solve problem 1.

6. For which of the problems above is Use Simpler
   Numbers an appropriate strategy to use? Explain.

Communicate ✓

For Rubric Scoring

Listen for information on how your work will be scored.

7. Each letter in the statements below represents one number
   in the box. What is the value of each letter? You may use a calculator.

   | 5 | 4.58 | 0.1764 | 4.7564 | 0.42 |

   $$C + A = D \qquad A^2 = E \qquad B - A < C$$

Decimals: Multiplication and Division

4

In this chapter you will:

Multiply and divide by powers of ten
Estimate and find products and quotients
Explore multiplying decimals by decimals
Apply properties of addition and
  multiplication to decimals
Use a calculator to investigate
  scientific notation
Solve multi-step problems

Critical Thinking Finding Together

The tenth term in a sequence is
1004.58. If the pattern is × 100, ÷ 10,
..., what is the first term? You may
use a calculator.

A Dividend Opinion

Said the Aliquant to the Aliquot,
"You're all used up, and I am not."
"Used up?" said the Aliquot. "Not a bit.
I happen to be a perfect fit.
You're a raveled thread. A wrong number.
You're about as useful as scrap lumber.
I slip into place like a mitered joint.
You hang out over your **decimal point**
Like a monkey asquat in a cuckoo's nest
With your tail adangle, self-impressed
By the way you twitch the thing about.
Stuck up about nothing but sticking out,
If I'm used up, you will discover
You're no fresh start. You're just left over
From nothing anyone would want."
Said the Aliquot to the Aliquant.

John Ciardi

121

Multiplying Decimals by 10, 100, and 1000

Lucinda used a calculator to multiply these decimals by 10, 100, and 1000. Study the patterns.

$10 \times 0.437 = 4.37$	$10 \times 0.82 = 8.2$	$10 \times 0.7 = 7$
$100 \times 0.437 = 43.7$	$100 \times 0.82 = 82$	$100 \times 0.7 = 70$
$1000 \times 0.437 = 437$	$1000 \times 0.82 = 820$	$1000 \times 0.7 = 700$

$10 \times 1.437 = 14.37$	$10 \times 1.82 = 18.2$	$10 \times 1.7 = 17$
$100 \times 1.437 = 143.7$	$100 \times 1.82 = 182$	$100 \times 1.7 = 170$
$1000 \times 1.437 = 1437$	$1000 \times 1.82 = 1820$	$1000 \times 1.7 = 1700$

▶ **To multiply a decimal by 10, 100, or 1000:**

- Count the number of zeros in the multiplier.

- Move the decimal point in the multiplicand to the *right* one place for each zero.

- Write as many zeros in the product as needed to place the decimal point correctly.

Study these examples.

$1\ 0 \times 0.5\,6 = 5.6$ 1 zero: Move 1 place to the right.

$1\ 0\ 0 \times 0.0\ 0\,4 = 0.4$ 2 zeros: Move 2 places to the right.

$1\ 0\ 0\ 0 \times 2.0\ 0\ 3 = 2\ 0\ 0\ 3$ 3 zeros: Move 3 places to the right.

$1\ 0\ 0 \times 0.8\ 0 = 8\ 0$ 2 zeros: Move 2 places to the right. Write 1 zero as a placeholder.

$1\ 0\ 0\ 0 \times 1\ 5.8\ 0\ 0 = 1\ 5{,}8\ 0\ 0$ 3 zeros: Move 3 places to the right. Write 2 zeros as placeholders.

Find the products. Use the patterns.

1. 10×0.814
100×0.814
1000×0.814

2. 10×0.041
100×0.041
1000×0.041

3. 10×5.02
100×5.02
1000×5.02

4. 10×0.3
100×0.3
1000×0.3

Multiply.

5. 10 × 0.1

6. 10 × 0.4

7. 10 × 0.77

8. 10 × 0.0049

9. 100 × 0.3

10. 100 × 0.05

11. 100 × 0.006

12. 100 × 0.1003

13. 1000 × 5.3

14. 1000 × 0.02

15. 1000 × 0.002

16. 1000 × 5.146

17. 10 × 0.0006

18. 1000 × 21.06

19. 100 × 19.41

20. 1000 × 12.0006

Find the products. Then write them in order from least to greatest.

21. a. 10 × 0.94

b. 100 × 0.092

c. 1000 × 0.0093

22. a. 100 × 0.05

b. 10 × 0.7

c. 1000 × 0.94

23. a. 1000 × 0.0062

b. 100 × 0.005

c. 10 × 0.042

24. a. 100 × 0.61

b. 100 × 0.70

c. 1000 × 0.0010

Find the missing factor.

Algebra

25. __?__ × 0.004 = 0.4

26. __?__ × 0.21 = 210

27. 100 × __?__ = 5

28. __?__ × 2.06 = 206

29. 1000 × __?__ = 8.77

30. 10 × __?__ = 0.02

PROBLEM SOLVING

31. Hesper oyucca whipplei is a plant that can grow 0.857 ft in one day. At that rate, how many feet can it grow in 10 days? in 100 days?

32. The largest tomato ever grown had a mass of 1.9 kg. The largest cabbage had a mass of 51.8 kg. Which mass is greater: 100 tomatoes or 10 cabbages?

 SHARE YOUR THINKING

Math Journal

33. Explain in your Math Journal.

a. What happens to a whole number such as 2300 when it is multiplied by 10, 100, and 1000; and

b. What happens to a decimal such as 0.42 when it is multiplied by 10, 100, and 1000.

Check your explanations on a calculator.

123

4-2 Estimating Decimal Products

Kevin rode his bike for 4.75 hours at an average speed of 18.6 miles per hour. About how far did he ride?

To find out about how far, estimate: 4.75 × 18.6.

▶ **To estimate a decimal product:**

- Round each factor to its greatest place.

  4.75 ⟶ 5
  18.6 ⟶ 20

- Multiply the rounded factors.

  5 × 20 = 100

  4.75 × 18.6 ≈ 100

  Both factors are rounded *up*.
  The actual product is *less than* 100.

Kevin rode about 100 miles.

Study these examples.

$$
\begin{array}{r}
1.329 \\
\times\, 33.29 \\
\end{array}
\longrightarrow
\begin{array}{r}
1 \\
\times\, 30 \\
\hline
30 \\
\end{array}
$$
Both factors are rounded down.

33.29 × 1.329 ≈ 30
The actual product is *greater than* 30.

$$
\begin{array}{r}
10.25 \\
\times\,\ 0.87 \\
\end{array}
\longrightarrow
\begin{array}{r}
1\,0 \\
\times\, 0.9 \\
\hline
9.0 \\
\end{array}
$$
One factor is rounded up, the other down.

0.87 × 10.25 ≈ 9
The estimate is *close to* the actual product.

Estimate each product. Tell whether the actual product is *greater than*, *less than*, or *close to* the estimated product.

1. 5.79
    × 3.7

2. 4.45
    × 6.2

3. 9.42
    × 6.6

4. 18.7
    × 5.8

5. 9.5
    × 0.63

6. 10.67
    × 0.84

7. 13.86
    × 0.74

8. 14.9
    × 0.72

9. 19.27
    × 1.8

10. 34.918
    × 5.2

11. 58.007
    × 12.39

12. 76.298
    × 8.547

Estimate the product.

13. 10.592   **14.** 14.6   **15.** 13.77   **16.** 584   **17.** 624
    ×   2.7   ×   0.5   ×   0.09   × 1.4   × 5.4

18. 3.28   **19.** 89.51   **20.** 47.4   **21.** 69.74   **22.** 12.8297
    ×   2.7   ×   3.6   ×   0.8   ×   6.6   ×   0.32

23. 10.6 × 23   **24.** 5.52 × 1.78   **25.** 0.9 × 13.6   **26.** 137 × 2.85

27. 6235 × 3.7   **28.** 2.8 × 31.89   **29.** 3.2 × 14.79   **30.** 0.7 × 103.95

31. 10.7 × 2.9 × 28.04   **32.** 1.5 × 2.8 × 12.1   **33.** 4.3 × 18.07 × 1.79

Estimate by rounding *one* factor to the nearest 10, 100, or 1000.
Use mental math.

34. 9.7 × 0.672   **35.** 10.2 × 5.6   **36.** 100.8 × 0.8   **37.** 96 × 1.235

38. 122 × 4.125   **39.** 10.3 × 17.7   **40.** 96 × 0.837   **41.** 997 × 14.5

PROBLEM SOLVING

42. Jack can swim 11.6 yards in a minute. If he can keep up that pace, about how far can he swim in 0.75 hour? in 2.5 hours?

43. Emma hiked for 3.7 hours at an average pace of 2.8 miles an hour. Did she reach a pond 12 miles from her starting point? Estimate to find out.

 **CHOOSE A COMPUTATION METHOD**

Is the product reasonable? Write *Yes* or *No*.
You may estimate or use a calculator.

44. 6.5 × 9.4 $\stackrel{?}{=}$ 61.1   **45.** 6.5 × 5.4 $\stackrel{?}{=}$ 3.51   **46.** 6.5 × 19.7 $\stackrel{?}{=}$ 128.05

47. 0.9 × 14.6 $\stackrel{?}{=}$ 131.4   **48.** 50.9 × 24.7 $\stackrel{?}{=}$ 1257.23   **49.** 0.9 × 100.17 $\stackrel{?}{=}$ 901.53

50. For which exercise(s) did you use a calculator to find an exact answer? Explain.

4-3 Multiplying Decimals by Whole Numbers

When Mariko arrived in London, she exchanged dollars for English pounds. If she had 48 dollars, and the exchange rate was 0.68 pounds for each dollar, how many pounds did she receive?

To find the number of pounds, multiply: $48 \times 0.68 = \underline{\ ?\ }$.

Estimate: $50 \times 0.7 = 35$
About 35 pounds

$48 \longrightarrow 50$	$100 \times 0.7 = 70$
$0.68 \longrightarrow 0.7$	$50 \times 0.7 = 35$

▶ **To multiply a decimal by a whole number:**

- Multiply as you would with whole numbers.
- Count the number of decimal places in the decimal factor.
- Mark off the *same number* of decimal places in the product.

Multiply as with whole numbers.

```
  0.6 8
×   4 8
  5 4 4
2 7 2
3 2 6 4
```

Write the decimal point in the product.

```
  0.6 8
×   4 8
  5 4 4    2 decimal places
2 7 2
3 2.6 4
```

Mariko received 32.64 pounds.

32.64 is close to the estimate of 35.

Study these examples.

```
  0.3 2 9
×       2    3 decimal places
  0.6 5 8
```

```
  1 2.7
×     9    1 decimal place
1 1 4.3
```

Write the decimal point in each product.

1.
```
    5.9
×     3
  1 7 7
```

2.
```
  1 2.9 2
×       7
  9 0 4 4
```

3.
```
  0.2 3 5
×       7
  1 6 4 5
```

4.
```
    9.2 7
×     1 5
1 3 9 0 5
```

5.
```
    2.4 6 3
×       2 6
  6 4 0 3 8
```

126

Estimate. Then multiply.

6.	7.	8.	9.	10.
0.9	0.7	0.59	0.47	0.32
× 22	× 79	× 43	× 21	× 73

11.	12.	13.	14.	15.
0.438	0.019	4.005	4.686	97.925
× 18	× 4	× 38	× 3	× 16

16. 4×0.6 17. 9×6.789 18. 25×0.45 19. 62×6.02

20. 15×45.79 21. 94×0.88 22. 7×0.17 23. 47×0.47

Multiplying Money Amounts

When multiplying money amounts by a whole number, mark off *two* decimal places in the product.

Estimate: $30 \times \$2 = \$60$

```
  $ 2.1 0
×     2 8
  1 6 8 0
  4 2 0
  $ 5 8.8 0
```

Write the dollar sign.

2 decimal places

Estimate. Then find the product.

24.	25.	26.	27.	28.
$8.32	$23.42	$8.27	$7.23	$23.45
× 9	× 7	× 52	× 43	× 62

PROBLEM SOLVING

Communicate

29. Which costs more, 7 lb of beef at $3.25 a pound or 12 lb of chicken at $1.79 a pound?

30. Ms. Lee bought 2.8 lb of pasta at $2.95 a pound. She has a $10 bill. Is this enough money for the purchase? Explain.

CONNECTIONS: SOCIAL STUDIES

31. In Paris, Sean received 6.2 francs for each dollar. He had 40 dollars. How many francs did he receive?

32. In Tokyo, Mel received 104.7 yen for each dollar. He had $100. How many yen did he receive?

Multiplying Decimals by Decimals

Remember: The place-value models can represent decimals.

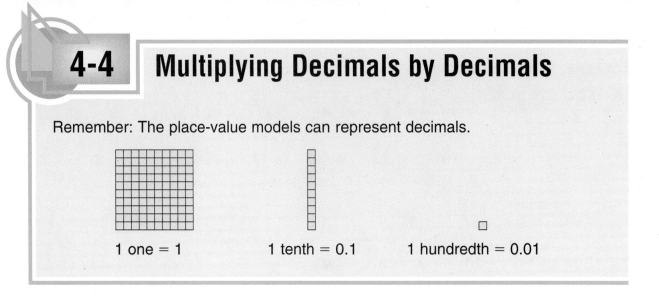

1 one = 1   1 tenth = 0.1   1 hundredth = 0.01

Multiply: 1.1 × 1.4 = ?

Hands-On Understanding

Materials Needed: flats, rods, units, paper, pencil

Step 1   Use 1 flat and 1 rod to show 1.1 vertically.

1.1

Step 2   Now place 4 rods to the right of the flat to show 1.4 horizontally.

1.4

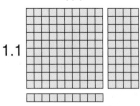

1.1

How many units (cubes) are needed to make a rectangle?

Step 3 Place units in your display to complete the rectangle.

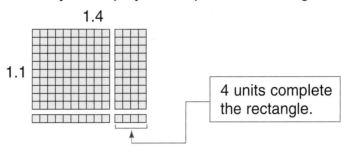

1.4

1.1

4 units complete the rectangle.

Step 4 How many ones are in your display? tenths? hundredths?

$1.1 \times 1.4 = 1$ one $+ 5$ tenths $+ 4$ hundredths
$= 1 + 0.5 + 0.04$
So $1.1 \times 1.4 = \underline{?}$

Communicate

1. How many decimal places (digits to the right of the decimal point) are there in 1.1? How many decimal places are there in 1.4?

2. How many decimal places are there in the product of 1.1 and 1.4?

   In this example, tenths $\times$ tenths $=$ hundredths.

3. When you multiply a decimal with tenths by another decimal with tenths will the product always be hundredths? Test several examples on your calculator to help you answer.

SHARE YOUR THINKING

Communicate

4. Use place-value models to show 1.3×2.4. Trade flats, rods, and units to regroup to display the correct product. Discuss your solution with the class.

129

4-5 More Multiplying Decimals

The distance from the beginning of the Right Angle Trail to the campground near the Blue Moose Brook is 14.3 km. Sid has hiked 0.75 of that distance. How far has he hiked?

To find how far Sid has hiked, multiply: 0.75×14.3 km = __?__

Estimate: $0.8 \times 10 = 8$ About 8 km

▶ **To multiply a decimal by a decimal:**

- Multiply as you would with whole numbers.
- Count the number of decimal places in *both* factors.
- Mark off the *same number* of decimal places in the product.

Multiply as with whole numbers.	Write the decimal point in the product.

```
      1 4.3
    × 0.7 5
      7 1 5
    1 0 0 1
    1 0 7 2 5
```

```
      1 4.3   ← 1 decimal place
    × 0.7 5   ← 2 decimal places
      7 1 5
    1 0 0 1
    1 0.7 2 5  ← 3 decimal places
```

Check.

Enter: 0.75 ✕ 14.3 ▤ ▭ 10.725

Sid has hiked 10.725 km.

Study these examples.

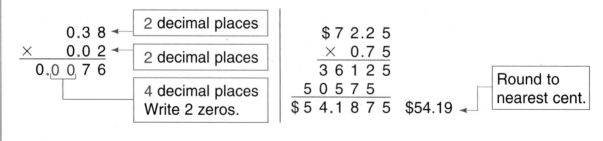

```
        0.3 8   ← 2 decimal places
    ×   0.0 2   ← 2 decimal places
      0.0 0 7 6
```
4 decimal places
Write 2 zeros.

```
      $ 7 2.2 5
    ×    0.7 5
      3 6 1 2 5
    5 0 5 7 5
    $ 5 4.1 8 7 5   $54.19
```
Round to nearest cent.

130

Multiply. Round to the nearest cent when necessary.

1. 0.6
   × 0.9

2. 0.14
   × 0.6

3. 7.35
   × 6.9

4. 0.629
   × 4.7

5. 9.702
   × 2.4

6. 0.43
   × 0.19

7. 0.61
   × 0.93

8. 0.163
   × 0.03

9. 0.911
   × 9.11

10. 0.414
    × 0.72

11. 13.5
    × 9.2

12. 0.20
    × 9.1

13. $8.05
    × 1.9

14. $9.20
    × 4.5

15. $10.50
    × 8

16. $59.50 × 2.4

17. 8.5 × 0.6

18. 4.12 × 1.8

19. 8.74 × 3.15

20. 9 × $56.95

21. 1.5 × 8.00

22. 6.2 × 9.5

23. 4.75 × $85

24. 2.3 × 0.2 × 5.1

25. 12.3 × 0.9 × 0.8

26. 2.7 × 19.5 × 0.5

Use estimation to check the products. Correct unreasonable products.

27. $0.8 \times 0.817 \overset{?}{=} 65.36$

28. $4.7 \times 2.6 \overset{?}{=} 12.22$

29. $6.4 \times 0.8 \overset{?}{=} 51.2$

30. $4.8 \times 15.94 \overset{?}{=} 7.6512$

31. $6.6 \times 48.3 \overset{?}{=} 31.878$

32. $0.94 \times 5.8 \overset{?}{=} 5.452$

PROBLEM SOLVING

33. Sadie hiked 37.6 km from the end of the Right Angle Trail to the beginning of the North Kaibab Trail. Sam hiked 0.4 as far. How much farther than Sam did Sadie hike?

34. Using a microscope, Ezra views some bacteria 0.003 mm in size. If the microscope magnifies objects 1000 times, how large do the bacteria appear?

 **SKILLS TO REMEMBER**

Find the quotient.

35. 50,000 ÷ 100

36. 2000 ÷ 10000

37. 80,000 ÷ 10

38. 60,000 ÷ 10,000

39. 40,000 ÷ 10,000

40. 75,000 ÷ 100

41. 25,000 ÷ 10

42. 618,500 ÷ 10

4-6 Dividing Decimals by 10, 100, and 1000

Eddie used a calculator to divide these decimals
by 10, 100, and 1000. Study the patterns.

637.4 ÷ 10 = 63.74	53.8 ÷ 10 = 5.38	8.7 ÷ 10 = 0.87
637.4 ÷ 100 = 6.374	53.8 ÷ 100 = 0.538	8.7 ÷ 100 = 0.087
637.4 ÷ 1000 = 0.6374	53.8 ÷ 1000 = 0.0538	8.7 ÷ 1000 = 0.0087

21.76 ÷ 10 = 2.176	6.15 ÷ 10 = 0.615	0.47 ÷ 10 = 0.047
21.76 ÷ 100 = 0.2176	6.15 ÷ 100 = 0.0615	0.47 ÷ 100 = 0.0047
21.76 ÷ 1000 = 0.02176	6.15 ÷ 1000 = 0.00615	0.47 ÷ 1000 = 0.00047

▶ **To divide a decimal by 10, 100, or 1000:**

- Count the number of zeros in the divisor.

- Move the decimal point to the *left* one place
  in the dividend for each zero in the divisor.

- Write zeros in the quotient as needed.

Study these examples.

6̖8.4 ÷ 1 0 = 6.8 4   1 zero: Move 1 place to the left.

2̖6 8.7 ÷ 1 0 0 = 2.6 8 7   2 zeros: Move 2 places to the left.

0̖3 2.5 ÷ 1 0 0 0 = 0.0 3 2 5   3 zeros: Move 3 places to the left.
Write 1 zero as a placeholder.

0̖0 1.8 2 ÷ 1 0 0 0 = 0.0 0 1 8 2   3 zeros: Move 3 places to the left.
Write 2 zeros as placeholders.

Find the quotients. Use the patterns.

1. 8329 ÷ 10	**2.** 724.8 ÷ 10	**3.** 56.39 ÷ 10	**4.** 2.8 ÷ 10
8329 ÷ 100	724.8 ÷ 100	56.39 ÷ 100	2.8 ÷ 100
8329 ÷ 1000	724.8 ÷ 1000	56.39 ÷ 1000	2.8 ÷ 1000

5. 4.27 ÷ 10	**6.** 8.1 ÷ 10	**7.** 0.6 ÷ 10	**8.** 0.18 ÷ 10
4.27 ÷ 100	8.1 ÷ 100	0.6 ÷ 100	0.18 ÷ 100
4.27 ÷ 1000	8.1 ÷ 1000	0.6 ÷ 1000	0.18 ÷ 1000

Divide.

9. $87 \div 10$ **10.** $4 \div 10$ **11.** $983 \div 10$ **12.** $0.63 \div 10$

13. $483 \div 100$ **14.** $29 \div 100$ **15.** $0.6 \div 100$ **16.** $7.33 \div 100$

17. $2814 \div 1000$ **18.** $874 \div 1000$ **19.** $71.02 \div 1000$ **20.** $0.5 \div 1000$

21. $0.02 \div 100$ **22.** $0.105 \div 10$ **23.** $30.8 \div 100$ **24.** $9.9 \div 10$

25. $849 \div 1000$ **26.** $3.9 \div 100$ **27.** $0.63 \div 10$ **28.** $0.17 \div 100$

29. $0.245 \div 100$ **30.** $5.628 \div 1000$ **31.** $9 \div 1000$ **32.** $19.95 \div 10$

Find the missing number.

33. $4.07 \div \underline{\ ?\ } = 0.0407$ **34.** $0.18 \div \underline{\ ?\ } = 0.018$ **35.** $22.8 \div \underline{\ ?\ } = 0.0228$

36. $\underline{\ ?\ } \div 100 = 56.7$ **37.** $\underline{\ ?\ } \div 10 = 0.07$ **38.** $\underline{\ ?\ } \div 1000 = 0.05$

Compare. Write <, =, or >.

39. $8 \div 100 \ \underline{\ ?\ }\ 80 \div 1000$ **40.** $5.6 \div 10 \ \underline{\ ?\ }\ 0.056 \div 100$

41. $17.5 \div 1000 \ \underline{\ ?\ }\ 175 \div 100$ **42.** $0.16 \div 100 \ \underline{\ ?\ }\ 16 \times 0.1$

43. $2.45 \times 100 \ \underline{\ ?\ }\ 24.5 \div 100$ **44.** $0.8 \times 1000 \ \underline{\ ?\ }\ 800 \div 1000$

PROBLEM SOLVING

45. Mr. Gleason divided 815.6 m of fencing into 100 equal sections. How long is each section?

46. Flying at a speed of 1000 km per hour, how long will it take to fly a distance of 6550 km? a distance of 12 500 km?

 SHARE YOUR THINKING

Communicate ✓

47. Work together with a classmate to solve the riddle. Explain your solution to the class.

I am a one-digit number.
The difference between multiplying me by 10 and dividing me by 10 is 49.5.
What number am I?

Patterning with Tenths, Hundredths, Thousandths

Examine these patterns for dividing numbers by 0.1, 0.01, and 0.001.

$$34 \div 0.1 = 340$$
$$34 \div 0.01 = 3400$$
$$34 \div 0.001 = 34{,}000$$

$$631.8 \div 0.1 = 6318$$
$$631.8 \div 0.01 = 63{,}180$$
$$631.8 \div 0.001 = 631{,}800$$

▶ **To divide by 0.1, 0.01, or 0.001:**

- Count the number of decimal places in the divisor.

- Move the decimal point to the *right* one place in the dividend for each decimal place in the divisor.

- Write zeros in the quotient as needed.

Study these examples.

$4.3\,6 \div 0.1 = 4\ 3.6$

1 decimal place in the divisor
Move 1 place to the right.

$4.3\ 6 \div 0.0\ 1 = 4\ 3\ 6$

2 decimal places in the divisor
Move 2 places to the right.

$4.3\ 6\ 0 \div 0.0\ 0\ 1 = 4\ 3\ 6\ 0$

3 decimal places in the divisor
Move 3 places to the right.
Write 1 zero.

When you divide by 0.1, 0.01, or 0.001, the quotient increases as the divisor decreases.

Find the quotients. Use the patterns.

1. $16 \div 0.1$
$16 \div 0.01$
$16 \div 0.001$

2. $329 \div 0.1$
$329 \div 0.01$
$329 \div 0.001$

3. $5.8 \div 0.1$
$5.8 \div 0.01$
$5.8 \div 0.001$

4. $27.6 \div 0.1$
$27.6 \div 0.01$
$27.6 \div 0.001$

Divide.

5. $84 \div 0.1$

6. $84 \div 0.01$

7. $84 \div 0.001$

8. $8.4 \div 0.01$

9. $97.8 \div 0.001$

10. $97.8 \div 0.01$

11. $97.8 \div 0.1$

12. $978 \div 0.1$

13. $237 \div 0.1$

14. $157.5 \div 0.1$

15. $42.23 \div 0.1$

16. $27.16 \div 0.01$

17. $82.06 \div 0.01$

18. $784.19 \div 0.01$

19. $2.5 \div 0.001$

20. $0.8 \div 0.001$

21. $0.72 \div 0.1$

22. $0.9 \div 0.01$

23. $188 \div 0.001$

24. $427.01 \div 0.01$

25. $56.56 \div 0.01$

26. $0.88 \div 0.1$

27. $1.56 \div 0.01$

28. $1 \div 0.001$

Compare. Write <, =, or >.

29. $23 \div 0.1$ ___?___ $23 \div 0.01$

30. $12.9 \div 0.01$ ___?___ $12.9 \div 0.001$

31. $15.4 \div 0.01$ ___?___ $15.4 \div 0.1$

32. $832 \div 0.1$ ___?___ $8.32 \div 0.1$

33. $5.9 \div 0.01$ ___?___ $59 \div 0.01$

34. $6.2 \div 0.01$ ___?___ $62 \div 0.1$

Compare the pattern in the first column with the pattern in the second column. Write a summary of what you observe.

Communicate ✓

35.
$0.637 \times 1000 = 637$
$0.637 \times 100 = 63.7$
$0.637 \times 10 = 6.37$
$0.637 \times 1 = 0.637$
$0.637 \times 0.1 = 0.0637$
$0.637 \times 0.01 = 0.00637$
$0.637 \times 0.001 = 0.000637$

$0.637 \div 1000 = 0.000637$
$0.637 \div 100 = 0.00637$
$0.637 \div 10 = 0.0637$
$0.637 \div 1 = 0.637$
$0.637 \div 0.1 = 6.37$
$0.637 \div 0.01 = 63.7$
$0.637 \div 0.001 = 637$

Find the missing number. Use the patterns above to help you.

36. $0.6 \times$ ___?___ $= 0.06$

37. $44 \div$ ___?___ $= 4400$

38. $8.6 \div$ ___?___ $= 8600$

39. $5.42 \div$ ___?___ $= 0.0542$

40. $97.7 \times$ ___?___ $= 0.0977$

41. ___?___ $\times 0.1 = 17.32$

PROBLEM SOLVING

42. How many dimes are in $18.60?

43. How many pennies are in $56?

4-8 Estimating Quotients

The fabric Hannah needs costs $4.65 per yard.
She has $23.50. About how many yards of
fabric can she buy?

To find about how many yards Hannah
can buy, estimate: $23.50 ÷ $4.65.

▶ **To estimate the quotient of two decimals
(or two money amounts):**

- Write compatible numbers for the
  dividend and the divisor.

  $23.50 ⟶ $24
  $4.65 ⟶ $4

- Divide.

  $24 ÷ $4 = 6

She can buy about 6 yards of fabric.

▶ Compare the dividend and the divisor to help
estimate quotients.

Dividend > Divisor	⟶	Quotient > 1
8 ÷ 0.16		8 ÷ 0.16 = 50

Dividend < Divisor	⟶	Quotient < 1
0.16 ÷ 8		0.16 ÷ 8 = 0.02

Estimate to place the decimal point in the quotient.

1. 29.52 ÷ 7.2 = 41   **2.** 18.7 ÷ 5.5 = 34   **3.** 49.6 ÷ 8 = 62

4. 38.13 ÷ 15.5 = 246   **5.** 40.18 ÷ 19.6 = 205   **6.** 225.15 ÷ 7.5 = 3002

7. 396.5 ÷ 12.2 = 325   **8.** 9.21 ÷ 7.5 = 1228   **9.** $37.75 ÷ 5 = $755

Estimate each quotient. Use compatible numbers.

10. $25.2 \div 6.1$ **11.** $18.3 \div 3.8$ **12.** $27.7 \div 4.4$ **13.** $34.1 \div 5.5$

14. $41.9 \div 8.6$ **15.** $54.3 \div 9.3$ **16.** $47.17 \div 6.88$ **17.** $38.8 \div 5.99$

18. $225.7 \div 6.8$ **19.** $182.8 \div 3.5$ **20.** $505.9 \div 52.7$ **21.** $798.2 \div 68.4$

22. $328 \div 15.9$ **23.** $885 \div 30.9$ **24.** $\$63.28 \div 4.4$ **25.** $\$96.78 \div \$9.50$

Compare. Write <, =, or >.

26. $8 \div 9 \underline{\ ?\ } 1$ **27.** $27.6 \div 7.4 \underline{\ ?\ } 1$ **28.** $14.9 \div 8.7 \underline{\ ?\ } 1$

29. $6.8 \div 18.9 \underline{\ ?\ } 1$ **30.** $1 \underline{\ ?\ } 0.7 \div 5.88$ **31.** $1 \underline{\ ?\ } 41.1 \div 0.999$

32. $1 \underline{\ ?\ } 1.28 \div 3.01$ **33.** $1 \underline{\ ?\ } 12.1 \div 0.894$ **34.** $1 \div 0.1 \underline{\ ?\ } 1$

Three Ways to Estimate Quotients

Front End	Compatible Numbers	Rounding
$35.5 \div 3.6$	$35.5 \div 3.6$	$35.5 \div 3.6$
$30 \div 3 = 10$	$36 \div 4 = 9$	$40 \div 4 = 10$

Complete. Estimate each quotient.

		Front End	Compatible Numbers	Rounding
35.	$39.2 \div 7.8$	?	?	?
36.	$48.6 \div 9.2$	?	?	?
37.	$60 \div 5.8$	?	?	?
38.	$152.8 \div 6.7$	?	?	?
39.	$\$225.50 \div 15.8$	?	?	?

PROBLEM SOLVING

40. Which method seems to give the most accurate estimate in each of exercises 35–39? Check with your calculator.

4-9 Dividing Decimals by Whole Numbers

Elena and five of her friends went out for pizza. The total bill was $18.66. They shared the bill equally. How much did each person pay?

To find the amount each person paid, divide: $18.66 ÷ 6 = __?__

Estimate: $18 ÷ 6 = $3   About $3

▶ **To divide a decimal by a whole number:**

Write the decimal point in the quotient directly above the decimal point in the dividend.	Divide as you would with whole numbers.	Check.

$$\begin{array}{r} . \\ 6\overline{)\$1\,8.6\,6} \end{array}$$

$$\begin{array}{r} \$3.1\,1 \\ 6\overline{)\$1\,8.6\,6} \\ -1\,8 \\ \hline 6 \\ -6 \\ \hline 0\,6 \\ -6 \\ \hline 0 \end{array}$$

$$\begin{array}{r} \$3.1\,1 \\ \times\quad\ 6 \\ \hline \$1\,8.6\,6 \end{array}$$

Write the dollar sign in the quotient.

Each person paid $3.11. $3.11 is close to the estimate of $3.

Study these examples.

$$\begin{array}{r} 0.4\,8 \\ 3\overline{)1.4\,4} \\ -1\,2 \\ \hline 2\,4 \\ -2\,4 \\ \hline 0 \end{array}$$

$$\begin{array}{r} 2.1\,9 \\ 4\overline{)8.7\,6} \\ -8 \\ \hline 0\,7 \\ -4 \\ \hline 3\,6 \\ -3\,6 \\ \hline 0 \end{array}$$

$$\begin{array}{r} 0.3\,7 \\ 2\,6\overline{)9.6\,2} \\ -7\,8 \\ \hline 1\,8\,2 \\ -1\,8\,2 \\ \hline 0 \end{array}$$

Write the letter of the correct answer.

1. $4\overline{)0.84}$ **a.** 2.1   **b.** 0.21   **c.** 21   **d.** 0.021

2. $2\overline{)2.912}$ **a.** 0.1456   **b.** 14.56   **c.** 1.456   **d.** 0.01456

Divide and check.

3. $67.2 \div 6$ **4.** $7.5 \div 3$ **5.** $49.32 \div 9$ **6.** $0.95 \div 5$

7. $21.60 \div 15$ **8.** $13.2 \div 22$ **9.** $0.784 \div 7$ **10.** $8.792 \div 4$

11. $62.1 \div 3$ **12.** $9.520 \div 7$ **13.** $\$77.20 \div 8$ **14.** $0.732 \div 6$

15. $5\overline{)99.5}$ **16.** $6\overline{)135.6}$ **17.** $7\overline{)\$17.85}$ **18.** $8\overline{)41.52}$

19. $12\overline{)\$34.80}$ **20.** $42\overline{)349.44}$ **21.** $4\overline{)0.8644}$ **22.** $5\overline{)0.8325}$

23. $2\overline{)0.9314}$ **24.** $5\overline{)\$50.25}$ **25.** $3\overline{)0.732}$ **26.** $4\overline{)\$24.12}$

27. $6\overline{)14.10}$ **28.** $3\overline{)0.1077}$ **29.** $8\overline{)0.016}$ **30.** $6\overline{)7.836}$

Compare. Write <, =, or >.

31. $0.57 \div 30$ _?_ $0.57 \div 3$ **32.** $92.4 \div 6$ _?_ $9.24 \div 6$

33. $4\overline{)48}$ _?_ $4\overline{)4.8}$ **34.** $5\overline{)0.015}$ _?_ $5\overline{)0.15}$

PROBLEM SOLVING

35. If 6 packages weigh 0.936 lb, what does 1 package weigh? What do 12 packages weigh?

36. Mary spent $.96 for 8 m of ribbon. What does 1 m of ribbon cost?

37. Irma wants to divide a bill of $48.24 equally among 8 people. What should each person pay?

38. Mr. Clark traveled 456.4 km in 14 days. If he traveled the same distance each day, how far did he travel in a day?

 **CALCULATOR ACTIVITY**

Find the average of each set of numbers.

39. 6.8, 4.9, 5.5, 7.2

40. 0.099, 0.2, 0.089, 0.12, 0.092, 0.108

41. $35.92, $37.16, $39, $33.95, $40.02

42. 4.8, 5, 4.5, 5.1, 4.75, 4.6, 5.25, 4.2, 4.0, 4.25

$\boxed{24.4}\; \div \; \boxed{4}\; = \; \boxed{}$

Dividing by a Decimal

Mrs. Martinez is shingling the roof of her house. Each shingle is 0.2 m wide. If the roof is 8.46 m wide, how many shingles can she put in each row?

To find the number of shingles, divide: 8.46 ÷ 0.2 = __?__

Estimate: 8 ÷ 0.2 = 40 ⟵ | 8 ÷ 0.1 = 8.0 = 80   80 ÷ 2 = 40 |
        About 40 shingles

▶ **To divide by a decimal:**

- Move the decimal point in the *divisor* to form a whole-number divisor.

  $0.2\overline{)8.4\ 6}$

- Move the decimal point in the *dividend* to the right the *same number* of places.

  $2\overline{)8.4\ 6}$

- Place the decimal point in the the quotient.

  $2\overline{)8\ \ 4.6}$

- Divide.

  $\begin{array}{r} 4\ 2.3 \\ 2\overline{)8\ 4.6} \end{array}$

- Check by multiplying.

  $\begin{array}{r} 4\ 2.3 \\ \times\ \ \ 0.2 \\ \hline 8.4\ 6 \end{array}$

Each row will have 42.3 shingles.   42.3 is close to the estimate of 40.

Study these examples.

$\begin{array}{r} 2\ 4.8 \\ 0.1\overline{)2.4\ 8} \end{array}$ $\begin{array}{r} 1\ 2.4 \\ 0.2\overline{)2.4\ 8} \end{array}$ $\begin{array}{r} 6.2 \\ 0.4\overline{)2.4\ 8} \end{array}$

140

Write the letter of the correct answer.

1. $0.3\overline{)1.2}$ **a.** 0.04   **b.** 0.4   **c.** 4   **d.** 40

2. $0.4\overline{)12.8}$ **a.** 320   **b.** 32   **c.** 3.2   **d.** 0.32

3. $0.2\overline{)1.8}$ **a.** 0.009   **b.** 0.09   **c.** 0.9   **d.** 9

4. $0.6\overline{)0.36}$ **a.** 0.06   **b.** 0.6   **c.** 6   **d.** 60

Divide and check.

5. $0.4\overline{)9.2}$ 6. $0.3\overline{)1.2}$ 7. $0.5\overline{)2.5}$ 8. $0.7\overline{)4.2}$

9. $0.8\overline{)5.6}$ 10. $0.9\overline{)4.5}$ 11. $0.6\overline{)74.4}$ 12. $0.2\overline{)8.4}$

13. $0.5\overline{)7.55}$ 14. $0.6\overline{)9.66}$ 15. $0.4\overline{)0.76}$ 16. $0.7\overline{)8.61}$

17. $92.4 \div 0.4$ 18. $6.3 \div 0.3$ 19. $257.2 \div 0.4$ 20. $0.96 \div 0.8$

21. $2.214 \div 0.9$ 22. $0.084 \div 0.3$ 23. $555.6 \div 0.6$ 24. $391.2 \div 0.4$

25. $0.7\overline{)95.13}$ 26. $0.3\overline{)53.34}$ 27. $0.6\overline{)2.88}$ 28. $0.8\overline{)7.68}$

29. $0.7\overline{)444.5}$ 30. $43.11 \div 0.3$ 31. $0.2\overline{)0.008}$ 32. $6.85 \div 0.5$

PROBLEM SOLVING

33. Mike is tiling a floor. If each tile is 0.3 m wide and the floor is 5.4 m wide, how many tiles will fit in each row?

34. Carlos cut a 25.8-ft length of rope into 0.6-ft segments. How many segments did he cut?

35. The perimeter of a square floor is 48.8 ft. How long is each side?

36. Yvette grew 13.68 in. in 12 months. On the average, how many inches did she grow per month?

 **CRITICAL THINKING**

Complete the pattern to find each quotient.

$24 \div 3 = 8$	**37.** $24 \div 0.3 = \underline{\ ?\ }$	**38.** $24 \div 0.03 = \underline{\ ?\ }$	**39.** $24 \div 0.003 = \underline{\ ?\ }$
$42 \div 6 = 7$	**40.** $42 \div 0.6 = \underline{\ ?\ }$	**41.** $42 \div 0.06 = \underline{\ ?\ }$	**42.** $42 \div 0.006 = \underline{\ ?\ }$

Decimal Divisors

One floor of a building is 3.75 m high. If the building is 116.25 m high and all floors are the same height, how many floors does the building have?

To find the number of floors, divide: $116.25 \div 3.75 = \underline{\ ?\ }$

Estimate: $120 \div 4 = 30$
                    About 30 floors

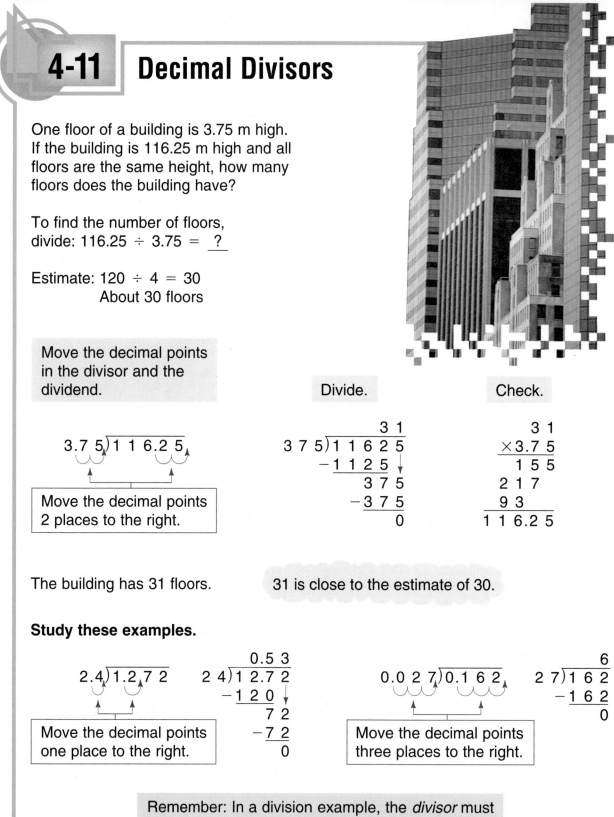

| Move the decimal points in the divisor and the dividend. | Divide. | Check. |

$$3.7\,5\,\overline{)\,1\,1\,6.2\,5}$$

Move the decimal points 2 places to the right.

```
              3 1
   3 7 5 )1 1 6 2 5
          -1 1 2 5 ↓
              3 7 5
            - 3 7 5
                  0
```

```
            3 1
         × 3.7 5
          1 5 5
        2 1 7
        9 3
      1 1 6.2 5
```

The building has 31 floors.   31 is close to the estimate of 30.

Study these examples.

$$2.4\,\overline{)\,1.2\,7\,2}$$

Move the decimal points one place to the right.

```
            0.5 3
   2 4 )1 2.7 2
        -1 2 0 ↓
            7 2
          - 7 2
              0
```

$$0.0\,2\,7\,\overline{)\,0.1\,6\,2}$$

Move the decimal points three places to the right.

```
              6
   2 7 )1 6 2
        -1 6 2
              0
```

Remember: In a division example, the *divisor* must always be a whole number.

Move the decimal points in the divisor and in the dividend to show the quotient.

1. $2.3\overline{)6.4\,6\,3}$ quotient: 2.8 1

2. $0.1\,9\overline{)0.1\,7\,4\,8}$ quotient: 0.9 2

3. $4\,6.1\overline{)2\,0\,7\,4.5}$ quotient: 4 5

4. $0.9\,2\overline{)2.8\,6\,1\,2}$ quotient: 3.1 1

5. $3.4\overline{)2.9\,2\,4}$ quotient: 0.8 6

6. $8\,2.1\overline{)2\,9\,5\,5.6}$ quotient: 3 6

Place the decimal point in the quotient.

7. $0.8\overline{)4.8\,2\,4}$ quotient: 6 0 3

8. $0.0\,1\,1\overline{)0.0\,9\,3\,5}$ quotient: 8 5

9. $0.0\,1\,2\overline{)0.0\,0\,1\,4\,4}$ quotient: 0 1 2

10. $1.5\overline{)0.0\,0\,4\,5}$ quotient: 0 0 0 3

11. $0.1\,8\overline{)0.0\,3\,6}$ quotient: 0 2

12. $0.0\,2\,4\overline{)0.0\,0\,1\,4\,4}$ quotient: 0 0 6

Divide and check.

13. $0.6\overline{)4.8}$

14. $2.4\overline{)0.48}$

15. $6.1\overline{)75.03}$

16. $3.2\overline{)2.080}$

17. $0.28\overline{)4.396}$

18. $0.75\overline{)0.7725}$

19. $0.07\overline{)3.5028}$

20. $0.08\overline{)1.9216}$

21. $6.9 \div 2.3$

22. $8.93 \div 4.7$

23. $0.78 \div 0.26$

24. $0.014 \div 0.07$

25. $\$1.25 \div 2.5$

26. $\$72.75 \div 0.75$

27. $0.20448 \div 0.24$

28. $0.12312 \div 0.36$

PROBLEM SOLVING

29. A stack of office memos ready for storage is 16.8 cm high. How many file boxes 2.1 cm high will be needed to store the memos?

30. One pound of ham contains 4.5 servings. It costs $2.59. What is the cost per serving? Round to the nearest cent.

FINDING TOGETHER

Place decimal points in the example to get the quotient described.

$$1794 \div 39 = 46$$

31. Quotient between (a) 0.4 and 0.5; (b) 4 and 5; (c) 4000 and 5000; (d) 400 and 500

4-12 Zeros in Division

Erin used 8.75 gal of gasoline to drive her car 210 mi. How many miles per gallon of gasoline did her car get?

To find the miles per gallon,
divide: $210 \div 8.75 = $ __?__

Estimate: $200 \div 8 = 25$ About 25 miles per gallon

Divide. It is sometimes necessary to write one or more zeros as placeholders *in the dividend* to complete the division.

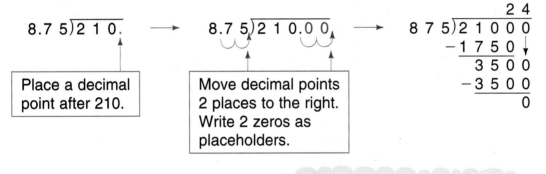

Place a decimal point after 210.

Move decimal points 2 places to the right. Write 2 zeros as placeholders.

Erin's car got 24 miles per gallon.

24 is close to the estimate of 25.

▶ If needed, write one or more zeros *in the quotient* to show the correct place value.

Write 1 zero in the quotient.

Write 2 zeros in the quotient.

Study this example.

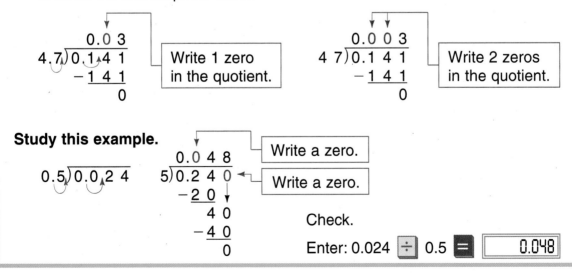

Write a zero.

Write a zero.

Check.

Enter: 0.024 ÷ 0.5 = 0.048

144

Divide. When needed, write zeros as placeholders in the dividend.

1. $0.4\overline{)0.2}$

2. $0.5\overline{)0.7}$

3. $0.8\overline{)0.5}$

4. $1.5\overline{)0.3}$

5. $0.8\overline{)1}$

6. $0.4\overline{)9}$

7. $2.5\overline{)6}$

8. $1.2\overline{)3}$

9. $0.05\overline{)0.7}$

10. $0.32\overline{)0.4}$

11. $0.08\overline{)0.7}$

12. $0.08\overline{)16}$

13. $6\overline{)3}$

14. $8\overline{)4}$

15. $0.2\overline{)0.03}$

16. $2.4\overline{)0.6}$

17. $0.7 \div 1.4$

18. $0.3 \div 2$

19. $0.03 \div 0.025$

20. $0.8 \div 0.032$

Divide. Write zeros in the quotient as needed.

21. $5\overline{)0.15}$

22. $4\overline{)0.36}$

23. $8\overline{)0.168}$

24. $80\overline{)0.8}$

25. $2.1\overline{)0.861}$

26. $6.2\overline{)0.372}$

27. $2.1\overline{)0.063}$

28. $0.6\overline{)0.036}$

29. $7\overline{)0.035}$

30. $9\overline{)0.414}$

31. $2\overline{)1.802}$

32. $9\overline{)0.099}$

33. $9.8\overline{)0.0196}$

34. $0.8\overline{)0.0328}$

35. $3.1\overline{)0.0279}$

36. $0.71\overline{)0.0142}$

37. $0.405 \div 0.5$

38. $0.352 \div 0.4$

39. $0.00092 \div 0.4$

40. $0.00042 \div 0.4$

41. $0.702 \div 9$

42. $0.0096 \div 3$

43. $4.32 \div 6$

44. $2.62 \div 8$

PROBLEM SOLVING

45. Forty laps around a track equal 2.5 km. How far is 1 lap around the track?

46. A wheel makes 1 turn in 0.7 second. What part of a turn can it make in 0.35 second?

47. Melons cost $.56 per pound. How many pounds can be bought with $5.60?

48. A can of juice costs $.48. How many cans can be bought with $12?

49. A greyhound runs at a speed of 39.35 miles per hour. How far will the greyhound run in 0.25 hour?

50. A baseball card is 0.65 mm thick. What is the height of 20 baseball cards? of 100 baseball cards?

51. Melinda bought 3.2 lb of cherries for $2.88. Cody paid $3.78 for 4.5 lb of cherries. Who paid more per pound? How much more?

4-13 Rounding Quotients

Mei bought 6 containers of apple juice for $1.90. To the nearest cent, what is the cost of each container of juice?

To find the cost of each container, divide: $1.90 ÷ 6 = ?

Estimate: $1.80 ÷ 6 = $.30

Rules for Rounding:

- Look at the digit to the right of the place to which you are rounding.

- If the digit is less than 5, round *down*. If the digit is 5 or more, round *up*.

Sometimes the division results in a remainder, no matter how many zeros are written in the dividend. You can **round** these quotients. In this problem, the quotient is rounded to the nearest cent (hundredths place).

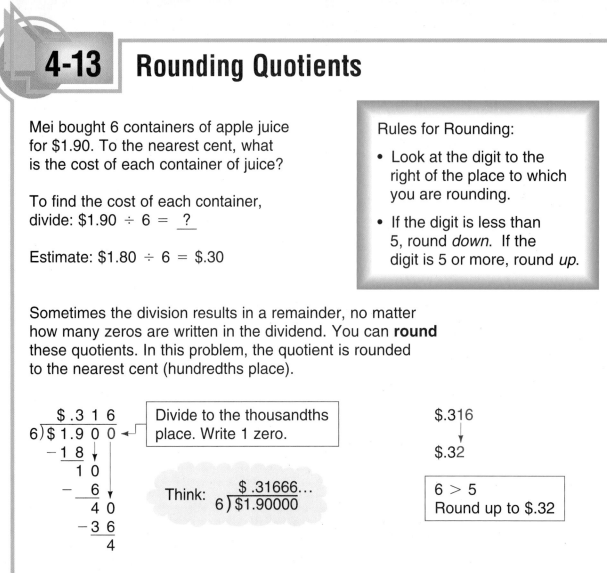

$$\begin{array}{r} \$ .3\ 1\ 6 \\ 6 \overline{)\ \$\ 1.9\ 0\ 0} \\ -1\ 8 \\ \hline 1\ 0 \\ -\ \ 6 \\ \hline 4\ 0 \\ -3\ 6 \\ \hline 4 \end{array}$$

Divide to the thousandths place. Write 1 zero.

Think: $\begin{array}{r} \$ .31666\ldots \\ 6 \overline{)\$1.90000} \end{array}$

$.316
↓
$.32

6 > 5
Round up to $.32

To the nearest cent, each container of juice costs $.32.

Study these examples.

Round to the nearest tenth.

$8 ÷ 3 = ?$

$$\begin{array}{r} 2.6\ 6 \\ 3 \overline{)8.0\ 0} \end{array}$$

Divide to hundredths. Write 2 zeros.

6 > 5 Round up. $8 ÷ 3 ≈ 2.7$

Round to the nearest thousandth.

$0.42 ÷ 0.19 = ?$

$$\begin{array}{r} 2.2\ 1\ 0\ 5 \\ 0.1\ 9 \overline{)0.4\ 2\ 0\ 0\ 0\ 0} \end{array}$$

Divide to ten thousandths. Write 4 zeros.

5 = 5 Round up. $0.42 ÷ 0.19 ≈ 2.211$

Divide. Round to the nearest tenth.

1. $6\overline{)8}$

2. $17\overline{)6}$

3. $9.2\overline{)20}$

4. $6.5\overline{)15}$

5. $2.3\overline{)0.4}$

6. $0.9\overline{)2.1}$

7. $3.1\overline{)6.5}$

8. $0.3\overline{)0.8}$

9. $0.4\overline{)0.85}$

10. $0.4\overline{)1.23}$

11. $0.03\overline{)0.11}$

12. $0.09\overline{)0.61}$

Divide. Round to the nearest hundredth or nearest cent.

13. $6\overline{)5}$

14. $3\overline{)22}$

15. $7\overline{)9.2}$

16. $4\overline{)1.5}$

17. $1.1\overline{)4.5}$

18. $1.5\overline{)0.4}$

19. $3.3\overline{)8.1}$

20. $0.7\overline{)4.5}$

21. $0.06\overline{)7.1}$

22. $0.07\overline{)9.3}$

23. $0.7\overline{)0.58}$

24. $0.3\overline{)0.71}$

25. $8\overline{)\$1.24}$

26. $6\overline{)\$8.23}$

27. $2\overline{)\$1.11}$

28. $3\overline{)\$5.19}$

Divide. Round to the nearest thousandth.

29. $6\overline{)0.4}$

30. $8\overline{)2.73}$

31. $3\overline{)7.055}$

32. $27\overline{)0.578}$

33. $0.3\overline{)0.61}$

34. $9.5\overline{)1808}$

35. $2.3\overline{)237}$

36. $0.07\overline{)0.4}$

PROBLEM SOLVING

37. A 32-oz box of cereal sells for $1.89. To the nearest cent, what is the price per ounce?

38. Juan can climb 3.7 km in 4 hours. To the nearest hundredth of a kilometer, how far can he climb in an hour?

39. Edna can run 5.5 km in 26 minutes. To the nearest tenth of a kilometer, what is her speed in kilometers per minute?

40. Mr. Shapiro used 14.7 gallons of gasoline to drive 392.7 miles. To the nearest tenth, what was his average number of miles per gallon?

 **SHARE YOUR THINKING**

Communicate

41. Use a newspaper advertisement to explain to a parent or another adult how to find *price per ounce* as in problem 37.

147

Working with Decimals

Algebra ✓

The properties of addition and multiplication of whole numbers are also true for decimals.

Commutative Property

	Think: "order."	
Addition		*Multiplication*
$0.6 + 1.9 = 1.9 + 0.6$		$0.6 \times 1.9 = 1.9 \times 0.6$

Associative Property

	Think: "grouping."	
Addition		*Multiplication*
$(0.4 + 0.6) + 0.8 = 0.4 + (0.6 + 0.8)$		$(0.4 \times 0.6) \times 0.8 = 0.4 \times (0.6 \times 0.8)$

Identity Property

	Think: "same."	
Addition		*Multiplication*
$0.7 + 0 = 0.7 \quad 0 + 0.7 = 0.7$		$0.7 \times 1 = 0.7 \quad 1 \times 0.7 = 0.7$

Zero Property of Multiplication

Think: "zero product."

$0.5 \times 0 = 0$ $0 \times 0.5 = 0$

Distributive Property of Multiplication over Addition

$$0.3 \times (0.8 + 0.2) = (0.3 \times 0.8) + (0.3 \times 0.2)$$

Name the property of addition or multiplication used.

1. $0.4 \times 0.6 = 0.6 \times 0.4$ **2.** $(1.3 + 0.7) + 5 = 1.3 + (0.7 + 5)$

3. $1.2 \times (0.4 + 1.1) = (1.2 \times 0.4) + (1.2 \times 1.1)$ **4.** $2.6 \times 1 = 2.6$

5. $0.48 \times c = 0.48$ **6.** $(n \times 0.5) \times 4 = n \times (0.5 \times 4)$

7. $12.6 + 0 = 12.6$ **8.** $4.5 \times (9.6 + 3.8) = (4.5 \times 9.6) + (4.5 \times 3.8)$

Complete. Use the properties of addition and multiplication.

9. $4 + 3.5 = 3.5 + \underline{\ ?\ }$ **10.** $2.77 \times 1 = \underline{\ ?\ }$ **11.** $0 + 12.3 = \underline{\ ?\ }$

12. $3.7 + \underline{\ ?\ } = 2.9 + 3.7$ **13.** $0.5 \times (2.9 + 3) = (0.5 \times 2.9) + (\underline{\ ?\ } \times 3)$

14. $(0.2 + 0.5) + 1.2 = 0.2 + (\underline{\ ?\ } + 1.2)$ **15.** $227.3 \times \underline{\ ?\ } = 0$

16. $0.04 \times \underline{\ ?\ } = 1.9 \times 0.04$ **17.** $(6.5 \times 2.7) \times 0.4 = \underline{\ ?\ } \times (2.7 \times 0.4)$

18. $1.03 \times n = n \times \underline{\ ?\ }$ **19.** $\underline{\ ?\ } \times (9.5 + \underline{\ ?\ }) = (2 \times 9.5) + (2 \times 0.9)$

Explain how to use the properties of addition and multiplication to make these computations easier. Then compute.   *Communicate*

20. $4 \times 9.7 \times 2.5$ **21.** $6.2 \times 3.7 \times 0$

22. $(6.7 \times 40) \times 2.5$ **23.** $(9.95 + 5.62) + 4.38$

24. $5 \times (2 + 1.5)$ **25.** $(50 \times 5.8) \times 2$

Using Multiplication to Divide

Study the pattern:
$$1 \div 0.5 = 2 \longrightarrow 1 \times 2 = 2$$
$$2 \div 0.5 = 4 \longrightarrow 2 \times 2 = 4$$
$$3 \div 0.5 = 6 \longrightarrow 3 \times 2 = 6$$

Dividing by 0.5 is the same as multiplying by 2.

Study the pattern:
$$1 \div 0.25 = 4 \longrightarrow 1 \times 4 = 4$$
$$2 \div 0.25 = 8 \longrightarrow 2 \times 4 = 8$$
$$3 \div 0.25 = 12 \longrightarrow 3 \times 4 = 12$$

Dividing by 0.25 is the same as multiplying by 4.

> How many $.50 are in....
> How many $.25 are in....

Divide. Use mental math.

26. $4 \div 0.5$ **27.** $6 \div 0.5$ **28.** $20 \div 0.5$ **29.** $4 \div 0.25$ **30.** $10 \div 0.25$

31. $25 \div 0.25$ **32.** $50 \div 0.5$ **33.** $40 \div 0.25$ **34.** $160 \div 0.5$ **35.** $225 \div 0.25$

 **CRITICAL THINKING**

36. Hannah is using her calculator to divide 8238.46 by 100, but the ÷ key is broken. Explain how she can find the answer by using a different calculator key.   *Communicate*

4-15 Scientific Notation

Scientists use **scientific notation** as a more compact and useful way to write very large numbers.

The Sun is about 93,000,000 miles from Earth.

▶ To write a number in scientific notation, write it as a *product* of two factors:

- One factor is a number greater than or equal to 1 but less than 10.

- The other factor is a power of 10, such as 10^2, 10^3, and so on.

> Scientific Notation
> 2.4×10^4
> 3×10^6 5.21×10^7

Write 93,000,000 in scientific notation.

- Move the decimal point to the *left* to get a number greater than or equal to 1 but less than 10.

  9 3 0 0 0 0 0

- Count the number of places the decimal point is moved. This is the power of ten.

  7 places moved. The power of 10 is 10^7.

- Drop the zeros to the right of the decimal. Write the product of the factors.

  9.3×10^7

So 93,000,000 equals 9.3×10^7.

Study these examples.

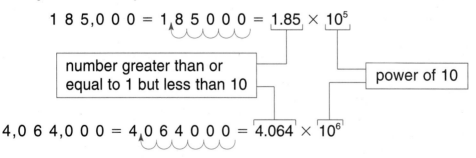

$1\,8\,5,0\,0\,0 = 1\,8\,5\,0\,0\,0 = 1.85 \times 10^5$

number greater than or equal to 1 but less than 10

power of 10

$4,0\,6\,4,0\,0\,0 = 4\,0\,6\,4\,0\,0\,0 = 4.064 \times 10^6$

Write in scientific notation.

1. 350,000
2. 475,000
3. 2,500,000
4. 4,650,000

5. 1,360,000
6. 87,000
7. 75,080
8. 82,000,000

9. 25,500,000
10. 477,000,000
11. 205,000
12. 1,825,500

13. 7,050,000
14. 405,050,000
15. 100,000,000
16. 9,000,000,000

Scientific Notation to Standard Form

Scientific Notation	Standard Form	
$3.6 \times 10^3 = 3.600 = 3600$		Move 3 places to the right.
$9.07 \times 10^4 = 9.0700 = 90,700$		Move 4 places to the right.

Write in standard form.

17. 3×10^2
18. 5×10^2
19. 6×10^3
20. 8×10^3

21. 3.5×10^3
22. 3.8×10^4
23. 4.04×10^5
24. 1.77×10^6

25. 8.03×10^5
26. 4.015×10^5
27. 6×10^8
28. 2.65×10^4

29. 2.165×10^6
30. 5.001×10^7
31. 4.323×10^5
32. 8.743×10^8

PROBLEM SOLVING

33. Mercury is about 60 000 000 km from the Sun. Write this number in scientific notation.

34. Light travels about 6 trillion miles in one year. Write this number in scientific notation.

 **CONNECTIONS: SCIENCE**

35. Research distances and dimensions in our solar system. Express the numbers using both standard form and scientific notation.

TECHNOLOGY

The Calculator and Scientific Notation

You can use a scientific calculator to find the standard numeral for a number written in scientific notation.

▶ Find the standard numeral for 3.2×10^4.

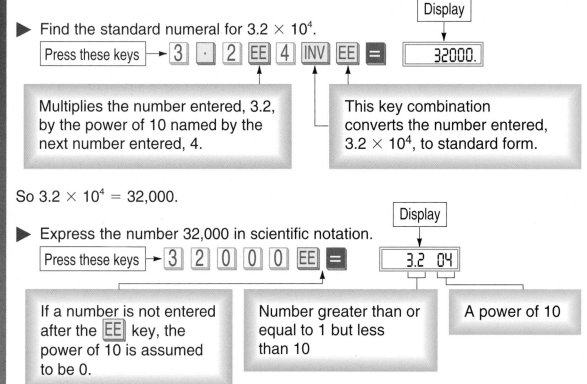

Press these keys ⟶ 3 · 2 EE 4 INV EE =

Display

32000.

Multiplies the number entered, 3.2, by the power of 10 named by the next number entered, 4.

This key combination converts the number entered, 3.2×10^4, to standard form.

So $3.2 \times 10^4 = 32{,}000$.

▶ Express the number 32,000 in scientific notation.

Press these keys ⟶ 3 2 0 0 0 EE =

Display

3.2 04

If a number is not entered after the EE key, the power of 10 is assumed to be 0.

Number greater than or equal to 1 but less than 10

A power of 10

So $32{,}000 = 3.2 \times 10^4$.

▶ Express the product of $(4.2 \times 10^3) \times (1.5 \times 10^5)$ in scientific notation.

Use the associative and commutative properties.

$(4.2 \times 1.5) \times (10^3 \times 10^5) = 6.3 \times 10^{(3 + 5)}$

Multiply the decimals.

Multiply the powers of 10.

Remember: When multiplying numbers with the same base, add the exponents.

So $(4.2 \times 10^3) \times (1.5 \times 10^5) = 6.3 \times 10^8$

Write the number in standard form.

1. 1.0×10^2

2. 1.0×10^3

3. 1.0×10^4

4. 1.0×10^5

5. 1.6×10^3

6. 4.5×10^5

7. 9.28×10^1

8. 6.09×10^7

9. 3×10^4

10. 9.9×10^6

11. 4.218×10^0

12. 1.07×10^1

Write the number in scientific notation.

13. 4100

14. 10,000,000

15. 100,000,000

16. 7999

17. 215

18. 79.8

19. 5.28

20. 5000

21. 1

22. 32,510

23. 100

24. 12,900

Find the product.

25. $(3.1 \times 10^3) \times (2.5 \times 10^2) = (3.1 \times 2.5) \times (10^3 \times 10^2) =$?

26. $(2.2 \times 10^5) \times (3 \times 10^1) = (2.2 \times 3) \times (10^5 \times 10^1) =$?

27. $(5.5 \times 10^4) \times (1.5 \times 10^4) =$?

28. $(3.85 \times 10^3) \times (1 \times 10^5) =$?

29. $(2.4 \times 10^6) \times (4.1 \times 10^4) =$?

30. $(6.57 \times 10^0) \times (1.2 \times 10^7) =$?

31. $(7.6 \times 10^0) \times (2.3 \times 10^0) =$?

32. $(5.2 \times 10^6) \times (1.5 \times 10^5) =$?

Solve. Use a calculator.

33. $1.6 \div 0.08 =$?
$16 \div 0.8 =$?
$160 \div 8 =$?

34. $4.9 \div 0.07 =$?
$49 \div 0.7 =$?
$490 \div 7 =$?

35. $3.6 \div 0.06 =$?
$36 \div 0.6 =$?
$360 \div 6 =$?

36. What do you notice about the quotients in exercises 33–35?

37. Describe the pattern in exercises 33–35.

38. Write the next number in the pattern: 0.2, 0.04, 0.008, ?

39. Describe the pattern in exercise 38.

4-17 | Problem Solving: Multi-Step Problem

Problem: Juanita is reading a 552-page book. She has already read 200 pages. If she reads 22 pages each day from now on, how many days will it take her to complete the book?

1 IMAGINE   Picture yourself in the problem.

2 NAME   *Facts:*   552-page book
She has read 200 pages.
She now reads 22 pages each day.

Question: How many days will it take Juanita to complete the book?

3 THINK   To find how long it will take to complete the book you must use two steps.

Step 1:   To find how many pages remain to be read, subtract:

$552 - 200 = \underline{\ ?\ }$ pages remaining

Step 2:   To find how many days it will take to complete the book, divide:

$\underline{\ ?\ }$ pages remaining $\div\ 22 = \underline{\ ?\ }$ number of days

4 COMPUTE   **Pages Remaining**   **Number of Days**

$552 - 200 = 352$

$$\begin{array}{r} 16 \\ 22\overline{)352} \\ -22\downarrow \\ \hline 132 \\ -132 \\ \hline \end{array}$$

It will take her 16 days to complete the book.

5 CHECK   Use opposite operations to check each step.

$$\begin{array}{r} 352 \\ +200 \\ \hline 552 \end{array} \qquad \begin{array}{r} 16 \\ \times 22 \\ \hline 32 \\ 32 \\ \hline 352 \end{array}$$

The answers check.

Solve. Watch for multi-step problems.

1. Mrs. Lopez bought 2 lb of apples at $1.09 a pound, 3 lb of oranges at $.89 a pound, and 3 lb of bananas at $.39 a pound. How much did the fruit cost her?

IMAGINE Picture Mrs. Lopez paying for fruit.

↓

NAME

Facts: 2 lb apples at $1.09 a pound
3 lb oranges at $.89 a pound
3 lb bananas at $.39 a pound

Question: How much did the fruit cost Mrs. Lopez?

↓

THINK To find the total cost you must use more than one step.

Step 1: To find the total cost for each fruit, multiply:
$2 \times \$1.09 = \underline{\ ?\ }$
$3 \times \$.89 = \underline{\ ?\ }$
$3 \times \$.39 = \underline{\ ?\ }$

Step 2: To find the total cost of all the fruit, add.

COMPUTE ⟶ **CHECK**

2. Dry cat food comes in two sizes: regular and jumbo. The regular size contains 8 oz and sells for $.96. The jumbo size has 12 oz and sells for $1.35. Which size is the better buy?

3. Judy's school has a 0.5-km track. One day at practice she ran around the track 9 times and then ran another 1 km in sprints. How far did she run that day?

4. One airplane traveled 2044 mi in 3.5 hours. Another traveled 3146 mi in 5.2 hours. Which airplane traveled faster? by how many miles an hour?

5. Natural Apple Sauce comes in 8-oz, 10-oz, and 16-oz jars. The 8-oz jar sells for $.54, the 10-oz jar for $.62, and the 16-oz jar for $1.00. Which size jar is most expensive per ounce?

6. Kim's mother is 3.5 times her age. Her father is 5 years older than her mother. Kim is 6 years older than her brother, who is 3. How old is Kim's father?

4-18 | Problem-Solving Applications

Solve each problem and explain the method you used.

1. Regina buys a bag of 2 dozen oranges for $2.99. What is the average cost per orange? Round to the nearest cent.

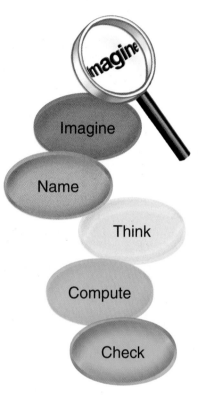

2. A bottle of iced tea costs $1.19. How much would a dozen bottles cost? a hundred bottles? a thousand bottles?

3. Cashews cost $3.98 a pound. Jake's bag weighed 2.7 pounds. Use estimation to find the cost of the cashews.

4. A 10.25-oz bag of pretzels costs $1.00. What is the cost per ounce?

5. Dried apricots cost $.29 per ounce. Mr. Carlson's bag of dried apricots weighs 18.8 ounces. How much will his bag of apricots cost?

6. Ted buys 100 packages of artificial sweetener. Each pack holds 0.035 oz of sweetener. How many ounces of sweetener does Ted buy?

7. How many ounces of cereal are in a package that includes this information on the nutrition label?
     serving size ....1.25 oz (1 cup)
     servings per package 8.2

8. Which of the following products costs less per ounce: 12 oz of peanut butter for $1.14 or 20 oz of peanut butter for $1.84?

9. Which is a better buy for Ms. Pinelli: a 1.9-lb cantaloupe for $1.69, or a 2.2-lb cantaloupe for $1.74?

10. A jar of peanuts costs $1.29. The peanuts cost $.129 per ounce. How many ounces of peanuts are in the jar?

Use a strategy from the list or another strategy you know to solve each problem.

USE THESE STRATEGIES:
Multi-Step Problem
Hidden Information
Extra Information
Use Simpler Numbers
Write a Number Sentence

11. Fine Foods sells a 2-lb wheel of cheese for $9.28. Stacey's Snacks sells the same cheese for $.27 per ounce. Which store has the better price?

12. Raphael pays $2.20 for 8 apples that weigh 2.5 lb. Kim buys 6 apples of the same kind that weigh 2 lb. How much does Kim spend?

13. A box of macaroni and cheese contains 2.4 servings. Each serving is 3.5 oz. How many ounces does the container hold?

14. Fine Foods sells a box of dried pineapple rings for $5.27. The rings cost $.62 per ounce. How many ounces are in the box?

15. Joe's prizewinning onion weighs 4.923 oz. A bag of 15 onions weighs 65.64 oz. How much lighter or heavier is Joe's onion than the average for this bag?

16. Pizza Delight comes in personal, average, and family sizes. The personal is 6 oz and sells for $1.09. The average is 12.5 oz and sells for $1.89. The family is 24 oz and sells for $2.79. Which is the best buy?

Use the chart for problems 17–20.

17. What is the price of the larger bottle of oregano?

18. How many more ounces does the large bag of wild rice contain than the small bag?

19. A restaurant needs 9 pounds of white rice. Will it be less expensive to buy one 5-lb bag and four 1-lb bags or two 5-lb bags?

20. Cindy buys 3 small boxes of raisins. How much less would she have had to spend to buy one large box instead?

Product	Size	Unit Price
Oregano	0.25 oz	$.48/oz
	0.5 oz	$.42/oz
White Rice	1 lb	$.09/oz
	5 lb	$.07/oz
Wild Rice	0.5 lb	$.37/oz
	3.25 lb	$.25/oz
Raisins	500 g	$5.48/kg
	1.2 kg	$5.20/kg

MAKE UP YOUR OWN

21. Use the chart to write a problem modeled on problem 19 above. Have a classmate solve it.

Chapter Review and Practice

Multiply.
(See pp. 122–131.)

1. 1000×0.02

2.
$$\begin{array}{r} 4.27 \\ \times\ \ \ 36 \\ \hline \end{array}$$

3.
$$\begin{array}{r} 0.695 \\ \times\ \ \ \ 12 \\ \hline \end{array}$$

4.
$$\begin{array}{r} 0.91 \\ \times\ \ 4.8 \\ \hline \end{array}$$

5. 0.4×0.002

6.
$$\begin{array}{r} 7.005 \\ \times\ \ 4.32 \\ \hline \end{array}$$

7.
$$\begin{array}{r} \$7.69 \\ \times\ \ \ \ \ 28 \\ \hline \end{array}$$

8.
$$\begin{array}{r} \$125.30 \\ \times\ \ \ \ \ \ \ 2.4 \\ \hline \end{array}$$

9. $3.55 \times \$46$

10.
$$\begin{array}{r} 0.927 \\ \times\ \ 0.08 \\ \hline \end{array}$$

11. $1.5 \times 0.66 \times 0.2$

Divide.
(See pp. 132–145.)

12. $0.79 \div 100$

13. $0.01\overline{)12.9}$

14. $16\overline{)154.4}$

15. $25\overline{)\$187.25}$

16. $0.084 \div 0.4$

17. $0.6\overline{)1.08}$

18. $4.8\overline{)2.544}$

19. $0.5\overline{)125}$

Divide. Round to the nearest hundredth or nearest cent.
(See pp. 146–147.)

20. $16 \div 6$

21. $0.11\overline{)5.9}$

22. $8\overline{)\$1.77}$

23. $9\overline{)\$57.59}$

Write in scientific notation.
(See pp. 150–151.)

24. $132{,}000$

25. 7796

26. $700{,}000{,}000{,}000$

Write in standard form.
(See pp. 150–151.)

27. 6×10^{3}

28. 5.4×10^{6}

29. 7.08×10^{5}

PROBLEM SOLVING
(See pp. 146–147, 154–157.)

30. Which costs more, 14.275 tons of coal at $85 a ton or 935 gallons of oil at $1.28 a gallon?

31. A deer runs at a speed of 27.7 miles per hour. How far does it run in 0.25 hour?

32. Kay can run 6 km in 26 minutes. Beth can run 4 km in 15.5 minutes. To the nearest tenth of a minute, how much faster does Beth run 1 kilometer?

(See *Still More Practice*, p. 509.)

PATTERNS: SEQUENCES

A **sequence** is a set of numbers given in a certain order. Each number is called a **term**. You can use a rule to find the next term in a sequence.

What is the rule for the sequence below?
What is the next term in the sequence?

1,   5,   9,   13,   17, . . . ◄—— . . . means the pattern continues indefinitely.
   +4   +4   +4   +4

Rule: Add 4   Next term: 17 + 4, or 21

What is the rule for the sequence below?
What is the next term in the sequence?

2.1,   4.2,   8.4,   16.8, . . .
   × 2   ×2   ×2

Rule: Multiply by 2   Next term: 16.8 × 2, or 33.6

Find the rule. Then use it to find the next term.

1. 10, 18, 26, . . .

2. 5, 20, 80, . . .

3. 106, 81, 56, . . .

4. 2.5, 5, 7.5, . . .

5. 0.4, 1.2, 3.6, . . .

6. 176.5, 17.65, 1.765, . . .

7. 0.125, 0.25, 0.5, . . .

8. 2, 0.4, 0.08, . . .

PROBLEM SOLVING

Use the sequence 1, 3, 5, 7, . . . for problems 9 and 10.

9. What is the sum of the first 2 terms? the first 3 terms? the first 4 terms? the first 5 terms?

10. Look at the sums you found. What pattern do you see? Use the pattern to predict the sum of the first 8 terms in the sequence. Check your prediction.

Check Your Mastery

Performance Assessment

Copy and complete each table. Describe any patterns that you see.

1.

n	n ÷ 0.5
2.5	
3.0	
3.5	

2.

x	x ÷ 0.2
0.08	
0.06	
0.04	

Find the product. Round products that are money amounts to the nearest cent.

3.   0.42
   $\times$ 56

4.   8.35
   $\times$ 6

5. 0.53×12.7

6.   0.39
   $\times 0.02$

7. $17.82
   $\times$ 47

8. $1.3 \times \$40.09$

Find the quotient.

9. $3\overline{)0.96}$

10. $4\overline{)\$14.24}$

11. $6.23 \div 0.001$

12. $0.7\overline{)7.91}$

13. $0.032\overline{)0.288}$

14. $0.0558 \div 6.2$

Rewrite in scientific notation or in standard form.

15. 576,000

16. 9,100,000

17. 9×10^4

18. 4.3×10^6

PROBLEM SOLVING   *Use a strategy or strategies you have learned.*

19. Sue can climb 7.6 km in 4.5 hours. To the nearest hundredth of a kilometer, how far can she climb in 1 hour?

20. Fred spent $10.69 on 9 used books. Fran spent $8.29 on 7 used books. Who spent less per book? Explain.

Number Theory and Fractions 5

You may have thought there was no mathematics in pizza. Well, there is. It turns out there is mathematics in plain cheese pizzas, sausage pizzas, pepperoni pizzas, pineapple pizzas, teriyaki pizzas, and avocado pizzas, just to name a few. (Sometimes, it's just not good to take mathematics too seriously.)

From *Math for Smarty Pants* by Marilyn Burns

In this chapter you will:

Investigate fractions, primes and composites
Compare, order and estimate fractions
Explore multiples and least common multiple
Relate fractions, mixed numbers, and decimals
Identify terminating and repeating decimals
Solve problems by finding a pattern

Critical Thinking/Finding Together

The number of slices in 1st giant pizza is a prime number between 10 and 20. The number of slices in 2nd and 3rd giant pizzas together is a multiple of 5. If the total number of slices is 32, how many slices are there in the 1st pizza?

5-1 Fractions

Mr. Paisley asked the students in his sixth grade mathematics class to represent the fraction $\frac{1}{6}$. He wrote the following reminder on the board:

A **fraction** is made up of:

$$\frac{\text{number of equal parts being considered}}{\text{number of equal parts in the whole}}$$

One student used pattern blocks to show $\frac{1}{6}$. A second student used a shaded diagram and a third student used a number line.

Pattern Blocks

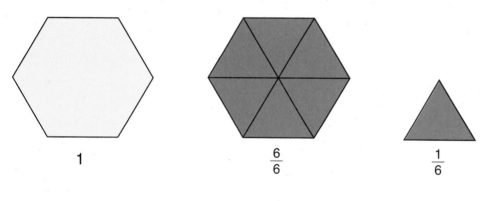

$$1 \qquad \frac{6}{6} \qquad \frac{1}{6}$$

Shaded Diagram

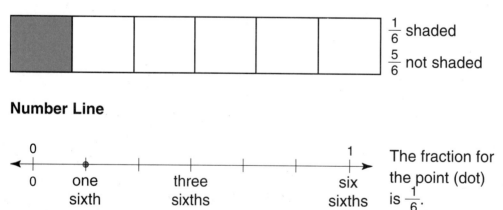

$\frac{1}{6}$ shaded

$\frac{5}{6}$ not shaded

Number Line

The fraction for the point (dot) is $\frac{1}{6}$.

Make a model for each fraction. Use pattern blocks, a shaded diagram, or a number line.

1. $\frac{1}{2}$ **2.** $\frac{1}{3}$ **3.** $\frac{1}{4}$ **4.** $\frac{2}{3}$ **5.** $\frac{3}{4}$ **6.** $\frac{3}{8}$

Write a fraction for each point.

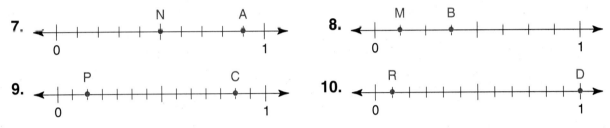

7.   N   A   0 ... 1

8.   M   B   0 ... 1

9.   P   C   0 ... 1

10.   R   D   0 ... 1

PROBLEM SOLVING

11. Of the 32 members of the Lansing Drama Club, 15 will build scenery, 8 will sew costumes, and the rest will make posters. What fractional part of the members will work on posters?

12. The school trophy case contains 10 awards. There are 4 plaques, 2 medals, and 4 ribbons. What fractional part is plaques? What fractional part is *not* ribbons?

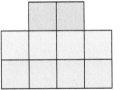

13. In the school store, there are 5 packs of typing paper, 7 packs of notebook paper, and 4 packs of computer paper. What fractional part is typing and notebook paper? What fractional part is *not* computer paper?

SHARE YOUR THINKING

Discuss with a small group of your classmates how to solve the following problems. Then complete each model.

Discuss ✓

14. Use hexagon and triangle pattern blocks as shown on page 162. Model the fraction $\frac{5}{12}$.

15. Use the same pattern blocks to model $\frac{1}{18}$.

163

5-2 Finding Equivalent Fractions

Fractions that name the same part of a set, region, or an object are called **equivalent fractions**. $\frac{1}{2}, \frac{2}{4}, \frac{3}{6}, \frac{4}{8}$ are equivalent.

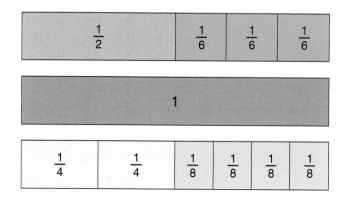

To find equivalent fractions, *multiply* or *divide* the numerator and the denominator by a fraction name for one (1) such as:

$$\frac{2}{2}, \frac{3}{3}, \frac{4}{4}, \frac{5}{5}$$

Study these examples.

$$\frac{5}{6} = \frac{?}{18}$$

| $6 \times 3 = 18$ |
| Multiply $\frac{5}{6}$ by $\frac{3}{3}$. |

$$\frac{5 \times ?}{6 \times 3} = \frac{?}{18}$$

$$\frac{5 \times 3}{6 \times 3} = \frac{15}{18}$$

$$\frac{5}{6} = \frac{15}{18}$$

$\frac{5}{6}$ and $\frac{15}{18}$ are equivalent fractions.

$$\frac{8}{32} = \frac{1}{?}$$

| $8 \div 8 = 1$ |
| Divide $\frac{8}{32}$ by $\frac{8}{8}$. |

$$\frac{8 \div 8}{32 \div ?} = \frac{1}{?}$$

$$\frac{8 \div 8}{32 \div 8} = \frac{1}{4}$$

$$\frac{8}{32} = \frac{1}{4}$$

$\frac{8}{32}$ and $\frac{1}{4}$ are equivalent fractions.

Complete.

1. $\dfrac{1}{3} = \dfrac{?}{6} = \dfrac{?}{9}$

2. $\dfrac{3}{4} = \dfrac{?}{8} = \dfrac{?}{12}$

3. $\dfrac{5}{8} = \dfrac{?}{16} = \dfrac{?}{24}$

4. $\dfrac{4}{5} = \dfrac{8}{?} = \dfrac{12}{?}$

5. $\dfrac{1}{6} = \dfrac{2}{?} = \dfrac{3}{?}$

6. $\dfrac{2}{7} = \dfrac{?}{14} = \dfrac{6}{?}$

164

In each exercise, which two figures show equivalent fractions? Explain your answer.

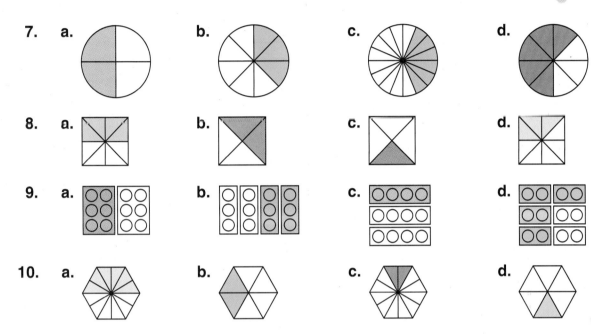

7. a.   b.   c.   d.

8. a.   b.   c.   d.

9. a.   b.   c.   d.

10. a.   b.   c.   d.

Write the missing number to complete the equivalent fraction.

11. $\frac{3}{4} = \frac{?}{12}$ 12. $\frac{1}{11} = \frac{?}{88}$ 13. $\frac{2}{9} = \frac{?}{81}$ 14. $\frac{2}{3} = \frac{?}{12}$ 15. $\frac{5}{7} = \frac{40}{?}$

16. $\frac{1}{8} = \frac{6}{?}$ 17. $\frac{1}{10} = \frac{3}{?}$ 18. $\frac{3}{11} = \frac{6}{?}$ 19. $\frac{9}{10} = \frac{?}{100}$ 20. $\frac{1}{25} = \frac{4}{?}$

21. $\frac{9}{30} = \frac{?}{10}$ 22. $\frac{4}{12} = \frac{?}{3}$ 23. $\frac{21}{28} = \frac{?}{4}$ 24. $\frac{40}{45} = \frac{?}{9}$ 25. $\frac{2}{6} = \frac{1}{?}$

26. $\frac{20}{30} = \frac{2}{?}$ 27. $\frac{9}{90} = \frac{1}{?}$ 28. $\frac{5}{25} = \frac{1}{?}$ 29. $\frac{44}{80} = \frac{?}{20}$ 30. $\frac{36}{42} = \frac{6}{?}$

31. $\frac{5}{14} = \frac{10}{?}$ 32. $\frac{4}{5} = \frac{?}{60}$ 33. $\frac{2}{7} = \frac{?}{49}$ 34. $\frac{49}{84} = \frac{7}{?}$ 35. $\frac{27}{30} = \frac{9}{?}$

CHALLENGE

36. I am equivalent to $\frac{2}{3}$. My numerator is 7 less than my denominator. What fraction am I?

37. I am equivalent to $\frac{36}{60}$. The sum of my numerator and my denominator is 24. What fraction am I?

5-3 Prime and Composite Numbers

▶ A **prime number** is a number greater than 1 that has *exactly two* factors, itself and 1.

Find all the factors of 11.

$1 \times 11 = 11$ Factors of 11: 1, 11

11 has exactly two factors, 11 and 1.
So 11 is a prime number.

Remember: 1 is *not* a prime number.

▶ A **composite number** is a number greater than 1 that has *more than two* factors.

Find all the factors of 25.

$1 \times 25 = 25$
$5 \times 5 = 25$ Factors of 25: 1, 5, 25

25 has more than two factors.
So 25 is a composite number.

Tell whether each number is *prime, composite,* or *neither.*

1. 24   **2.** 35   **3.** 2   **4.** 9   **5.** 19

6. 21   **7.** 33   **8.** 11   **9.** 1   **10.** 0

11. 51   **12.** 26   **13.** 81   **14.** 100   **15.** 41

Copy these statements in your Math Journal. Then tell whether each statement is True or False. Explain.

Math Journal

16. Any whole number, except 0 and 1, is either prime or composite.

17. No composite number is an even number.

18. All prime numbers are odd numbers.

Make and complete a chart like the one below for the numbers 1–20. Use the chart for exercises 19–26.

Identify the number(s).

19. prime numbers

20. composite numbers

21. have exactly three factors

22. has only one factor

23. have six factors

24. Which number is a factor of all of the numbers?

25. Which numbers have both 2 and 3 as factors?

26. Which numbers have both 2 and 5 as factors?

	Factors	Number of Factors	Prime or Composite
1	1	1	neither
2	1, 2	?	prime
3	1, 3	?	
17	?	?	?
18	?	?	?
19	?	?	?
20	1, 2, 4, 5, 10, 20	?	?

 FINDING TOGETHER

Six is called a **perfect number** because it is the sum of all its factors, not including itself.

The factors of 6 are 1, 2, 3, 6.
1 + 2 + 3 = 6

27. Find another perfect number. Extend the chart above to help you.

28. Are any prime numbers also perfect numbers? Explain.

29. Use your computer or reference books in the school library to define these two types of numbers: (a) deficient numbers; and (b) abundant numbers. Discuss your results with the class.

Discuss ✓

5-4 Prime Factorization

Every composite number can be written as the product of prime factors. This is called **prime factorization**.

Find the prime factorization of 36.

Use a **factor tree** to find the prime factors:

- Start with the composite number.

- Choose *any* 2 factors.

- Continue factoring until all the branches show *prime* numbers.

- Arrange the prime factors in order from least to greatest.

36

③ × 12 ← Composite Factor again.

③ × ③ × 4 ← Composite Factor again.

③ × ③ × ② × ②

2 × 2 × 3 × 3

The prime factorization of 36 is 2 × 2 × 3 × 3.

Using exponents, the prime factorization of 36 can be expressed as follows: $2^2 \times 3^2$

$2 \times 2 = 2^2$ and
$3 \times 3 = 3^2$

No matter which 2 factors you begin with, the prime factorization will always be the same.

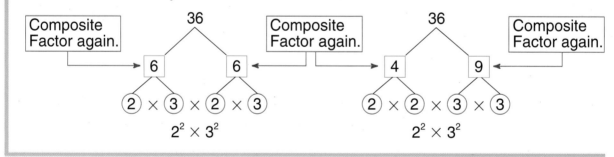

Composite Factor again. → 36

6   6 ← Composite Factor again.

② × ③ × ② × ③

$2^2 \times 3^2$

Composite Factor again. → 36

4   9 ← Composite Factor again.

② × ② × ③ × ③

$2^2 \times 3^2$

Make a factor tree for each.

1. 45   **2.** 64   **3.** 72   **4.** 88   **5.** 48

Express each in exponent form.

6. 5 × 2 × 2 × 5   **7.** 2 × 2 × 2 × 3   **8.** 11 × 2 × 11 × 5

9. 7 × 3 × 2 × 7 × 3   **10.** 5 × 13 × 5 × 5   **11.** 2 × 7 × 2 × 7

Find the prime factorization and write in exponent form.

12. 32   **13.** 24   **14.** 50   **15.** 125   **16.** 63

17. 72   **18.** 44   **19.** 60   **20.** 100   **21.** 96

Prime Factorization Using Divisibility Rules

You can use the divisibility rules to help you find the prime factorization of larger numbers.

If	Divisible by:
the number is even	2
the digit sum is divisible by 3	3
the last two digits are divisible by 4	4
the number ends in 5 or 0	5
the digit sum is divisible by 9	9
the number ends in 0	10

Find the prime factorization of 9450.

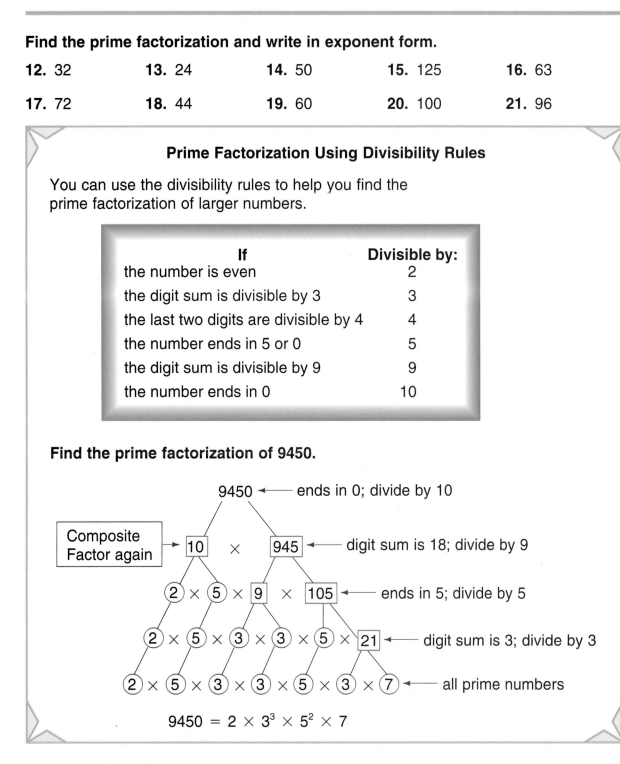

9450 ← ends in 0; divide by 10

Composite Factor again → 10 × 945 ← digit sum is 18; divide by 9

2 × 5 × 9 × 105 ← ends in 5; divide by 5

2 × 5 × 3 × 3 × 5 × 21 ← digit sum is 3; divide by 3

2 × 5 × 3 × 3 × 5 × 3 × 7 ← all prime numbers

$9450 = 2 \times 3^3 \times 5^2 \times 7$

Find the prime factorization. Use a calculator to check.

22. 95   **23.** 114   **24.** 153   **25.** 390   **26.** 504

27. 189   **28.** 225   **29.** 540   **30.** 1215   **31.** 2916

5-5 Greatest Common Factor

The **greatest common factor** (**GCF**) of two or more numbers is the greatest number that is a factor of the numbers.

Find the greatest common factor of 8, 12, and 20.

To find the GCF:

- List all the factors of each number.

  8: 1, 2, 4, 8
  12: 1, 2, 3, 4, 6, 12
  20: 1, 2, 4, 5, 10, 20

- Find the common or shared factors.

  Common factors:
  1, 2, 4

- Choose the greatest common factor.

  The greatest common factor is 4.

The GCF of 8, 12, and 20 is 4.

4 divides evenly into 8, 12, and 20.

Write all the common factors for each set of numbers.

1. 8 and 24   **2.** 10 and 30   **3.** 15 and 35   **4.** 12 and 18

5. 16 and 20   **6.** 12 and 24   **7.** 30 and 18   **8.** 45 and 20

9. 4, 6, and 8   **10.** 6, 9, and 12   **11.** 5, 12, and 14   **12.** 6, 14, and 22

Find the GCF of each pair of numbers.

13. 6 and 12   **14.** 12 and 36   **15.** 8 and 10   **16.** 6 and 14

17. 9 and 30   **18.** 8 and 36   **19.** 24 and 42   **20.** 7 and 40

Find the GCF of each set of numbers.

21. 8, 24, and 32   **22.** 5, 30, and 35   **23.** 15, 30, and 45

Finding the GCF Using Prime Factorization

Find the GCF of 27 and 54.

Use factor trees to find the prime factorization of each number.

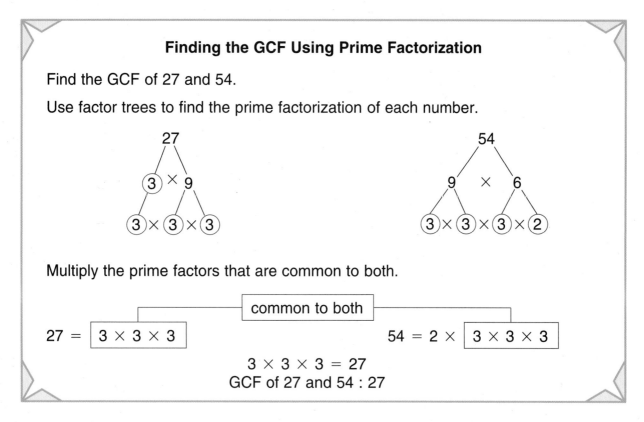

Multiply the prime factors that are common to both.

27 = $\boxed{3 \times 3 \times 3}$ 54 = 2 × $\boxed{3 \times 3 \times 3}$

common to both

$3 \times 3 \times 3 = 27$
GCF of 27 and 54 : 27

Find the GCF. Use prime factorization.

24. 48 and 56   **25.** 64 and 96   **26.** 36 and 72   **27.** 80 and 100

28. 45 and 75   **29.** 39 and 104   **30.** 48 and 84   **31.** 100 and 125

32. 14, 49, and 70   **33.** 48, 80, and 112   **34.** 18, 54, and 90

Find the pairs of numbers.

35. Between 10 and 20 that have 6 as their GCF

36. Between 12 and 18 that have 4 as their GCF

37. Between 15 and 30 that have 5 as their GCF

38. Between 16 and 24 that have 8 as their GCF

 SHARE YOUR THINKING   Communicate

39. You can *divide* to find the greatest common factor (GCF) of two or more numbers. Write a short note to your teacher to tell how you can use division to find the GCF in exercise 32 above.

Fractions in Simplest Form

A fraction is in **simplest form**, or **lowest terms**, when the numerator and denominator have no common factor other than 1.

Cass and Arti surveyed 32 classmates to find out when they did their homework. The results showed that 20 out of 32 ($\frac{20}{32}$) of the students did homework after dinner. Rename this fraction in simplest form.

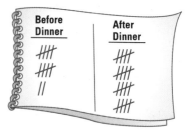

Before Dinner	After Dinner
⑴⑴⑴ ⑴⑴⑴ ⑴⑴	⑴⑴⑴ ⑴⑴⑴ ⑴⑴⑴ ⑴⑴⑴

To rename a fraction as an equivalent fraction in simplest form:

- Find the GCF of the numerator and the denominator.

$\frac{20}{32}$ → factors of 20: 1, 2, 4, 5, 10, 20
→ factors of 32: 1, 2, 4, 8, 16, 32

GCF of 20 and 32: 4

- Divide the numerator and the denominator by their GCF.

$$\frac{20}{32} = \frac{20 \div 4}{32 \div 4} = \frac{5}{8}$$

The simplest form of $\frac{20}{32}$ is $\frac{5}{8}$.

Write the letter of the GCF of the numerator and the denominator of each fraction.

1. $\frac{3}{6}$ **a.** 1   **b.** 6   **c.** 3   **d.** 18

2. $\frac{7}{8}$ **a.** 7   **b.** 1   **c.** 8   **d.** 14

3. $\frac{10}{12}$ **a.** 10   **b.** 12   **c.** 2   **d.** 1

4. $\frac{25}{45}$ **a.** 25   **b.** 9   **c.** 5   **d.** 1

5. $\frac{80}{100}$ **a.** 20   **b.** 10   **c.** 50   **d.** 2

6. $\frac{11}{132}$ **a.** 1   **b.** 11   **c.** 132   **d.** 12

Is the fraction in lowest terms? Write *Yes* or *No*.

7. $\frac{2}{3}$
8. $\frac{1}{8}$
9. $\frac{4}{8}$
10. $\frac{5}{10}$
11. $\frac{3}{10}$
12. $\frac{1}{12}$

13. $\frac{7}{21}$
14. $\frac{12}{25}$
15. $\frac{10}{18}$
16. $\frac{6}{21}$
17. $\frac{12}{18}$
18. $\frac{5}{24}$

19. $\frac{16}{27}$
20. $\frac{9}{12}$
21. $\frac{14}{35}$
22. $\frac{24}{34}$
23. $\frac{17}{36}$
24. $\frac{18}{72}$

Rename each as a fraction in simplest form.

25. $\frac{5}{10}$
26. $\frac{15}{40}$
27. $\frac{16}{48}$
28. $\frac{3}{18}$
29. $\frac{16}{20}$
30. $\frac{9}{36}$

31. $\frac{5}{55}$
32. $\frac{12}{16}$
33. $\frac{20}{50}$
34. $\frac{21}{49}$
35. $\frac{12}{24}$
36. $\frac{12}{30}$

37. $\frac{12}{44}$
38. $\frac{30}{55}$
39. $\frac{14}{42}$
40. $\frac{14}{18}$
41. $\frac{5}{35}$
42. $\frac{20}{32}$

43. $\frac{14}{20}$
44. $\frac{16}{24}$
45. $\frac{20}{32}$
46. $\frac{9}{36}$
47. $\frac{6}{27}$
48. $\frac{16}{28}$

PROBLEM SOLVING Write each answer as a fraction in simplest form.

49. Lions spend about 20 hours a day sleeping. What part of their day do lions spend sleeping? What part of their day are they awake?

50. At the circus, 128 of the 160 animals are *not* lions. What part of the animals are lions?

 **CRITICAL THINKING**

Write *sometimes*, *always*, or *never*.

51. A fraction with 1 as a numerator is in simplest form.

52. A fraction with a prime number in the numerator is in simplest form.

173

Mixed Numbers and Improper Fractions

When you compute with fractions, you must be able to express **mixed numbers** as **improper fractions**, and vice versa.

▶ **To rename a mixed number as an improper fraction:**

- Multiply the whole number by the denominator.

- Add the product to the numerator.

- Write the sum over the denominator.

$$3\frac{1}{2} = \frac{(2 \times 3) + 1}{2} = \frac{7}{2}$$

improper fraction

Remember: An **improper fraction** has a numerator *equal to* or *greater than* its denominator.

▶ **To rename an improper fraction as a whole number or as a mixed number:**

- Divide the numerator by the denominator.

- If there is a remainder, write it over the denominator and express in simplest form.

$$\frac{38}{4} = 4\overline{)38}\;\;^{9\;\text{R2}}$$

$$\frac{38}{4} = 9\frac{2}{4}$$

$$= 9\frac{1}{2}$$

Remember: Read $9\frac{1}{2}$ as nine *and* one half.

Study these examples.

$$\frac{18}{9} = 9\overline{)18}\;\;^{2} \longrightarrow \frac{18}{9} = 2 \qquad 10\frac{5}{6} = \frac{(6 \times 10) + 5}{6} = \frac{60 + 5}{6} = \frac{65}{6}$$

Write the word name for each mixed number.

1. $7\frac{1}{2}$ 2. $8\frac{4}{5}$ 3. $5\frac{1}{10}$ 4. $22\frac{1}{8}$ 5. $1\frac{1}{20}$ 6. $11\frac{11}{12}$

Write as a mixed number.

7. eleven and one fourth

8. nine and nine tenths

9. sixteen and three fifths

10. thirty and two thirds

11. twenty and fifteen sixteenths

12. twenty-one and seven tenths

Express each mixed number as an improper fraction.

13. $4\frac{1}{4}$ 14. $2\frac{1}{2}$ 15. $1\frac{3}{8}$ 16. $2\frac{1}{8}$ 17. $3\frac{4}{5}$ 18. $5\frac{2}{7}$

19. $1\frac{1}{9}$ 20. $1\frac{1}{10}$ 21. $11\frac{1}{3}$ 22. $12\frac{1}{2}$ 23. $15\frac{1}{4}$ 24. $12\frac{2}{7}$

Express each improper fraction as a whole number or a mixed number in simplest form.

25. $\frac{6}{5}$ 26. $\frac{9}{7}$ 27. $\frac{11}{8}$ 28. $\frac{5}{3}$ 29. $\frac{14}{2}$ 30. $\frac{48}{8}$

31. $\frac{12}{8}$ 32. $\frac{15}{9}$ 33. $\frac{44}{6}$ 34. $\frac{92}{10}$ 35. $\frac{88}{6}$ 36. $\frac{110}{5}$

SHARE YOUR THINKING

37. Each of the mixed numbers in the box below must be simplified. Work with a classmate to decide why. Discuss with the class. Then simplify each mixed number.

Discuss ✓

$$15\frac{10}{24} \qquad 15\frac{24}{24} \qquad 15\frac{25}{24}$$

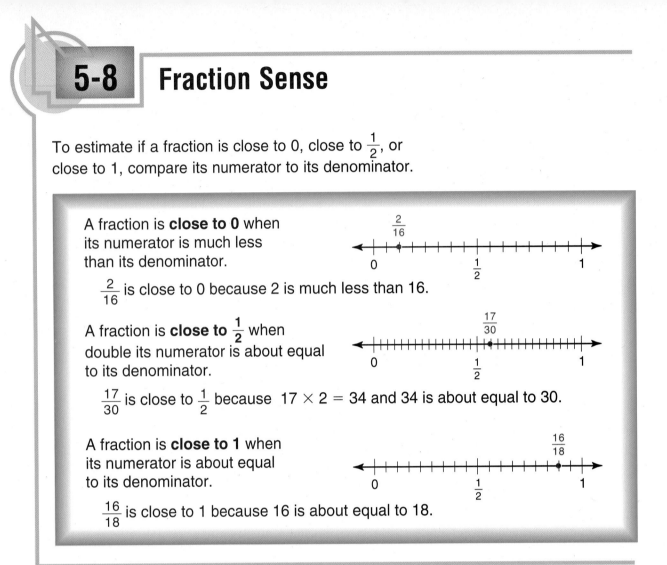

5-8 Fraction Sense

To estimate if a fraction is close to 0, close to $\frac{1}{2}$, or close to 1, compare its numerator to its denominator.

A fraction is **close to 0** when its numerator is much less than its denominator.

$\frac{2}{16}$ is close to 0 because 2 is much less than 16.

A fraction is **close to $\frac{1}{2}$** when double its numerator is about equal to its denominator.

$\frac{17}{30}$ is close to $\frac{1}{2}$ because $17 \times 2 = 34$ and 34 is about equal to 30.

A fraction is **close to 1** when its numerator is about equal to its denominator.

$\frac{16}{18}$ is close to 1 because 16 is about equal to 18.

Write the fraction that names each point.
Tell whether the fraction is close to 0, $\frac{1}{2}$, or 1.

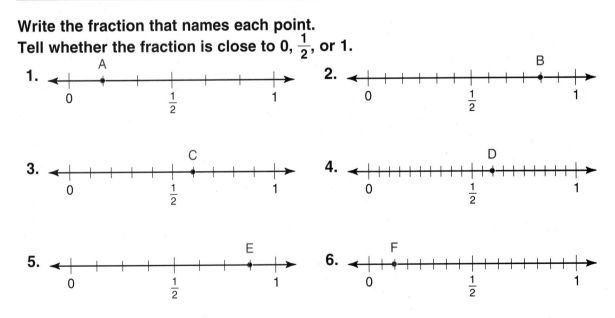

1. A

2. B

3. C

4. D

5. E

6. F

Tell whether the fraction is close to 0, $\frac{1}{2}$, or 1.

7. $\frac{1}{8}$ **8.** $\frac{2}{15}$ **9.** $\frac{6}{7}$ **10.** $\frac{13}{15}$ **11.** $\frac{7}{13}$ **12.** $\frac{8}{15}$

13. $\frac{27}{60}$ **14.** $\frac{7}{90}$ **15.** $\frac{12}{25}$ **16.** $\frac{91}{100}$ **17.** $\frac{106}{200}$ **18.** $\frac{37}{40}$

19. $\frac{20}{38}$ **20.** $\frac{11}{200}$ **21.** $\frac{5}{61}$ **22.** $\frac{42}{90}$ **23.** $\frac{39}{40}$ **24.** $\frac{15}{32}$

25. $\frac{13}{27}$ **26.** $\frac{17}{28}$ **27.** $\frac{1}{4}$ **28.** $\frac{3}{4}$ **29.** $\frac{30}{40}$ **30.** $\frac{25}{100}$

Complete. Write a fraction that is close to 0.

31. $\frac{?}{12}$ **32.** $\frac{?}{20}$ **33.** $\frac{?}{9}$ **34.** $\frac{1}{?}$ **35.** $\frac{7}{?}$ **36.** $\frac{12}{?}$

Complete. Write a fraction that is close to $\frac{1}{2}$.

37. $\frac{?}{7}$ **38.** $\frac{?}{25}$ **39.** $\frac{?}{15}$ **40.** $\frac{12}{?}$ **41.** $\frac{9}{?}$ **42.** $\frac{?}{42}$

Complete. Write a fraction that is close to 1.

43. $\frac{?}{7}$ **44.** $\frac{?}{30}$ **45.** $\frac{?}{14}$ **46.** $\frac{35}{?}$ **47.** $\frac{24}{?}$ **48.** $\frac{?}{100}$

Is the fraction a little more than $\frac{1}{2}$? Write *Yes* or *No*. Explain.

49. $\frac{14}{25}$ **50.** $\frac{11}{24}$ **51.** $\frac{16}{30}$ **52.** $\frac{17}{35}$ **53.** $\frac{52}{110}$ **54.** $\frac{61}{122}$

SKILLS TO REMEMBER

**Write the product that comes out of the machine.
Input these numbers.**

55. 7 **56.** 8 **57.** 9

58. 10 **59.** 11 **60.** 12

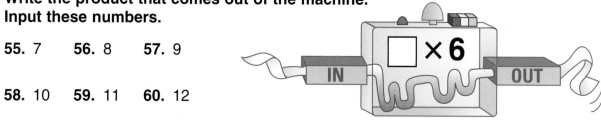

177

5-9 Least Common Multiple

The planet Jupiter takes 12 years to make one complete revolution around the Sun. The planet Saturn takes 30 years to make one complete revolution around the Sun. If an astronomer looks at both planets at 9:00 P.M. tonight, how many years will it be before they have the exact same position in the sky?

To find the number of years, you must find the **least common multiple (LCM)** of 12 and 30.

▶ The least common multiple of two or more numbers is the least number, except 0, that is a common multiple of both (or all) of the numbers.

Multiples of 12: 12, 24, 36, 48, 60, 72, . . .

Multiples of 30: 30, 60, 90, 120, 150, 180, . . .

> Extend the list until you find a common multiple of the numbers.

The least common multiple (LCM) of 12 and 30 is 60.

So Jupiter and Saturn will have the exact same position in the sky in 60 years.

Find the LCM of each pair or set of numbers.

1. 3, 4	**2.** 2, 5	**3.** 3, 6	**4.** 8, 24	**5.** 12, 15
6. 4, 10	**7.** 1, 9	**8.** 6, 5	**9.** 12, 10	**10.** 40, 16
11. 3, 4, 6	**12.** 1, 6, 7	**13.** 4, 5, 10	**14.** 4, 6, 8	
15. 5, 6, 12	**16.** 3, 9, 12	**17.** 8, 12, 36	**18.** 10, 18, 72	

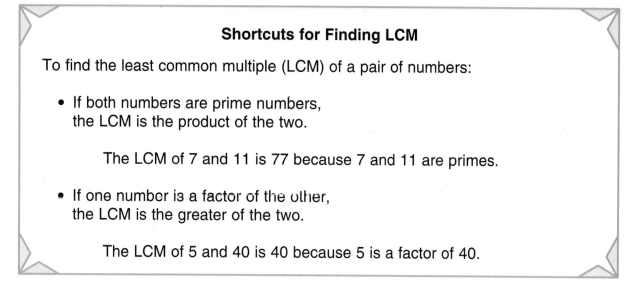

Shortcuts for Finding LCM

To find the least common multiple (LCM) of a pair of numbers:

- If both numbers are prime numbers,
  the LCM is the product of the two.

  The LCM of 7 and 11 is 77 because 7 and 11 are primes.

- If one number is a factor of the other,
  the LCM is the greater of the two.

  The LCM of 5 and 40 is 40 because 5 is a factor of 40.

Find the LCM of each pair of numbers.

19. 3, 7	**20.** 2, 3	**21.** 7, 21	**22.** 3, 9	**23.** 12, 4	**24.** 10, 5
25. 7, 2	**26.** 11, 5	**27.** 3, 15	**28.** 16, 32	**29.** 1, 9	**30.** 12, 1
31. 7, 56	**32.** 8, 40	**33.** 12, 48	**34.** 13, 39	**35.** 11, 17	**36.** 13, 15

 CALCULATOR ACTIVITY

Find the LCM of each pair of numbers.   35 and 42

- Determine the GCF using
  prime factorization.

  $35 = 5 \times 7$
  $42 = 2 \times 3 \times 7$
  GCF of 35 and 42 is 7.

- Calculate the product of
  the two numbers.

  ③ ⑤ ✕ ④ ② ▬   ☐ 1470.

- Divide the product
  by the GCF.

  ① ④ ⑦ ⓪ ÷ ⑦ ▬   ☐ 210.

- The quotient is the LCM.

  The LCM of 35 and 42 is 210.

37. 36 and 63   **38.** 40 and 45   **39.** 30 and 72   **40.** 27 and 36

Comparing Fractions

Compare: $\frac{21}{50}$ $\underline{\ ?\ }$ $\frac{42}{50}$ $\frac{13}{14}$ $\underline{\ ?\ }$ $\frac{11}{14}$

▶ When fractions have *like* denominators, compare their numerators.

$\frac{21}{50}$ $\underline{\ ?\ }$ $\frac{42}{50}$ ⟶ 21 < 42 $\frac{13}{14}$ $\underline{\ ?\ }$ $\frac{11}{14}$ ⟶ 13 > 11

$\frac{21}{50} < \frac{42}{50}$ $\frac{13}{14} > \frac{11}{14}$

▶ When fractions have *unlike* denominators, rename them
to get their *least common denominator*. Then compare.

> The **least common denominator** (**LCD**) of two or more fractions
> is the least common multiple (LCM) of their denominators.

Compare: $\frac{5}{6}$ $\underline{\ ?\ }$ $\frac{4}{5}$

To compare fractions using the LCD:

- Find the LCD.

  The LCD of $\frac{5}{6}$ and $\frac{4}{5}$ is 30.

  $\frac{5}{6}$ Multiples of 6: 6, 12, 18, 24, 30, . . .

  $\boxed{\text{LCM}}$

  $\frac{4}{5}$ Multiples of 5: 5, 10, 15, 20, 25, 30, . . .

- Rename the original fractions
  with the LCD as the
  new denominator.

  $\frac{5}{6} = \frac{?}{30}$ ⟶ $\frac{5 \times 5}{6 \times 5}$ ⟶ $\frac{25}{30}$

  $\frac{4}{5} = \frac{?}{30}$ ⟶ $\frac{4 \times 6}{5 \times 6}$ ⟶ $\frac{24}{30}$

- Compare the numerators.

  25 > 24, so $\frac{25}{30} > \frac{24}{30}$.

So $\frac{5}{6} > \frac{4}{5}$.

Compare. Write <, =, or >.

1. $\frac{7}{8}$? $\frac{5}{8}$

2. $\frac{9}{20}$? $\frac{9}{20}$

3. $\frac{14}{30}$? $\frac{26}{30}$

4. $\frac{17}{21}$? $\frac{10}{21}$

5. $\frac{12}{7}$? $\frac{16}{7}$

6. $\frac{9}{8}$? $\frac{8}{8}$

7. $\frac{22}{6}$? $\frac{32}{6}$

8. $\frac{19}{19}$? $\frac{20}{19}$

Find the least common denominator (LCD) of each pair of fractions.

9. $\frac{3}{5}$ and $\frac{1}{4}$

10. $\frac{3}{4}$ and $\frac{1}{10}$

11. $\frac{7}{8}$ and $\frac{5}{6}$

12. $\frac{1}{2}$ and $\frac{2}{3}$

13. $\frac{1}{12}$ and $\frac{3}{24}$

14. $\frac{1}{3}$ and $\frac{4}{9}$

15. $\frac{5}{7}$ and $\frac{12}{49}$

16. $\frac{2}{5}$ and $\frac{4}{7}$

17.–24. Rename each pair of fractions in exercises 9–16 so they have the LCD as their denominator.

Compare. Write <, =, or >.

25. $\frac{1}{4}$? $\frac{7}{16}$

26. $\frac{7}{10}$? $\frac{3}{5}$

27. $\frac{4}{21}$? $\frac{1}{7}$

28. $\frac{6}{14}$? $\frac{2}{7}$

29. $\frac{3}{5}$? $\frac{5}{8}$

30. $\frac{4}{7}$? $\frac{6}{9}$

31. $\frac{7}{12}$? $\frac{9}{15}$

32. $\frac{10}{25}$? $\frac{7}{10}$

33. $\frac{11}{16}$? $\frac{11}{16}$

34. $\frac{4}{5}$? $\frac{12}{15}$

35. $\frac{11}{20}$? $\frac{9}{15}$

36. $\frac{11}{21}$? $\frac{22}{42}$

MENTAL MATH

Compare. Write < or >. Look for fractions close to 0, $\frac{1}{2}$, or 1.

37. $\frac{6}{11}$? $\frac{5}{6}$ $\longrightarrow$ $\frac{6}{11}$ is close to $\frac{1}{2}$; $\frac{5}{6}$ is close to 1; $\frac{6}{11} < \frac{5}{6}$

38. $\frac{10}{9}$? $\frac{5}{8}$

39. $\frac{1}{11}$? $\frac{6}{13}$

40. $\frac{15}{31}$? $\frac{2}{30}$

41. $\frac{16}{17}$? $\frac{3}{7}$

42. $\frac{11}{20}$? $\frac{9}{20}$

43. $\frac{3}{32}$? $\frac{21}{32}$

44. $\frac{7}{15}$? $\frac{8}{7}$

45. $\frac{15}{16}$? $\frac{17}{16}$

Ordering Fractions

Write these fractions in order from least to greatest: $\frac{2}{3}, \frac{3}{4}, \frac{5}{8}$.

To order fractions with unlike denominators, rename them so that the denominators are the same.

- Find the LCD.

$$\frac{2}{3} \qquad \frac{3}{4} \qquad \frac{5}{8} \qquad \boxed{\text{LCD is 24.}}$$

- Rename each fraction using the LCD.

$$\frac{2}{3} = \frac{16}{24}, \quad \frac{3}{4} = \frac{18}{24}, \quad \frac{5}{8} = \frac{15}{24}$$

- Compare the numerators and write the fractions in order.

$$\frac{15}{24} < \frac{16}{24} < \frac{18}{24}$$

From least to greatest: $\frac{5}{8}, \frac{2}{3}, \frac{3}{4}$ From greatest to least: $\frac{3}{4}, \frac{2}{3}, \frac{5}{8}$

Study these examples.

Order from greatest to least.

$$2\frac{1}{8}, 2\frac{3}{4}, \frac{19}{8}$$
$$\downarrow \qquad \downarrow \qquad \downarrow$$
$$2\frac{1}{8}, 2\frac{6}{8}, 2\frac{3}{8}$$

$$\frac{19}{8} = 8\overline{)19}^{2\frac{3}{8}}$$

From greatest to least: $2\frac{3}{4}, \frac{19}{8}, 2\frac{1}{8}$

$$6\frac{5}{6}, 6\frac{3}{5}, 6\frac{2}{3}$$
$$\downarrow \qquad \downarrow \qquad \downarrow \qquad \boxed{\text{LCD is 30.}}$$
$$6\frac{25}{30}, 6\frac{18}{30}, 6\frac{20}{30}$$

From greatest to least: $6\frac{5}{6}, 6\frac{2}{3}, 6\frac{3}{5}$

Write in order from least to greatest.

1. $\frac{4}{5}, \frac{7}{10}, \frac{3}{4}$

2. $\frac{5}{12}, \frac{3}{8}, \frac{5}{6}$

3. $\frac{1}{4}, \frac{2}{5}, \frac{3}{8}$

4. $\frac{7}{9}, \frac{2}{3}, \frac{1}{6}$

5. $\frac{3}{5}, \frac{7}{10}, \frac{3}{8}$

6. $\frac{4}{5}, \frac{5}{6}, \frac{13}{15}$

7. $\frac{2}{11}, \frac{3}{4}, \frac{1}{2}$

8. $\frac{11}{24}, \frac{5}{8}, \frac{7}{12}$

9. $1\frac{2}{3}, 1\frac{5}{12}, 1\frac{5}{8}$

10. $3\frac{8}{10}, 3\frac{5}{6}, 3\frac{2}{3}$

11. $9\frac{3}{5}, 9\frac{5}{8}, 9\frac{7}{10}$

12. $7\frac{2}{9}, 7\frac{1}{3}, 7\frac{3}{4}$

Write in order from greatest to least.

13. $\frac{7}{12}, \frac{1}{2}, \frac{2}{3}$

14. $\frac{1}{4}, \frac{1}{3}, \frac{1}{5}$

15. $\frac{2}{5}, \frac{7}{10}, \frac{1}{3}$

16. $\frac{7}{12}, \frac{4}{5}, \frac{9}{10}$

17. $5\frac{4}{5}, 5\frac{3}{4}, 5\frac{7}{8}$

18. $2\frac{2}{3}, 2\frac{3}{4}, 2\frac{4}{5}$

19. $\frac{17}{18}, \frac{7}{9}, \frac{2}{3}$

20. $\frac{3}{7}, \frac{1}{2}, \frac{3}{14}$

21. $\frac{21}{9}, \frac{12}{9}, \frac{9}{12}$

22. $\frac{7}{6}, \frac{14}{5}, \frac{31}{10}$

23. $1\frac{2}{15}, \frac{18}{15}, 1\frac{4}{15}$

24. $\frac{21}{9}, 1\frac{5}{9}, \frac{8}{3}$

PROBLEM SOLVING

25. Tony saw three pumpkins labeled $5\frac{3}{8}$ lb, $5\frac{1}{4}$ lb, and $5\frac{5}{16}$ lb. Which pumpkin was the heaviest? Explain.

26. In a standing broad jump contest three students reached these distances: Patty, $6\frac{5}{12}$ ft; Hank, $6\frac{3}{4}$ ft; and Terry, $6\frac{5}{6}$ ft. Who won the contest? Explain.

27. If you put a jar $12\frac{3}{4}$ inches tall into a carton $12\frac{7}{12}$ inches high, will the jar stick out? Explain.

28. Which of these model cars can fit into the box? Explain.

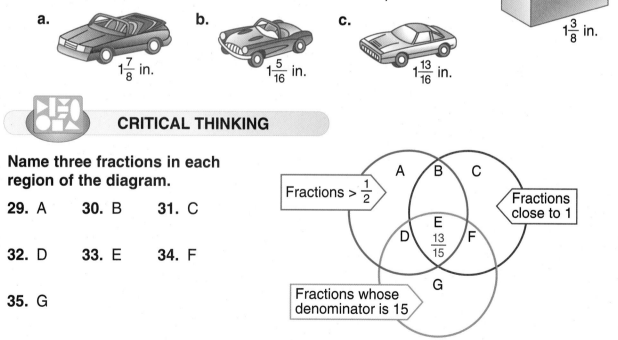

a. $1\frac{7}{8}$ in.

b. $1\frac{5}{16}$ in.

c. $1\frac{13}{16}$ in.

$1\frac{3}{8}$ in.

CRITICAL THINKING

Name three fractions in each region of the diagram.

29. A

30. B

31. C

32. D

33. E

34. F

35. G

Fractions > $\frac{1}{2}$

Fractions close to 1

Fractions whose denominator is 15

E $\frac{13}{15}$

5-12 Fractions, Mixed Numbers, Decimals

Fractions and mixed numbers with denominators that are *powers of ten* can be renamed as decimals. The word names are the same.

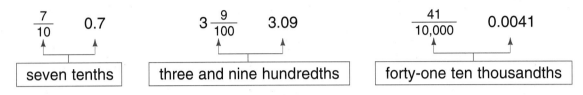

$\frac{7}{10}$ 0.7

seven tenths

$3\frac{9}{100}$ 3.09

three and nine hundredths

$\frac{41}{10,000}$ 0.0041

forty-one ten thousandths

To rename fractions and mixed numbers with denominators that are powers of ten as decimals:

- Read the given fraction. $\frac{8}{100}$ → eight hundredths

- Determine the decimal place.   hundredths: *two* decimal places

- Write an equivalent decimal.   eight hundredths = 0.08

Study these examples.

$6\frac{3}{10}$ → six and three tenths   (tenths: *one* decimal place) → 6.3

$\frac{41}{100}$ → forty-one hundredths   (hundredths: *two* decimal places) → 0.41

$\frac{16}{1000}$ → sixteen thousandths   (thousandths: *three* decimal places) → 0.016

$\frac{201}{10,000}$ → two hundred one ten thousandths   (ten thousandths: *four* decimal places) → 0.0201

Choose the equivalent decimal.

1. $\frac{7}{100}$ **a.** 0.700   **b.** 0.07   **c.** 0.007   **d.** 700.7

2. $13\frac{28}{100}$ **a.** 13.28   **b.** 13.028   **c.** 13.0028   **d.** 0.1328

3. $\frac{109}{1000}$ **a.** 100.9   **b.** 0.0109   **c.** 0.109   **d.** 109.001

184

Choose the equivalent fraction.

4. 0.02
 a. $\frac{2}{100}$
 b. $\frac{2}{10}$
 c. $\frac{2}{1000}$
 d. $\frac{1}{2}$

5. 0.1005
 a. $1\frac{5}{10}$
 b. $\frac{1005}{10,000}$
 c. $\frac{1005}{1000}$
 d. $\frac{1005}{100}$

6. 4.12
 a. $4\frac{12}{10}$
 b. $4\frac{12}{1000}$
 c. $4\frac{12}{100}$
 d. $\frac{412}{1000}$

Write the word name. Then write the equivalent decimal or fraction.

7. $\frac{9}{10}$
 8. $\frac{35}{100}$
 9. $\frac{81}{1000}$
 10. $\frac{71}{10,000}$
 11. $6\frac{2}{10}$

12. $16\frac{19}{100}$
 13. $4\frac{12}{1000}$
 14. $9\frac{417}{1000}$
 15. $\frac{2}{1000}$
 16. $\frac{56}{10,000}$

17. 0.87
 18. 0.022
 19. 0.0563
 20. 0.1578
 21. 7.52

22. 8.009
 23. 16.573
 24. 37.069
 25. 5.0005
 26. 11.0011

Renaming Improper Fractions as Decimals

To rename improper fractions with denominators that are powers of ten as decimals:

- Rename the improper fraction as a mixed number.

$$\frac{628}{100} \longrightarrow 100)\overline{628} \quad 6\frac{28}{100}$$

- Rename the mixed number as a decimal.

$$6\frac{28}{100} = 6.28$$

Write the equivalent decimal or whole number.

27. $\frac{25}{10}$
 28. $\frac{420}{100}$
 29. $\frac{372}{10}$
 30. $\frac{4620}{100}$
 31. $\frac{5390}{1000}$

32. $\frac{1472}{1000}$
 33. $\frac{7000}{1000}$
 34. $\frac{20,000}{10,000}$
 35. $\frac{34,000}{10,000}$
 36. $\frac{79,500}{10,000}$

Fractions: Renaming as Decimals

Write $\frac{3}{4}$ as a decimal.

▶ **To rename a fraction as a decimal:**

- Divide the numerator by the denominator.

$$4\overline{)3}$$

- Place a decimal point after the numerator and in the quotient.

$$4\overline{)3.}$$

- Divide. Add zeros as needed.

$$4\overline{)\underset{0.7\ 5}{3.0\ 0}}$$

So $\frac{3}{4}$ = 0.75.

Write $9\frac{1}{16}$ as a decimal.

▶ **To rename a mixed number as a decimal:**

- Separate the mixed number into a whole number part and a fraction part.

$$9\frac{1}{16} = 9 + \frac{1}{16}$$

- Rename the fraction part as a decimal.

$$\frac{1}{16} \longrightarrow 16\overline{)\underset{0.0\ 6\ 2\ 5}{1.0\ 0\ 0\ 0}}$$

- Add the whole number part and the decimal.

$$9 + 0.0\ 6\ 2\ 5 = 9.0\ 6\ 2\ 5$$

So $9\frac{1}{16}$ = 9.0625.

Write each fraction as a decimal.

1. $\frac{2}{5}$

2. $\frac{1}{2}$

3. $\frac{1}{4}$

4. $\frac{4}{5}$

5. $\frac{3}{8}$

6. $\frac{6}{15}$

7. $\frac{9}{20}$

8. $\frac{3}{50}$

9. $\frac{1}{20}$

10. $\frac{1}{25}$

11. $\frac{22}{50}$

12. $\frac{19}{20}$

Write each mixed number as a decimal.

13. $7\frac{4}{5}$ **14.** $15\frac{1}{2}$ **15.** $28\frac{3}{5}$ **16.** $12\frac{5}{8}$ **17.** $7\frac{3}{100}$

18. $9\frac{4}{1000}$ **19.** $11\frac{3}{50}$ **20.** $80\frac{3}{4}$ **21.** $44\frac{11}{20}$ **22.** $61\frac{1}{8}$

Shortcut for Renaming Fractions

Write $\frac{12}{25}$ as a decimal.

When a fraction has a denominator that is a factor of 10, 100, 1000, . . . :

Think: 25 is a factor of 100.

- Rename as an equivalent fraction with a denominator that is a power of ten.

$$\frac{12 \times 4}{25 \times 4} = \frac{48}{100}$$

- Read the fraction. Then write the decimal.

$$\frac{48}{100}$$

forty-eight hundredths

0.48

So $\frac{12}{25} = 0.48$.

Write each fraction or mixed number as a decimal.
Use the shortcut whenever possible.

23. $\frac{3}{20}$ **24.** $\frac{7}{25}$ **25.** $7\frac{9}{10}$ **26.** $8\frac{3}{10}$ **27.** $\frac{32}{50}$

28. $\frac{11}{25}$ **29.** $\frac{5}{16}$ **30.** $\frac{25}{32}$ **31.** $9\frac{12}{20}$ **32.** $10\frac{3}{25}$

PROBLEM SOLVING Write each answer as a decimal.

33. Ann has nine tenths of a dollar. How much money does she have?

34. Allan has three fifths of a dollar. How much money does he have?

35. Roy has one and three fourths dollars. How much more money does he need to have $3.25?

36. Dong has three and two fifths dollars and Rita has four and one fourth dollars. How much more money does Rita have?

Write 0.35 as a fraction in simplest form.

To rename a decimal as a fraction:

- Read the given decimal.   0.35 ⟶ thirty-five hundredths

- Determine the denominator   The denominator is 100.
  of the fraction.

- Write an equivalent fraction.   thirty-five hundredths $= \frac{35}{100}$

- Simplify if necessary. $\frac{35}{100} = \frac{35 \div 5}{100 \div 5} = \frac{7}{20}$

So $0.35 = \frac{7}{20}$.

Study this example.

Write 9.008 as a mixed number in simplest form.

$$9.008 \longrightarrow \text{nine and eight thousandths} \longrightarrow 9\frac{8}{1000}$$

$$\frac{8}{1000} = \frac{8 \div 8}{1000 \div 8} = \frac{1}{125}$$

So $9.008 = 9 + \frac{1}{125} = 9\frac{1}{125}$.

Complete.

1. $0.63 = \frac{?}{100}$ **2.** $0.05 = \frac{5}{?} = \frac{?}{20}$ **3.** $0.259 = \frac{?}{1000}$

4. $0.750 = \frac{750}{?} = \frac{?}{4}$ **5.** $8.7 = 8\frac{?}{10}$ **6.** $4.09 = 4\frac{9}{?}$

7. $2.627 = 2\frac{?}{1000}$ **8.** $5.500 = 5\frac{?}{1000} = 5\frac{?}{2}$ **9.** $38.03 = 38\frac{?}{?}$

Write each decimal as a fraction in simplest form.

10. 0.9   **11.** 0.07   **12.** 0.43   **13.** 0.77   **14.** 0.003

15. 0.127   **16.** 0.45   **17.** 0.36   **18.** 0.675   **19.** 0.325

20. 0.0033   **21.** 0.0009   **22.** 0.441   **23.** 0.101   **24.** 0.0500

Write each decimal as a mixed number in simplest form.

25. 1.09   **26.** 5.7   **27.** 11.31   **28.** 12.1   **29.** 2.5

30. 8.4   **31.** 9.16   **32.** 6.35   **33.** 1.055   **34.** 3.004

35. 6.0005   **36.** 8.0010   **37.** 3.375   **38.** 2.95   **39.** 20.0750

Copy and complete.

	Decimal	**Words**	**Fraction in Simplest Form**
40.	4.7	four and seven tenths	?
41.	3.88	?	?
42.	?	two hundred and six thousandths	?
43.	?	one thousand eleven ten thousandths	?
44.	101.003	?	?

CONNECTIONS: SCIENCE

Rename each decimal as a fraction or mixed number in simplest form.

45. A tortoise travels 0.7 mile per hour.

46. An elephant can run at a speed of 24.5 miles per hour.

47. The height of a zebra may be 1.55 meters.

48. The height of a flower may be 0.44 meter.

49. The length of a boa constrictor may be 7.8 meters.

Terminating and Repeating Decimals

Alberto plays baseball for the Piney Creek Wildcats.
In his first 30 times at bat, he gets 10 hits.
Write his batting average as a decimal.

Rename $\frac{10}{30}$ as an equivalent decimal.

$$3\,0\overline{)1\,0.0\,0\,0\,0}^{0.3\,3\,3\,3\,\ldots}$$

No matter how many zeros you write in the dividend,
the division just keeps on going. When the same
digits repeat in the quotient, you have a
repeating decimal.

Alberto's batting average is 0.3333 . . . , or 0.333
when rounded to the nearest thousandth.

▶ Every fraction can be renamed as a **terminating decimal**
or a **repeating decimal**.

A terminating decimal has a
definite number of decimal places.
When you divide, the remainder
is 0.

$$\frac{5}{8} \rightarrow 8\overline{)5.0\,0\,0}^{0.6\,2\,5}$$

$$\frac{5}{8} = 0.6\,2\,5$$

A repeating decimal has one
or more digits that repeat
indefinitely. When you divide,
the remainder is never zero.

$$\frac{3}{11} \rightarrow 1\,1\overline{)3.0\,0\,0\,0\,0}^{0.2\,7\,2\,7\,2\,\ldots}$$

$$\frac{3}{11} = 0.2\,7\,2\,7\,\ldots$$

▶ Repeating decimals may be written with a *bar* over
the digit or digits that repeat.

0.2727 . . . = 0.$\overline{27}$ ◀── ⎫ bar

5.13636 . . . = 5.13$\overline{6}$ ◀── ⎫ bar

Think: The digits 2 and 7
repeat indefinitely.

Think: The digits 3 and 6
repeat indefinitely. The
digit 1 does not repeat.

Rewrite each repeating decimal with a bar over the part that repeats.

1. 0.66666 . . .
2. 0.11111 . . .
3. 0.45454 . . .
4. 0.09090 . . .

5. 0.83333 . . .
6. 0.26666 . . .
7. 2.384848 . . .
8. 5.13232 . . .

Write each repeating decimal showing eight decimal places.

9. $0.\overline{1}$
10. $0.\overline{12}$
11. $0.1\overline{4}$
12. $0.2\overline{8}$

13. $5.\overline{3}$
14. $12.0\overline{6}$
15. $7.2\overline{7}$
16. $13.2\overline{17}$

Rename each fraction as a terminating or repeating decimal.

17. $\frac{1}{8}$
18. $\frac{13}{20}$
19. $\frac{5}{11}$
20. $\frac{1}{3}$
21. $\frac{3}{4}$

22. $\frac{2}{9}$
23. $\frac{7}{16}$
24. $\frac{5}{12}$
25. $\frac{11}{18}$
26. $\frac{1}{16}$

Rename each mixed number as a terminating or repeating decimal.

27. $4\frac{2}{5}$
28. $6\frac{1}{4}$
29. $12\frac{2}{3}$
30. $15\frac{2}{3}$
31. $1\frac{3}{8}$

32. $121\frac{1}{9}$
33. $33\frac{1}{3}$
34. $5\frac{5}{16}$
35. $28\frac{21}{36}$
36. $11\frac{13}{25}$

CALCULATOR ACTIVITY

Math Journal

Copy and complete each table. Explain the patterns in your Math Journal.

	Fraction	Decimal		Fraction	Decimal
37.	$\frac{1}{9}$ $\boxed{1} \div \boxed{9} \boxed{=}$	$0.111111\overline{1}$	41.	$\frac{1}{11}$	?
38.	$\frac{2}{9}$	?	42.	$\frac{2}{11}$	?
39.	$\frac{3}{9}$	?	43.	$\frac{3}{11}$	?
40.	$\frac{4}{9}$	?	44.	$\frac{4}{11}$	?

5-16 Problem Solving: Find a Pattern

Problem: Tim gave all his baseball cards to Walter. On the first day he gave him 1 card. On each day after that he gave him 3 times the number he had given him the day before. At the end of 4 days, Tim had given away all the cards he had. How many did he give to Walter in all?

1 IMAGINE Picture yourself giving away cards.

2 NAME *Facts:* first day—1 card

each day after—3 times the number he gave the day before

after 4 days—all cards are given away

Question: How many cards did Tim give to Walter?

3 THINK To find how many cards Tim gave Walter, make a table to list the number of cards given away and the total. Multiply the number he gave away the day before by 3. To find the total, add the amount given away to the total of the day before. Look for a pattern.

4 COMPUTE

Day	1st	2nd	3rd	4th
Walter received	1	3 × 1 3	3 × 3 9	3 × 9 27
Tim's total	1	4	13	40

Tim gave Walter 40 baseball cards.

5 CHECK Begin with 40 cards and subtract the number Tim gave away each day.
Do you have 0 left on the 4th day? Yes.

$$40 - 1 - 3 - 9 - 27 \overset{?}{=} 0$$
$$39 - 3 - 9 - 27 \overset{?}{=} 0$$
$$36 - 9 - 27 \overset{?}{=} 0$$
$$27 - 27 = 0$$

The answer checks.

Solve each problem. Find a pattern to help you.

1. In a science experiment, Joel discovered that his record of the changes in a liquid's temperature formed a pattern. In each of the first 3 minutes the temperature increased 1.5° F; in each of the next 2 minutes it decreased 0.75° F. Then this pattern repeated itself. If Joel started measuring the temperature at 50° F, how long would it take the temperature to reach 62° F?

IMAGINE   Picture yourself working the experiment.

NAME   *Facts:*   each of first 3 min–increase 1.5° F
                         each of next 2 min–decrease 0.75° F
                         starting temperature–50° F

      Question: How long will it take the temperature to reach 62° F?

THINK   To find how many minutes it will take to reach 62° F, make a table to list the time and degrees increased or decreased. Look for a pattern.

      COMPUTE ⟶ **CHECK**

2. Alice makes a necklace with 24 red and white beads. If she creates a pattern of 1 red and 3 white beads, how many red beads will she use? how many white beads?

3. Find the next three terms in this sequence:
$\frac{1}{8}, \frac{1}{2}, \frac{3}{8}, \frac{3}{4}, \frac{5}{8}$, 1. What is the pattern?

4. Hector caught 2 fish on Monday, 4 on Tuesday, 8 on Wednesday, 16 on Thursday, and so on. Following this pattern, how many fish did he catch on Saturday?

5. Crystal builds a tower out of blocks for her little brother. She uses 7 blocks. The edge of each block is $1\frac{1}{4}$ in. shorter than the edge of the block under it. If the bottom block is $9\frac{3}{4}$ in. on each edge, how long is the edge of the top block?

5-17 Problem-Solving Applications

Solve each problem and explain the method you used.

1. Stella paints this pattern. What fraction, in simplest form, names the shaded region?

2. Dom has these tubes of paint: $\frac{1}{4}$ oz crimson, $\frac{2}{3}$ oz burnt sienna, $\frac{2}{5}$ oz black, and $\frac{6}{9}$ oz magenta. Which two tubes have the same amount of paint?

3. Milly has $\frac{19}{4}$ oz of white paint. Is this more than 5 ounces?

4. Julio's favorite brushes are the following lengths: $\frac{15}{2}$ in., $7\frac{1}{3}$ in., $7\frac{5}{9}$ in., and $\frac{31}{4}$ in. How would he arrange the brushes in order from shortest to longest?

5. One sheet of watercolor paper is 0.01 in. thick. Write 0.01 as a fraction.

6. Jeremiah has finished $\frac{5}{8}$ of his painting. Write this fraction as a decimal.

7. Stella's newest painting has an area of 156.25 in.^{2} Write this decimal as a mixed number.

8. Becky has $3\frac{3}{4}$ quarts of paint thinner. She also has 7.5 pints of turpentine. Does she have more paint thinner or turpentine?

Use the chart for problems 9–11.

9. Which two pencils are the same length?

10. What are the longest and shortest pencils?

11. Which pencils are close to 5 in. in length?

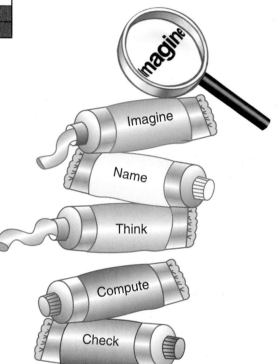

Color	Length
red	$6\frac{3}{4}$ in.
orange	7.8 in.
yellow	$5\frac{2}{5}$ in.
green	$\frac{11}{5}$ in.
blue	$7\frac{9}{10}$ in.
indigo	5.4 in.
violet	$\frac{40}{9}$ in.

Use a strategy from the list or another strategy you know to solve each problem.

12. A painting is $\frac{1}{4}$ in. longer than it is wide. Its length is 8.2 in. The frame is 2.7 in. thick. What is the width of the painting?

13. James folds a sheet of drawing paper in fourths, then in thirds, and then in half. Estimate into how many parts his paper is divided. Check your answer by following the folds.

14. Gary uses $\frac{3}{8}$ of a tube of raw sienna to paint a fall landscape. He also uses $\frac{2}{5}$ oz of cadmium red for the same picture. How much more raw sienna than cadmium red does he use?

15. Joanne paints these three pictures. If she continues the pattern in a fourth picture, what fractional part of that picture will be shaded?

16. Every third day Fran goes to calligraphy class. Every fourth day she goes to pottery class. On March 1 Fran attends both classes. How many days that month will the 2 classes fall on the same day?

17. Gesso boards are advertised at 3 boards for $14.20. How much would Danielle pay for one gesso board?

Use the circle graph for problems 18–20.

18. What part of Abby's artworks are oil paintings?

19. What type of art makes up $\frac{1}{5}$ of Abby's work?

20. Abby did 5 charcoals this week. When these charcoals are added to the data from the graph, what part of her work will be charcoals?

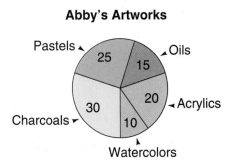

Abby's Artworks

Pastels 25
Oils 15
Charcoals 30
Watercolors 10
Acrylics 20

MAKE UP YOUR OWN

21. Write a problem modeled on problem 17. Have a classmate solve it.

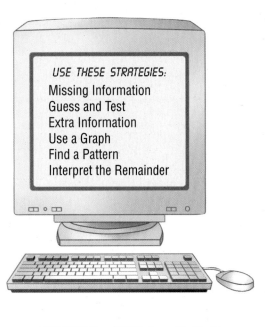

Chapter Review and Practice

Write the missing number to complete the equivalent fraction. *(See pp. 164–165.)*

1. $\frac{2}{3} = \frac{?}{9}$

2. $\frac{3}{4} = \frac{9}{?}$

3. $\frac{20}{90} = \frac{?}{9}$

4. $\frac{15}{45} = \frac{?}{3}$

Find the prime factorization and write in exponent form. *(See pp. 168–169.)*

5. 28

6. 30

7. 75

8. 84

Find the GCF of each pair of numbers. *(See pp. 170–171.)*

9. 3 and 27

10. 12 and 48

11. 21 and 35

Find the LCM of each pair of numbers. *(See pp. 178–179.)*

12. 3 and 5

13. 6 and 18

14. 4 and 15

Rename each as indicated. *(See pp. 172–175, 184–191.)*

15. $\frac{9}{45}$ in simplest form

16. $4\frac{2}{3}$ as an improper fraction

17. $7\frac{3}{8}$ as a decimal

18. 0.45 as a fraction in simplest form

19. $\frac{5}{16}$ as a decimal

20. 9.6 as a mixed number in simplest form

Compare. Write <, =, or >. *(See pp. 180–181.)*

21. $\frac{19}{21}$? $\frac{13}{21}$

22. $\frac{7}{12}$? $\frac{3}{4}$

23. $\frac{6}{16}$? $\frac{3}{8}$

Write in order from least to greatest. *(See pp. 182–183.)*

24. $\frac{2}{3}, \frac{5}{6}, \frac{3}{5}$

25. $\frac{2}{9}, \frac{1}{3}, \frac{3}{4}$

26. $8\frac{5}{6}, 8\frac{7}{12}, 8\frac{3}{4}$

PROBLEM SOLVING *(See pp. 162–163, 176–177.)*

27. Of 24 paintings, 6 are landscapes, 9 are oils, and the rest are watercolors. What fractional part are watercolors?

28. Jamie must choose a fraction strip that is close to $\frac{1}{2}$ from $\frac{5}{6}, \frac{4}{7},$ or $\frac{4}{11}$ strips. Which should he choose? Explain.

(See *Still More Practice,* p. 510.)

THE SIEVE OF ERATOSTHENES

A Greek mathematician named Eratosthenes created a method for finding prime numbers. The method is called the **Sieve of Eratosthenes**. You can use it to find all of the prime numbers between 1 and 100.

1	2	3	4	5	6	7	8	9	10
11	12	13	14	15	16	17	18	19	20
21	22	23	24	25	26	27	28	29	30
31	32	33	34	35	36	37	38	39	40
41	42	43	44	45	46	47	48	49	50
51	52	53	54	55	56	57	58	59	60
61	62	63	64	65	66	67	68	69	70
71	72	73	74	75	76	77	78	79	80
81	82	83	84	85	86	87	88	89	90
91	92	93	94	95	96	97	98	99	100

Copy the chart above. Then use it to complete the following.

1. Cross out 1, because it is neither prime nor composite.

2. Circle 2, the first prime number. Cross out every multiple of 2.

3. Circle 3, the second prime number. Cross out every multiple of 3, including those already crossed out.

4. Circle 5, the third prime number. Cross out every multiple of 5, including those already crossed out.

5. Circle 7, the fourth prime number. Cross out every multiple of 7, including those already crossed out. Circle the remaining numbers. The circled numbers are prime numbers.

6. Tell whether each number is prime or composite:

   a. 13   **b.** 37   **c.** 49   **d.** 57   **e.** 59   **f.** 84

7. Find prime numbers that complete each statement.

   a. 69 = _?_ × _?_   **b.** _?_ + _?_ = 66

   c. 91 = _?_ × _?_   **d.** _?_ − _?_ = 76

197

Check Your Mastery

Performance Assessment

Find a pair of numbers, if any, for each description.

1. Between 9 and 25 that have 8 as their GCF

2. Between 0 and 9 that have 1 as their GCF

3. Between 0 and 9 that have 8 as their GCF and their LCM

4. LCM is between 100 and 200 and their GCF is 64

Write the missing number to complete the equivalent fraction.

5. $\frac{5}{6} = \frac{?}{12}$

6. $\frac{10}{15} = \frac{2}{?}$

7. $\frac{14}{100} = \frac{7}{?}$

Tell whether each number is *prime* or *composite*.

8. 5

9. 21

10. 17

Find the prime factorization.

11. 34

12. 42

13. 90

Rename each as indicated.

14. $\frac{16}{48}$ in simplest form

15. $5\frac{1}{2}$ as an improper fraction

16. $7\frac{1}{8}$ as a decimal

17. 0.64 as a fraction in simplest form

18. $\frac{3}{7}$ as a decimal

19. $12\frac{1}{3}$ as a decimal

PROBLEM SOLVING   *Use a strategy or strategies you have learned.*

20. Lou Ann must choose a melon from ones that weigh $4\frac{2}{5}$ lb, $4\frac{3}{10}$ lb, or $4\frac{1}{2}$ lb. If she wants to choose the heaviest, which melon should she choose? Explain.

21. Dawind makes a quilt pattern with 45 blue and yellow patches. If he uses 2 blue patches for every 3 yellow patches, how many blue patches will he use? how many yellow patches?

Cumulative Review II

Choose the best answer.

1. Find the value of n.

   $n = 13,024 \div 32$

   a. 40 R7
   b. 407
   c. 407 R25
   d. not given

2. $\begin{array}{r} 4550 \\ \times\ 240 \\ \hline \end{array}$

   a. 10,920
   b. 109,200
   c. 10,920,000
   d. not given

3. 1000×0.6

   a. 0.600
   b. 60
   c. 600
   d. not given

4. Estimate.

   8.7×19.52

   a. 30
   b. 80
   c. 90
   d. 180

5. $3 \times 0.4 \times 0.8$

   a. 0.096
   b. 0.96
   c. 9.6
   d. not given

6. $\begin{array}{r} 21.7 \\ \times\ 0.04 \\ \hline \end{array}$

   a. 0.868
   b. 8.68
   c. 86.8
   d. not given

7. $0.413 \div 0.01$

   a. 0.41300
   b. 4.13
   c. 41.3
   d. not given

8. Estimate.

   $218.7 \div 3.9$

   a. 5
   b. 50
   c. 90
   d. 700

9. $0.7\overline{)3.934}$

   a. 0.0562
   b. 0.562
   c. 5.62
   d. not given

10. $0.07875 \div 0.75$

    a. 0.105
    b. 1.5
    c. 10.5
    d. not given

11. Which is the GCF of 18 and 30?

    a. 3
    b. 6
    c. 9
    d. 90

12. Which is equivalent to $2\frac{2}{5}$?

    a. 2.2
    b. 2.4
    c. 2.5
    d. 2.6

13. Which is ordered greatest to least?

    a. $\frac{7}{10}, \frac{7}{8}, \frac{2}{3}$
    b. $\frac{9}{10}, \frac{2}{5}, \frac{2}{3}$
    c. $\frac{5}{6}, \frac{3}{4}, \frac{1}{2}$
    d. none of these

14. Which is equivalent to $3\frac{1}{4}$?

    a. $\frac{4}{3}$
    b. $\frac{13}{3}$
    c. $\frac{7}{4}$
    d. $\frac{13}{4}$

15. Which will give a terminating decimal as a quotient?

    a. $1 \div 2$
    b. $1 \div 3$
    c. $1 \div 7$
    d. $1 \div 9$

16. Rename $1\frac{2}{3}$ as a repeating decimal.

    a. $1.\overline{2}$
    b. $1.\overline{3}$
    c. $1.\overline{6}$
    d. $1.\overline{7}$

Ongoing Assessment II

For Your Portfolio

Solve each problem. Explain the steps and the strategy
or strategies you used for each. Then choose one from
problems 1–4 for your Portfolio.

1. Chico has four and three tenths
dollars and Alma has nine and
three fourths dollars. Alma has
how much more money than Chico?

2. Naomi added 0.25 to the
difference she obtained from
subtracting 1.19 from 3.23.
What number did Naomi
end up with?

3. Which number is 10^5 more than
$(4 \times 10^4) + (3 \times 10^3) +$
$(2 \times 10^2) + (1 \times 10)$?

   a. 43,260   b. 53,210
   c. 143,210   d. 1,432,100

4. Which number is (3.67×10^3)
times (2.35×10^7)?

   a. 1.32×10^4 b. 6.02×10^{10}
   c. 8.6245×10^4 d. 8.6245×10^{10}

Tell about it.

5. Write a note to your teacher to explain how you
found the answer to problem 4. Be as specific
as you can. Were you able to eliminate some
answers right away? Did you estimate?

Communicate ✓

For Rubric Scoring

Listen for information on how your work will be scored.

6. Fill in the boxes using the digits 2, 3, 4, 5 so that the
product satisfies the four clues below. Each digit
2, 3, 4, 5 should be used only once. You may use a
calculator to help you.

Clue #1: Product is between 10 and 16.

Clue #2: Three is such a lonely number!

Clue #3: Answer must contain the digits 1, 2, 5, and 7.

Clue #4: The hundredths digit in the answer is even.

Fractions: Addition and Subtraction

$\frac{4}{5}$

$\frac{3}{4}$

6

$\frac{3}{8}$

$\frac{9}{12}$

$\frac{1}{2}$

Where is math in dinnertime?

One whole pizza pie:
Two toppings to choose.
Three slices with peppers: 3/8.
Four slices with pepperoni: 1/2.
(How many slices are plain?)

From *Math in the Bath* by
Sara Atherlay

In this chapter you will:

Explore addition properties
Estimate, add, and subtract fractions
Use the fraction calculator to rename
  and compute
Solve problems by working backwards

Critical Thinking/Finding Together

Work with a classmate to write a pizza
problem similar to the problem above.
Use your favorite pizza toppings. Challenge
student pairs to solve the problem.

Addition Properties: Fractions

Algebra

The following properties of addition are true for whole numbers.

Commutative	Associative	Identity
$a + b = b + a$	$(a + b) + c = a + (b + c)$	$a + 0 = a$

In this lesson, you will decide if they are also true when
a, b, and c are fractions.

True or false: $\frac{3}{8} + \frac{1}{8} = \frac{1}{8} + \frac{3}{8}$?

 ## Hands-On Understanding

Materials Needed: fraction strips, paper, pencil

Step 1
Use fraction strips to show $\frac{3}{8} + \frac{1}{8}$ as shown at the right.

How many $\frac{1}{8}$s are there in all?
What is the sum of $\frac{3}{8}$ and $\frac{1}{8}$?

Step 2
Use fraction strips to show $\frac{1}{8} + \frac{3}{8}$.
Use blue for $\frac{1}{8}$ and red for $\frac{3}{8}$.
How many $\frac{1}{8}$s are there in all?
What is the sum of $\frac{1}{8}$ and $\frac{3}{8}$?

Step 3
Compare the strips you made for $\frac{3}{8} + \frac{1}{8}$ and for $\frac{1}{8} + \frac{3}{8}$.

Is the total number of $\frac{1}{8}$s the same for both strips?
Is $\frac{3}{8} + \frac{1}{8} = \frac{1}{8} + \frac{3}{8}$ true or false? Explain.

Communicate

Step 4
Draw other fraction strips with different denominators that are the same length as the strip you made for $\frac{3}{8} + \frac{1}{8}$.
Which one of your strips shows the answer in simplest form? Explain.

Step 5 Make two fraction strips to show $\frac{4}{8} + \frac{6}{8}$ and $\frac{6}{8} + \frac{4}{8}$.

Complete: $\frac{4}{8} + \frac{6}{8} = \frac{?}{8}$ $\frac{6}{8} + \frac{4}{8} = \frac{?}{8}$

Step 6 What is the sum of $\frac{4}{8}$ and $\frac{6}{8}$ or $\frac{6}{8}$ and $\frac{4}{8}$? Draw two fraction strips that show this sum in simplest form.

Communicate

1. Choose the one that shows the sum of $\frac{2}{8} + \frac{4}{8}$ or $\frac{4}{8} + \frac{2}{8}$.

   a. $\frac{6}{16}$ **b.** $\frac{6}{8}$ **c.** $\frac{8}{16}$

2. Choose the one that shows the sum of $\frac{2}{8} + \frac{4}{8}$ or $\frac{4}{8} + \frac{2}{8}$ in simplest form.

   a. $\frac{2}{3}$ **b.** $\frac{3}{8}$ **c.** $\frac{3}{4}$

3. Discuss with the class why the commutative property of addition is true for fractions.

Discuss

SHARE YOUR THINKING

4. Use fraction strips to show that the associative property of addition is true for fractions.

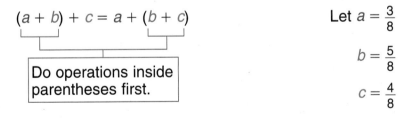

$(a + b) + c = a + (b + c)$

Do operations inside parentheses first.

Let $a = \frac{3}{8}$

$b = \frac{5}{8}$

$c = \frac{4}{8}$

5. Use fraction strips to show that the identity property of addition is true for fractions.

   $a + 0 = a$ Let $a = \frac{4}{8}$.

6. Show each sum in exercises 4 and 5 to the class. Be sure to display strips that show the sum in simplest form.

Estimating Sums and Differences

A package of cheese weighs $\frac{9}{16}$ pound. Brittany needs 1 pound of cheese. If she buys two packages, will she have enough cheese?

To find the approximate amount of cheese in two packages, estimate: $\frac{9}{16}$ lb $+$ $\frac{9}{16}$ lb.

▶ **To estimate the sum (or difference) of fractions:**

- Round each fraction to 0, $\frac{1}{2}$, or 1.

- Add (or subtract) the rounded numbers.

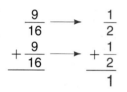

$$\frac{9}{16} \longrightarrow \frac{1}{2}$$
$$+\frac{9}{16} \longrightarrow +\frac{1}{2}$$
$$\overline{1}$$

> $\frac{9}{16}$ is a little more than $\frac{1}{2}$.
> So $\frac{9}{16} + \frac{9}{16}$ is a little more than 1.

Brittany will have enough cheese.

Study these examples.

Tell whether the sum or difference is close to 0, $\frac{1}{2}$, or 1.

$$\frac{1}{5} + \frac{3}{7}$$
$$\downarrow \quad \downarrow$$
$$0 + \frac{1}{2} = \frac{1}{2}$$

So $\frac{1}{5} + \frac{3}{7}$ is close to $\frac{1}{2}$.

$$\frac{7}{12} - \frac{1}{9}$$
$$\downarrow \quad \downarrow$$
$$\frac{1}{2} - 0 = \frac{1}{2}$$

So $\frac{7}{12} - \frac{1}{9}$ is close to $\frac{1}{2}$.

$$\frac{11}{12} - \frac{8}{10}$$
$$\downarrow \quad \downarrow$$
$$1 - 1 = 0$$

So $\frac{11}{12} - \frac{8}{10}$ is close to 0.

Match each exercise to its estimated sum or difference in the box.

1. $\dfrac{1}{5} + \dfrac{8}{9}$ 2. $\dfrac{6}{7} + \dfrac{11}{12}$

3. $\dfrac{9}{10} - \dfrac{5}{8}$ 4. $\dfrac{6}{13} - \dfrac{8}{18}$

a. $1 + 1 = 2$	b. $1 - \dfrac{1}{2} = \dfrac{1}{2}$
c. $\dfrac{1}{2} - \dfrac{1}{2} = 0$	d. $0 + 1 = 1$

Estimate the sum or difference.

5. $\dfrac{1}{11} + \dfrac{4}{9}$ 6. $\dfrac{15}{16} - \dfrac{1}{10}$ 7. $\dfrac{2}{9} + \dfrac{5}{6}$ 8. $\dfrac{11}{12} + \dfrac{12}{14}$

9. $\dfrac{7}{15} - \dfrac{1}{10}$ 10. $\dfrac{18}{20} - \dfrac{13}{24}$ 11. $\dfrac{3}{11} - \dfrac{1}{6}$ 12. $\dfrac{1}{9} + \dfrac{3}{10}$

13. $\dfrac{9}{10} + \dfrac{1}{6} + \dfrac{3}{8}$ 14. $\dfrac{1}{9} + \dfrac{1}{7} + \dfrac{1}{2}$ 15. $\dfrac{15}{16} + \dfrac{5}{8} + \dfrac{4}{9} + \dfrac{3}{25}$

Estimating with Mixed Numbers

To estimate the sum (or difference) of mixed numbers:

- Round each mixed number to the nearest whole number.
- Add (or subtract) the rounded numbers.

$$8\dfrac{1}{5} \longrightarrow 8$$
$$+9\dfrac{5}{8} \longrightarrow +10$$
$$\overline{\phantom{+9\dfrac{5}{8}}} \quad \overline{18} \leftarrow \text{estimated sum}$$

$$15\dfrac{1}{2} \longrightarrow 16$$
$$-9\dfrac{9}{10} \longrightarrow -10$$
$$\overline{\phantom{-9\dfrac{9}{10}}} \quad \overline{6} \leftarrow \text{estimated difference}$$

Estimate the sum or difference.

16. $7\dfrac{2}{3}$ $+4\dfrac{3}{4}$

17. $9\dfrac{1}{3}$ $-3\dfrac{5}{12}$

18. $16\dfrac{1}{8}$ $+13\dfrac{8}{9}$

19. $12\dfrac{1}{2}$ $-4\dfrac{7}{10}$

20. $10\dfrac{1}{6}$ $-9\dfrac{8}{9}$

21. $15\dfrac{3}{4} - \dfrac{9}{10}$ 22. $19\dfrac{2}{15} + \dfrac{6}{7}$ 23. $12\dfrac{3}{5} + \dfrac{10}{12} + 9\dfrac{8}{15}$

PROBLEM SOLVING

24. Antonio needs at least 15 pounds of chicken for a dinner party. He buys three packages: $3\dfrac{1}{4}$ lb, $4\dfrac{3}{7}$ lb, and $5\dfrac{2}{3}$ lb. Will this be enough chicken? Explain.

Communicate

6-3 Adding Fractions

Felix taped together three pieces of paper. They measured $\frac{9}{16}$ in., $\frac{7}{8}$ in., and $\frac{3}{4}$ in. How long was the taped piece?

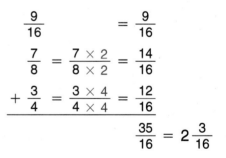

$\frac{9}{16}$ in. $\frac{7}{8}$ in. $\frac{3}{4}$ in.

To find the length of the taped piece, add: $\frac{9}{16}$ in. $+ \frac{7}{8}$ in. $+ \frac{3}{4}$ in. $=$ _?_

Estimate: $\frac{9}{16} + \frac{7}{8} + \frac{3}{4} \longrightarrow \frac{1}{2} + 1 + 1 = 2\frac{1}{2}$

About $2\frac{1}{2}$ in. long

▶ **To add fractions:**

- Find the least common denominator (LCD) of the fractions.

  Multiples of 16: 16, 32, 48, 64, . . .
  Multiples of 8: 8, 16, 24, 32, . . .
  Multiples of 4: 4, 8, 12, 16, . . .
  The LCD is 16.

  > The (LCD) of $\frac{9}{16}$, $\frac{7}{8}$, and $\frac{3}{4}$ is the least common multiple (LCM) of the denominators.

- Rename each fraction as an equivalent fraction with the LCD as the denominator.

- Add. Express the sum in simplest form.

$$\frac{9}{16} = \frac{9}{16}$$

$$\frac{7}{8} = \frac{7 \times 2}{8 \times 2} = \frac{14}{16}$$

$$+ \frac{3}{4} = \frac{3 \times 4}{4 \times 4} = \frac{12}{16}$$

$$\frac{35}{16} = 2\frac{3}{16}$$

Check.

Enter: 9 ⁄ 16 + 7 ⁄ 8 + 3 ⁄ 4 = ▮ 35/16 ▮

The taped piece of paper was $2\frac{3}{16}$ in. long.

Find the LCD for each set of fractions.

1. $\dfrac{2}{3}, \dfrac{3}{4}$ 2. $\dfrac{5}{12}, \dfrac{5}{6}$ 3. $\dfrac{7}{8}, \dfrac{1}{6}$ 4. $\dfrac{1}{2}, \dfrac{1}{4}, \dfrac{1}{8}$ 5. $\dfrac{1}{3}, \dfrac{5}{9}, \dfrac{1}{2}$

Add. Estimate to help you.

6. $\dfrac{1}{2}$
$+\dfrac{1}{3}$

7. $\dfrac{1}{4}$
$+\dfrac{2}{5}$

8. $\dfrac{1}{8}$
$+\dfrac{5}{6}$

9. $\dfrac{1}{12}$
$+\dfrac{1}{3}$

10. $\dfrac{4}{5}$
$+\dfrac{1}{20}$

11. $\dfrac{8}{15}$
$+\dfrac{1}{3}$

12. $\dfrac{7}{9}$
$+\dfrac{1}{2}$

13. $\dfrac{5}{7}$
$+\dfrac{3}{8}$

14. $\dfrac{3}{10}$
$+\dfrac{1}{6}$

15. $\dfrac{2}{9}$
$+\dfrac{7}{12}$

16. $\dfrac{3}{8}$
$+\dfrac{5}{24}$

17. $\dfrac{4}{9}$
$+\dfrac{1}{5}$

18. $\dfrac{1}{4}$
$\dfrac{1}{3}$
$+\dfrac{1}{2}$

19. $\dfrac{11}{20}$
$\dfrac{2}{5}$
$+\dfrac{1}{2}$

20. $\dfrac{1}{6}$
$\dfrac{1}{9}$
$+\dfrac{1}{9}$

21. $\dfrac{3}{4}$
$\dfrac{1}{6}$
$+\dfrac{1}{2}$

22. $\dfrac{3}{20}$
$\dfrac{1}{5}$
$+\dfrac{3}{10}$

23. $\dfrac{5}{6}$
$\dfrac{5}{8}$
$+\dfrac{7}{24}$

Compare. Use <, =, or >. Use a fraction calculator to check your answers.

24. $\dfrac{1}{5} + \dfrac{3}{10}$ __?__ $\dfrac{1}{2}$ 25. $\dfrac{4}{7} + \dfrac{1}{2}$ __?__ $\dfrac{13}{14}$ 26. $\dfrac{1}{6} + \dfrac{4}{9}$ __?__ $\dfrac{2}{3}$ 27. $\dfrac{7}{9} + \dfrac{1}{10}$ __?__ 1

PROBLEM SOLVING

28. Hector skied $\dfrac{7}{16}$ mi and then $\dfrac{5}{8}$ mi. How far did he ski in all?

29. One jogger ran $\dfrac{4}{5}$ mi and another ran $\dfrac{1}{10}$ mi. How far did they run in all?

30. In water, sound travels about $\dfrac{9}{10}$ mi in a second. How far will it travel in 2 seconds?

31. Three fifths of the 2nd floor is used for hallways and $\dfrac{3}{20}$ for offices. What part of the floor is used for both?

SKILLS TO REMEMBER

Write as a mixed number in simplest form.

32. $\dfrac{18}{4}$ 33. $\dfrac{11}{5}$ 34. $\dfrac{29}{6}$ 35. $\dfrac{76}{8}$ 36. $\dfrac{57}{9}$ 37. $\dfrac{85}{20}$

6-4 Adding Mixed Numbers

Ms. Posio owns a stock which advanced $1\frac{3}{8}$ points on Monday and $2\frac{1}{8}$ points on Tuesday. What was the total advance for the two days?

To find the number of points the stock advanced, add: $1\frac{3}{8} + 2\frac{1}{8} = \underline{\ ?\ }$

Estimate: $1\frac{3}{8} + 2\frac{1}{8}$

$1 + 2 = 3$ About 3 points

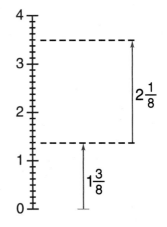

▶ **To add mixed numbers with *like* denominators:**

- Add the fractions.
- Add the whole numbers.
- Express the sum in simplest form.

$$\begin{array}{r} 1\frac{3}{8} \\ +\,2\frac{1}{8} \\ \hline 3\frac{4}{8} = 3\frac{1}{2} \end{array}$$ ← simplest form

The total advance of the stock was $3\frac{1}{2}$ points.

▶ **To add mixed numbers with *unlike* denominators:**

- Find the LCD of the fractions.
- Rename each fraction as an equivalent fraction with the LCD as the denominator.
- Add. Express the sum in simplest form.

$$\begin{array}{r} 14\frac{1}{10} = 14\frac{3}{30} \\ +\,26\frac{1}{15} = 26\frac{2}{30} \\ \hline = 40\frac{5}{30} = 40\frac{1}{6} \end{array}$$ ←

LCD of 10 and 15 is 30.

simplest form

Study these examples.

$$\begin{array}{r} 17 \\ +\ 9\frac{3}{16} \\ \hline 26\frac{3}{16} \end{array}$$ ← Add whole numbers. Bring down the fraction.

$$\begin{array}{r} 11\frac{5}{8} = 11\frac{5}{8} \\ +\ 5\frac{3}{4} = 5\frac{6}{8} \\ \hline 16\frac{11}{8} = 16 + 1\frac{3}{8} = 17\frac{3}{8} \end{array}$$

Rename the sum.

208

Rename each in simplest form.

1. $7\frac{6}{9}$ 2. $6\frac{8}{18}$ 3. $9\frac{15}{15}$ 4. $5\frac{8}{20}$ 5. $16\frac{12}{8}$ 6. $22\frac{14}{9}$

Complete.

7. $4\frac{1}{6} = 4\frac{?}{24}$
   $+3\frac{1}{4} - 3\frac{?}{24}$
   $\overline{\quad 7\frac{?}{24} = 7\frac{?}{12}}$

8. $\quad\ 8\frac{4}{5}$
   $+\ 9\frac{1}{5}$
   $\overline{\ 17\frac{?}{?} = \ \underline{?}}$

9. $7\frac{7}{20} = 7\frac{?}{?}$
   $+4\frac{4}{5} = \ 4\frac{16}{?}$
   $\overline{\quad 11\frac{?}{20} = 12\frac{?}{20}}$

Add. Estimate to help you.

10. $6\frac{2}{9}$
    $+3\frac{2}{9}$

11. $6\frac{1}{7}$
    $+8\frac{3}{7}$

12. $4\frac{1}{6}$
    $+2\frac{3}{8}$

13. $1\frac{5}{6}$
    $+2\frac{1}{3}$

14. $2\frac{2}{5}$
    $+3\frac{1}{10}$

15. $6\frac{2}{3}$
    $+7\frac{2}{5}$

16. 14
    $+\ 7\frac{5}{9}$

17. $9\frac{3}{8}$
    $+4$

18. $8\frac{7}{12}$
    $+\ \frac{5}{12}$

19. $16\frac{1}{8}$
    $+\ 7\frac{7}{8}$

20. $12\frac{7}{10}$
    $+23\frac{7}{30}$

21. $25\frac{7}{18}$
    $+15\frac{1}{6}$

22. $3\frac{7}{8}$
    $+3\frac{1}{2}$

23. $8\frac{3}{4}$
    $+6\frac{1}{3}$

24. $6\frac{11}{16}$
    $+12\frac{3}{4}$

25. $18\frac{3}{4}$
    $+20\frac{2}{3}$

26. $10\frac{9}{20}$
    $+\ 8\frac{3}{4}$

27. $15\frac{5}{6}$
    $+12\frac{7}{9}$

28. $9\frac{3}{7} + 6\frac{2}{7} + 4\frac{1}{7}$ 29. $6\frac{1}{2} + 3\frac{1}{3} + 4\frac{5}{6}$ 30. $4\frac{3}{5} + 2\frac{3}{10} + 1\frac{1}{2}$

PROBLEM SOLVING

31. It takes $1\frac{2}{3}$ gal for paint repairs in the den and $1\frac{1}{2}$ gal for the kitchen. How much paint is that in all?

32. A butcher sold packages of meat weighing $1\frac{2}{3}$ lb and $5\frac{3}{4}$ lb. What was the total weight of the meat?

MENTAL MATH

Add. Look for sums of 1.

33. $\frac{1}{2} + \frac{3}{4} + \frac{1}{2}$ 34. $5\frac{1}{4} + 6\frac{2}{3} + 11\frac{3}{4}$ 35. $\frac{1}{4} + 7\frac{3}{4} + \frac{1}{5}$

36. $2\frac{6}{8} + 9 + 3\frac{1}{4}$ 37. $\frac{2}{3} + 1\frac{1}{5} + 7\frac{5}{15}$ 38. $8\frac{1}{6} + 9\frac{1}{7} + 2\frac{5}{6}$

6-5 | Subtracting Fractions

The chart shows the fractional part of family income spent on food for five countries. How much greater is the fractional part for India than for the U.S.A.?

Family Income Spent on Food	
India	$\frac{11}{20}$
China	$\frac{1}{2}$
Mexico	$\frac{1}{3}$
Japan	$\frac{1}{5}$
U.S.A.	$\frac{1}{10}$

To find how much greater,

subtract: $\frac{11}{20} - \frac{1}{10} = \underline{\ ?\ }$

Estimate: $\frac{11}{20} - \frac{1}{10} \longrightarrow \frac{1}{2} - 0 = \frac{1}{2}$

About $\frac{1}{2}$ greater

▶ **To subtract fractions:**

- Find the LCD of the fractions.

- Rename each fraction as an equivalent fraction with the LCD as the denominator.

- Subtract. Express the difference in simplest form.

$$\begin{array}{rcl} \frac{11}{20} & = & \frac{11}{20} \\ -\frac{1}{10} = \frac{1 \times 2}{10 \times 2} & = & \frac{2}{20} \\ \hline & & \frac{9}{20} \end{array}$$

LCD of 20 and 10 is 20.

simplest form

Check.

Enter: 11 / 20 − 1 / 10 = $\boxed{9/20}$

The fractional part of India's family income spent on food is $\frac{9}{20}$ greater than that of the U.S.A.

Subtract. Estimate to help you.

1. $\frac{3}{4}$ $-\frac{5}{8}$
2. $\frac{5}{6}$ $-\frac{1}{2}$
3. $\frac{7}{10}$ $-\frac{1}{5}$
4. $\frac{1}{5}$ $-\frac{1}{25}$
5. $\frac{1}{2}$ $-\frac{3}{10}$
6. $\frac{4}{5}$ $-\frac{1}{15}$

7. $\frac{3}{4}$ $-\frac{2}{3}$
8. $\frac{7}{8}$ $-\frac{5}{6}$
9. $\frac{1}{2}$ $-\frac{2}{5}$
10. $\frac{6}{7}$ $-\frac{1}{2}$
11. $\frac{3}{8}$ $-\frac{1}{10}$
12. $\frac{4}{5}$ $-\frac{3}{4}$

Find the difference. Use a fraction calculator to check your answers.

13. $\dfrac{7}{8} - \dfrac{4}{5}$ **14.** $\dfrac{3}{10} - \dfrac{7}{30}$ **15.** $\dfrac{4}{11} - \dfrac{8}{22}$ **16.** $\dfrac{25}{48} - \dfrac{3}{8}$

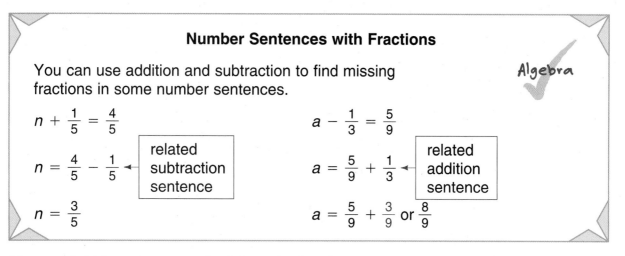

Number Sentences with Fractions

You can use addition and subtraction to find missing fractions in some number sentences.

Algebra ✓

$n + \dfrac{1}{5} = \dfrac{4}{5}$

$n = \dfrac{4}{5} - \dfrac{1}{5}$ ← related subtraction sentence

$n = \dfrac{3}{5}$

$a - \dfrac{1}{3} = \dfrac{5}{9}$

$a = \dfrac{5}{9} + \dfrac{1}{3}$ ← related addition sentence

$a = \dfrac{5}{9} + \dfrac{3}{9}$ or $\dfrac{8}{9}$

Use a related sentence to find the missing fraction or whole number.

17. $n + \dfrac{1}{8} = \dfrac{7}{8}$ **18.** $y - \dfrac{1}{6} = \dfrac{1}{6}$ **19.** $m + \dfrac{1}{3} = \dfrac{1}{2}$ **20.** $z - \dfrac{1}{9} = \dfrac{1}{18}$

21. $p - \dfrac{5}{6} = \dfrac{3}{8}$ **22.** $t + \dfrac{3}{5} = \dfrac{5}{6}$ **23.** $c - 0 = \dfrac{3}{5}$ **24.** $d + 0 = \dfrac{5}{12}$

25. $f - 3\dfrac{1}{2} = 3\dfrac{1}{2}$ **26.** $g - 1\dfrac{1}{4} = \dfrac{3}{4}$ **27.** $\dfrac{2}{3} = r - \dfrac{1}{3}$ **28.** $\dfrac{7}{8} = b + \dfrac{7}{16}$

PROBLEM SOLVING Use the chart on page 210 for problems 29–30.

29. How much greater is the fractional part for China than for Japan?

30. How much greater is the fractional part for India than for Mexico?

31. Mr. Baumbach plans to leave his estate to his four children. One gets $\dfrac{1}{4}$ of his estate, the second gets $\dfrac{1}{16}$, and the third gets $\dfrac{3}{8}$. How much does the fourth child get? Explain your answer.

Communicate ✓

MAKE UP YOUR OWN

Write a word problem with two fractions so that the:

32. sum is $\dfrac{1}{2}$ **33.** difference is $\dfrac{1}{8}$ **34.** difference is $\dfrac{12}{45}$

Subtracting Mixed Numbers

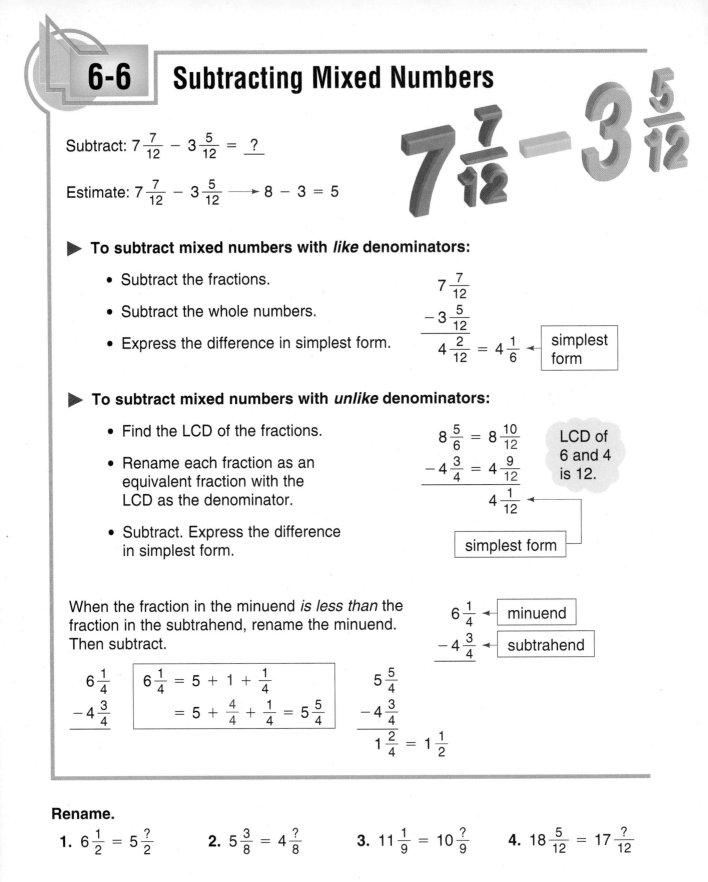

Subtract: $7\frac{7}{12} - 3\frac{5}{12} = \underline{\ ?\ }$

Estimate: $7\frac{7}{12} - 3\frac{5}{12} \longrightarrow 8 - 3 = 5$

▶ **To subtract mixed numbers with _like_ denominators:**

- Subtract the fractions.

- Subtract the whole numbers.

- Express the difference in simplest form.

$$
\begin{array}{r}
7\frac{7}{12} \\
-3\frac{5}{12} \\
\hline
4\frac{2}{12} = 4\frac{1}{6}
\end{array}
$$

← simplest form

▶ **To subtract mixed numbers with _unlike_ denominators:**

- Find the LCD of the fractions.

- Rename each fraction as an equivalent fraction with the LCD as the denominator.

- Subtract. Express the difference in simplest form.

$$
\begin{array}{r}
8\frac{5}{6} = 8\frac{10}{12} \\
-4\frac{3}{4} = 4\frac{9}{12} \\
\hline
4\frac{1}{12}
\end{array}
$$

LCD of 6 and 4 is 12.

simplest form

When the fraction in the minuend *is less than* the fraction in the subtrahend, rename the minuend. Then subtract.

$$
\begin{array}{r}
6\frac{1}{4} \\
-4\frac{3}{4}
\end{array}
$$

← minuend

← subtrahend

$$
\begin{array}{r}
6\frac{1}{4} \\
-4\frac{3}{4}
\end{array}
\qquad
\boxed{
\begin{aligned}
6\frac{1}{4} &= 5 + 1 + \frac{1}{4} \\
&= 5 + \frac{4}{4} + \frac{1}{4} = 5\frac{5}{4}
\end{aligned}
}
\qquad
\begin{array}{r}
5\frac{5}{4} \\
-4\frac{3}{4} \\
\hline
1\frac{2}{4} = 1\frac{1}{2}
\end{array}
$$

Rename.

1. $6\frac{1}{2} = 5\frac{?}{2}$ **2.** $5\frac{3}{8} = 4\frac{?}{8}$ **3.** $11\frac{1}{9} = 10\frac{?}{9}$ **4.** $18\frac{5}{12} = 17\frac{?}{12}$

Subtract. Estimate to help you.

5. $6\frac{4}{9}$ $-4\frac{1}{9}$ **6.** $7\frac{5}{8}$ $-3\frac{3}{8}$ **7.** $5\frac{3}{4}$ $-1\frac{3}{4}$ **8.** $2\frac{9}{10}$ $-2\frac{9}{10}$ **9.** $9\frac{1}{8}$ $-3\frac{5}{8}$ **10.** $11\frac{5}{7}$ $-\ 9\frac{6}{7}$

11. $9\frac{1}{3}$ $-1\frac{5}{6}$ **12.** $13\frac{1}{6}$ $-\ 9\frac{3}{4}$ **13.** $10\frac{2}{5}$ $-\ 3\frac{7}{10}$ **14.** $11\frac{1}{4}$ $-\ 6\frac{2}{3}$ **15.** $15\frac{2}{15}$ $-\ 6\frac{4}{5}$ **16.** $13\frac{1}{12}$ $-\ 4\frac{3}{8}$

Subtracting Mixed Numbers and Whole Numbers

$9\frac{3}{4}$
-3
$\overline{6\frac{3}{4}}$

Subtract whole numbers. Bring down the fraction.

$9\quad = 8\frac{4}{4}$
$-3\frac{3}{4} = 3\frac{3}{4}$
$\overline{\qquad 5\frac{1}{4}}$

$9 = 8 + 1$
$= 8 + \frac{4}{4}$
$= 8\frac{4}{4}$

Compute. Estimate to help you.

17. $7\frac{3}{5}$ -4 **18.** $11\frac{1}{8}$ $-\ 6$ **19.** $14\frac{5}{9}$ $-\ 8\frac{2}{9}$ **20.** $6\frac{7}{12}$ $-2\frac{1}{12}$ **21.** $8\frac{1}{2}$ $-1\frac{1}{2}$ **22.** $10\frac{4}{7}$ $-\ 7\frac{4}{7}$

23. $7\frac{1}{5}$ $-4\frac{3}{5}$ **24.** $8\frac{1}{8}$ $-2\frac{7}{8}$ **25.** $8\frac{2}{3}$ $-1\frac{5}{6}$ **26.** $6\frac{1}{5}$ $-2\frac{3}{4}$ **27.** $12\frac{1}{4}$ $-10\frac{1}{3}$ **28.** $15\frac{1}{6}$ $-\ 6\frac{3}{8}$

29. $10 - 1\frac{1}{4}$ **30.** $9 - 1\frac{1}{9}$ **31.** $7\frac{1}{2} - 7\frac{1}{3}$ **32.** $15\frac{1}{5} - 15\frac{1}{6}$

33. $9\frac{1}{5} - 3\frac{5}{8}$ **34.** $8\frac{7}{9} - \frac{5}{8}$ **35.** $24\frac{1}{3} - 3\frac{9}{10}$ **36.** $18 - 3\frac{5}{6}$

37. $8\frac{1}{6} - 3\frac{3}{4} + 2\frac{1}{2}$ **38.** $9\frac{1}{12} - 5\frac{3}{8} - 1\frac{3}{4}$ **39.** $6\frac{1}{2} + 7\frac{1}{3} - 8\frac{1}{4}$

PROBLEM SOLVING

40. There are $16\frac{1}{3}$ yd of material on a bolt. If $5\frac{3}{4}$ yd are used, how much material is left on the bolt?

41. Tricia usually works 40 hours a week. Last week she was absent $6\frac{1}{4}$ hours. How many hours did she work?

Maurice made a solution for a chemistry experiment by mixing $1\frac{3}{4}$ pt of colored water with $\frac{1}{2}$ pt of lemon juice. How many pints of solution did he make?

To find the amount of solution, add: $1\frac{3}{4}$ pt $+ \frac{1}{2}$ pt $= \underline{\ ?\ }$

Maurice and his friend Beth both used mental math and the associative property of addition to compute.

Maurice $\qquad\qquad 1\frac{3}{4} = 1\frac{1}{4} + \frac{1}{2}$ $\qquad$ *Beth* $\qquad\qquad \frac{1}{2} = \frac{1}{4} + \frac{1}{4}$

$$1\frac{3}{4} + \frac{1}{2} = 1\frac{1}{4} + \frac{1}{2} + \frac{1}{2}$$
$$= 1\frac{1}{4} + \left(\frac{1}{2} + \frac{1}{2}\right)$$
$$= 1\frac{1}{4} + 1$$
$$= 2\frac{1}{4}$$

associative property

$$1\frac{3}{4} + \frac{1}{2} = 1\frac{3}{4} + \frac{1}{4} + \frac{1}{4}$$
$$= \left(1\frac{3}{4} + \frac{1}{4}\right) + \frac{1}{4}$$
$$= 2 + \frac{1}{4}$$
$$= 2\frac{1}{4}$$

Maurice made $2\frac{1}{4}$ pt of solution.

Study these examples.

$$6\frac{1}{3} + 4\frac{4}{5} + 5\frac{2}{3} = 6\frac{1}{3} + 5\frac{2}{3} + 4\frac{4}{5} \longleftarrow \boxed{\text{commutative property}}$$
$$= \left(6\frac{1}{3} + 5\frac{2}{3}\right) + 4\frac{4}{5} \longleftarrow \boxed{\text{associative property}}$$
$$= 12 + 4\frac{4}{5} = 16\frac{4}{5}$$

$$7\frac{5}{8} + \left(8\frac{1}{2} - 8\frac{1}{2}\right) = 7\frac{5}{8} + 0 \longleftarrow$$
$$= 7\frac{5}{8} \longleftarrow$$

identity property

> Remember: Commutative means "order."
> Associative means "grouping."
> Identity means "same."

Name the property of addition.

1. $5\frac{1}{2} + 3\frac{1}{4} = 3\frac{1}{4} + 5\frac{1}{2}$ **2.** $7\frac{1}{8} + 0 = 7\frac{1}{8}$

3. $1\frac{1}{6} + (1\frac{1}{5} + 1\frac{1}{4}) = (1\frac{1}{6} + 1\frac{1}{5}) + 1\frac{1}{4}$

4. $0 + 10\frac{1}{10} = 10\frac{1}{10}$ **5.** $0 + 2\frac{2}{3} = 2\frac{2}{3} + 0$

6. $(9\frac{1}{6} + \frac{5}{6}) + \frac{1}{2} = 9\frac{1}{6} + (\frac{5}{6} + \frac{1}{2})$ **7.** $1\frac{1}{2} + b - b + 1\frac{1}{2}$

8. $2\frac{1}{4} + b = 2\frac{1}{4}$ **9.** $a + (b + \frac{1}{2}) = (a + b) + \frac{1}{2}$

Add. Use mental math and the properties of addition.

10. $\frac{1}{2} + \frac{1}{4}$ **11.** $\frac{1}{2} + \frac{1}{2}$ **12.** $\frac{1}{4} + \frac{1}{4}$ **13.** $\frac{1}{2} + \frac{3}{4}$

14. $5\frac{1}{2} + 2\frac{3}{4}$ **15.** $\frac{3}{4} + 11\frac{3}{4}$ **16.** $\frac{1}{2} + \frac{1}{8}$ **17.** $\frac{5}{8} + \frac{1}{2}$

18. $\frac{3}{8} + \frac{3}{4}$ **19.** $10\frac{3}{8} + \frac{5}{8}$ **20.** $8\frac{1}{4} + 9\frac{1}{2}$ **21.** $12\frac{1}{4} + 6\frac{7}{8}$

Compute. Explain how you used the properties of addition to help you.

22. $5\frac{2}{5} + 3\frac{3}{5} + 6\frac{1}{4}$ **23.** $8\frac{1}{8} + 4\frac{1}{4} + 5\frac{7}{8}$

24. $0 + 11\frac{2}{5}$ **25.** $9\frac{1}{6} + 0$

26. $7\frac{1}{2} + 6 + 4\frac{1}{4}$ **27.** $3\frac{3}{4} + 2\frac{1}{2} + 11$

28. $12\frac{1}{4} + 5\frac{1}{8} + 2\frac{1}{2}$ **29.** $9\frac{1}{2} + 4\frac{3}{7} + 1\frac{1}{4}$

30. $(\frac{1}{2} + 2\frac{3}{5}) + 1\frac{1}{2}$ **31.** $8\frac{1}{4} + (4\frac{1}{9} + \frac{3}{4})$

32. $3\frac{1}{2} + (1\frac{1}{4} - 1\frac{1}{4})$ **33.** $(2\frac{1}{5} - 2\frac{1}{5}) + (3\frac{2}{3} + 1\frac{1}{4})$

MENTAL MATH

Subtract by "adding on" in steps.

34. $7 - 2\frac{1}{4}$ $2\frac{1}{4} + \boxed{\frac{3}{4}} = 3 \longrightarrow 3 + \boxed{4} = 7 \longrightarrow$ Difference: $4\frac{3}{4}$

35. $6 - 3\frac{1}{2}$ **36.** $8 - 2\frac{2}{3}$ **37.** $2\frac{5}{8} - 1\frac{7}{8}$ **38.** $8\frac{9}{14} - 6\frac{11}{14}$

TECHNOLOGY

Computing Fractions and Decimals

You can use the $\boxed{\text{F⇆D}}$ key on a fraction calculator to rename a fraction or mixed number as a decimal or a decimal as a fraction or mixed number.

▶ Rename $1\frac{1}{4}$ as a decimal.

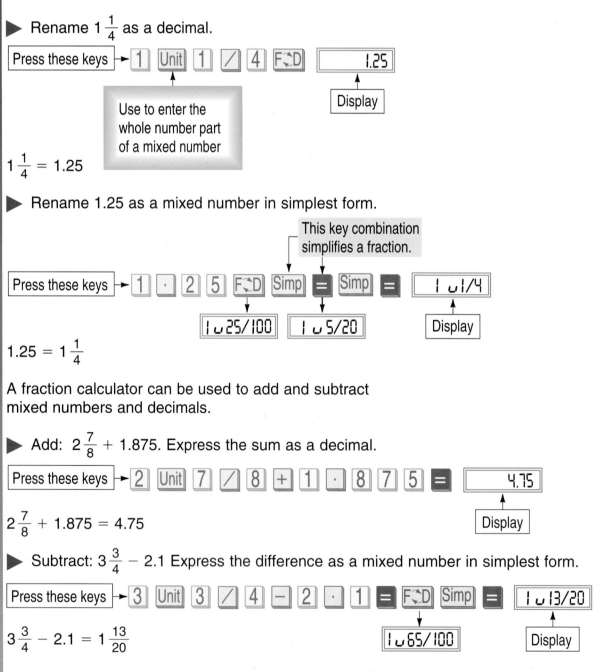

Press these keys → $\boxed{1}$ $\boxed{\text{Unit}}$ $\boxed{1}$ $\boxed{/}$ $\boxed{4}$ $\boxed{\text{F⇆D}}$ 1.25

Display

Use to enter the whole number part of a mixed number

$1\frac{1}{4} = 1.25$

▶ Rename 1.25 as a mixed number in simplest form.

This key combination simplifies a fraction.

Press these keys → $\boxed{1}$ $\boxed{.}$ $\boxed{2}$ $\boxed{5}$ $\boxed{\text{F⇆D}}$ $\boxed{\text{Simp}}$ $\boxed{=}$ $\boxed{\text{Simp}}$ $\boxed{=}$ 1 ⌴ 1/4

1 ⌴ 25/100   1 ⌴ 5/20   Display

$1.25 = 1\frac{1}{4}$

A fraction calculator can be used to add and subtract mixed numbers and decimals.

▶ Add: $2\frac{7}{8} + 1.875$. Express the sum as a decimal.

Press these keys → $\boxed{2}$ $\boxed{\text{Unit}}$ $\boxed{7}$ $\boxed{/}$ $\boxed{8}$ $\boxed{+}$ $\boxed{1}$ $\boxed{.}$ $\boxed{8}$ $\boxed{7}$ $\boxed{5}$ $\boxed{=}$ 4.75

$2\frac{7}{8} + 1.875 = 4.75$ Display

▶ Subtract: $3\frac{3}{4} - 2.1$ Express the difference as a mixed number in simplest form.

Press these keys → $\boxed{3}$ $\boxed{\text{Unit}}$ $\boxed{3}$ $\boxed{/}$ $\boxed{4}$ $\boxed{-}$ $\boxed{2}$ $\boxed{.}$ $\boxed{1}$ $\boxed{=}$ $\boxed{\text{F⇆D}}$ $\boxed{\text{Simp}}$ $\boxed{=}$ 1 ⌴ 13/20

$3\frac{3}{4} - 2.1 = 1\frac{13}{20}$ 1 ⌴ 65/100   Display

216

Write each as a decimal.

1. $\frac{1}{8}$
2. $\frac{2}{5}$
3. $\frac{9}{12}$
4. $3\frac{11}{20}$
5. $2\frac{3}{5}$

Write each as a fraction or mixed number in simplest form.

6. 0.8
7. 0.25
8. 1.7
9. 0.375
10. 5.8

Add or subtract. Write each as a decimal.

11. $\frac{3}{8} + 2.4$
12. $5.3 + 2\frac{1}{4}$
13. $8.2 - 1\frac{1}{5}$

14. $\frac{5}{8} + 0.54$
15. $\frac{7}{10} - 0.063$
16. $9.85 - 5\frac{1}{4}$

Add or subtract. Write each as a fraction or mixed number in simplest form.

17. $2.4 + 1\frac{1}{4}$
18. $1\frac{5}{8} - 0.75$
19. $3\frac{3}{5} - 2.9$

20. $31.45 - 24\frac{1}{5}$
21. $20\frac{3}{4} + 0.01$
22. $8\frac{1}{2} + 2.63$

Solve. For exercises 23, 25, and 27, use the $\boxed{\text{F⟺D}}$ key to write each as a decimal. Look for a pattern.

23. $\boxed{\dfrac{1}{9} \quad \dfrac{2}{9} \quad \dfrac{3}{9} \quad \dfrac{4}{9} \quad \dfrac{5}{9}}$

24. Predict the decimal equivalent for $\frac{8}{9}$.

25. $\boxed{\dfrac{1}{11} \quad \dfrac{2}{11} \quad \dfrac{3}{11} \quad \dfrac{4}{11} \quad \dfrac{5}{11}}$

26. Predict the decimal equivalent for $\frac{10}{11}$.

27. $\boxed{\dfrac{1}{33} \quad \dfrac{2}{33} \quad \dfrac{3}{33} \quad \dfrac{4}{33} \quad \dfrac{5}{33}}$

28. Predict the decimal equivalent for $\frac{20}{33}$.

29. What type of decimal is expressed in exercises 23–28?

30. If $\frac{1}{15} = 0.0666667$, then $\frac{2}{15} = $ _?_ .

31. If $\frac{3}{15} = 0.2$, then $\frac{6}{15} = $ _?_ .

32. Predict the decimal equivalents for $\frac{8}{15}$ and $\frac{9}{15}$.

6-9 Problem Solving: Working Backwards

Problem: Mrs. Kline bought a bolt of material. She used $4\frac{3}{4}$ yd to make a dress and $2\frac{3}{8}$ yd for a jacket. After buying $1\frac{1}{8}$ yd more she had $6\frac{1}{2}$ yd left to make two pairs of slacks. How much material was on the bolt she bought?

1 IMAGINE Draw and label a picture.

2 NAME *Facts:* She used $4\frac{3}{4}$ yd and $2\frac{3}{8}$ yd.
She bought $1\frac{1}{8}$ yd more.
She had $6\frac{1}{2}$ yd left.

Question: How much material was on the bolt she bought?

3 THINK Look at the picture you drew and labeled. Begin with the amount of material she had left ($6\frac{1}{2}$ yd) and *work backwards*. Subtract the amount she had bought and add the amounts she used to find the first amount of material she bought:

$$\underline{\ ?\ } - 4\frac{3}{4}\text{ yd} - 2\frac{3}{8}\text{ yd} + 1\frac{1}{8}\text{ yd} = 6\frac{1}{2}\text{ yd}$$
$$\qquad\text{used}\qquad\text{used}\qquad\text{bought}\quad\text{left}$$

$$6\frac{1}{2}\text{ yd} - 1\frac{1}{8}\text{ yd} + 2\frac{3}{8}\text{ yd} + 4\frac{3}{4}\text{ yd} = \underline{\ ?\ }$$

4 COMPUTE Add or subtract in order from left to right.

$$6\frac{1}{2}\text{ yd} - 1\frac{1}{8}\text{ yd} + 2\frac{3}{8}\text{ yd} + 4\frac{3}{4}\text{ yd} = \underline{\ ?\ } \qquad 6\frac{4}{8} - 1\frac{1}{8} = 5\frac{3}{8}$$

$$5\frac{3}{8}\text{ yd} + 2\frac{3}{8}\text{ yd} + 4\frac{3}{4}\text{ yd} = \underline{\ ?\ } \qquad 5\frac{3}{8} + 2\frac{3}{8} = 7\frac{6}{8}$$

$$7\frac{3}{4}\text{ yd} + 4\frac{3}{4}\text{ yd} = \underline{\ ?\ } \qquad 7\frac{3}{4} + 4\frac{3}{4} = 11\frac{6}{4} = 12\frac{2}{4}$$

$$12\frac{2}{4}\text{ yd} = 12\frac{1}{2}\text{ yd}$$

Mrs. Kline originally bought $12\frac{1}{2}$ yd of material.

5 CHECK Begin with the first amount and compute by working *forward*:

$$12\frac{1}{2}\text{ yd} - 4\frac{3}{4}\text{ yd} - 2\frac{3}{8}\text{ yd} + 1\frac{1}{8}\text{ yd} = 6\frac{1}{2}\text{ yd}$$

Solve. Use the strategy of Working Backwards.

1. After losing $2\frac{1}{2}$ lb in each of the first 2 weeks
   of March and gaining $1\frac{3}{8}$ lb in each of the next
   2 weeks, Ted's prizewinning piglet weighed
   110 lb. How much did his piglet weigh on March 1?

IMAGINE   Create a mental picture.

NAME   *Facts:* $2\frac{1}{2}$ lb lost each of the first 2 weeks

                        $1\frac{3}{8}$ lb gained each of the next 2 weeks

                        110 lb is its final weight.

         Question: How much did the piglet weigh March 1?

THINK   Begin with the piglet's final weight of 110 lb and *work
         backwards.* Subtract the pounds it gained and add
         the pounds it lost to find its weight on March 1.

$$110 \text{ lb} - 1\frac{3}{8} \text{ lb} - 1\frac{3}{8} \text{ lb} + 2\frac{1}{2} \text{ lb} + 2\frac{1}{2} \text{ lb}$$

COMPUTE   →   **CHECK**

2. At the end of one school day, Ms. Dinger had 17 crayons left.
   She remembered giving out 14 crayons in the morning, getting
   back 12 crayons at recess, and giving out 11 crayons after
   lunch. How many crayons did Ms. Dinger have at the start
   of the day?

3. Jason was given his allowance on Sunday. On Monday
   he bought a book for $2.95. On Tuesday Kurt paid Jason
   the $3.50 he owed him. Jason now has $6.05. How much
   was his allowance?

4. Lee wrote a 2-digit number. She divided it by 9,
   added 24, and doubled the result. Her final answer
   was 64. What number did Lee write?

5. Rita bought some peaches. She used $3\frac{2}{3}$ lb to make peach
   cobbler. Then she used $\frac{5}{6}$ lb in fruit salad. After her neighbor
   gave her $2\frac{1}{2}$ lb from her tree, she had $3\frac{1}{4}$ lb. How many
   pounds of peaches did she buy?

Problem-Solving Applications

Solve each problem and explain the method used.

1. Ms. Carson's class makes silk flowers for a craft fair. A silk rose is $12\frac{1}{2}$ in. long. A silk lily is 15 in. long. How much longer is the lily?

2. Marissa cuts petals out of red ribbon. One petal is $\frac{9}{16}$ in. long, another is $\frac{13}{16}$ in. long, and a third is $1\frac{5}{16}$ in. long. How many inches of ribbon does she need to cut for all three petals?

3. Paul cuts wire stems for the flowers. He has a piece of wire 12 in. long. He cuts a stem $9\frac{1}{2}$ in. long. How much wire does he have left?

4. Paul uses $\frac{2}{3}$ yard of green tape to wrap one stem and $\frac{3}{4}$ yard of green tape to wrap another. How much tape does he use in all?

5. Gloria made a wreath using $4\frac{2}{3}$ yards of green ribbon and $5\frac{3}{4}$ yards of yellow ribbon. How much ribbon did she use?

6. José made a flower arrangement that was $18\frac{5}{16}$ in. tall. It was a little too tall, so he cut $2\frac{1}{2}$ in. off the stems. How tall was the finished arrangement?

7. At the fair, $\frac{1}{5}$ of the class worked the booth. Another $\frac{2}{3}$ of the class had created the flowers. The remaining part of the class decorated the booth. What part of the class decorated the booth?

8. Mr. McCauley's class made wicker baskets for the craft fair. Each basket is $6\frac{1}{8}$ in. tall and $2\frac{5}{16}$ in. wider than it is tall. How wide are the baskets?

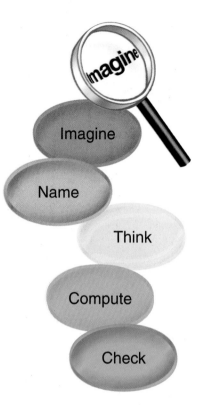

Imagine

Name

Think

Compute

Check

Use a strategy from the list or another strategy you know to solve each problem.

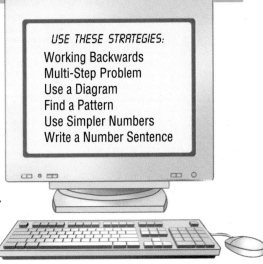

USE THESE STRATEGIES:
Working Backwards
Multi-Step Problem
Use a Diagram
Find a Pattern
Use Simpler Numbers
Write a Number Sentence

9. Dan started with a 36-in. strip of wicker. He cut two $11\frac{5}{8}$-in. pieces. Does he have enough left over to cut two $6\frac{1}{2}$-in. pieces?

10. Ben cut wicker strips to make a basket. He cut two $8\frac{3}{4}$-in. strips and one $6\frac{1}{2}$-in. strip from one long piece. He had $\frac{5}{8}$ in. of wicker left over. How long was the original piece of wicker?

11. Write the next five numbers in this series:
$\frac{23}{24}, \frac{11}{12}, \frac{21}{24}, \frac{5}{6}, \frac{19}{24}, \frac{3}{4}$.

12. Mr. Cortez spent $10.25 of his money on wicker each week for 2 weeks. Next he collected $4.65 and $6.70 from students. Then he had $45.53 to spend. How much money did he have originally?

13. Nedra made a basket handle using three pieces of wicker. The first piece was 9 in. long, the second was $\frac{1}{2}$ in. longer than the first, and the third was $\frac{3}{4}$ in. shorter than the second. How much wicker did she use?

Use the diagram for problems 14–16.

14. How much wider is the thick wicker than the thin?

15. Jason makes a basket using the medium width wicker. The basket is as tall as 8 strips of the wicker. Is the basket taller than 4 inches?

16. Arlene's basket is 9 strips tall. The pattern is thin, thick, thick, thin, thick, thick, and so on. How tall is Arlene's basket?

Widths of Wicker

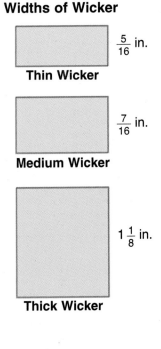

$\frac{5}{16}$ in.
Thin Wicker

$\frac{7}{16}$ in.
Medium Wicker

$1\frac{1}{8}$ in.
Thick Wicker

MAKE UP YOUR OWN

17. Plan a wicker basket of your own. Then write a problem about it. Have a classmate solve it.

221

Chapter Review and Practice

Estimate the sum or difference. *(See pp. 204–205.)*

1. $\dfrac{5}{9} + \dfrac{3}{7}$ **2.** $\dfrac{14}{15} - \dfrac{7}{8}$ **3.** $9\dfrac{1}{6} - 7\dfrac{1}{2}$ **4.** $12\dfrac{9}{10} + 11\dfrac{6}{7}$

Add or subtract. Estimate to help you. *(See pp. 206–213.)*

5. $\begin{array}{r} \dfrac{1}{2} \\ +\dfrac{1}{3} \\ \hline \end{array}$ **6.** $\begin{array}{r} 4\dfrac{1}{8} \\ +6\dfrac{3}{8} \\ \hline \end{array}$ **7.** $\begin{array}{r} \dfrac{1}{3} \\ -\dfrac{1}{4} \\ \hline \end{array}$ **8.** $\begin{array}{r} 4\dfrac{5}{7} \\ -1\dfrac{2}{7} \\ \hline \end{array}$ **9.** $\begin{array}{r} \dfrac{1}{9} \\ +\dfrac{2}{3} \\ \hline \end{array}$ **10.** $\begin{array}{r} \dfrac{3}{5} \\ +\dfrac{3}{4} \\ \hline \end{array}$

11. $\begin{array}{r} 6\dfrac{5}{6} \\ +1\dfrac{1}{8} \\ \hline \end{array}$ **12.** $\begin{array}{r} 3\dfrac{1}{2} \\ +9\dfrac{2}{9} \\ \hline \end{array}$ **13.** $\begin{array}{r} 10\dfrac{5}{6} \\ +19 \\ \hline \end{array}$ **14.** $\begin{array}{r} 13 \\ +8\dfrac{4}{7} \\ \hline \end{array}$ **15.** $\begin{array}{r} 5 \\ -1\dfrac{1}{2} \\ \hline \end{array}$ **16.** $\begin{array}{r} 10 \\ -2\dfrac{3}{4} \\ \hline \end{array}$

17. $14\dfrac{3}{10} - 1\dfrac{1}{3}$ **18.** $1\dfrac{3}{8} + 19\dfrac{2}{3}$ **19.** $10 - 1\dfrac{1}{10}$ **20.** $4\dfrac{3}{7} - 1\dfrac{1}{6}$

21. $4\dfrac{1}{4} + 1\dfrac{1}{2} + 2\dfrac{3}{8}$ **22.** $\dfrac{1}{8} + 3\dfrac{2}{3} + 3\dfrac{7}{8}$ **23.** $15\dfrac{3}{8} - 9\dfrac{7}{8}$

(See pp. 210–211.)

Use a related sentence to find the missing fraction or whole number.

24. $b + \dfrac{1}{3} = \dfrac{2}{3}$ **25.** $n - \dfrac{1}{5} = \dfrac{4}{5}$ **26.** $m - \dfrac{1}{4} = \dfrac{1}{4}$ **27.** $a + \dfrac{1}{6} = \dfrac{4}{6}$

28. $f - 1\dfrac{1}{3} = \dfrac{1}{9}$ **29.** $t + \dfrac{1}{8} = \dfrac{3}{5}$ **30.** $\dfrac{2}{7} = r + \dfrac{2}{7}$ **31.** $\dfrac{4}{9} = v - \dfrac{1}{27}$

Compute. Use mental math and the properties of addition. *(See pp. 214–215.)*

32. $\dfrac{1}{2} + \dfrac{1}{2}$ **33.** $\dfrac{1}{4} + \dfrac{1}{4}$ **34.** $5\dfrac{1}{6} + \left(3\dfrac{1}{2} - 3\dfrac{1}{2}\right)$

35. $0 + 9\dfrac{3}{5}$ **36.** $7\dfrac{1}{4} + 3\dfrac{1}{2}$ **37.** $\left(\dfrac{2}{3} + 5\dfrac{1}{2}\right) + 5\dfrac{1}{3}$

PROBLEM SOLVING

(See pp. 204–205, 220–221.)

38. XYZ stock advanced $\dfrac{2}{3}$ point on Monday and $\dfrac{2}{3}$ point on Tuesday. What was the total advance?

39. Angela ran $5\dfrac{3}{10}$ mi yesterday and $6\dfrac{1}{2}$ mi today. How many miles in all did she run?

40. Michael has three lengths of ribbon: $14\dfrac{1}{2}$ yd, $25\dfrac{1}{3}$ yd, and $9\dfrac{7}{8}$ yd. Estimate the total length.

41. Grace's cat weighs 15 lb. The cat loses $4\dfrac{3}{4}$ lb. How much does her cat weigh then?

(See Still More Practice, p. 511.)

LOGIC: STATEMENTS AND NEGATIONS

In logic, the **negation** of a statement is formed
by denying the original statement. When a statement
is true, its negation is false. When a statement
is false, its negation is true.

Statement		**Negation**	
A square is round.	False	A square is not round.	True
Seven is an odd number.	True	Seven is not an odd number.	False
A square has 5 sides.	False	No squares have 5 sides.	True

**Tell whether each is a negation of the statement:
"A triangle has 4 sides." Write *Yes* or *No.***

1. Some triangles have 4 sides.

2. A triangle does not have 4 sides.

3. No triangles have 4 sides.

4. A square has 4 sides.

**Tell whether the statement is *True* or *False*. Then write the
negation of the statement and tell whether it is *True* or *False.***

5. All squares have 4 sides.

6. Sixteen is a prime number.

7. No circles have 3 sides.

8. No prime numbers are even.

9. A fraction cannot be renamed
as a decimal.

10. An odd number is not
divisible by 4.

11. Fractions can be added if they
are like fractions.

12. The sum of a fraction and
zero is not zero.

Write a statement and its negation for each description.

13. A statement about adding fractions or mixed numbers.
Statement: True; Negation: False

14. A statement about subtracting fractions or mixed numbers.
Statement: False; Negation: True

223

Performance Assessment

Solve the problems and explain your methods.

1. Eduardo's fraction calculator is not working correctly. When he enters a fraction, the calculator adds $\frac{1}{4}$, and on the next step it subtracts $2\frac{1}{2}$, and then continues the same pattern without stopping. Here are the first four numbers that Eduardo got:

   $8\frac{5}{8}, 6\frac{1}{8}, 6\frac{3}{8}, 3\frac{7}{8}, \underline{\ ?\ }, \underline{\ ?\ }$

   What are the fifth and sixth numbers in this sequence? With what number did Eduardo start?

Add or subtract. Estimate to help you.

2. $\begin{array}{r} \frac{1}{3} \\ +\frac{1}{4} \\ \hline \end{array}$

3. $\begin{array}{r} 2\frac{1}{5} \\ +4\frac{3}{5} \\ \hline \end{array}$

4. $\begin{array}{r} \frac{1}{2} \\ -\frac{1}{3} \\ \hline \end{array}$

5. $\begin{array}{r} 9\frac{5}{7} \\ -8\frac{2}{7} \\ \hline \end{array}$

6. $\begin{array}{r} \frac{5}{8} \\ +\frac{1}{6} \\ \hline \end{array}$

7. $\begin{array}{r} \frac{7}{12} \\ +\frac{13}{24} \\ \hline \end{array}$

8. $\begin{array}{r} 8\frac{3}{5} \\ +6\frac{1}{3} \\ \hline \end{array}$

9. $\begin{array}{r} 5\frac{5}{7} \\ +4 \\ \hline \end{array}$

10. $\begin{array}{r} \frac{5}{6} \\ -\frac{1}{5} \\ \hline \end{array}$

11. $\begin{array}{r} 4\frac{3}{8} \\ -1\frac{1}{7} \\ \hline \end{array}$

12. $1\frac{1}{2} + 2\frac{1}{3} + 1\frac{5}{6}$

13. $12 - 1\frac{1}{8}$

14. $15\frac{9}{10} - 14\frac{1}{3}$

Use a related sentence to find the missing fraction or whole number.

15. $a + \frac{1}{4} = \frac{1}{2}$

16. $t + 1\frac{1}{2} = 3\frac{1}{2}$

17. $n - \frac{3}{8} = \frac{5}{16}$

Compute. Use the addition properties when possible.

18. $2\frac{4}{5} + (1\frac{1}{3} - 1\frac{1}{3})$

19. $(\frac{1}{4} + 5\frac{3}{8}) + 3\frac{3}{4}$

20 $2\frac{1}{2} + 4 + 1\frac{1}{4}$

PROBLEM SOLVING   *Use a strategy or strategies you have learned.*

21. Luciano bought two bags of apples weighing $5\frac{1}{4}$ lb and $8\frac{5}{8}$ lb. What was the total weight of the two bags of apples?

22. From a board 4 ft long, a piece $1\frac{2}{3}$ ft is cut off. What is the length of the board that is left?

Fractions: Multiplication and Division

7

We are just about to go home when Rebecca remembers the special birthday cupcakes her mom made.

There are 24 KIDS in the class.
Rebecca has 24 CUPCAKES.

X So what's the problem?

Rebecca wants Mrs. Fibonacci to have a cupcake, too.

Everyone is going crazy trying to figure out what fraction of a cupcake each person will get.

I'm the first to figure out the answer.

I raise my hand and tell Mrs. Fibonacci I'm allergic to cupcakes.

EVERYONE (24) believes me and gets ONE (1) cupcake. NO ONE (0) has to figure out fractions.

From *Math Curse* by
Jon Scieszka and Lane Smith

In this chapter you will:

Investigate properties of multiplication and reciprocals
Explore the meaning of division of fractions
Estimate and find products and quotients
Apply the order of operations to fractions
Work with fractions and money
Solve problems by using a diagram

Critical Thinking/Finding Together

In a group of 50 people, $\frac{3}{5}$ are male and $\frac{1}{5}$ of the males wear glasses. How many of the males wear glasses? females?

$\frac{1}{3}$

$\frac{1}{3}$

$\frac{1}{3}$

$\frac{1}{3}$

Multiplying Fractions by Fractions

Ms. Amazing is a great magician. She can take a red rope that is $\frac{3}{8}$ yd long and change $\frac{2}{3}$ of it into a blue rope. How long is the blue part of the rope?

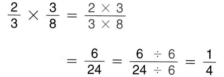

To find how long the blue part of the rope is, multiply: $\frac{2}{3} \times \frac{3}{8}$ yd = _?_

To multiply a fraction by a fraction:

- Multiply the numerators. Then multiply the denominators.

- Write the product in simplest form.

$$\frac{2}{3} \times \frac{3}{8} = \frac{2 \times 3}{3 \times 8}$$

$$= \frac{6}{24} = \frac{6 \div 6}{24 \div 6} = \frac{1}{4}$$

The blue part of the rope is $\frac{1}{4}$ yd long.

To multiply a fraction by a fraction, using cancellation:

- Divide any numerator and any denominator by their GCF. (This is called **cancellation**.)

- Cancel until the numerators and denominators have no common factor other than 1.

- Multiply the numerators. Multiply the denominators. The product will be in simplest form.

$2 \div 2 = 1 \qquad 3 \div 3 = 1$

$$\frac{2}{3} \times \frac{3}{8} = \frac{\overset{1}{\cancel{2}}}{\underset{1}{\cancel{3}}} \times \frac{\overset{1}{\cancel{3}}}{\underset{4}{\cancel{8}}}$$

$3 \div 3 = 1 \qquad 8 \div 2 = 4$

$$= \frac{1 \times 1}{1 \times 4}$$

$$= \frac{1}{4}$$

Study these examples.

$$\frac{4}{15} \times \frac{5}{6} = \frac{\overset{2}{\cancel{4}}}{\underset{3}{\cancel{15}}} \times \frac{\overset{1}{\cancel{5}}}{\underset{3}{\cancel{6}}} = \frac{2 \times 1}{3 \times 3} = \frac{2}{9}$$

$$\frac{7}{12} \times \frac{6}{7} = \frac{\overset{1}{\cancel{7}}}{\underset{2}{\cancel{12}}} \times \frac{\overset{1}{\cancel{6}}}{\underset{1}{\cancel{7}}} = \frac{1 \times 1}{2 \times 1} = \frac{1}{2}$$

Complete.

1. $\dfrac{5}{6} \times \dfrac{7}{9} = \dfrac{5 \times ?}{6 \times 9}$

$\phantom{\dfrac{5}{6} \times \dfrac{7}{9}} = \dfrac{?}{?}$

2. $\dfrac{9}{10} \times \dfrac{20}{21} = \dfrac{\overset{3}{\cancel{9}}}{\underset{1}{\cancel{10}}} \times \dfrac{\overset{?}{\cancel{20}}}{\underset{?}{\cancel{21}}}$

$\phantom{\dfrac{9}{10} \times \dfrac{20}{21}} = \dfrac{3 \times ?}{1 \times ?} = \dfrac{?}{?}$

Tell how you can cancel to simplify the multiplication. Then find the product.

3. $\dfrac{1}{2} \times \dfrac{2}{3}$　　**4.** $\dfrac{3}{5} \times \dfrac{1}{6}$　　**5.** $\dfrac{5}{12} \times \dfrac{3}{4}$　　**6.** $\dfrac{2}{9} \times \dfrac{6}{8}$　　**7.** $\dfrac{16}{20} \times \dfrac{10}{12}$

Multiply. Cancel whenever possible.

8. $\dfrac{3}{4} \times \dfrac{1}{5}$　　**9.** $\dfrac{5}{8} \times \dfrac{1}{3}$　　**10.** $\dfrac{3}{7} \times \dfrac{5}{6}$　　**11.** $\dfrac{7}{18} \times \dfrac{3}{5}$　　**12.** $\dfrac{8}{10} \times \dfrac{1}{2}$

13. $\dfrac{12}{20} \times \dfrac{5}{6}$　　**14.** $\dfrac{7}{8} \times \dfrac{3}{4}$　　**15.** $\dfrac{5}{9} \times \dfrac{2}{3}$　　**16.** $\dfrac{3}{4} \times \dfrac{2}{9}$　　**17.** $\dfrac{7}{12} \times \dfrac{1}{7}$

18. $\dfrac{9}{16} \times \dfrac{4}{5}$　　**19.** $\dfrac{4}{21} \times \dfrac{1}{8}$　　**20.** $\dfrac{14}{18} \times \dfrac{2}{3}$　　**21.** $\dfrac{24}{50} \times \dfrac{10}{12}$　　**22.** $\dfrac{1}{9} \times \dfrac{1}{10}$

23. $\dfrac{9}{10} \times \dfrac{1}{2}$　　　**24.** $\dfrac{5}{8} \times \dfrac{2}{3} \times \dfrac{7}{10}$　　　**25.** $\dfrac{3}{4} \times \dfrac{1}{6} \times \dfrac{2}{5}$

PROBLEM SOLVING

26. A recipe calls for $\dfrac{1}{2}$ cup of butter. If Harold wants to make $\dfrac{1}{2}$ of the recipe, how much butter should he use?

27. Ebony lifts weights $\dfrac{3}{4}$ hour a day. Adam lifts weights $\dfrac{1}{2}$ as long. How long does Adam lift weights?

28. In water, sound travels approximately $\dfrac{9}{10}$ of a mile per second. How far does sound travel in $\dfrac{1}{3}$ second?

29. A can holds $\dfrac{7}{8}$ qt of water. How much water is in the can when it is $\dfrac{2}{3}$ full?

30. Of the students in the sophomore class, $\dfrac{2}{5}$ have cameras; $\dfrac{1}{4}$ of the students with cameras join the photography club. What fraction of the students in the sophomore class joins the photography club? What fraction does not?

7-2 Multiplying Fractions and Whole Numbers

Nancy uses $\frac{2}{3}$ ft of silk to make a bow for a package. How much silk is needed for 4 bows?

$$0 \qquad \frac{1}{3} \qquad \frac{2}{3} \qquad \frac{3}{3} \qquad \frac{4}{3} \qquad \frac{5}{3} \qquad \frac{6}{3} \qquad \frac{7}{3} \qquad \frac{8}{3}$$

To find the amount of silk needed,
multiply: $4 \times \frac{2}{3}$ ft = $\underline{\ ?\ }$

To multiply a fraction and a whole number:

- Rename the whole number as an improper fraction with a denominator of 1.

$$4 \times \frac{2}{3} = \frac{4}{1} \times \frac{2}{3}$$

- Multiply the numerators. Then multiply the denominators.

$$= \frac{4 \times 2}{1 \times 3}$$

- Write the product in simplest form.

$$= \frac{8}{3} = 2\frac{2}{3}$$

Nancy needs $2\frac{2}{3}$ ft of silk.

Study these examples.

Multiply: $3 \times \frac{1}{4} = \underline{\ ?\ }$

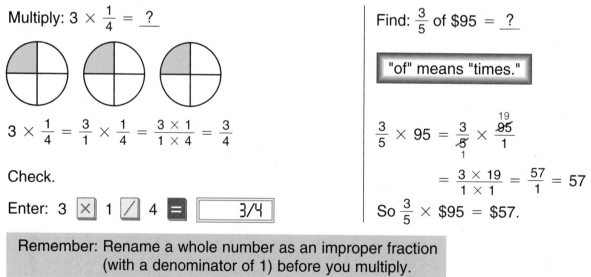

$$3 \times \frac{1}{4} = \frac{3}{1} \times \frac{1}{4} = \frac{3 \times 1}{1 \times 4} = \frac{3}{4}$$

Check.

Enter: 3 $\boxed{\times}$ 1 $\boxed{/}$ 4 $\boxed{=}$ $\boxed{3/4}$

Find: $\frac{3}{5}$ of \$95 = $\underline{\ ?\ }$

"of" means "times."

$$\frac{3}{5} \times 95 = \frac{3}{\overset{}{\underset{1}{\cancel{5}}}} \times \frac{\overset{19}{\cancel{95}}}{1}$$

$$= \frac{3 \times 19}{1 \times 1} = \frac{57}{1} = 57$$

So $\frac{3}{5} \times \$95 = \$57$.

Remember: Rename a whole number as an improper fraction (with a denominator of 1) before you multiply.

Multiply.

1. $4 \times \frac{3}{4}$
2. $10 \times \frac{3}{5}$
3. $18 \times \frac{1}{3}$
4. $24 \times \frac{1}{12}$

5. $25 \times \frac{4}{5}$
6. $20 \times \frac{3}{10}$
7. $9 \times \frac{3}{4}$
8. $27 \times \frac{1}{2}$

9. $\frac{5}{6} \times 18$
10. $\frac{7}{9} \times 45$
11. $\frac{4}{5} \times 12$
12. $\frac{3}{7} \times 9$

Find the product.

13. $\frac{1}{2}$ of 12
14. $\frac{1}{8}$ of 40
15. $\frac{2}{3}$ of 9
16. $\frac{3}{4}$ of 44

17. $\frac{1}{9}$ of 3
18. $\frac{3}{8}$ of 4
19. $\frac{5}{6}$ of 20
20. $\frac{3}{8}$ of 18

21. $\frac{2}{5}$ of \$20
22. $\frac{5}{6}$ of \$36
23. $\frac{1}{2}$ of \$7.50
24. $\frac{2}{3}$ of \$2.70

PROBLEM SOLVING

25. David is making 12 flags for the parade. Each flag requires $\frac{2}{3}$ yd of material. How many yards of material are needed?

26. Rori budgets $\frac{3}{10}$ of her \$540 weekly income for rent. How much money is budgeted for rent each week?

27. Sociologists have determined that $\frac{2}{5}$ of the people in the world are shy. A personnel manager is interviewing 150 people. How many of these people might be shy?

28. A mathematics exam contains 75 questions. Amos answers $\frac{4}{5}$ of the questions correctly. How many questions does he answer correctly?

29. Of the 24 players on the football team, $\frac{1}{4}$ are first-year players and $\frac{1}{3}$ are second-year players. How many of the players are in their first or second year?

 **SHARE YOUR THINKING**

Communicate

30. Using a calculator, which of these procedures can you use to find $\frac{7}{8}$ of 44: multiply 7 by 44, then divide by 8 or divide 7 by 8, then multiply by 44? Share your conclusion with your teacher.

31. Find these products on a calculator: $\frac{1}{8}$ of 12; $75 \times \frac{3}{10}$; $\frac{5}{6} \times \$1000$.

7-3 Properties and the Reciprocal

Algebra ✓

Multiplication properties may help you multiply fractions.

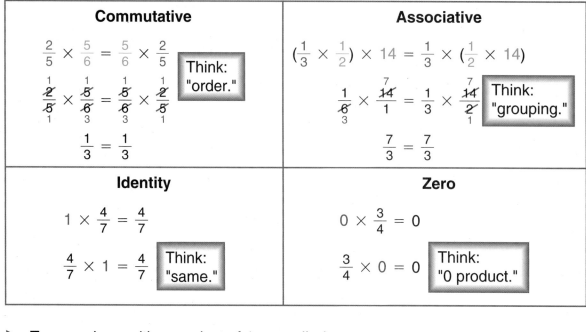

Commutative	Associative
$\frac{2}{5} \times \frac{5}{6} = \frac{5}{6} \times \frac{2}{5}$ Think: "order."	$(\frac{1}{3} \times \frac{1}{2}) \times 14 = \frac{1}{3} \times (\frac{1}{2} \times 14)$
$\frac{1}{\cancel{2}} \times \frac{1}{\cancel{5}} = \frac{\cancel{5}}{\cancel{6}} \times \frac{\cancel{2}}{\cancel{5}}$	$\frac{1}{\cancel{6}} \times \frac{\cancel{14}}{1} = \frac{1}{3} \times \frac{\cancel{14}}{\cancel{2}}$ Think: "grouping."
$\frac{1}{3} = \frac{1}{3}$	$\frac{7}{3} = \frac{7}{3}$

Identity	Zero
$1 \times \frac{4}{7} = \frac{4}{7}$	$0 \times \frac{3}{4} = 0$
$\frac{4}{7} \times 1 = \frac{4}{7}$ Think: "same."	$\frac{3}{4} \times 0 = 0$ Think: "0 product."

▶ Two numbers with a product of 1 are called **reciprocals** of each other.

$\frac{3}{4} \times \frac{4}{3} = \frac{\overset{1}{\cancel{3}}}{\cancel{4}} \times \frac{\overset{1}{\cancel{4}}}{\cancel{3}} = \frac{1}{1} = 1$ ◄ $\frac{3}{4}$ and $\frac{4}{3}$ are reciprocals.

$6 \times \frac{1}{6} = \frac{\overset{1}{\cancel{6}}}{1} \times \frac{1}{\cancel{6}} = \frac{1}{1} = 1$ ◄ 6 and $\frac{1}{6}$ are reciprocals.

To find the reciprocal of a number:

- Write the number as a fraction. $10 = \frac{10}{1}$

- *Invert* the fraction by exchanging the position of the numerator and the denominator. $\frac{10}{1} \times \frac{1}{10}$

- Check if the product of the numbers is 1. $\frac{10}{1} \times \frac{1}{10} = \frac{\overset{1}{\cancel{10}}}{1} \times \frac{1}{\cancel{10}} = \frac{1}{1} = 1$

10 and $\frac{1}{10}$ are reciprocals. Their product is 1.

Complete. Use the properties of multiplication.

1. $\frac{1}{5} \times \frac{3}{4} = \frac{3}{4} \times \underline{\ ?\ }$

2. $\frac{3}{8} \times \underline{\ ?\ } = \frac{3}{8}$

3. $\underline{\ ?\ } \times \frac{5}{6} = \frac{5}{6}$

4. $\frac{1}{2} \times \underline{\ ?\ } = 0$

5. $\frac{1}{3} \times 0 = \underline{\ ?\ } \times \frac{1}{3}$

6. $\frac{7}{10} \times \underline{\ ?\ } = \frac{2}{3} \times \frac{7}{10}$

7. $\frac{1}{4} \times (\frac{1}{5} \times \frac{1}{6}) - (\frac{1}{4} \times \underline{\ ?\ }) \times \frac{1}{6}$

8. $\underline{\ ?\ } \times (14 \times \frac{1}{3}) = (\frac{1}{2} \times 14) \times \frac{1}{3}$

Find the missing reciprocal in each number sentence.

9. $4 \times \underline{\ ?\ } = 1$

10. $7 \times \underline{\ ?\ } = 1$

11. $\frac{1}{8} \times \underline{\ ?\ } = 1$

12. $\frac{1}{12} \times \underline{\ ?\ } = 1$

13. $24 \times \underline{\ ?\ } = 1$

14. $100 \times \underline{\ ?\ } = 1$

15. $\frac{4}{5} \times \underline{\ ?\ } = 1$

16. $\frac{8}{9} \times \underline{\ ?\ } = 1$

Write the reciprocal of each number.

17. 9

18. 15

19. $\frac{1}{7}$

20. $\frac{11}{12}$

21. $\frac{7}{4}$

22. $\frac{12}{5}$

Explain how using properties or reciprocals can make these computations easier. Then compute.

23. $\frac{3}{5} \times 14 \times \frac{1}{2}$

24. $\frac{3}{4} \times \frac{7}{12} \times 0$

25. $(\frac{1}{4} \times \frac{7}{8}) \times 16$

26. $(46 \times \frac{1}{9}) \times 9$

27. $\frac{7}{8} \times (\frac{8}{7} \times 33)$

28. $\frac{4}{5} \times (9 \times \frac{5}{4})$

29. $\frac{5}{7} \times \frac{5}{8} \times \frac{14}{25}$

30. $14 \times \frac{14}{15} \times 15$

31. $\frac{3}{7} \times 9 \times 21$

SKILLS TO REMEMBER

Write each whole number or mixed number as an improper fraction.

32. 5

33. 12

34. $1\frac{1}{2}$

35. $3\frac{1}{4}$

36. $5\frac{5}{6}$

37. $8\frac{7}{8}$

7-4 Multiplying Mixed Numbers

The weight of water is $62\frac{1}{2}$ lb per cubic foot. What is the weight of $2\frac{4}{5}$ cubic feet of water?

To find the weight of $2\frac{4}{5}$ cubic feet of water, multiply: $2\frac{4}{5} \times 62\frac{1}{2}$ lb $= \underline{\ ?\ }$

Estimate: $2\frac{4}{5} \times 62\frac{1}{2} \longrightarrow 3 \times 63 = 189$
About 189 lb

To multiply with mixed numbers:

- Rename both factors as improper fractions.

$$2\frac{4}{5} \times 62\frac{1}{2} = \frac{14}{5} \times \frac{125}{2}$$

- Cancel where possible.

$$= \frac{\overset{7}{\cancel{14}}}{\underset{1}{\cancel{5}}} \times \frac{\overset{25}{\cancel{125}}}{\underset{1}{\cancel{2}}}$$

- Multiply the numerators. Then multiply the denominators.

$$= \frac{7 \times 25}{1 \times 1}$$

- If the product is an improper fraction, rename it as a whole or mixed number.

$$= \frac{175}{1} = 175$$

The weight of $2\frac{4}{5}$ cubic feet of water is 175 lb.

175 lb is close to the estimate of 189 lb.

Study these examples.

$$\frac{2}{3} \times 3\frac{3}{8} = \frac{2}{3} \times \frac{27}{8} \qquad \text{Estimate: } 1 \times 3 = 3$$

$$= \frac{\overset{1}{\cancel{2}}}{\underset{1}{\cancel{3}}} \times \frac{\overset{9}{\cancel{27}}}{\underset{4}{\cancel{8}}} = \frac{9}{4} = 2\frac{1}{4}$$

$$9 \times 5\frac{2}{3} = \frac{9}{1} \times \frac{17}{3} \qquad \text{Estimate: } 9 \times 6 = 54$$

$$= \frac{\overset{3}{\cancel{9}}}{1} \times \frac{17}{\underset{1}{\cancel{3}}} = \frac{51}{1} = 51$$

Estimate each product.

1. $7\frac{2}{3} \times 4\frac{1}{4}$

2. $9\frac{1}{5} \times 3\frac{7}{12}$

3. $8 \times 9\frac{7}{8}$

4. $7\frac{1}{2} \times 10$

5. $\frac{5}{6} \times 32\frac{4}{7}$

6. $75\frac{1}{10} \times \frac{3}{5}$

7. $12\frac{1}{4} \times 5\frac{1}{4}$

8. $15\frac{7}{8} \times 1\frac{5}{9}$

Multiply. Estimate to help you.

9. $2\frac{2}{3} \times 1\frac{1}{2}$ **10.** $2\frac{1}{4} \times 1\frac{1}{3}$ **11.** $1\frac{2}{5} \times 3\frac{3}{4}$ **12.** $3\frac{3}{4} \times 1\frac{1}{5}$

13. $1\frac{1}{2} \times 6$ **14.** $3\frac{1}{3} \times 6$ **15.** $10 \times 1\frac{2}{5}$ **16.** $24 \times 1\frac{1}{6}$

17. $\frac{2}{3} \times 2\frac{1}{2}$ **18.** $\frac{3}{4} \times 2\frac{2}{3}$ **19.** $3\frac{1}{7} \times 4\frac{2}{3}$ **20.** $2\frac{2}{5} \times 3\frac{1}{6}$

21. $2\frac{1}{10} \times \frac{6}{7}$ **22.** $5\frac{5}{8} \times \frac{5}{9}$ **23.** $1\frac{1}{6} \times 9$ **24.** $3\frac{1}{8} \times 12$

25. $8\frac{1}{6} \times 3\frac{3}{7}$ **26.** $3\frac{1}{9} \times 2\frac{1}{7}$ **27.** $\frac{3}{4}$ of $2\frac{2}{3}$ **28.** $\frac{5}{9}$ of $2\frac{1}{4}$

29. $6 \times 5\frac{3}{5} \times 1\frac{2}{3}$ **30.** $6\frac{2}{3} \times 7 \times 1\frac{1}{5}$ **31.** $2\frac{1}{6} \times 5\frac{1}{3} \times 1\frac{7}{8}$

Compare. Write <, =, or >.

32. $2\frac{1}{2} \times 3\frac{1}{4} \underline{\quad?\quad} 2\frac{1}{4} \times 3\frac{1}{2}$ **33.** $1\frac{2}{3} \times 3\frac{1}{4} \underline{\quad?\quad} 3\frac{1}{4} \times 1\frac{2}{3}$

34. $3\frac{3}{5} \times 1\frac{1}{2} \underline{\quad?\quad} 2\frac{1}{2} \times 1\frac{3}{4}$ **35.** $6\frac{1}{4} \times 2\frac{1}{4} \underline{\quad?\quad} 3\frac{1}{2} \times 4\frac{1}{8}$

Find the value of _n_. Use the properties of multiplication.   *Algebra* ✓

36. $n \times 1 = 3\frac{1}{2}$ **37.** $n \times 4\frac{1}{5} = 4\frac{1}{5} \times 5$ **38.** $1\frac{1}{3} \times n = 0$

39. $(n \times \frac{1}{2}) \times 4 = \frac{1}{3} \times (\frac{1}{2} \times 4)$ **40.** $5\frac{1}{2} \times 0 = n \times 5\frac{1}{2}$

PROBLEM SOLVING

41. One serving of meat is about $3\frac{1}{2}$ oz. A person needs 2 servings a day for proper nutrition. How many ounces of meat is this?

42. Round steak contains $3\frac{1}{2}$ servings per pound. How many servings are there in 10 lb of round steak?

43. The weight of water is $62\frac{1}{2}$ lb per cubic foot. What is the weight of $5\frac{1}{3}$ cubic feet of water?

44. A long-playing record makes $33\frac{1}{3}$ revolutions per minute. If it plays for 42 min, how many revolutions does it make?

Each loop in a spring takes $\frac{3}{8}$ in. of wire.
How many loops can be made from 3 in. of wire?

To find how many loops can be made,
divide: $3 \div \frac{3}{8} = \underline{\ ?\ }$

Think: How many $\frac{3}{8}$s are in 3?

You can use a **diagram** to help you divide 3 by $\frac{3}{8}$.

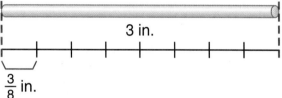

3 in.

$\frac{3}{8}$ in.

Count the number of $\frac{3}{8}$ in. units. There are 8 units. So $3 \div \frac{3}{8} = 8$.

Eight loops can be made from 3 in. of wire.

Study these examples.

Write a division sentence for each diagram.

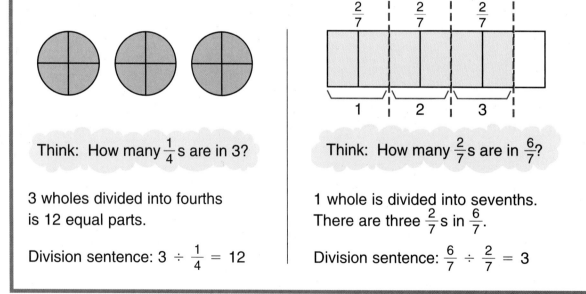

Think: How many $\frac{1}{4}$s are in 3?

3 wholes divided into fourths
is 12 equal parts.

Division sentence: $3 \div \frac{1}{4} = 12$

Think: How many $\frac{2}{7}$s are in $\frac{6}{7}$?

1 whole is divided into sevenths.
There are three $\frac{2}{7}$s in $\frac{6}{7}$.

Division sentence: $\frac{6}{7} \div \frac{2}{7} = 3$

Use the diagram to find each.

1.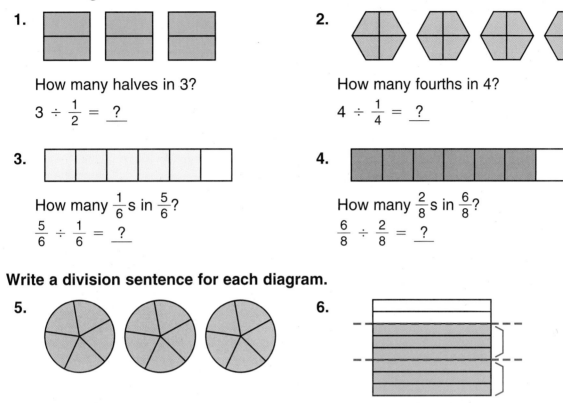

How many halves in 3?

$3 \div \frac{1}{2} = \underline{?}$

2.

How many fourths in 4?

$4 \div \frac{1}{4} = \underline{?}$

3.

How many $\frac{1}{6}$s in $\frac{5}{6}$?

$\frac{5}{6} \div \frac{1}{6} = \underline{?}$

4.

How many $\frac{2}{8}$s in $\frac{6}{8}$?

$\frac{6}{8} \div \frac{2}{8} = \underline{?}$

Write a division sentence for each diagram.

5.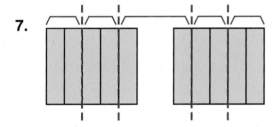

6.

7.

8.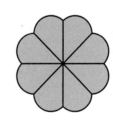

Draw a diagram to show each division sentence.

9. $4 \div \frac{1}{2}$ **10.** $3 \div \frac{1}{3}$ **11.** $\frac{4}{5} \div \frac{1}{5}$ **12.** $\frac{8}{10} \div \frac{2}{10}$

13. $2 \div \frac{2}{6}$ **14.** $3 \div \frac{3}{7}$ **15.** $1 \div \frac{1}{6}$ **16.** $1 \div \frac{3}{6}$

SHARE YOUR THINKING

17. Draw diagrams to help you divide by the unit fractions $\frac{1}{2}, \frac{1}{3}, \frac{1}{5}$ in these exercises: (a) $2 \div \frac{1}{2}$; (b) $1 \div \frac{1}{3}$; (c) $3 \div \frac{1}{5}$.

18. Write a rule in your Math Journal that tells what happens when you divide a whole number by a unit fraction.

Math
Journal

Dividing Fractions by Fractions

Eduardo discovered a short way to divide fractions by examining some division sentences and the related multiplication sentences.

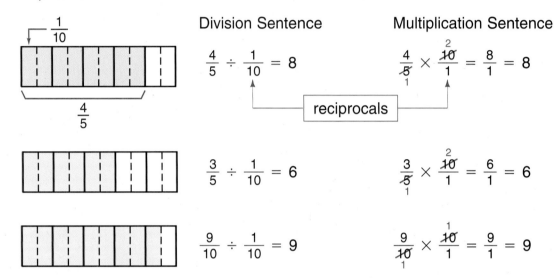

	Division Sentence	Multiplication Sentence
	$\dfrac{4}{5} \div \dfrac{1}{10} = 8$	$\dfrac{4}{\cancel{5}} \times \dfrac{\cancel{10}^{2}}{1} = \dfrac{8}{1} = 8$
	$\dfrac{3}{5} \div \dfrac{1}{10} = 6$	$\dfrac{3}{\cancel{5}} \times \dfrac{\cancel{10}^{2}}{1} = \dfrac{6}{1} = 6$
	$\dfrac{9}{10} \div \dfrac{1}{10} = 9$	$\dfrac{9}{\cancel{10}} \times \dfrac{\cancel{10}^{1}}{1} = \dfrac{9}{1} = 9$

reciprocals

Eduardo concluded that **dividing by a fraction** gives the same result as **multiplying by its reciprocal**.

▶ **To divide a fraction by a fraction:**

- Multiply by the reciprocal of the divisor.

- Cancel where possible.

- If the product is an improper fraction, rename it as a whole or mixed number.

reciprocals

$$\dfrac{3}{4} \div \dfrac{1}{16} = \dfrac{3}{4} \times \dfrac{16}{1}$$

$$= \dfrac{3}{\cancel{4}_{1}} \times \dfrac{\cancel{16}^{4}}{1}$$

$$= \dfrac{12}{1} = 12$$

Study these examples.

$$\dfrac{7}{20} \div \dfrac{1}{5} = \dfrac{7}{20} \times \dfrac{5}{1}$$

Reciprocal of $\dfrac{1}{5}$ is 5 or $\dfrac{5}{1}$.

$$= \dfrac{7}{\underset{4}{\cancel{20}}} \times \dfrac{\cancel{5}^{1}}{1} = \dfrac{7}{4} = 1\dfrac{3}{4}$$

$$\dfrac{11}{14} \div \dfrac{11}{12} = \dfrac{11}{14} \times \dfrac{12}{11}$$

$$= \dfrac{\cancel{11}^{1}}{\underset{7}{\cancel{14}}} \times \dfrac{\cancel{12}^{6}}{\cancel{11}_{1}}$$

$$= \dfrac{6}{7}$$

Complete.

1. $\dfrac{9}{13} \div \dfrac{3}{5} = \dfrac{9}{13} \times \dfrac{5}{3} = \underline{?}$

2. $\dfrac{12}{25} \div \dfrac{3}{10} = \dfrac{12}{25} \times \dfrac{10}{3} = \underline{?}$

3. $\dfrac{3}{7} \div \dfrac{1}{14} = \dfrac{3}{7} \times \dfrac{?}{?} = \underline{?}$

4. $\dfrac{1}{8} \div \dfrac{1}{16} = \dfrac{?}{?} \times \dfrac{16}{1} = \underline{?}$

Divide. Draw a diagram to help you.

5. $\dfrac{1}{2} \div \dfrac{1}{4}$

6. $\dfrac{2}{5} \div \dfrac{1}{10}$

7. $\dfrac{1}{4} \div \dfrac{1}{16}$

8. $\dfrac{1}{2} \div \dfrac{1}{10}$

9. $\dfrac{7}{8} \div \dfrac{1}{8}$

10. $\dfrac{5}{6} \div \dfrac{1}{6}$

11. $\dfrac{6}{8} \div \dfrac{3}{8}$

12. $\dfrac{6}{16} \div \dfrac{2}{16}$

Divide.

13. $\dfrac{3}{5} \div \dfrac{3}{4}$

14. $\dfrac{2}{3} \div \dfrac{3}{4}$

15. $\dfrac{6}{7} \div \dfrac{3}{5}$

16. $\dfrac{1}{4} \div \dfrac{1}{5}$

17. $\dfrac{5}{8} \div \dfrac{5}{8}$

18. $\dfrac{2}{5} \div \dfrac{2}{5}$

19. $\dfrac{5}{24} \div \dfrac{5}{12}$

20. $\dfrac{6}{13} \div \dfrac{3}{26}$

21. $\dfrac{2}{9} \div \dfrac{1}{3}$

22. $\dfrac{1}{8} \div \dfrac{1}{5}$

23. $\dfrac{16}{25} \div \dfrac{3}{5}$

24. $\dfrac{9}{28} \div \dfrac{3}{7}$

25. $\dfrac{14}{15} \div \dfrac{8}{9}$

26. $\dfrac{9}{10} \div \dfrac{6}{7}$

27. $\dfrac{1}{6} \div \dfrac{1}{11}$

28. $\dfrac{1}{11} \div \dfrac{1}{6}$

PROBLEM SOLVING

29. How many $\dfrac{1}{16}$s are there in $\dfrac{3}{8}$?

30. How many $\dfrac{1}{100}$s are there in $\dfrac{1}{10}$?

31. Eric cut a $\dfrac{1}{2}$ ft piece of wood into $\dfrac{1}{8}$ ft strips. How many strips did he make?

32. Karen cut a piece of leather $\dfrac{3}{4}$ yd long into $\dfrac{1}{16}$ yd strips. How many strips did she make?

33. Ms. Appell bought $\dfrac{1}{3}$ bushel of apples. She used $\dfrac{3}{4}$ of the apples to make applesauce. What part of a bushel did she use for applesauce?

34. Which operation did you use to solve problem 33? Explain why you chose this operation.

Communicate ✓

237

Estimation in Division

You can use estimation to determine if the quotient
of two fractions is **less than 1** or **greater than 1**.

Study these division examples:

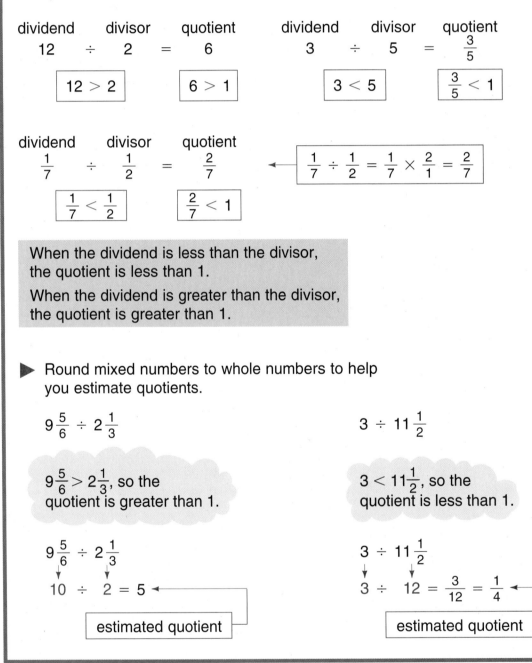

| dividend | | divisor | | quotient |
| 12 | ÷ | 2 | = | 6 |

| 12 > 2 | | 6 > 1 |

| dividend | | divisor | | quotient |
| 3 | ÷ | 5 | = | $\frac{3}{5}$ |

| 3 < 5 | | $\frac{3}{5}$ < 1 |

| dividend | | divisor | | quotient |
| $\frac{1}{7}$ | ÷ | $\frac{1}{2}$ | = | $\frac{2}{7}$ |

$$\frac{1}{7} \div \frac{1}{2} = \frac{1}{7} \times \frac{2}{1} = \frac{2}{7}$$

| $\frac{1}{7} < \frac{1}{2}$ | | $\frac{2}{7} < 1$ |

When the dividend is less than the divisor,
the quotient is less than 1.

When the dividend is greater than the divisor,
the quotient is greater than 1.

▶ Round mixed numbers to whole numbers to help
you estimate quotients.

$$9\frac{5}{6} \div 2\frac{1}{3}$$

$9\frac{5}{6} > 2\frac{1}{3}$, so the
quotient is greater than 1.

$$9\frac{5}{6} \div 2\frac{1}{3}$$
$$10 \div 2 = 5$$

estimated quotient

$$3 \div 11\frac{1}{2}$$

$3 < 11\frac{1}{2}$, so the
quotient is less than 1.

$$3 \div 11\frac{1}{2}$$
$$3 \div 12 = \frac{3}{12} = \frac{1}{4}$$

estimated quotient

238

Tell whether the quotient is *less than 1* or *greater than 1*.

1. $\frac{6}{7} \div \frac{3}{7}$

2. $\frac{2}{5} \div \frac{4}{5}$

3. $\frac{1}{3} \div \frac{1}{10}$

4. $\frac{1}{15} \div \frac{1}{12}$

5. $\frac{3}{7} \div \frac{3}{11}$

6. $\frac{4}{5} \div \frac{4}{9}$

7. $\frac{2}{3} \div \frac{3}{4}$

8. $\frac{7}{8} \div \frac{5}{6}$

9. $\frac{5}{9} \div \frac{7}{18}$

10. $\frac{17}{36} \div \frac{5}{12}$

11. $\frac{11}{12} \div \frac{3}{7}$

12. $\frac{4}{9} \div \frac{9}{10}$

13. $2 \div \frac{3}{7}$

14. $4\frac{1}{10} \div \frac{9}{10}$

15. $1\frac{1}{2} \div 1\frac{1}{3}$

16. $3\frac{5}{6} \div 1\frac{14}{15}$

Estimate. Round each mixed number to the nearest whole number.

17. $8\frac{1}{3} \div 1\frac{5}{6}$

18. $9\frac{3}{4} \div 4\frac{3}{4}$

19. $11 \div 1\frac{7}{8}$

20. $17 \div 2\frac{1}{4}$

21. $6\frac{1}{8} \div 9\frac{2}{5}$

22. $3\frac{8}{9} \div 12\frac{1}{9}$

23. $11\frac{1}{2} \div \frac{11}{12}$

24. $\frac{7}{8} \div 9\frac{1}{2}$

Compare. Write $<$ or $>$. Use estimation to help you.

25. $3 \div \frac{1}{12}$ __?__ 1

26. $\frac{1}{2} \div \frac{3}{7}$ __?__ 1

27. $2 \div \frac{1}{10}$ __?__ $\frac{1}{10} \div 2$

28. $9 \div 3$ __?__ $9 \div 3\frac{1}{3}$

29. $\frac{7}{8} \div 1\frac{1}{2}$ __?__ $1\frac{1}{2} \div \frac{7}{8}$

30. $10\frac{5}{6} \div 2\frac{3}{8}$ __?__ $9\frac{5}{6} \div 1\frac{3}{8}$

PROBLEM SOLVING

31. If $2\frac{7}{9}$ yd of material is cut into 5 pieces of the same length, about how long is each piece?

32. A piece of wire $\frac{3}{5}$ yd long is to be cut into 6 pieces of the same length. About how long is each piece?

33. About how much will each person get if 10 people share $12\frac{1}{2}$ qt of strawberries equally?

 **CRITICAL THINKING**

34. When the dividend stays the same and the divisor increases, what happens to the quotient?

35. Explain your answer to exercise 34 in your Math Journal. Is your conclusion true for whole numbers, decimals, and fractions?

Math Journal

Dividing Whole Numbers

At a party, $\frac{3}{4}$ gal of yogurt is equally divided among 6 people. How much is each person's serving?

To find the amount of each serving, divide: $\frac{3}{4}$ gal ÷ 6 = ___?___

Estimate: $\frac{3}{4} \div 6 \longrightarrow 1 \div 6 = \frac{1}{6}$

About $\frac{1}{6}$ gal

To divide with whole numbers and fractions:

- Rename the whole number as an improper fraction with a denominator of 1.

$$\frac{3}{4} \div 6 = \frac{3}{4} \div \frac{6}{1}$$

Think: Divide $\frac{3}{4}$ into 6 equal parts.

- Multiply by the reciprocal of the divisor. Cancel where possible.

$$= \frac{\overset{1}{\cancel{3}}}{4} \times \frac{1}{\underset{2}{\cancel{6}}}$$

- Multiply the numerators. Then multiply the denominators.

$$= \frac{1 \times 1}{4 \times 2}$$

- Write the quotient in simplest form.

$$= \frac{1}{8}$$

Each person's serving is $\frac{1}{8}$ gal.

Study these examples.

Divide: $8 \div \frac{5}{6} =$ ___?___

$$8 \div \frac{5}{6} = \frac{8}{1} \div \frac{5}{6}$$

$$= \frac{8}{1} \times \frac{6}{5}$$

$$= \frac{48}{5} = 9\frac{3}{5} \leftarrow \boxed{\text{simplest form}}$$

Check.

Enter: 8 ÷ 5 / 6 = 48/5

Divide: $\frac{1}{2} \div 10 =$ ___?___

$$\frac{1}{2} \div 10 = \frac{1}{2} \div \frac{10}{1}$$

$$= \frac{1}{2} \times \frac{1}{10}$$

$$= \frac{1}{20} \leftarrow \boxed{\text{simplest form}}$$

Check.

Enter: 1 / 2 ÷ 10 = 1/20

Complete. Use a fraction calculator to check your answers.

1. $6 \div \frac{1}{4} = \frac{6}{1} \div \frac{?}{?}$

   $= \frac{6}{1} \times \frac{?}{?} = \underline{?}$

2. $\frac{1}{3} \div 8 = \frac{1}{3} \div \frac{?}{?}$

   $= \frac{1}{3} \times \frac{?}{?} = \underline{?}$

Divide. Estimate to help you.

3. $3 \div \frac{1}{3}$

4. $4 \div \frac{1}{5}$

5. $2 \div \frac{3}{5}$

6. $4 \div \frac{5}{6}$

7. $\frac{1}{4} \div 6$

8. $\frac{1}{3} \div 5$

9. $\frac{3}{4} \div 9$

10. $\frac{5}{7} \div 10$

11. $9 \div \frac{3}{7}$

12. $36 \div \frac{6}{7}$

13. $27 \div \frac{3}{5}$

14. $8 \div \frac{8}{9}$

15. $\frac{5}{6} \div 10$

16. $\frac{4}{9} \div 8$

17. $\frac{1}{8} \div 8$

18. $\frac{1}{10} \div 10$

19. $100 \div \frac{20}{21}$

20. $40 \div \frac{8}{17}$

21. $\frac{11}{12} \div 22$

22. $\frac{7}{15} \div 42$

Compare. Write <, =, or >.

23. $16 \div \frac{8}{9} \underline{\ ?\ } 10 \div \frac{1}{2}$

24. $\frac{1}{3} \div 10 \underline{\ ?\ } 10 \div \frac{1}{3}$

25. $8 \div \frac{3}{4} \underline{\ ?\ } 6 \div \frac{3}{4}$

26. $\frac{1}{2} \div 14 \underline{\ ?\ } \frac{1}{4} \div 7$

27. $9 \times \frac{2}{7} \underline{\ ?\ } 9 \div \frac{2}{7}$

28. $\frac{11}{13} \div 9 \underline{\ ?\ } \frac{11}{13} \times \frac{1}{9}$

PROBLEM SOLVING

29. How many $\frac{3}{8}$-ft boards can be cut from a 6-ft board?

30. What is the quotient when $\frac{3}{7}$ is divided by 9?

31. A child's shirt requires $\frac{5}{6}$ yd of fabric. How many shirts can be made from 25 yd of fabric?

32. How many $\frac{2}{3}$-cup sugar bowls can be filled from 10 cups of sugar?

33. A $\frac{1}{2}$-ton weight is to be lifted by 5 people. How many pounds must each person lift? (*Hint:* 1 ton = 2000 pounds.)

34. After driving 240 mi, $\frac{3}{5}$ of a trip was completed. How long was the total trip? How many miles were left to drive?

241

7-9 Dividing a Mixed Number

A theatre owner has a $1\frac{5}{6}$-hour movie as her feature presentation. How many times each day can the movie be shown if the projection time available each day is $7\frac{1}{3}$ hours?

To find how many times the movie can be shown, divide: $7\frac{1}{3} \div 1\frac{5}{6} = \underline{\ ?\ }$

Think: How many $1\frac{5}{6}$s are in $7\frac{1}{3}$?

Estimate: $7\frac{1}{3} \div 1\frac{5}{6}$ $7 \div 2 = 3\frac{1}{2}$

To divide with mixed numbers:

- Rename both mixed numbers as improper fractions.

- Multiply by the reciprocal of the divisor. Cancel where possible.

- Multiply the numerators. Then multiply the denominators.

- Write the quotient in simplest form.

$$7\frac{1}{3} \div 1\frac{5}{6} = \frac{22}{3} \div \frac{11}{6}$$

$$= \frac{\overset{2}{\cancel{22}}}{\underset{1}{\cancel{3}}} \times \frac{\overset{2}{\cancel{6}}}{\underset{1}{\cancel{11}}}$$

$$= \frac{2 \times 2}{1 \times 1}$$

$$= \frac{4}{1} = 4$$

The movie can be shown 4 times.

Study these examples.

Divide: $1\frac{1}{2} \div \frac{1}{6} = \underline{\ ?\ }$

$$1\frac{1}{2} \div \frac{1}{6} = \frac{3}{2} \div \frac{1}{6}$$

$$= \frac{3}{\underset{1}{\cancel{2}}} \times \frac{\overset{3}{\cancel{6}}}{1}$$

$$= \frac{9}{1} = 9$$

Divide: $6 \div 4\frac{2}{3} = \underline{\ ?\ }$

> Rename both numbers as improper fractions.

$$6 \div 4\frac{2}{3} = \frac{6}{1} \div \frac{14}{3}$$

$$= \frac{\overset{3}{\cancel{6}}}{1} \times \frac{3}{\underset{7}{\cancel{14}}}$$

$$= \frac{9}{7} = 1\frac{2}{7}$$

Complete.

1. $10\frac{1}{2} \div 1\frac{1}{2} = \frac{21}{2} \div \frac{?}{?}$

$\quad\quad = \frac{21}{2} \times \frac{?}{?} = \underline{\ ?\ }$

2. $4\frac{2}{3} \div 6 = \frac{?}{3} \div \frac{?}{1}$

$\quad\quad = \frac{?}{3} \times \frac{1}{?} = \underline{\ ?\ }$

Divide. Estimate to help you.

3. $4\frac{2}{3} \div 2\frac{1}{3}$

4. $4\frac{4}{5} \div 1\frac{1}{5}$

5. $3\frac{1}{4} \div 1\frac{1}{2}$

6. $5\frac{5}{6} \div 2\frac{2}{3}$

7. $\frac{5}{8} \div 1\frac{1}{4}$

8. $\frac{2}{3} \div 1\frac{3}{4}$

9. $5\frac{3}{4} \div \frac{2}{3}$

10. $2\frac{5}{6} \div \frac{3}{4}$

11. $1\frac{2}{3} \div 6$

12. $1\frac{4}{5} \div 3$

13. $6 \div 2\frac{1}{4}$

14. $12 \div 1\frac{1}{7}$

15. $2\frac{1}{3} \div \frac{1}{6}$

16. $3\frac{1}{2} \div \frac{7}{10}$

17. $5\frac{1}{4} \div \frac{11}{16}$

18. $11\frac{2}{3} \div \frac{7}{8}$

19. $\frac{9}{10} \div 9$

20. $\frac{5}{6} \div 10$

21. $\frac{3}{8} \div \frac{9}{14}$

22. $\frac{4}{7} \div \frac{20}{21}$

23. $5\frac{2}{3} \div 1\frac{1}{3}$

24. $5\frac{1}{7} \div 2\frac{1}{7}$

25. $32 \div 1\frac{1}{7}$

26. $18 \div 2\frac{1}{4}$

27. $5\frac{5}{8} \div 18$

28. $9\frac{3}{4} \div 26$

29. $20\frac{1}{4} \div 90$

30. $12\frac{1}{2} \div 50$

31. $84 \div 5\frac{1}{4}$

32. $26 \div 3\frac{1}{2}$

33. $2\frac{2}{5} \div 1\frac{7}{10}$

34. $4\frac{1}{5} \div 4\frac{2}{3}$

PROBLEM SOLVING

35. A curtain requires $2\frac{3}{5}$ yd of material. How many curtains can be made from 39 yd of material?

36. An insect walks about $22\frac{1}{2}$ ft per hour. At this rate, how long will it take the insect to walk 90 ft?

37. How many boards $1\frac{1}{4}$ ft long can be cut from a board $9\frac{7}{8}$ ft long? What part of a $1\frac{1}{4}$ ft board is left over?

CHALLENGE

Divide.

38. $21 \div \frac{3}{16} \div 2\frac{2}{3}$

39. $1\frac{1}{3} \div 5 \div 2\frac{3}{5}$

40. $3\frac{1}{3} \div 2\frac{1}{2} \div 1\frac{1}{3}$

Order of Operations Using Fractions

You can use the order of operations rules to simplify mathematical expressions containing fractions.

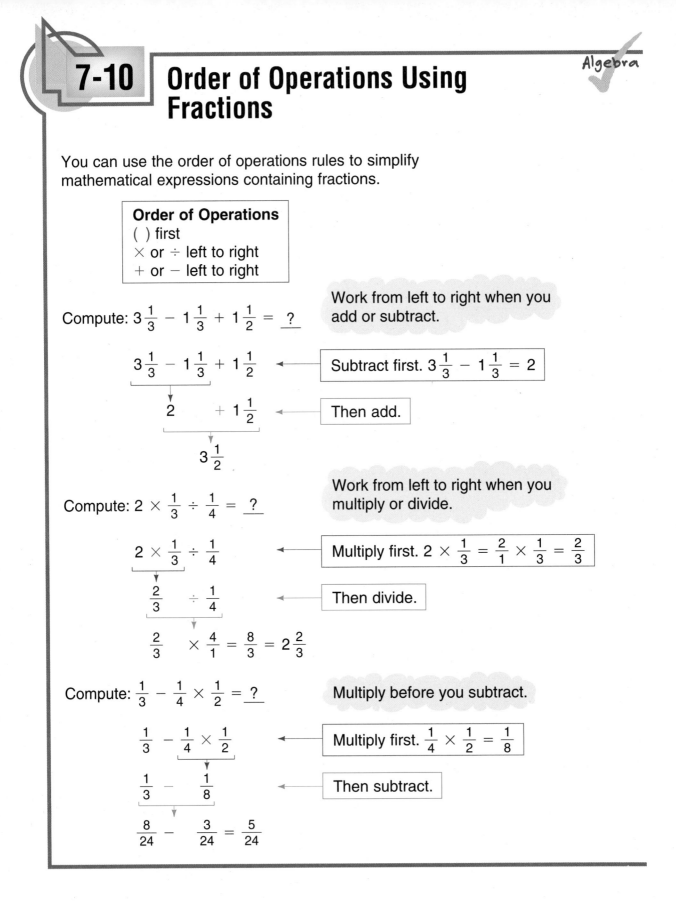

Order of Operations
( ) first
× or ÷ left to right
+ or − left to right

Compute: $3\frac{1}{3} - 1\frac{1}{3} + 1\frac{1}{2} = \underline{\ ?\ }$

Work from left to right when you add or subtract.

$3\frac{1}{3} - 1\frac{1}{3} + 1\frac{1}{2}$ ← Subtract first. $3\frac{1}{3} - 1\frac{1}{3} = 2$

$2 \quad + 1\frac{1}{2}$ ← Then add.

$3\frac{1}{2}$

Compute: $2 \times \frac{1}{3} \div \frac{1}{4} = \underline{\ ?\ }$

Work from left to right when you multiply or divide.

$2 \times \frac{1}{3} \div \frac{1}{4}$ ← Multiply first. $2 \times \frac{1}{3} = \frac{2}{1} \times \frac{1}{3} = \frac{2}{3}$

$\frac{2}{3} \quad \div \frac{1}{4}$ ← Then divide.

$\frac{2}{3} \quad \times \frac{4}{1} = \frac{8}{3} = 2\frac{2}{3}$

Compute: $\frac{1}{3} - \frac{1}{4} \times \frac{1}{2} = \underline{\ ?\ }$

Multiply before you subtract.

$\frac{1}{3} - \frac{1}{4} \times \frac{1}{2}$ ← Multiply first. $\frac{1}{4} \times \frac{1}{2} = \frac{1}{8}$

$\frac{1}{3} - \frac{1}{8}$ ← Then subtract.

$\frac{8}{24} - \frac{3}{24} = \frac{5}{24}$

Compute.

1. $1\frac{1}{4} + 3\frac{1}{4} + 2$

2. $5\frac{7}{8} - 1\frac{3}{8} - 2$

3. $\frac{3}{5} - \frac{1}{5} + 6$

4. $2\frac{4}{7} + \frac{3}{7} - 1$

5. $\frac{1}{2} \times \frac{3}{4} \times \frac{1}{8}$

6. $\frac{3}{5} \times 5 \times \frac{1}{2}$

7. $6 \times \frac{1}{2} \div \frac{1}{4}$

8. $9 \times \frac{1}{3} \div \frac{1}{2}$

9. $\frac{5}{6} + \frac{1}{6} - \frac{1}{2}$

10. $\frac{4}{9} - \frac{1}{9} + \frac{2}{3}$

11. $\frac{1}{8} + \frac{1}{2} \times 16$

12. $1\frac{2}{3} - 6 \times \frac{1}{6}$

Expressions with Parentheses

Compute: $(1\frac{1}{3} \times \frac{6}{7}) \div \frac{1}{7} = \underline{\ ?\ }$

$(\frac{4}{\underset{1}{\cancel{3}}} \times \frac{\overset{2}{\cancel{6}}}{7}) \div \frac{1}{7} = \frac{8}{7} \div \frac{1}{7}$

> Compute within parentheses first.

$= \frac{8}{\underset{1}{\cancel{7}}} \times \frac{\overset{1}{\cancel{7}}}{1} = \frac{8}{1} = 8$

Compute.

13. $(1\frac{1}{4} \times 4) - \frac{1}{3}$

14. $(8 \div 1\frac{1}{3}) + 6$

15. $(1\frac{2}{3} \times 1\frac{1}{2}) \div 5$

16. $(10 \div 1\frac{2}{3}) \times \frac{7}{8}$

17. $\frac{4}{9} \times (1\frac{1}{2} + 1\frac{3}{4})$

18. $1\frac{1}{3} \times (2\frac{1}{2} - 1\frac{1}{4})$

MENTAL MATH

Use compatible numbers to estimate each product.

19. $\frac{5}{6}$ of $43 \longrightarrow \frac{5}{6}$ of 42 $\qquad \frac{1}{6} \times \frac{42}{1} = 7$, so $\frac{5}{6}$ of $42 = 5 \times 7 = 35$.

> estimate

20. $\frac{1}{5}$ of 34

21. $\frac{2}{3}$ of 25

22. $\frac{3}{4}$ of 198

23. $\frac{2}{5}$ of 139

24. $\frac{1}{4}$ of $95

25. $\frac{1}{8}$ of $76

26. $\frac{1}{5}$ of $64.90

27. $\frac{1}{3}$ of $3.95

Fractions with Money

A package of copier paper costs $1.50.
Leroy bought a package for $\frac{3}{4}$ of the cost.
How much did Leroy pay?

To determine how much Leroy paid,
find: $\frac{3}{4}$ of $1.50 = \underline{\ ?\ }$

Estimate: $\frac{3}{4}$ of $1.50 \longrightarrow \frac{3}{4}$ of $1.60

$\frac{1}{4}$ of $1.60 = \$.40$, so $\frac{3}{4}$ of $1.60 = \$1.20$.

$\frac{3}{4}$ of $1.50 = \frac{3}{4} \times \$1.50$ "of" means "times."

$= \frac{3}{4} \times \frac{\$1.50}{1}$

$= \frac{3 \times \$1.50}{4 \times 1} = \frac{\$4.50}{4}$

$\quad\quad\quad = \$1.13$

$$\begin{array}{r} \$1.125 = \$1.13 \\ 4\overline{)\$4.500} \end{array}$$

Round to the nearest cent.

Leroy paid $1.13. $1.13 is close to the estimate of $1.20.

Study this example.

Divide: $8.75 $\div$ $1\frac{1}{4}$

$\$8.75 \div 1\frac{1}{4} = \frac{\$8.75}{1} \div \frac{5}{4} = \frac{\$8.75}{1} \times \frac{4}{5}$ Multiply by the reciprocal.

Rename as improper fractions.

$= \frac{\overset{\$1.75}{\cancel{\$8.75}}}{1} \times \frac{4}{\underset{1}{\cancel{5}}} = \frac{\$7.00}{1} = \$7.00$

Compute. Round to the nearest cent when necessary.

1. $\frac{1}{2}$ of $46
2. $\frac{1}{5}$ of $85
3. $\frac{1}{3}$ of $6.09
4. $\frac{1}{4}$ of $8.32

5. $\frac{3}{4}$ of $70
6. $\frac{2}{5}$ of $86
7. $\frac{2}{3}$ of $21.50
8. $\frac{3}{8}$ of $16.50

9. $3.50 ÷ $3\frac{1}{2}$
10. $5.50 ÷ $1\frac{2}{3}$
11. $36.75 ÷ $3\frac{3}{4}$
12. $11.20 ÷ $1\frac{1}{3}$

13. $14.90 ÷ $2\frac{1}{2}$
14. $11.40 ÷ $1\frac{1}{5}$
15. $6.65 ÷ $1\frac{3}{4}$
16. $56 ÷ $\frac{7}{8}$

PROBLEM SOLVING

17. Hiro wants to sell the bicycle he bought originally for $220 for $\frac{3}{5}$ of that price. What is the selling price of the bicycle?

18. Mary Ann bought a typewriter marked $350 for $\frac{3}{4}$ of the price. How much did she pay?

19. Joni paid $8.75 for a $3\frac{1}{2}$-square-foot rug. How much is that per square foot?

20. John bought a roast that weighed $4\frac{1}{2}$ lb for $12.60. How much is that per pound?

21. Dennis spent $\frac{1}{4}$ of his $18 weekly allowance. How much money does he have left?

22. A $35 dress in a store is marked "$\frac{1}{4}$ off." What is the new price of the dress?

23. Mr. Bucks has $44,000 to share with his three children and five grandchildren. He gives the first child $\frac{1}{2}$, the second child $\frac{1}{5}$, and the third child $\frac{1}{8}$ of the money. How much money is left for the grandchildren? What fractional part of the money is that?

 **CONNECTIONS: HISTORY**

24. Shampoo that cost $.90 in 1967 costs about $3\frac{5}{9}$ times as much today. What is today's cost?

25. A bicycle that cost $180 in 1975 costs about $2\frac{1}{2}$ times as much today. What is today's cost?

247

7-12 Problem Solving: Use a Diagram

Algebra ✓

Problem: Of 27 bowling balls for rent, 12 are black and 24 are speckled. What part of the bowling balls are both black and speckled?

1 IMAGINE Picture a shelf of bowling balls for rent.

2 NAME *Facts:*   27 bowling balls
                    12 are black
                    24 are speckled

Question: What part of the bowling balls are both black and speckled?

3 THINK Make a Venn diagram and list the facts.
Let A represent black bowling balls.
Let B represent speckled bowling balls.
Let C represent black speckled bowling balls.

12 Black   **24 Speckled**

A   C   B

To find the number of bowling balls that are both black and speckled (C), subtract 27 from the sum of 12 and 24. $(12 + 24) - 27 = C$

Then write a fraction to show the number of black speckled bowling balls out of all the balls.

$\dfrac{C}{27} = ?$

$\dfrac{C}{27}$ $\dfrac{\text{number of black speckled}}{\text{total number of bowling balls}}$

4 COMPUTE
$(12 + 24) - 27 = C$
    $36 \quad - 27 = 9$

$\dfrac{9}{27} = \dfrac{1}{3}$

One third of the bowling balls are both black and speckled.

Black   **Speckled**
  12   **24**

$\begin{array}{c} 12 \\ -9 \\ \hline 3 \end{array}$ 9 $\begin{array}{c} 24 \\ -9 \\ \hline 15 \end{array}$

5 CHECK Add the number of bowling balls in each region:

$A + B + C \overset{?}{=} 27$ $A + C \overset{?}{=} 12$ $C + B \overset{?}{=} 24$

$3 + 15 + 9 \overset{?}{=} 27$ $3 + 9 \overset{?}{=} 12$ $9 + 15 \overset{?}{=} 24$

$\qquad\qquad 27 = 27$ $12 = 12$ $24 = 24$

Your answer checks.

Solve by using a Venn diagram.

1. The school paper lists the names of the 18 baseball players, 20 volleyball players, and 16 soccer players. One person belongs to all three teams. One third of the baseball players and 4 of the volleyball players also belong to the soccer team. How many students play only soccer?

Baseball 18   **Volleyball** 20

Soccer 16

IMAGINE Draw and label a diagram.

NAME *Facts:*   18 baseball players
20 volleyball players
16 soccer players
1 student plays on all three teams.
$\frac{1}{3}$ of the baseball players also play soccer.
4 volleyball players also play soccer.

Question: How many students play only soccer?

THINK To find the number who play both soccer and baseball, multiply: $\frac{1}{3} \times 18 = 6$

Use the Venn diagram and subtract to find the number in each overlapping region.
$6 - 1 = \underline{\ ?\ }$ and $4 - 1 = \underline{\ ?\ }$

Then solve for the number who play only soccer.
$16 - (1 + ? + ?) = \underline{\ ?\ }$ soccer only

COMPUTE ⟶ **CHECK**

2. There are 26 shops at the minimall. One third of the 12 shops that provide services also provide goods. How many shops provide only goods?

3. At a buffet table, ham, chicken, and beef were being served. Of the 200 guests, 70 ate ham, 100 ate chicken, 85 ate beef, 25 ate ham and beef, and 30 ate chicken and ham. One tenth of the guests ate all three. How many guests ate only ham?

4. While on vacation 50 people could opt to fish, scuba dive, and/or water ski. Of the group, 35 went fishing, 32 went scuba diving, 14 tried water skiing, 21 tried both fishing and scuba diving, 4 tried only water skiing, and 10 did all three. How many went only fishing? only scuba diving?

249

Problem-Solving Applications

Solve each problem and explain the method you used.

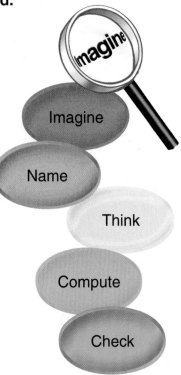

1. Hanley's Farm Stand sets out $\frac{2}{5}$ bushel of apples. One half of the apples are sold in the first hour the stand is open. What part of a bushel is left at the end of the first hour?

2. Solomon can pick $\frac{5}{8}$ bushel of grapes each hour. How many bushels can he pick in 7 hours?

3. An apple cake recipe calls for $2\frac{2}{3}$ cups of apple slices. Each apple supplies about $\frac{2}{3}$ cup of slices. How many apples are needed to make the cake?

4. Geraldine picked $4\frac{1}{2}$ quarts of strawberries. Lonnie picked $\frac{3}{4}$ as much as Geraldine. How many quarts of strawberries did Lonnie pick?

5. A melon weighs $3\frac{1}{3}$ lb. Jimmy cuts the fruit into 8 equal pieces. How much does each piece weigh?

6. There are 48 packages of blueberries on sale. Of these $\frac{1}{4}$ are wild. The rest of the blueberries are cultivated. How many packages of cultivated blueberries are on sale?

7. Pat picked $10\frac{1}{2}$ pints of raspberries and packaged them in half-pint containers. How many containers did he use?

8. A pound of plums costs $1.05. How much do $3\frac{1}{5}$ lb cost?

9. How many $\frac{2}{5}$-lb slices can be cut from a 15-lb watermelon?

10. How many $\frac{1}{2}$-gal containers can be filled from a $25\frac{1}{2}$-gal keg of cider?

11. Maya picks $\frac{3}{4}$ bushel of peaches in $\frac{2}{3}$ hour. How many bushels can she pick in one hour?

Use a strategy from the list or another strategy you know to solve each problem.

USE THESE STRATEGIES:
Multi-Step Problem
Use Simpler Numbers
Hidden Information
Use a Diagram
Working Backwards

12. There is an average of 18 cherries in a 4-oz container. How many cherries are in a 1-lb container?

13. Hanley's has $\frac{7}{8}$ bushel of peaches when the farm stand opens in the morning. By noon, $\frac{1}{4}$ of the peaches are left. What part of the bushel of peaches was sold in the morning?

14. The stand sells $\frac{1}{2}$ quart of berries for $1.49. Do $2\frac{3}{4}$ quarts of berries cost more than $10?

15. Hanley's has 130 lb of grapes, which are sold in $2\frac{1}{2}$-lb bags. How many bags of grapes can be prepared for the stand?

16. A customer buys a $2\frac{5}{8}$-lb melon for $1.05. Would a $4\frac{1}{2}$-lb melon cost more than $2?

17. By 11 A.M. Kathy had sold $\frac{1}{6}$ of the 5-lb bags of pears. Between 11 A.M. and 4 P.M. she sold 2 dozen more bags. If she had 11 bags left at 4 P.M., how many bags did she have when the stand opened?

Use the diagram for problems 18–20.

Anne, Bill, Carol, Derek, and Emmy each bought berries.

18. Who bought only strawberries?

19. How many people bought raspberries?

20. Who bought both strawberries and raspberries?

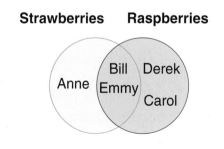

Strawberries   Raspberries

Anne   Bill   Derek
Emmy   Carol

MAKE UP YOUR OWN

21. Make up your own problem modeled on problem 16 above. Then have a classmate solve it.

Chapter Review and Practice

Multiply. Cancel whenever possible. *(See pp. 226–229.)*

1. $\frac{5}{6} \times \frac{3}{4}$ 2. $\frac{4}{5} \times \frac{3}{7}$ 3. $\frac{1}{4} \times 48$ 4. $\frac{6}{11} \times \frac{33}{42}$ 5. $18 \times \frac{2}{3}$

6. $3\frac{3}{7} \times 3\frac{1}{2}$ 7. $1\frac{3}{8} \times 1\frac{3}{5}$ 8. $4\frac{2}{7} \times \frac{2}{5}$ 9. $\frac{2}{9}$ of 3   10. $3\frac{3}{5} \times 5\frac{1}{2}$

Complete. Name the property of multiplication used. *(See pp. 230–231.)*

11. $\underline{\ ?\ } \times (8 \times \frac{2}{3}) = (\frac{1}{4} \times 8) \times \frac{2}{3}$

12. $\underline{\ ?\ } \times \frac{3}{5} = \frac{3}{5}$ 13. $\frac{7}{11} \times \frac{11}{14} = \frac{11}{14} \times \underline{\ ?\ }$

Write the reciprocal of each number. *(See pp. 230–231.)*

14. $\frac{2}{3}$ 15. 32   16. $\frac{7}{8}$ 17. $\frac{11}{4}$ 18. $\frac{5}{6}$

Estimate. Then multiply. *(See pp. 232–233.)*

19. $4\frac{1}{5} \times 2\frac{2}{3}$ 20. $2\frac{2}{5} \times 6\frac{1}{2}$ 21. $2\frac{4}{7} \times 3\frac{1}{2}$ 22. $3\frac{1}{3} \times 2\frac{7}{10}$

Draw a diagram to show each division sentence. *(See pp. 234–235.)*

23. $6 \div \frac{1}{3}$ 24. $4 \div \frac{4}{7}$ 25. $\frac{1}{3} \div \frac{1}{9}$

Estimate. Then divide. *(See pp. 234–243, 246–247.)*

26. $\frac{5}{6} \div \frac{2}{3}$ 27. $\frac{5}{12} \div \frac{1}{6}$ 28. $8 \div 3\frac{1}{5}$ 29. $4 \div 1\frac{2}{5}$ 30. $10\frac{2}{5} \div 2\frac{1}{6}$

Compute. *(See pp. 244–245.)*

31. $\frac{1}{3}$ of $48   32. $\frac{3}{5}$ of $12.75   33. $36.40 \div 1\frac{1}{7}$ 34. $8.00 \div \frac{4}{5}$

PROBLEM SOLVING *(See pp. 246–247, 250–251.)*

35. Terry had $11.60. He spent $\frac{3}{4}$ of it. How much money did he have left?

36. Elisa buys $6\frac{1}{2}$ lb of apples for $8.38. What is the cost of one pound of apples?

(See *Still More Practice*, p. 512.)

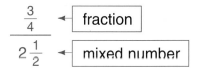

COMPLEX FRACTIONS

A **complex fraction** has a fraction or mixed number in its numerator, its denominator, or both.

$$\frac{\frac{2}{5}}{10} \leftarrow \text{fraction} \qquad \frac{32}{\frac{4}{5}} \leftarrow \text{fraction} \qquad \frac{\frac{3}{4}}{2\frac{1}{2}} \leftarrow \text{fraction / mixed number}$$

To simplify a complex fraction, divide the numerator by the denominator.

$$\frac{\frac{2}{5}}{10} = \frac{2}{5} \div 10 = \frac{2}{5} \div \frac{10}{1} = \frac{2}{5} \times \frac{1}{10} = \frac{1}{25}$$

$$\frac{32}{\frac{4}{5}} = 32 \div \frac{4}{5} = \frac{32}{1} \div \frac{4}{5} = \frac{32}{1} \times \frac{5}{4} = \frac{40}{1} = 40$$

$$\frac{\frac{3}{4}}{2\frac{1}{2}} = \frac{3}{4} \div 2\frac{1}{2} = \frac{3}{4} \div \frac{5}{2} = \frac{3}{4} \times \frac{2}{5} = \frac{3}{10}$$

Simplify. Use a fraction calculator to check your answers. (*Hint:* See page 216 for calculator keys to use.)

1. $\dfrac{\frac{4}{5}}{\frac{5}{9}}$
 2. $\dfrac{\frac{4}{7}}{5}$
 3. $\dfrac{\frac{7}{8}}{6}$
 4. $\dfrac{2}{\frac{3}{5}}$
 5. $\dfrac{\frac{1}{5}}{\frac{3}{4}}$
 6. $\dfrac{\frac{2}{3}}{\frac{2}{5}}$

7. $\dfrac{\frac{2}{3}}{8}$
 8. $\dfrac{4}{\frac{8}{9}}$
 9. $\dfrac{8}{\frac{2}{5}}$
 10. $\dfrac{\frac{3}{7}}{\frac{9}{10}}$
 11. $\dfrac{\frac{8}{9}}{4}$
 12. $\dfrac{\frac{4}{5}}{\frac{3}{8}}$

13. $\dfrac{21}{\frac{7}{8}}$
 14. $\dfrac{\frac{5}{9}}{3}$
 15. $\dfrac{\frac{7}{10}}{\frac{5}{12}}$
 16. $\dfrac{12}{\frac{4}{9}}$
 17. $\dfrac{\frac{15}{16}}{\frac{3}{8}}$
 18. $\dfrac{14}{\frac{2}{7}}$

19. $\dfrac{\frac{9}{10}}{\frac{3}{10}}$
 20. $\dfrac{12}{\frac{6}{7}}$
 21. $\dfrac{2\frac{2}{3}}{4}$
 22. $\dfrac{6}{3\frac{3}{5}}$
 23. $\dfrac{9}{4\frac{1}{2}}$
 24. $\dfrac{2\frac{3}{4}}{3\frac{1}{3}}$

Check Your Mastery

Performance Assessment

Find the error or errors in each computation. Explain.

1. $\frac{3}{8} \times (10 \times \frac{8}{3})$

$= \frac{3}{8} \times (\frac{1}{10} \times \frac{8}{3})$

$= \frac{3}{8} \times (\frac{8}{3} \times \frac{1}{10})$

$= (\frac{3}{8} \times \frac{8}{3}) \times \frac{1}{10}$

$= 1 \times \frac{1}{10} = \frac{1}{10}$?

2. $\frac{4}{5} \div \frac{5}{4} \times 0$

$= (\frac{4}{5} \div \frac{5}{4}) \times 0$

$= 1 \times 0$

$= 1$?

3. $19 \times \frac{19}{20} \times \frac{20}{19}$

$= (19 \times \frac{19}{20}) \times (19 \times \frac{20}{19})$

$= (\frac{1}{19} \times \frac{19}{20}) \times (\frac{19}{1} \times \frac{20}{19})$

$= \frac{1}{20} \times 20$

$= 1$?

Multiply or divide. Estimate to help you.

4. $\frac{4}{5} \times \frac{1}{7}$

5. $\frac{3}{4} \times \frac{4}{9}$

6. $\frac{3}{20} \times \frac{14}{15}$

7. $\frac{7}{8} \times \frac{6}{35} \times \frac{5}{9}$

8. $\frac{2}{3} \times 24$

9. $5 \times \frac{2}{5}$

10. $\frac{1}{4}$ of 16

11. $\frac{7}{10}$ of $20

12. $5\frac{2}{5} \times 3\frac{1}{3}$

13. $1\frac{1}{2} \times 1\frac{2}{3}$

14. $12 \times 2\frac{5}{6}$

15. $6\frac{2}{3} \times 5 \times 1\frac{1}{5}$

16. $\frac{2}{9} \div \frac{4}{9}$

17. $\frac{7}{8} \div \frac{5}{16}$

18. $9 \div \frac{1}{2}$

19. $\frac{8}{9} \div 4$

20. $7\frac{1}{2} \div 3\frac{3}{4}$

21. $2\frac{1}{7} \div 2\frac{1}{2}$

22. $7\frac{5}{7} \div \frac{9}{14}$

23. $9 \div 1\frac{1}{3}$

Compute. Use the order of operations rules.

24. $\frac{5}{12} - \frac{1}{12} + \frac{2}{3}$

25. $\frac{1}{4} + \frac{1}{2} \times 12$

26. $(8 \div 1\frac{1}{3}) \times \frac{3}{8}$

PROBLEM SOLVING   *Use a strategy or strategies you have learned.*

27. If $5\frac{1}{3}$ yd of material is cut into 15 pieces of the same length, about how long is each piece?

28. Ms. Kirby drove 140 mi in $3\frac{1}{2}$ hours. What was her average speed in miles per hour?

Cumulative Test I

Choose the best answer.

1. Which shows the standard form of 2 trillion, 14 million, 800 thousand?

a. 2,014,800   **b.** 2,014,000,800,000
c. 2,014,800,000   **d.** 2,000,014,800,000

2. Which shows the decimal 0.8741 rounded to its greatest nonzero place?

a. 1   **b.** 0.9
c. 0.874   **d.** 0.87

3. Estimate.

9,879,632
+   763,986

a. 9,700,000
b. 10,700,000
c. 11,700,000
d. 9,000,000,000

4. $4.56 + $.56 + $44

a. $5.56
b. $49.12
c. $104.56
d. not given

5. 2,729,000 − 409,026

a. 2,320,026
b. 2,320,974
c. 2,320,984
d. not given

6. 68 − 0.054

a. 0.014
b. 67.46
c. 67.946
d. not given

7. Which shows greatest to least?

a. 1.88; 1.8; 1.08; 1.008

b. 1.8; 1.88; 1.08; 1.008

c. 1.008; 1.88; 1.08; 1.8

d. none of these

8. Which shows least to greatest?

a. $2\frac{2}{5}, 2\frac{2}{3}, 2\frac{1}{4}$

b. $2\frac{2}{3}, 2\frac{2}{5}, 2\frac{1}{4}$

c. $2\frac{1}{4}, 2\frac{2}{3}, 2\frac{2}{5}$

d. none of these

9.
$$\frac{11}{12}$$
$$-\frac{2}{3}$$

a. $\frac{1}{4}$

b. $\frac{5}{6}$

c. $1\frac{1}{9}$

d. not given

10. Find the value of m:

$$m - \frac{2}{3} = \frac{4}{5}$$

a. $\frac{2}{15}$

b. $\frac{3}{4}$

c. $\frac{8}{15}$

d. $1\frac{7}{15}$

11. How much more than 2×10^3 is 2500?

a. 500   **b.** 2300

c. 3000   **d.** 5500

12. Which is greater than 3 but less than $6\frac{1}{2}$?

a. $5 + 1\frac{3}{4}$ **b.** $1\frac{1}{2} + 1\frac{1}{2}$

c. $9\frac{1}{9} - 5\frac{5}{12}$ **d.** $20 - 17\frac{1}{3}$

Compute. Estimate to help you.

13. 1000×0.285

14. $\begin{array}{r} 8.005 \\ \times\ \ 5.32 \\ \hline \end{array}$

15. $119\overline{)190,400}$

16. $6.2\overline{)0.0558}$

17. $\dfrac{5}{6} + \dfrac{7}{8}$

18. $\dfrac{2}{3} \times \dfrac{9}{10}$

19. $4\dfrac{1}{5} \times 2\dfrac{1}{2}$

20. $8 \div 2\dfrac{2}{3}$

21. $\dfrac{3}{8} \times \$21.60$

22. $10\dfrac{2}{5} \div 2\dfrac{1}{6}$

23. $\left(\dfrac{3}{4} + 2\dfrac{1}{5}\right) + \dfrac{1}{4}$

24. $\dfrac{1}{4} + \dfrac{1}{3} \times 9$

PROBLEM SOLVING
Solve each problem. Then explain the method you used.

25. Keewhan's fish tank holds 15 gallons of water. One gallon of water weighs 8.33 lb. What is the weight of the water if the fish tank is filled to the top?

26. Connie had $82.50. She spent $\dfrac{4}{5}$ of it on a shirt. About how much did she have left?

27. Heather drove 25.4 km in 4 days. To the nearest tenth of a kilometer, what was the average distance driven per day?

28. A theater has 675 seats. There are three times as many seats in a row as there are rows. How many rows and how many seats are there?

29. Paulo needs $5\dfrac{1}{3}$ yd of material to make a curtain of a certain size. How many such curtains can he make if he has 78 yd of material? What fractional part of another curtain does Paulo have left?

For Rubric Scoring

Listen for information on how your work will be scored.

30. A factory makes CD players. The table below shows that it produces 29 CD players during the first 4 days of production.

Day (*d*)	1	2	3	4
Number of CD players (*n*)	8	15	22	29

a. What pattern do you see in the 1st row of the table? 2nd row?

1st: | *d* | 1 | 2 | 3 | 4 |

2nd: | *n* | 8 | 15 | 22 | 29 |

b. Predict the number of CD players produced in the first 12 days of production. Upon what do you base your prediction?

c. Write an expression to show how you can get the numbers in the 2nd row of the table from the numbers in the 1st row.

Statistics and Probability

8

Lunch Time

Oh, for a piece of papaya,
or a plate of beef lo mein—

Oh, for a bowl of Irish stew,
or fresh paella from Spain—

Oh, for a forkful of couscous,
or a chunk of Jarlsberg cheese—

Oh for some lasagna,
or a bowl of black-eyed peas—

Of all the tasty foods
That I would love to try,

I sit here and wonder
why, oh why,
Mama packed me
this liverwurst on rye.

Lee Bennett Hopkins

In this chapter you will:

Explore types of graphs
Learn about range, mean, median, and mode
Survey, collect, organize, report, and
   interpret data
Investigate stem-and-leaf plots, box-and-
   whiskers, and double bar and line graphs
Recognize misleading statistics
Make predictions based on probability
   of simple and compound events
Learn about computer graphs
Solve problems by making an organized list

Critical Thinking/Finding Together

Research each of the foods mentioned in the
poem. Make tree diagrams to find the number
of two-food combinations you can make.

Update your skills. See page 17.

8-1 Graphing Sense

The students in Ms. Pleau's sixth grade mathematics class recorded the temperature at 6:00 P.M. for twenty consecutive days. The data the class collected is shown in the table.

Temperatures (°F)			
20	30	22	25
30	25	25	10
19	25	22	15
25	26	21	15
20	28	20	18

Make a line plot of the temperature data.

 Hands-On Understanding

Materials Needed: straightedge, paper, pencil, graph paper

Step 1 Draw a horizontal number line at the bottom of a sheet of graph paper.

Step 2 Write a zero at one end of the line plot and the greatest temperature at the other end.

Step 3 Use multiples of 5 for the scale of your plot. Mark evenly spaced intervals on the line.

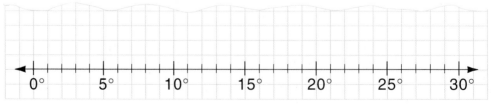

Mark an X on the line plot for each temperature.

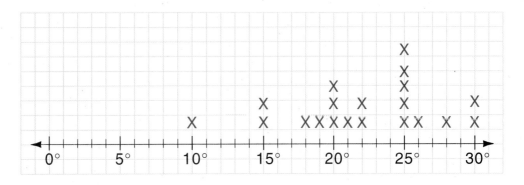

Communicate

1. What do the five Xs above 25° tell you?

2. Around which two temperatures does the data seem to cluster (group)?

   Discuss ✓

3. Work with your classmates to draw line plots of the temperature data using a scale interval of 2 and then 10. Which scale — 2, 5, or 10 — seems to work best for the data? Explain.

4. How is a line plot like a bar graph? How is it different?

PROJECT

5. Record temperature data for twenty consecutive days at 6:00 P.M. for your town or city.

   • Make a line plot of the data.

   • Make another appropriate graph to show the same data.

   Discuss ✓

   • Display and discuss the graphs with the class.

8-2 Surveys

Iris wanted to know whether the students in her school think that it is a good idea to make the school day longer. She decided to conduct a **survey**.

To conduct a survey:

- Write and ask questions to determine the opinions on the topic.

- Record the responses.

- Organize the data in a table or graph.

Since Iris could not talk to every student in the school, her math teacher, Mr. Smith, suggested that she use a **sample** of the school population. Iris' sample included the same number of students from each grade, the same number of boys as girls, students from various ethnic backgrounds, and so on.

> A sample represents a larger population.

Iris asked this question of 30 students: *Do you think the school day should be longer?* Answer *Yes, No,* or *Not Sure.*

Responses: Yes—8 students   No—18 students   Not Sure—4 students

She presented her findings in both a pictograph and a bar graph.

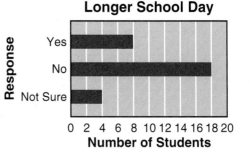

From her survey, Iris predicted that most students in her school do not think that the school day should be longer.

The bar graph shows the results of a survey about favorite types of movies. Use the graph to answer the questions.

1. How many people chose comedy?

2. How many more people chose science fiction than adventure?

3. How many people in all were surveyed?

4. What fractional part of those surveyed chose western?

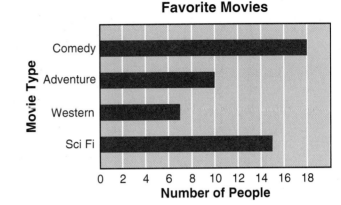

Favorite Movies

The pictograph shows the results of a survey about favorite drinks. Use the graph to answer the questions.

5. Which drinks were chosen by fewer than 8 people?

6. How many of those surveyed did not choose soda or iced tea as their favorite?

7. What fractional part of those surveyed chose either lemonade or soda?

8. You are ordering drinks for 200 people. Based on the results of the survey, how many of each drink would you order?

Favorite Drinks	
Soda	🍾 🍾 🍾
Lemonade	🍾 🍾
Apple juice	🍾 🍾 🍾
Orange juice	🍾 🍾
Iced tea	🍾
Key: Each 🍾 = 4 votes	

Communicate ✓

Tell whether each location would be an appropriate place for the survey. Write *Yes* or *No*. Explain your answer.

9. Favorite food; Italian restaurant

10. Favorite holiday; shopping mall

11. Favorite sport; football game

12. Favorite music; retirement home

 **CRITICAL THINKING**

13. Nancy makes predictions from survey data that she collects. Name three things that can affect the accuracy of Nancy's predictions. Discuss your ideas with the class.

Discuss ✓

8-3 Collecting Data

Daria's gym teacher wanted to find out how fast most sixth graders could run. Thirty students ran the 100-meter dash. The recorded times, in seconds, are shown on the board.

14.2	14.4	13.9	14.9	14.4	13.8	14.4	14.3	13.8	14.2
14.6	13.7	15.1	13.9	14.0	13.8	14.4	15.4	15.2	13.8
14.4	14.1	14.2	13.7	14.2	14.4	14.6	14.0	15.1	13.9

▶ Daria collected these data. She organized them in a **frequency table**. She grouped the data by time and used **tally marks** to record the number of students whose running times fell within each range of times.

Running Times: 100-m Dash		
Speed (seconds)	Tally	Total
13.6–14.0	⊬⊬⊬⊬ I	11
14.1–14.5	⊬⊬⊬⊬ II	12
14.6–15.0	III	3
15.1–15.5	IIII	4

Daria examined the frequency table. She concluded that most students could run the race in 14.5 seconds or less.

▶ Rico used this frequency table to record the same data.

Speed (seconds)	13.7	13.8	13.9	14.0	14.1	14.2	14.3	14.4	14.6	14.9	15.1	15.2	15.4
Tally	II	IIII	III	II	I	IIII	I	⊬⊬ I	II	I	II	I	I
Total	2	4	3	2	1	4	1	6	2	1	2	1	1

He concluded that 14.4 seconds was the time most frequently run for the 100-meter dash.

Copy and complete the frequency table.

	Distances: Standing Long Jump		
	Distance (meters)	Tally	Total
1.	1.1–1.3	⊬⊬ I	?
2.	1.4–1.6	?	11
3.	1.7–1.9	?	5
4.	2.0–2.2	II	?
5.	2.3–2.5	II	?

Use the frequency table at the bottom of page 262 for exercises 6–9.

6. How many long jumps were recorded?

7. Within which range do most jumps fall?

8. How many more students jumped from 1.4 m to 1.6 m than from 1.7 m to 1.9 m?

9. Write a conclusion about the long-jump data shown in the table.

Some students in the sixth grade at Owens School participated in the softball throw. The numerical data below show the distances thrown, in meters.

10. Organize the data in an ungrouped frequency table.

11. How many softball throws were recorded?

Softball Throw: Grade 6							
21	20	28	21	24	20	22	28
20	28	26	24	21	23	28	26
24	23	29	20	23	20	21	28

12. Which distance was thrown most often? Which was thrown exactly 4 times?

13. Write a conclusion about the data in the table.   *Communicate* ✓

Some Owens School sixth graders participated in the 50-meter dash. The times, in seconds, are shown.

14. Organize the data in an ungrouped frequency table.

15. Which of the times was run least often? most often?

50-Meter Dash: Grade 6									
6.9	6.8	7.0	6.9	7.3	6.7	6.9	7.0	7.1	7.0
7.0	7.2	6.8	7.1	6.9	6.9	7.2	6.8	7.1	6.9

16. How many students ran the race in 6.8 seconds?

17. What fractional part of the runners ran the distance in 7 seconds or less?

CONNECTIONS: PHYSICAL EDUCATION

18. Conduct a softball throw (distances to nearest meter or nearest yard) with the students in your mathematics or physical education class (as in exercises 10–13 above).   *Communicate* ✓

   a. Collect the data and make a frequency table.

   b. Write the three best conclusions you can make about the data in your frequency table.

Range, Mean, Median, and Mode

Gloria exercises daily by doing sit-ups. She keeps a record of her progress and uses **statistics** to describe the progress.

Day	S	M	T	W	Th	F	S
Number of Sit-ups	28	30	30	37	35	40	45

- The **range** of this set of data is the difference between the greatest number and the least number of sit-ups.

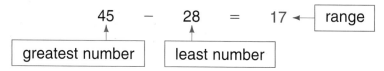

- The **mean** of this set of data is the *average* number of sit-ups. To find the mean, add the numbers and then divide by the number of numbers.

$$\frac{28 + 30 + 30 + 37 + 35 + 40 + 45}{7} = \frac{245}{7} = 35 \leftarrow \boxed{\text{mean}}$$

- The **median** of this set of data is the middle number when the data are listed in order from least to greatest.

28, 30, 30, 35, 37, 40, 45

$\uparrow$

$\boxed{\text{median}}$

> If there is an even number of numbers in the set of data, the median is the mean (average) of the two middle scores.
>
> 12, 13, 14, 15   Median $= \dfrac{13 + 14}{2} = 13.5$

- The **mode** of this set of data is the number that occurs most frequently.

28, 30, 30, 37, 35, 40, 45

$\boxed{\text{mode}}$

> Some sets of data have *no* mode: 29, 33, 35, 31, 30, 32
>
> Some have more than one mode: 28, 31, 31, 29, 36, 29, 35
> The modes are 29 and 31.

Find the range, mean, median, and mode for each set of data.

1. 2, 2, 3, 5, 7, 8, 10, 19

2. 70, 110, 90, 70, 60

3. 104, 94, 97, 99, 95, 90, 93

4. 83, 85, 108, 81, 74, 89, 87, 107

5. 1.9, 2.7, 1.4, 12.3, 5.7

6. 2.2, 2.2, 2.9, 2.2, 2.6

7. $5.25, $6.50, $4.90, $5.75

8. $4.50, $4.95, $4.80, $6.25, $4.25

PROBLEM SOLVING

9. The range of Meg's 5 test scores is 22. Her scores on tests 1–4 were: 81, 87, 70, and 83. What was her score on test 5?

10. Tito wants to have a mean (average) test score of 85. His scores on 3 tests were: 81, 80, and 85. What must his next test score be to meet his goal?

11. Suki's test scores were: 86, 96, 52, 86, 86, and 79. Would mean, median, or mode best describe these scores? Explain.

12. Sal's Shoes sold 8 size 4s, 12 size 5s, 11 size 7s, and 9 size 8s. Would mean, median, or mode best describe these data? Explain.

Box-and-Whisker Plot

A **box-and-whisker plot** uses both the range and the median to analyze data. It shows the **extremes** (greatest and least numbers), the **upper quartile** (median of the upper half of the data), and the **lower quartile** (median of the lower half of the data).

Scores		
50	60	80
100	90	70
	110	

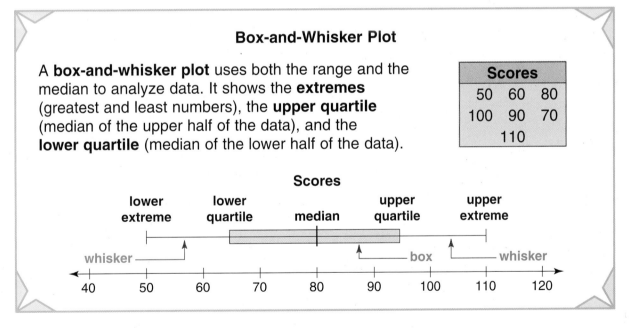

Use the box-and-whisker plot for exercises 13–16.

13. What is the median score?

14. What are the extremes?

15. What fractional part of the scores fall within the box?

16. What fractional part of the scores fall along each whisker?

265

The table below shows some of the tallest buildings in Baltimore, Maryland. You can organize and display the same data in a **stem-and-leaf plot**.

Building	Number of Floors
U.S. Fidelity & Guaranty	40
NationsBank Bldg.	34
Harbour Court	28
250 W. Pratt St.	26
6 St. Paul Place	37
Commerce Place	30
Union Trust Tower	24
IBM Tower	28
World Trade Center	32
Tremont Plaza Hotel	37
Charles Center South	26

To make a stem-and-leaf plot:

- Look at the front-end digits. Write them in order. They are the *stems*.

Stem	
2	
3	
4	

- The *leaves* are the ones digits. Write each leaf to the right of its stem.

Stem	Leaf				
2	8	6	4	8	6
3	4	7	0	2	7
4	0				

- Rewrite the leaves in order from least to greatest.

Stem	Leaf				
2	4	6	6	8	8
3	0	2	4	7	7
4	0				

Range: 40 − 24 = 16
Median: Count to find the sixth (middle) number, 30.
Mode: There are 3 modes—26, 28, and 37.

The stem-and-leaf plot shows the heights (number of floors) of some of the tallest buildings in Charlotte, North Carolina.

Use the plot to answer the questions.

Stem	Leaf
6	0
4	0 2
3	2 2 2
2	4 7

1. The heights of how many buildings are shown?

2. How many floors are there in the tallest building?

3. How many buildings have 40 or more floors?

4. What are the range, the median, and the mode of the data?

The data shows the heights (number of floors) of some tall buildings in Denver, Colorado.

56	54	52	43	41	40
36	35	31	32	34	42
32	28	26			

Copy and complete the stem-and-leaf plot. Then answer the questions.

5.
Stem	Leaf
5	? ? ?
4	? ? ? ?
3	? ? ? ? ? ?
2	? ?

6. What are the range, the median, and the mode of the data?

7. Write a statement that summarizes what the plot shows.

CONNECTIONS: HISTORY

The data shows the ages of the first 21 United States presidents at the time of their inaugurations.

57	61	57	57	58	57	61
54	68	51	49	64	50	48
65	52	56	46	54	49	50

Make a stem-and-leaf plot of the data. Then answer the questions.

8. What are the range, the median, and the mode of the data?

9. At the time of inauguration, how many of the 21 presidents were less than 60 years old?

10. Write a letter to your history or social studies teacher to tell what you have just learned about the ages of the first 21 United States presidents. Use your stem-and-leaf plot to explain what you have learned.

Communicate ✓

8-6 Working with Graphs and Statistics

Use the graphs to answer each question.

1. Write a brief summary to tell what is shown by the pictograph. What data is shown by the graph? What are some conclusions you can make about the data? Share your summary with the class.

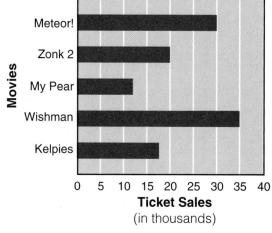

Average Attendance: Home Games	
Highmount Eagles	⚾ ⚾ ⚾ ⚾
Pine Hill Hens	⚾ ⚾
Ashoken Ants	⚾ ⚾ ⚾
Millerville Moles	⚾ ⚾ ⚾ ⚾ ⚾
Key: Each ⚾ = 100 people	

2. Write a brief summary to tell what is shown by the bar graph. Share your summary with the class.

3. What other type of graph could you use to display the data? Explain.

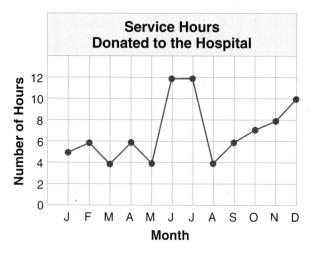

Weekend Movie Attendance

Movies / Ticket Sales (in thousands)

4. Write a brief summary to tell what is shown by the line graph. Share your summary with the class.

5. What trends, if any, do you see in the data?

6. Could you use a bar graph to display the information shown in the pictograph above? the line graph above? Explain.

Service Hours Donated to the Hospital

Number of Hours / Month

Use the information in the table for exercises 7–11.

Annual Salaries: F.H. Murphy Company				
$20,000	$30,000	$325,000	$28,000	$32,000
$26,000	$30,000	$26,000	$28,000	$30,000

7. Make a frequency table of the salaries of F.H. Murphy Company.

8. Use your frequency table from exercise 7 to find the mean, median, and mode of the salary data.

9. Which statistic—mean, median, or mode—most accurately describes the average annual salary of F.H. Murphy Company employees? Discuss your choice with the class.

Discuss

10. How would the mean, median, and mode be affected if the greatest salary at F.H. Murphy Company were changed to $3,000,000 and the other salaries all remained the same? Compute the new mean, median, and mode.

11. Which statistic—mean, median, or mode—is most affected by extreme values? Explain your answer using:
    a. a set of numbers that contains one number that is much greater than all of the other numbers in the set; and
    b. a set of numbers that contains one number that is much less than all of the other numbers in the set.

Communicate

 CRITICAL THINKING

12. You are comparing the numbers of home runs hit by five intramural baseball teams during the 1999 season of the Ashe County Baseball League. Which type of graph would you use? Explain.

13. Which type of graph would you use to show changes in base-path lengths from the beginning of the league in 1943 to the present day? Explain.

14. If you assume that the annual salaries of baseball coaches at city high schools are about the same, which statistic—mean, median, or mode—would you use to describe the average annual salary? Explain.

8-7 Making Line Graphs

Ms. Li's students are studying traffic patterns in the neighborhood. They recorded hourly traffic through one intersection from 1:00 P.M. to 8:00 P.M.

Vehicles in Intersection							
Time	1-2	2-3	3-4	4-5	5-6	6-7	7-8
Vehicles	368	325	288	390	505	478	352

Maria displayed the results in a **line graph**.

▶ **To make a line graph:**

- Draw horizontal and vertical axes on grid paper.

- Use the data from the table to choose an appropriate scale.

  Think: Choose intervals of 50.

- Draw and label the scale on the vertical axis. Start at 0 and label equal intervals.

- Label the horizontal axis. List each time period.

- Locate the points on the graph. Mark each with a dot.

- Connect the points with line segments.

- Write a title for your graph.

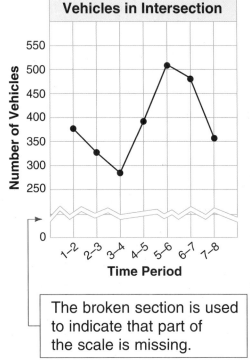

The broken section is used to indicate that part of the scale is missing.

Use the line graph above for exercises 1–4.

1. What data are shown on the horizontal axis? the vertical axis?

2. During which time period was the traffic flow the greatest?

3. Which time periods show an increase over the previous time period? Which show a decrease? Why was the 5:00–6:00 P.M. traffic flow so great?

4. How would *increasing* the size of the intervals on the vertical axis change the appearance of the graph? What effect would *decreasing* the interval size have?

Copy and complete the graph to show the data in the table.

5.

School Fair Profits	
Year	Profit (in dollars)
1997	850
1996	740
1995	700
1994	620
1993	525
1992	585

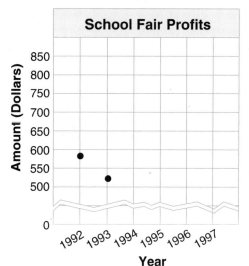

Use the completed line graph for exercises 6–9.

6. What does each interval on the vertical scale represent?

7. What trend does the graph show? Estimate the profit for 1998.

8. By how much did profits increase from 1995 to 1997?

9. What is the range of the profits? What is the mean (average) profit?

Make a line graph for each set of data.

10.

New Book Club Members						
Week	1	2	3	4	5	6
Members	45	58	82	81	97	103

11.

Bottles Recycled						
Month	Jan.	Feb.	Mar.	Apr.	May	Jun.
Bottles	186	152	106	92	124	199

12.

Park Clean-up Volunteers							
Day	S	M	T	W	Th	F	S
Volunteers	17	26	51	48	30	96	22

13.

Kathy's Scoring Average						
Year	1992	1993	1994	1995	1996	1997
Points	12.5	14.8	11.7	15.5	17.9	16.0

PROJECT

14. Record the amount of time you spend, in minutes, either talking on the telephone or reading each day for one week.
- Make a line graph.
- Summarize the information from the graph (including the range of the data).
- Describe any trends.

Communicate

Analyzing Line Graphs

In a recent survey of orders at pizza restaurants, it was discovered that of every 50 pizzas ordered, people will request extra cheese on 9 of them.

You can use a line graph to compare the two quantities:

- number of pizzas ordered
- number of pizzas with extra cheese

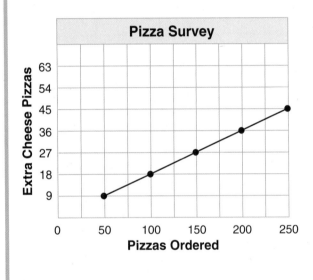

As the number of pizzas ordered increases, the number of pizzas with extra cheese also increases.

From left to right, the line slopes upward.

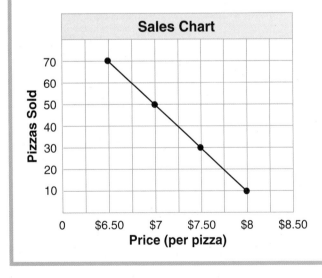

Mr. Fleury, owner of Pizzas Unlimited, makes this line graph to predict what will happen to sales if pizza prices are increased.

As the price per pizza increases, the number of pizzas sold decreases.

From left to right, the line slopes downward.

PROBLEM SOLVING

Communicate

1. A certain clock loses 3 minutes every 12 hours as shown in the table. Draw a line graph of the data and determine if the line slopes upward or downward. Explain why this happens.

Number of Hours	12	24	36	48	60
Time Lost (minutes)	3	6	9	12	15

2. The table shows the monthly cable fee and the number of subscribers (in thousands). Draw a line graph of the data and determine if the line slopes upward or downward. Explain why this happens.

Monthly Cable Fee	$50	$40	$30	$20	$10
Number of Subscribers (thousands)	2	4	6	8	10

3. As an experiment, Mr. Fleury of Pizzas Unlimited charges four different prices for the same pizza at four different stores that he manages. One week's results are shown in the table. Draw a line graph of this data.

Price (per pizza)	$6.50	$7.00	$7.50	$8.00
Pizzas sold	75	62	20	10

4. Compare your line graph in problem 3 with the line graph at the bottom of page 272. Discuss with your class why one graph is a "smooth" line while the other is not. Is Mr. Fleury's prediction about the effect of price increases on sales reasonable? Explain.

CRITICAL THINKING

Algebra

5. As *b* increases, what happens to *c*?

6. As *c* decreases, what happens to *b*?

7. If *c* = 75, what is the value of *b*?

8. Predict the value of *b* if *c* = 165.

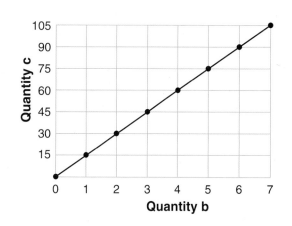

8-9 Double Line and Double Bar Graphs

Sam's Shirts sells T-shirts and sweatshirts. The double line graph shows how many of each were in stock during a 9-month period.

A **double line graph** compares two sets of data over time.

Each set is graphed separately, but on the same grid.

The *key* explains which set of data is shown by each line.

The graph shows, for example, that on January 1 there were 275 T-shirts and 225 sweatshirts in stock.

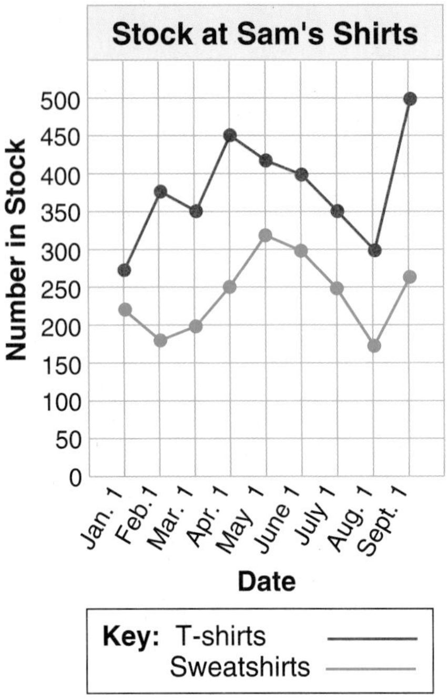

Stock at Sam's Shirts

Key: T-shirts ———
       Sweatshirts ———

Use the double line graph above for exercises 1–6.

1. How many more T-shirts than sweatshirts were in stock on July 1?

2. What can you say about the stock of sweatshirts between February 1 and May 1? Explain your answer.

    Communicate

3. What can you say about the stock of T-shirts from April 1 to August 1? Explain your answer.

4. When was the difference between the stock of T-shirts and sweatshirts the greatest? On what date was it the least?

5. When was the total stock of T-shirts and sweatshirts the greatest? When was the total stock the least?

6. Why might sales of both T-shirts and sweatshirts increase greatly from August 1 to September 1? Explain.

    Communicate

A **double bar graph** compares two related sets of data.

Solve. Use the double bar graph at the right.

Daily Shirt Sales at Sam's

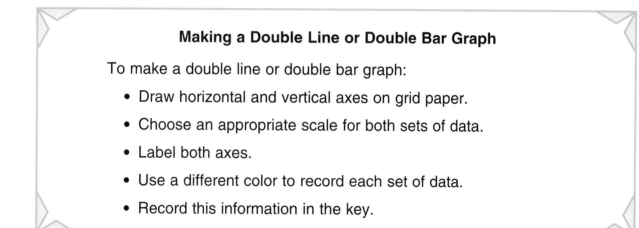

7. How many more T-shirts than sweatshirts were sold on Monday?

8. On which day were the most T-shirts sold? the most sweatshirts?

9. Which day had the greatest number of total sales?

10. On which day was there the least difference between T-shirt sales and sweatshirt sales?

11. How would you summarize the information about shirt sales shown by the graph?

Making a Double Line or Double Bar Graph

To make a double line or double bar graph:

- Draw horizontal and vertical axes on grid paper.

- Choose an appropriate scale for both sets of data.

- Label both axes.

- Use a different color to record each set of data.

- Record this information in the key.

Draw the graphs as indicated in exercises 12 and 13.

12. Make a double line graph for this table of Sam's stock.

Date	Jeans	Shorts
Oct. 1	525	175
Nov. 1	425	225
Dec. 1	450	210
Jan. 1	475	200
Feb. 1	480	185

13. Make a double bar graph for this table of favorite T-shirt colors.

Favorite Color	Girls	Boys
White	4	7
Red	8	9
Blue	12	9
Green	11	6
Yellow	7	11

Interpreting Circle Graphs

Shapiro's Marketing conducted a survey of music preferences on the basis of store sales of CDs.

A **circle graph** shows how parts are related to a whole. The circle graph at the right shows the fraction of the 200 people surveyed who purchased each type of music.

Music Preferences

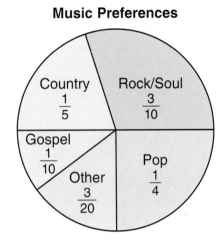

To find out how many of the 200 people chose country music, find $\frac{1}{5}$ of 200.

$$\frac{1}{5} \times \frac{200}{1} \longrightarrow \frac{1}{\overset{}{\underset{1}{5}}} \times \frac{\overset{40}{200}}{1} = 40$$

So 40 people chose country music.

Remember: A circle graph shows *parts of a whole*.
The parts may be given as fractions or percents.

Use the circle graph above for exercises 1-7.

	Type of Music	Fraction	Number of People
1.	Country	$\frac{1}{5}$	40
2.	Rock/Soul	?	?
3.	Pop	?	?
4.	Gospel	?	?
5.	Other	?	?

6. Together, what part of the people surveyed chose either pop or country? pop or rock/soul? pop, country, or rock/soul?

7. Shapiro's conducted the same survey last month with 250 people and got the same fractions. Will the circle graph look the same or different? Explain.

Solve. Use the circle graph at the right.

8. There are 64 students in the sixth grade at Whitman School. How many favor mystery books?

9. How many sixth graders chose science fiction books?

10. What fractional part of the sixth graders prefer books that are *not* science fiction?

11. How many more sixth graders chose mystery books than sports books?

12. Which two types of books do one fourth of the sixth graders favor? Which two types do three fourths of the sixth graders favor?

13. How would the circle graph look different if 8 fewer sixth graders chose mystery books and chose science fiction books instead? Explain.

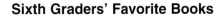

Sixth Graders' Favorite Books

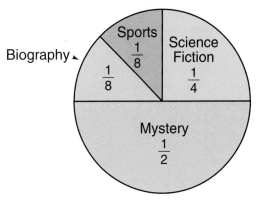

Solve. Use the circle graph at the right.

14. Tiger is Ed's cat. How many hours each day does he spend eating?

15. How many more hours does Tiger spend sleeping than he does playing?

16. To which two activities does Tiger devote the same amount of time? How much time?

Tiger's Day

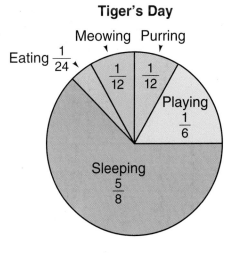

    CRITICAL THINKING

17. How many times greater is Tiger's sleeping time than his eating time?

18. On graph paper, construct a different kind of appropriate graph that shows the number of hours (not the fractional part) that Tiger spends on each activity in one week (7 days). Give your graph a title and label all of its parts.

8-11 Probability

For each of the spinners, there are five possible results, or **outcomes**: 1, 3, 5, 7, 9.

With Spinner A, each number has the same chance of occurring. The outcomes are **equally likely**.

With Spinner B, the outcomes are **not equally likely**. You are more likely to land on 1 than to land on 9.

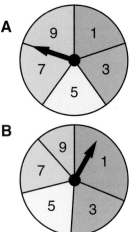

▶ The **probability** of a given event, E, is the chance that it will occur in an experiment. For equally likely outcomes, probability is given as a *ratio* by the formula:

$$P(E) = \frac{\text{number of favorable outcomes}}{\text{number of possible outcomes}}$$

Read: Probability of event E

$$P(7) = \frac{\text{number of favorable outcomes}}{\text{number of possible outcomes}} = \frac{1}{5} \leftarrow \boxed{\text{There is one 7.}}$$

$$P(\text{not } 7) = \frac{\text{number of favorable outcomes}}{\text{number of possible outcomes}} = \frac{4}{5} \leftarrow \boxed{\begin{array}{l}\text{There are four}\\\text{favorable outcomes:}\\\text{1, 3, 5, 9.}\end{array}}$$

$P(7) + P(\text{not } 7) = 1: \frac{1}{5} + \frac{4}{5} = 1$

These events are **complementary**.

▶ $P(5 \text{ or } 7)$ is an example of a probability statement for a **combined event**—probability of spinning a 5 *or* probability of spinning a 7.

Use Spinner A.

To find the probability of a combined event containing "or":

- Find the probability of each event.
- Add the two probabilities.

$P(5 \text{ or } 7) = P(5) + P(7)$
$= \frac{1}{5} + \frac{1}{5} = \frac{2}{5}$

278

For each experiment, list the possible outcomes. Then write whether the outcomes are *equally likely* or *not equally likely*.

1. Toss a marker on the board.

1	2
3	4

2. Spin the spinner.

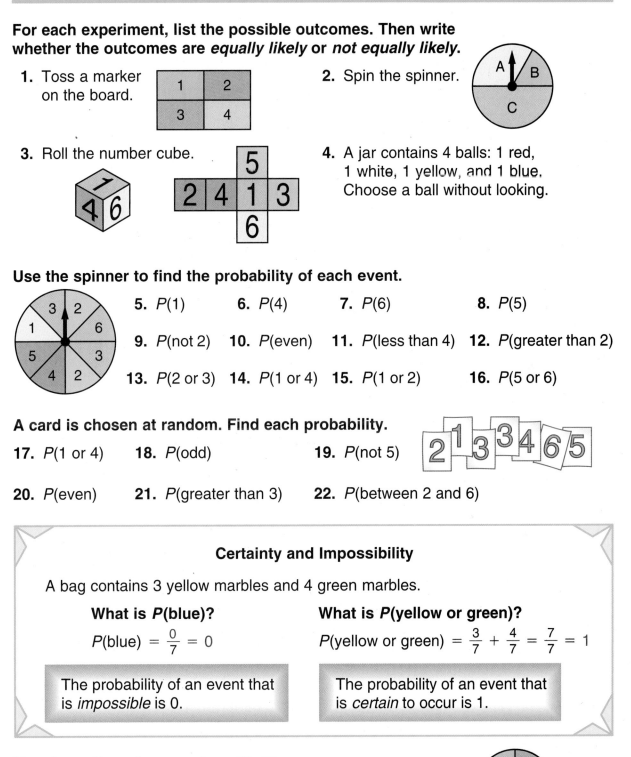

3. Roll the number cube.

4. A jar contains 4 balls: 1 red, 1 white, 1 yellow, and 1 blue. Choose a ball without looking.

Use the spinner to find the probability of each event.

5. *P*(1)   6. *P*(4)   7. *P*(6)   8. *P*(5)

9. *P*(not 2)   10. *P*(even)   11. *P*(less than 4)   12. *P*(greater than 2)

13. *P*(2 or 3)   14. *P*(1 or 4)   15. *P*(1 or 2)   16. *P*(5 or 6)

A card is chosen at random. Find each probability.

17. *P*(1 or 4)   18. *P*(odd)   19. *P*(not 5)

20. *P*(even)   21. *P*(greater than 3)   22. *P*(between 2 and 6)

Certainty and Impossibility

A bag contains 3 yellow marbles and 4 green marbles.

What is *P*(blue)?

$P(\text{blue}) = \frac{0}{7} = 0$

The probability of an event that is *impossible* is 0.

What is *P*(yellow or green)?

$P(\text{yellow or green}) = \frac{3}{7} + \frac{4}{7} = \frac{7}{7} = 1$

The probability of an event that is *certain* to occur is 1.

Use the spinner for exercises 23–26.

23. What is *P*(even)?   24. What is *P*(odd)?

25. Find *P*(less than 2).   26. Find *P*(between 1 and 7).

8-12 Compound Events

In probability, a **compound event** involves two or more events that are considered as a single event. One event follows the other. $P(A, B)$ means P(1st event, 2nd event).

(A, B) means A followed by B.

You can use **tree diagrams** to list all the possible outcomes for a compound event.

Experiment 1 Draw a card. Replace it. Draw a second card. How many outcomes are there? What is $P(1, 7)$?

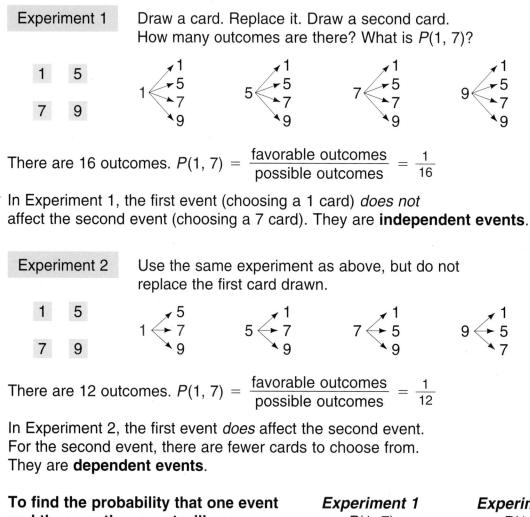

There are 16 outcomes. $P(1, 7) = \dfrac{\text{favorable outcomes}}{\text{possible outcomes}} = \dfrac{1}{16}$

In Experiment 1, the first event (choosing a 1 card) *does not* affect the second event (choosing a 7 card). They are **independent events**.

Experiment 2 Use the same experiment as above, but do not replace the first card drawn.

There are 12 outcomes. $P(1, 7) = \dfrac{\text{favorable outcomes}}{\text{possible outcomes}} = \dfrac{1}{12}$

In Experiment 2, the first event *does* affect the second event. For the second event, there are fewer cards to choose from. They are **dependent events**.

To find the probability that one event and then another event will occur:

- Find the probability of each event.

- *Multiply* the two probabilities.

Experiment 1
$P(1, 7)$
$P(1, 7) = \dfrac{1}{4} \times \dfrac{1}{4}$
$= \dfrac{1}{16}$

Experiment 2
$P(1, 7)$
$P(1, 7) = \dfrac{1}{4} \times \dfrac{1}{3}$
$= \dfrac{1}{12}$

Draw tree diagrams. List all possible outcomes.

1. Toss a penny and roll a 1-6 number cube.

2. Spin the spinner and choose a marble without looking.

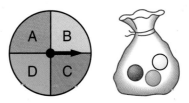

Find each probability. Use the experiment in exercise 1.

3. P(H, 3)   4. P(H, even)   5. P(T, 2 or 3)   6. P(T, odd)   7. P(H, 1, 2, or 3)

Spin the spinners.

8. List all possible outcomes. Use tree diagrams.

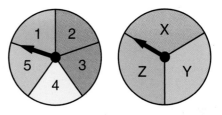

Find each probability. Use the experiment in exercise 8.

9. P(1, X)   10. P(2 or 3, Z)   11. P(odd, Y)   12. P(even, X)

Draw tree diagrams to find each number of possible outcomes.

13. You are choosing a movie and a restaurant among 3 movies and 4 restaurants.

14. You are choosing an outfit from a selection of 4 shirts, 2 pairs of pants, and 2 sweaters.

Find the probability: (a) if the first choice is replaced; and (b) if the first choice is not replaced.

Experiment: Choose a pair of socks from a drawer containing 3 yellow (Y) pairs, 2 white (W) pairs, and 1 red (R) pair. Then choose a second pair.

15. P(R, Y)   16. P(Y, R)   17. P(Y, W)   18. P(R, Y or W)

SKILLS TO REMEMBER

Multiply.

19. $\frac{1}{8} \times 3200$ 20. $\frac{3}{4} \times 120$ 21. $\frac{7}{10} \times 1000$ 22. $\frac{5}{6} \times 3600$

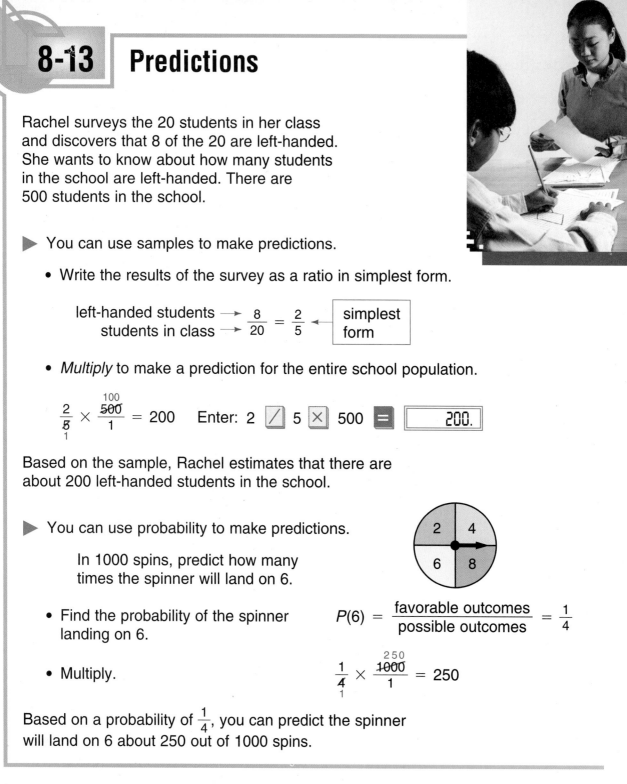

8-13 Predictions

Rachel surveys the 20 students in her class and discovers that 8 of the 20 are left-handed. She wants to know about how many students in the school are left-handed. There are 500 students in the school.

▶ You can use samples to make predictions.

- Write the results of the survey as a ratio in simplest form.

$$\text{left-handed students} \rightarrow \frac{8}{20} = \frac{2}{5} \leftarrow \boxed{\text{simplest form}}$$
$$\text{students in class} \rightarrow$$

- *Multiply* to make a prediction for the entire school population.

$$\frac{2}{\underset{1}{5}} \times \frac{\overset{100}{\cancel{500}}}{1} = 200 \quad \text{Enter: } 2 \boxed{/} 5 \boxed{\times} 500 \boxed{=} \boxed{200.}$$

Based on the sample, Rachel estimates that there are about 200 left-handed students in the school.

▶ You can use probability to make predictions.

In 1000 spins, predict how many times the spinner will land on 6.

- Find the probability of the spinner landing on 6.

$$P(6) = \frac{\text{favorable outcomes}}{\text{possible outcomes}} = \frac{1}{4}$$

- Multiply.

$$\frac{1}{\underset{1}{4}} \times \frac{\overset{250}{\cancel{1000}}}{1} = 250$$

Based on a probability of $\frac{1}{4}$, you can predict the spinner will land on 6 about 250 out of 1000 spins.

In 2000 spins, predict the number of times the spinner above would land on each of the following.

1. number > 5

2. number < 4

3. number between 2 and 8

PROBLEM SOLVING

4. In a sample of 100 people who saw a movie, 56 said they enjoyed it. How many people would you expect to say they enjoyed the movie out of a group of 5000 people?

5. Out of 20 batteries tested, 1 was found to be defective. Out of 1000 batteries, how many would you expect to be defective?

6. In a sample of 50 people in the stadium, 7 were wearing baseball caps. How many of the 50,000 fans would you predict to be wearing baseball caps?

7. Norma reached into a jar filled with jelly beans and pulled out 25 beans. Five of them were red. If there are 600 beans in the jar, how many would you expect to be red?

8. How many heads would you expect to get if you tossed a dime 100 times? Why?

9. You roll a 1–6 number cube 1000 times. How many rolls of 5 or 6 would you expect to get?

10. In 1000 spins, predict the number of times the spinner will land on A.

11. In 1000 spins, predict the number of times the spinner will land on B or C.

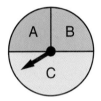

The bar graph shows the music preferences of 50 people.

12. In a population of 1000 people, how many would you expect to choose each type of music?

13. Would a survey of 100 people give the same results? Explain.

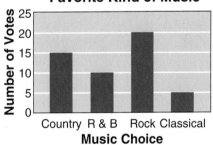

PROJECT

14. Conduct a survey of the students in your school (or grade) to find out how many are right-handed, left-handed or ambidextrous (use both hands).

  a. Determine whom you will sample and the questions you will ask.

  b. Show the data you collect in an appropriate graph.

  c. Make predictions for the entire school (or grade).

Communicate

Misleading Graphs and Statistics

The way data are presented in a graph can affect the impression the graph makes. Examine the two graphs below.

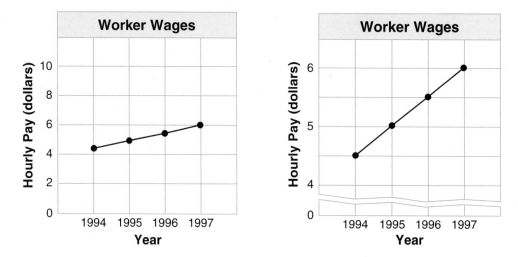

Both graphs show the same information about worker hourly pay. Although both show that wages have increased, the graph at the right gives the impression that wages have increased more rapidly. What causes this?

Look at the vertical scale of each graph. The *expanded scale* on the graph at the right creates the impression of a faster rate of increase.

Study this example.

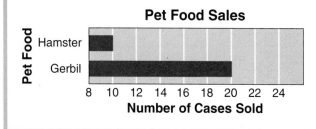

This graph is *misleading.* Although gerbil food sales are *twice* those of hamster food sales, the graph gives the impression that the difference is much greater.

Use the graphs above for exercises 1–2.

1. It is time for a new labor contract. Which line graph would you use if you represented labor? Which would you use if you represented management? Explain.

2. Why does the bar graph give the impression that gerbil food sales are about 6 times as great as hamster food sales? Draw an accurate bar graph for the data.

The table and graphs show the number of exercise videos sold during a 5-month period.

Month	Jan.	Feb.	Mar.	Apr.	May
Sales (hundreds)	7.5	10	12	14	16

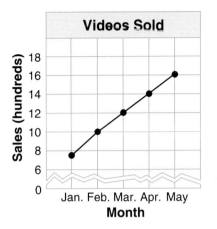

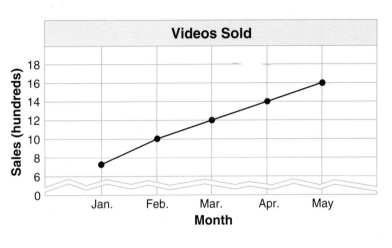

3. Do both graphs show the same data?

4. Why do you get a different impression about the data from the two graphs?

5. Which graph would someone use who wants to convince you that the sales of exercise videos have risen dramatically?

The graph at the right shows sales of pet-training videos.

6. How many pet-training videos were sold in 1996? in 1997?

7. What is misleading about this graph?

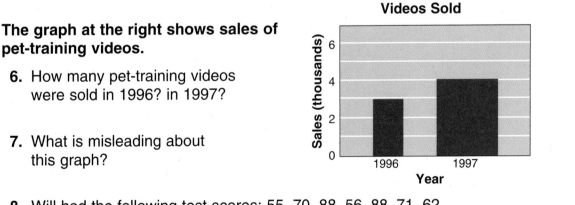

8. Will had the following test scores: 55, 70, 88, 56, 88, 71, 62. He told his friend, "My most typical score was 88" (the mode). Discuss with the class why Will's statement is misleading.

Discuss ✓

9. An advertiser said "Nine out of 10 dentists agree. Glomb works best." Discuss with the class why this statement might be misleading.

TECHNOLOGY

Computer Graphing

You can use a computer program to graph data.
The program will do most of the work once
the data and information about the graph are entered.

Use the data below to draw a bar and a line graph.

Calculator Sales–Units Sold

	1st Quarter	2nd Quarter	3rd Quarter	4th Quarter
model X90	510	760	830	970
model X110	670	410	320	540
model JZII	410	400	350	330

▶ Answer the following questions
to create a bar graph.

- Type of graph? **bar**
- Graph which column(s)? **2nd Quarter**
- Vertical or horizontal bars? **Vertical**
- Graph title? **2nd Quarter Sales**
- Vertical label? **Units Sold**
- Horizontal label? **Model**
- Vertical scale? **automatic**

> Most programs automatically
> choose an appropriate scale.

▶ Answer the following questions
to create a line graph.

- Type of graph? **line**
- Graph which column(s)? **all**
- Graph title? **Sales**
- Vertical label? **Units Sold**
- Horizontal label? **Quarter**
- Labels for key? **X90**
  X110
  JZII
- Vertical scale? **automatic**

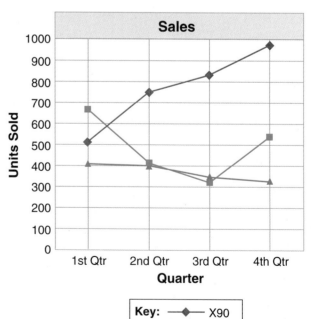

Solve. Use the data and the bar graph on page 286.

1. How many calculators does each unit on the vertical scale represent?

2. What is the range of sales represented on the graph?

3. How many units of the best-selling calculator sold in the 2nd quarter? 3rd quarter?

4. Which two calculator models were about equally popular in the 2nd quarter?

5. What is the mean number of calculators sold for each quarter?

6. What are the median and mode for the set of data over the 4 quarters?

Solve. Use the line graph on page 286.

7. How do you know which line represents model JZII?

8. Which line represents model X90? X110?

9. Which model's sales steadily increased over the 4 quarters?

10. The company wants to discontinue one line of calculators. Which model should the company discontinue? Explain why.

Use the data below to solve each.

Slow-Pitch Softball—Division 2—Games Won

	1995	1996	1997
Mudhens	36	34	36
Cougars	28	22	18
Sand Fleas	36	38	44

11. The Cougars' manager wants to make a bar graph of the team's wins from 1995 to 1997. Answer the bar graph questions on page 286. Then draw the graph.

12. The Division 2 president wants to make a line graph of the games won by the Division 2 teams from 1995–1997. Answer the line graph questions on page 286. Then draw the graph.

13. What is the mean number of wins for each team from 1995 to 1997?

14. What are the median and mode for the set of data over the 3 years?

8-16 Problem Solving: Organized List

Problem: Frank designed this math game for his class. It is played by tossing a marker twice on the square and finding the sum of the two addends on which the marker lands. How many different sums are possible?

1 IMAGINE Picture yourself playing Frank's game.

2 NAME

Facts: A marker is tossed 2 times.
Possible addends — $\frac{1}{2}$, $\frac{1}{4}$, $2\frac{1}{4}$, $1\frac{1}{2}$

Question: How many different sums are possible?

3 THINK Since a marker is tossed twice, you can get the same addend or 2 different addends. To find the possible sums, make an *organized list* using the addends. Eliminate extra addends that have the same sum.

> Remember: A marker can land twice on the same addend.

4 COMPUTE

Addend	$\frac{1}{2}$	$\frac{1}{2}$	$\frac{1}{2}$	$\frac{1}{2}$	$\cancel{\frac{1}{4}}$	$\frac{1}{4}$	$\frac{1}{4}$	$\frac{1}{4}$
Addend	$\frac{1}{2}$	$\frac{1}{4}$	$2\frac{1}{4}$	$1\frac{1}{2}$	$\cancel{\frac{1}{2}}$	$\frac{1}{4}$	$2\frac{1}{4}$	$1\frac{1}{2}$
Sum	1	$\frac{3}{4}$	$2\frac{3}{4}$	2	$\cancel{\frac{3}{4}}$	$\frac{1}{2}$	$2\frac{1}{2}$	$1\frac{3}{4}$

$$\frac{1}{2} + \frac{1}{4} = \frac{1}{4} + \frac{1}{2}$$

Addend	$\cancel{2\frac{1}{4}}$	$\cancel{2\frac{1}{4}}$	$2\frac{1}{4}$	$2\frac{1}{4}$	$\cancel{1\frac{1}{2}}$	$\cancel{1\frac{1}{2}}$	$\cancel{1\frac{1}{2}}$	$1\frac{1}{2}$
Addend	$\cancel{\frac{1}{2}}$	$\cancel{\frac{1}{4}}$	$2\frac{1}{4}$	$1\frac{1}{2}$	$\cancel{\frac{1}{2}}$	$\cancel{\frac{1}{4}}$	$\cancel{2\frac{1}{4}}$	$1\frac{1}{2}$
Sum	$\cancel{2\frac{3}{4}}$	$\cancel{2\frac{1}{2}}$	$4\frac{1}{2}$	$3\frac{3}{4}$	$\cancel{2}$	$\cancel{1\frac{3}{4}}$	$\cancel{3\frac{3}{4}}$	3

There are 10 different sums possible.

5 CHECK Use a tree diagram to be sure all possible combinations are listed.

$\frac{1}{2} + \square$
$\frac{1}{2}$ $\frac{1}{4}$ $2\frac{1}{4}$ $1\frac{1}{2}$

$2\frac{1}{4} + \square$
$\frac{1}{2}$ $\frac{1}{4}$ $2\frac{1}{4}$ $1\frac{1}{2}$

$\frac{1}{4} + \square$
$\frac{1}{2}$ $\frac{1}{4}$ $2\frac{1}{4}$ $1\frac{1}{2}$

$1\frac{1}{2} + \square$
$\frac{1}{2}$ $\frac{1}{4}$ $2\frac{1}{4}$ $1\frac{1}{2}$

Solve. Make an organized list to help you.

1. The nursery has 6 evergreen trees to be used in landscaping the park. There are 4 different areas where the trees can be put. In how many ways can the trees be placed so that each area has at least one tree?

IMAGINE   Draw a picture of trees and a park.

NAME   *Facts:*   6 evergreen trees
4 different areas

Question: In how many ways can the trees be placed so that each area has at least one tree?

THINK   Make an organized list of 4 different park areas and trees to go into each.
(Make as many combinations as possible with 6 trees in 4 areas.)

COMPUTE ⟶ **CHECK**

2. How many different four-digit numbers can you make using the digits 0, 1, 2, and 3 if repetition of a digit is not permitted?

3. Jason has a yellow shirt, a pink shirt, and a blue shirt; a pair of black slacks and a pair of tan slacks; a pin-striped sport coat and a black sport coat. How many different three-piece outfits can he make?

4. How many different combinations of numbers and letters can be made from spinning a dial marked 4, 7, and 9 and spinning a dial marked *A, B, C, D,* and *E?*

5. In a bowling game you have 2 chances to knock down the 10 pins. How many different ways can the pins be knocked down if with every 2 tries all 10 pins are knocked down?

6. How many different three-digit numbers can you make using the digits 0, 1, and 2 if repetition of digits *is* permitted?

289

Problem-Solving Applications

Solve each problem and explain the method you used.

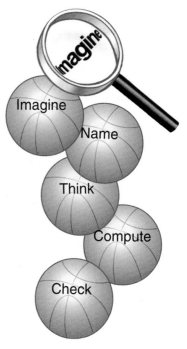

1. In their last 8 basketball games, the Johnston Jump Shots scored these points: 85, 62, 74, 71, 81, 65, 81, and 57. What is their mean score per game?

Imagine

Name

Think

Compute

Check

2. What are the range, the median, and the mode of scores for the Jump Shots' last 8 games?

3. Make a bar graph to show these scores.

4. In their previous 8 games the Jump Shots scored these points: 72, 74, 81, 88, 58, 60, 67, and 68. Make a stem-and-leaf plot using all 8 scores. What are the range, median, and mode of the scores?

5. The Jump Shots have won the coin toss at the beginning of their last three games. What is the probability that they will win the coin toss at the beginning of their next game?

6. There are 2 red, 2 orange, and 1 brown practice balls in the equipment locker. The coach gives out 2 balls at the beginning of practice. What is the probability of her choosing a red ball and a brown ball if the first ball is replaced? If the first ball is *not* replaced?

Use the line graph for problems 7–9.

7. Which team had the greater mean score in February?

8. Which team had a mean score of 62 in March?

9. Over the four months, which team had the greater mean score?

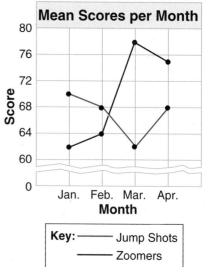

Use a strategy from the list or another strategy you know to solve each problem.

USE THESE STRATEGIES:
Write a Number Sentence
Organized List
Missing Information
Use a Graph
Multi-Step Problem

10. There are 5 teams in the local basketball league: the Jump Shots, the Zoomers, the Victors, the Hoopsters, and the Towers. Each season, every team plays every other team twice. How many games are played in a season?

11. The Victors won twice as many games as the Hoopsters. How many games did the Hoopsters win?

12. Janine scored the following points in the first 5 games: 24, 29, 20, 28, and 19. How many points must Janine score in the sixth game to keep her median and mean scores the same?

Use the bar graph for problems 13–15.

13. Which team scored more points in the first game?

14. Which game had the greatest point spread between the winning and losing scores?

15. Which team won 3 out of 4 of the play-off games?

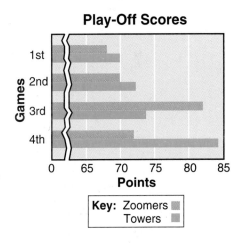

Play-Off Scores

Games: 1st, 2nd, 3rd, 4th

Points: 0, 65, 70, 75, 80, 85

Key: Zoomers
Towers

Use the circle graph for problems 16–18.

16. Which two players together scored one fourth of the points in Game 4?

17. What part of the team's points did Janine score?

18. What was the mean score per player for the players in game 4?

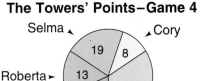

The Towers' Points – Game 4

Selma 19, Cory 8, Roberta 13, Connie 16, Janine 28

MAKE UP YOUR OWN

19. Suppose the five girls in the circle graph each had to choose 1 of 6 numbers for their team uniforms. Draw a tree diagram to show the possible uniform numbers each girl can choose.

Chapter Review and Practice

Use the bar graph to answer the questions. *(See pp. 258–261.)*

1. How many students chose basketball?

2. How many fewer students chose football than soccer?

3. How many students in all were surveyed?

4. What fractional part of those surveyed did *not* choose baseball?

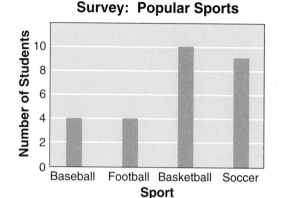

Survey: Popular Sports

Find the range, mean, median, and mode for each set of data. *(See pp. 264–265.)*

5. 93, 95, 118, 91, 84, 93, 112

6. $3.75, $3.95, $4.80, $4.86

Make a double line graph for this table. *(See pp. 274–275.)*

7.

Doris Ann's Weekend Jobs					
Weekend	**1**	**2**	**3**	**4**	**5**
Frame Making	15	10	10	9	10
Frame Painting	3	11	15	14	7

PROBLEM SOLVING
Find the probability. *(See pp. 278–281.)*

Experiment: A box contains 3 red marbles, 1 black marble, and 2 white marbles. Choose marbles at random.

8. *P*(red)

9. *P*(black)

10. *P*(white)

11. *P*(green)

12. *P*(red or black)

13. *P*(not black)

14. *P*(not red)

15. *P*(marble)

16. Choose one marble. Replace it. Choose another marble. What is *P*(red, white)?

17. Choose one marble. Do not replace it. Choose another marble. What is *P*(black, red)?

(See *Still More Practice,* p. 513.)

TRIPLE LINE AND BAR GRAPHS

A **triple line graph** and a **triple bar graph** are used
to compare three sets of data. Each set of data is
graphed separately but on the same grid.

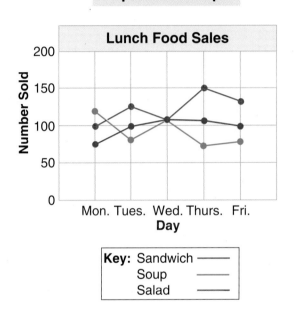

Triple Line Graph

Lunch Food Sales

Key: Sandwich ———
Soup ———
Salad ———

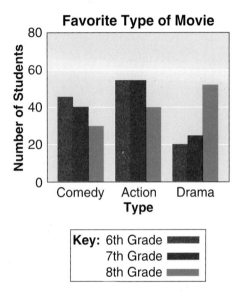

Triple Bar Graph

Favorite Type of Movie

Key: 6th Grade ▬▬
7th Grade ▬▬
8th Grade ▬▬

Use the graphs above for exercises 1–8.

1. On which day were the most
sandwiches sold?

2. On which two days were the
same number of salads sold?

3. On which day did the three
foods have equal sales?

4. On Monday, which food was
most popular? least popular?

5. Which type of movie was least
preferred by Grade 6?

6. In which grade did the fewest
students prefer comedy?

7. Which type of movie was liked
by the same number of 6th
and 7th graders?

8. In which grade did the
greatest number of students
prefer drama?

Performance Assessment

Use the pictograph for problems 1–2.

1. If the number of dogs owned as pets by 6th graders increases by 6 and the number owned by 5th graders decreases by 5, how many total dogs are owned by the students in all four grades?

2. Write and solve a problem using the pictograph.

Numbers of Dogs Owned as Pets	
3rd grade	🐶 🐶 🐶
4th grade	🐶 🐶 🐶 🐶 🐶
5th grade	🐶 🐶 🐶
6th grade	🐶 🐶 🐶 🐶 🐶

Key: 🐶 = 2 dogs

Use the table for exercises 3–5.

3. Organize the data in a frequency table.

4. Find the range and mode of the data.

Pages Read: Grade 6

31	20	38	31	24	30	12	38
20	18	36	34	21	33	38	36
34	23	39	30	43	30	31	28

5. Make a line plot and a stem-and-leaf plot of the data.

Name the graph described.

6. Uses small pictures or symbols.

7. Shows changes over time.

8. Shows parts of a whole.

Make a double line graph.

9. **Number of Books Sold**

Month	Jan.	Feb.	Mar.	Apr.
History	100	125	100	90
Psychology	175	180	170	120

PROBLEM SOLVING   *Use a strategy or strategies you have learned.*

10. Use spinner A. Find each probability: $P(1 \text{ or } 4)$, $P(\text{not } 2)$, $P(9)$.

11. Spin A and then spin B. List all possible outcomes.

12. Use spinner B. In 600 spins, predict the number of times the spinner will land on 6, 7, or 8.

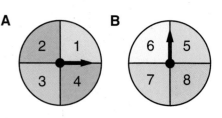

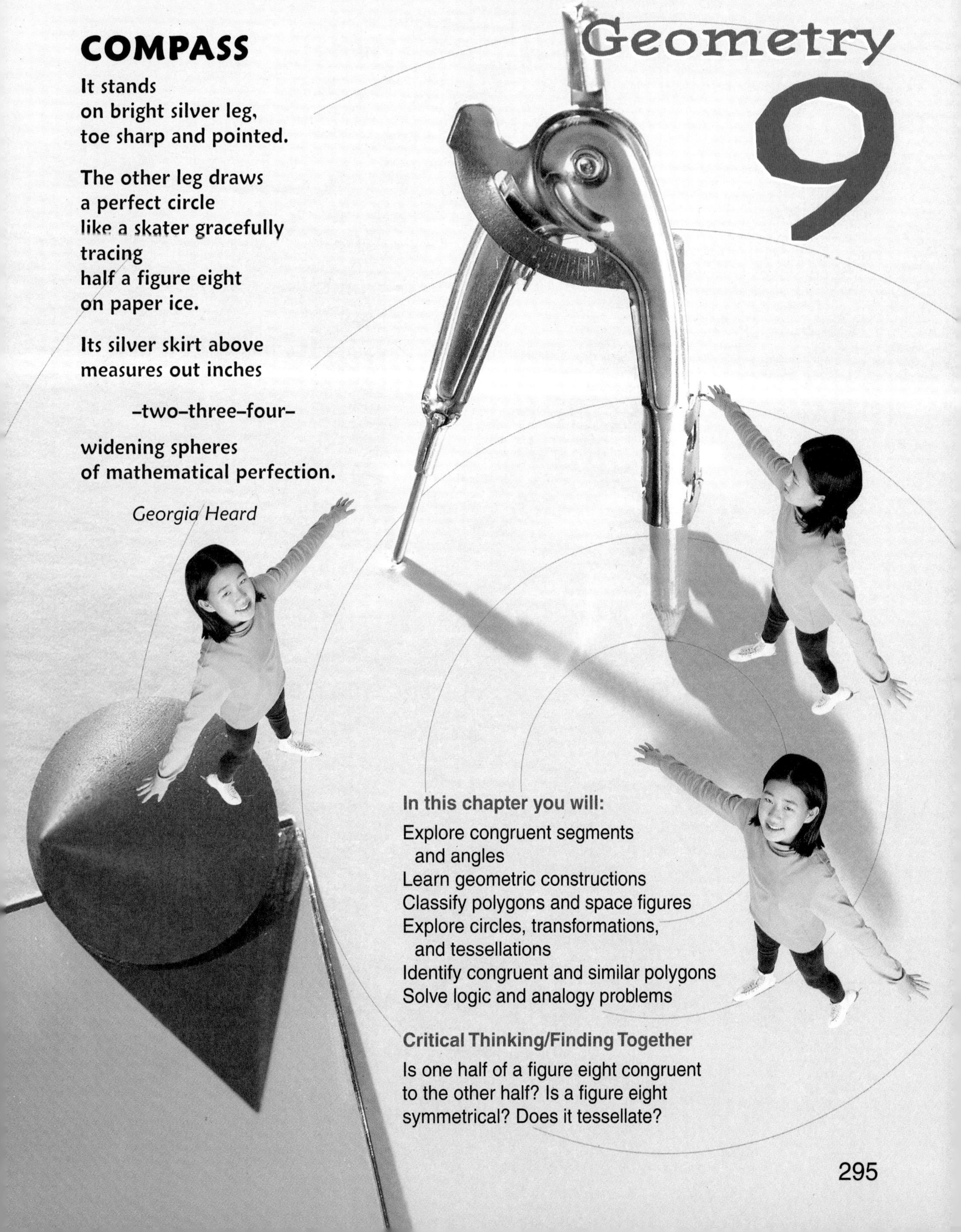

COMPASS

It stands
on bright silver leg,
toe sharp and pointed.

The other leg draws
a perfect circle
like a skater gracefully
tracing
half a figure eight
on paper ice.

Its silver skirt above
measures out inches

   –two–three–four–

widening spheres
of mathematical perfection.

Georgia Heard

Geometry
9

In this chapter you will:

Explore congruent segments
  and angles
Learn geometric constructions
Classify polygons and space figures
Explore circles, transformations,
  and tessellations
Identify congruent and similar polygons
Solve logic and analogy problems

Critical Thinking/Finding Together

Is one half of a figure eight congruent
to the other half? Is a figure eight
symmetrical? Does it tessellate?

Update your skills. See page 20.

9-1 Congruent Segments and Angles

Line segments that have the same length are **congruent**. Angles are congruent if they have the same measure (same angle opening).

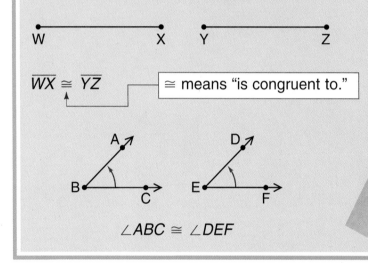

$\overline{WX} \cong \overline{YZ}$

≅ means "is congruent to."

$\angle ABC \cong \angle DEF$

Record your answers on a sheet of paper.

Hands-On Understanding

Materials Needed: straightedge, compass, paper, pencil, tracing paper

| Step 1 | To decide if $\overline{PQ}$ is congruent to $\overline{RS}$, trace $\overline{PQ}$ onto a sheet of paper. |

| Step 2 | Place your tracing on top of $\overline{RS}$. |

| Step 3 | Compare the length of your tracing with the length of $\overline{RS}$. |

Does your tracing of $\overline{PQ}$ exactly cover $\overline{RS}$? Is $\overline{PQ}$ congruent to $\overline{RS}$? Explain your answer.

| **Step 4** | You can use a compass to decide if $\overline{PQ}$ and $\overline{RS}$ are congruent. First, put the compass point at point P and the compass pencil at point Q, as shown. |

| **Step 5** | Keep the same compass opening. Pick up your compass and place the compass point at point R and the compass pencil at point S.

Do the two segments have the same compass opening? Do the two segments have the same length? Are they congruent? |

| **Step 6** | You can use a tracing to decide if $\angle JKL$ and $\angle STU$ are congruent. Trace $\angle JKL$ onto a sheet of paper. |

| **Step 7** | Place your tracing on top of $\angle STU$. |

| **Step 8** | Compare the opening of the angle on your tracing with the opening of $\angle STU$.

Do the two angles have the same opening? Is $\angle JKL$ congruent to $\angle STU$? Explain. |

Communicate

Which pairs of line segments or angles are congruent? Explain.

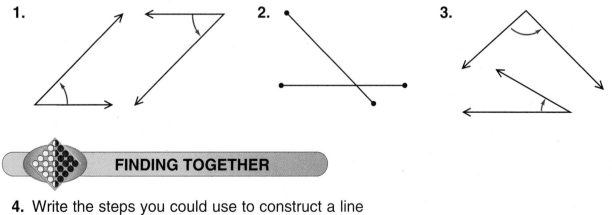

1. 2. 3.

FINDING TOGETHER

4. Write the steps you could use to construct a line segment congruent to $\overline{AB}$. Trace $\overline{AB}$ onto a sheet of paper and do the construction.

297

9-2 Constructing Perpendicular Lines

Lines (line segments or rays) that intersect to form four congruent angles are **perpendicular**.

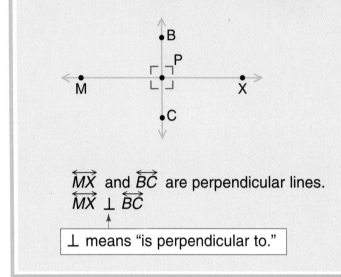

$\overleftrightarrow{MX}$ and $\overleftrightarrow{BC}$ are perpendicular lines.
$\overleftrightarrow{MX} \perp \overleftrightarrow{BC}$

$\perp$ means "is perpendicular to."

Construct a line perpendicular to the given line.

Hands-On Understanding

Materials Needed: straightedge, compass, paper, pencil

| Step 1 | Copy the line onto your own paper. Mark a point S on the line. |

| Step 2 | Use your compass. Put the compass point on *S* and use the compass pencil to draw an arc that intersects (crosses) the line at any point *R*. Rotate the compass so the pencil also draws an arc at a point *T* on the line. |

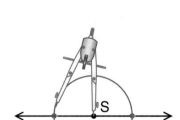

| Step 3 | Write *R* and *T* to label the two points. |

298

Step 4	Open the compass to more than half the length of $\overline{RT}$.

Step 5	*From point R*, rotate the compass so the pencil draws an arc (above the line).

Step 6	Keep the same compass opening. From point T, rotate the compass so the pencil draws another arc (above the line) that intersects the first arc.

Step 7	Write N to label the point where the two arcs intersect.

Step 8	Draw $\overleftrightarrow{NS}$. Line NS is perpendicular to line RT at point S. In symbols, you write $\overleftrightarrow{NS} \perp \overleftrightarrow{RT}$. It is also true that $\overleftrightarrow{RT} \perp \overleftrightarrow{NS}$.

Communicate

1. Describe the four angles formed by two perpendicular lines. What type of angles are they?

Discuss ✓

2. Explain to your parents or another adult how to construct perpendicular lines. Use Steps 1–8 to help you.

CRITICAL THINKING

3. Draw a line and pick two points on the line. Construct a perpendicular line at each point. What do you discover?

9-3 Measuring and Drawing Angles

An angle is formed by two rays with a common endpoint. The rays are the **sides** of the angle. The common endpoint is the **vertex** of the angle. You can use a protractor to measure the number of units or **degrees** (°) in an angle.

▶ **To measure ∠XYZ:**

- Place the center mark of the protractor on the **vertex** of the angle, Y, with $\overrightarrow{YX}$ pointing to 0°.

- Read the measure of the angle where $\overrightarrow{YZ}$ crosses the protractor.

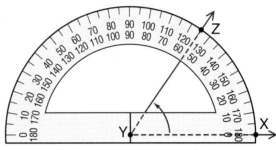

∠XYZ measures 55°.

You can use a protractor to draw an angle of a given number of degrees.

▶ **To draw an angle of 140°:**

- Draw a base ray, $\overrightarrow{MN}$.

- Place the center mark of the protractor on M with $\overrightarrow{MN}$ pointing to 0°.

- Mark P at 140°.

- Draw $\overrightarrow{MP}$.

Mark P at 140° point of protractor.

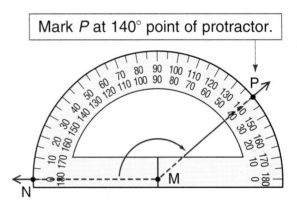

Give the measure of each angle.

1. ∠AOB
2. ∠AOC
3. ∠AOD
4. ∠GOB
5. ∠GOF
6. ∠EOD

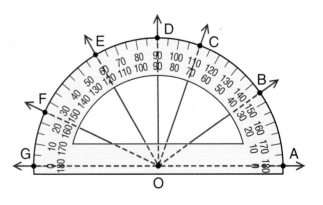

Use a protractor to draw an angle of the given measure.

7. 40°   **8.** 75°   **9.** 90°   **10.** 135°   **11.** 5°   **12.** 180°

Estimate the measure of each angle. Then use a protractor to find the exact measure.

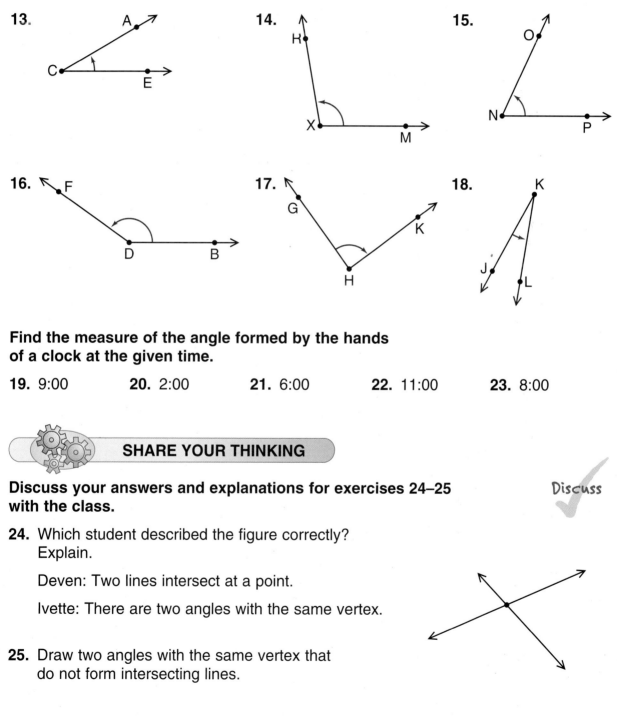

13.

14.

15.

16.

17.

18.

Find the measure of the angle formed by the hands of a clock at the given time.

19. 9:00   **20.** 2:00   **21.** 6:00   **22.** 11:00   **23.** 8:00

SHARE YOUR THINKING

Discuss your answers and explanations for exercises 24–25 with the class.

Discuss

24. Which student described the figure correctly? Explain.

Deven: Two lines intersect at a point.

Ivette: There are two angles with the same vertex.

25. Draw two angles with the same vertex that do not form intersecting lines.

Classifying Angles

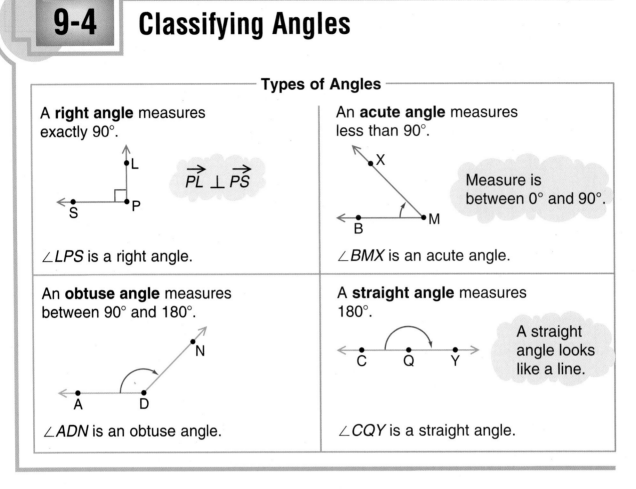

Types of Angles

A **right angle** measures exactly 90°.

∠LPS is a right angle.

$\overrightarrow{PL} \perp \overrightarrow{PS}$

An **acute angle** measures less than 90°.

Measure is between 0° and 90°.

∠BMX is an acute angle.

An **obtuse angle** measures between 90° and 180°.

∠ADN is an obtuse angle.

A **straight angle** measures 180°.

A straight angle looks like a line.

∠CQY is a straight angle.

The measure of an angle is given. Write whether the angle is *right*, *acute*, *obtuse*, or *straight*.

1. 45° **2.** 90° **3.** 4° **4.** 112° **5.** 175° **6.** 180°

Classify each angle as *right*, *acute*, *obtuse*, or *straight*. Use a protractor to check.

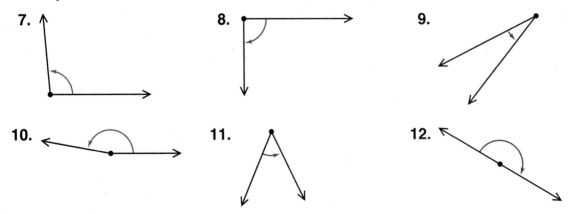

7.

8.

9.

10.

11.

12.

Pairs of Angles

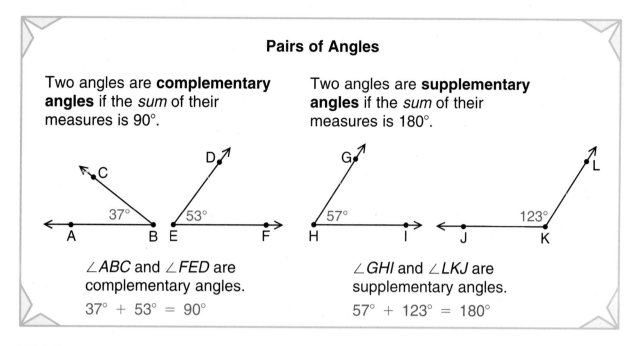

Two angles are **complementary angles** if the *sum* of their measures is 90°.

∠ABC and ∠FED are complementary angles.

37° + 53° = 90°

Two angles are **supplementary angles** if the *sum* of their measures is 180°.

∠GHI and ∠LKJ are supplementary angles.

57° + 123° = 180°

Which are the measures of complementary angles? supplementary angles? neither?

13. 60°, 30°

14. 130°, 50°

15. 113°, 67°

16. 110°, 90°

17. 90°, 90°

18. 179°, 1°

19. 45°, 45°

20. 97°, 93°

Write *True* or *False*. If false, explain why.   Communicate ✓

21. The vertex of ∠CAD is C.

22. ∠DAB is a right angle.

23. $\overrightarrow{CA}$ is a part of $\overleftrightarrow{CM}$.

24. ∠BAC is an obtuse angle.

25. $\overleftrightarrow{AM}$ and $\overleftrightarrow{MA}$ name the same ray.

26. ∠CAD and ∠MAB are complementary angles.

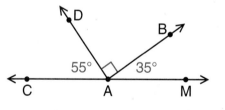

Draw the angles described.

27. Two complementary angles that share a common side

28. Two supplementary angles that are congruent and share a common side

303

Elmira is making a class pennant. How can she use a compass to **construct** an angle congruent to the angle drawn?

To construct an angle _DEF_ congruent to angle _ABC_:

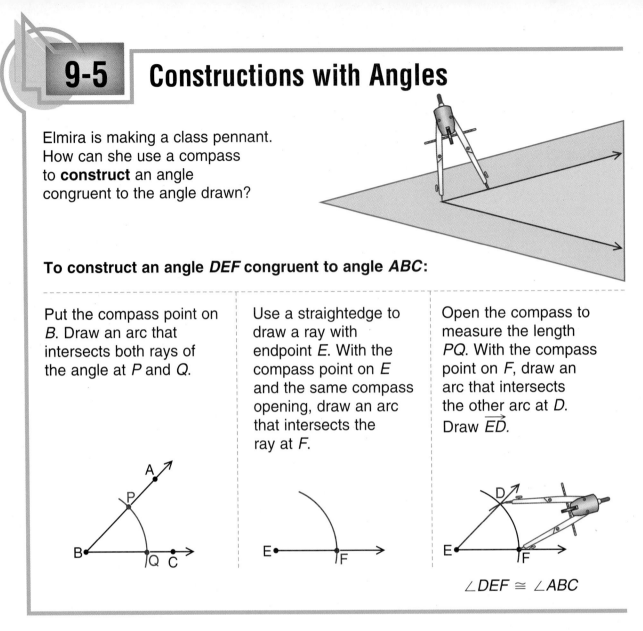

Put the compass point on _B_. Draw an arc that intersects both rays of the angle at _P_ and _Q_.

Use a straightedge to draw a ray with endpoint _E_. With the compass point on _E_ and the same compass opening, draw an arc that intersects the ray at _F_.

Open the compass to measure the length _PQ_. With the compass point on _F_, draw an arc that intersects the other arc at _D_. Draw $\overrightarrow{ED}$.

$\angle DEF \cong \angle ABC$

Use unlined paper to trace each of the angles. Then construct an angle congruent to each.

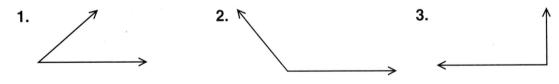

1.

2.

3.

Construct the indicated angles.

4. Draw an angle with each given measure: 70° and 135°. Construct an angle congruent to each.

Angle Bisector

An **angle bisector** is a ray that divides the angle into two congruent parts. You can use a compass and a straightedge to **bisect** an angle.

To construct the bisector of angle *XYZ*:

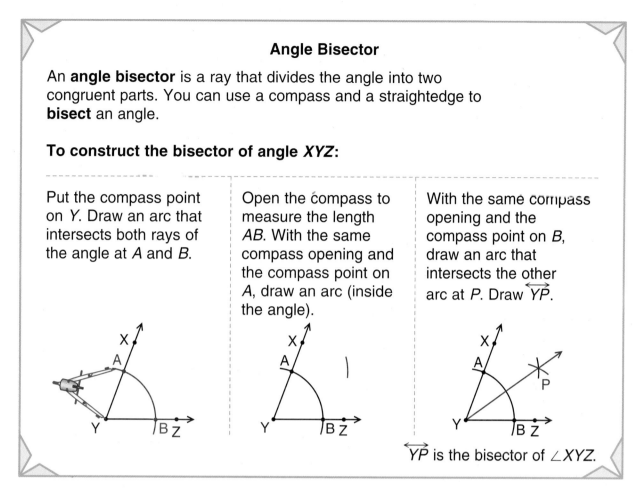

Put the compass point on *Y*. Draw an arc that intersects both rays of the angle at *A* and *B*.

Open the compass to measure the length *AB*. With the same compass opening and the compass point on *A*, draw an arc (inside the angle).

With the same compass opening and the compass point on *B*, draw an arc that intersects the other arc at *P*. Draw $\overleftrightarrow{YP}$.

$\overleftrightarrow{YP}$ is the bisector of ∠*XYZ*.

Name the bisector of each angle and the two congruent angles formed.

5.

6.
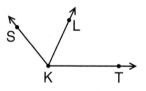

7.
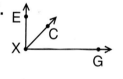

Draw each angle described. Then bisect it.

8. acute angle

9. obtuse angle

10. right angle

CHALLENGE

11. Draw an angle with a measure of 160°. Construct an angle with $\frac{1}{4}$ of this measure. (*Hint:* Each angle formed by a bisector has a measure of $\frac{1}{2}$ the measure of the original angle.) Share your results with the class.

Communicate

Update your skills. See page 22.

9-6 Polygons

When all of the sides of a polygon are congruent and all of the angles are congruent, the polygon is a **regular polygon**.

> A regular triangle is an equilateral triangle. A regular quadrilateral is a square.

A regular octagon has 8 congruent sides and 8 congruent angles.

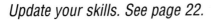

Examples	
Regular	**Not Regular**

▶ A **diagonal** of a polygon is a line segment that connects two vertices and is *not* a side.

$\overline{AC}, \overline{AD}, \overline{BD}, \overline{BE}, \overline{CE}$ are diagonals of pentagon *ABCDE*.

Every pentagon, whether it is regular or not regular, has five diagonals.

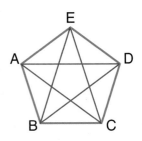

▶ You can draw some of the diagonals of a polygon to help you decide if the polygon is **convex** or **concave**.

convex polygon — diagonals do not have any points outside the polygon; *JKLMNO* is a convex hexagon.

concave polygon — one or more diagonals have points outside the polygon; *PRSTUV* is a concave hexagon. Diagonal $\overline{PU}$ is outside the hexagon.

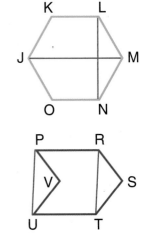

Tell whether the polygon is regular or *not* regular. Then tell whether it is convex or concave. Explain your answers.

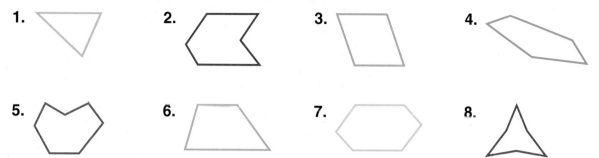

1.   2.   3.   4.

5.   6.   7.   8.

Copy and complete this table. Draw each polygon.

	Figure (draw)	Name	Number of Sides	Number of Vertices
9.	?	heptagon	7	?
10.	?	nonagon	?	9
11.	?	decagon	10	?

Find the number of diagonals in each polygon.

12. pentagon   13. hexagon   14. heptagon   15. octagon

16. In your answers for exercises 12–15, does it matter if the polygons are convex or concave? Explain.

Communicate ✓

Draw the polygon described. Then draw and name its diagonals.

17. rhombus *JKLM*   18. regular quadrilateral *WXYZ*

19. concave pentagon *ABCDE*   20. regular triangle *NRT*

SKILLS TO REMEMBER

Algebra ✓

Find the missing number.

21. $c + 98 = 180 \longrightarrow c = 180 - 98 \longrightarrow c = 82$ 22. $a + 110 = 180$

23. $b + 275 = 360$ 24. $310 + a = 360$ 25. $165 + c = 180$

Classifying Triangles

Triangles are classified by the lengths of their sides and by the measures of their angles.

Sides

equilateral triangle (A)—all sides congruent
isosceles triangle (B)—two sides congruent
scalene triangle (C)—no sides congruent

Angles

acute triangle (D)—three acute angles
obtuse triangle (E)—one obtuse angle
right triangle (F)—one right angle

▶ The **sum of the measures** of the angles of any triangle is **180°**. You can use mental math or a number sentence to find the measure of one angle when you know the measures of the other two.

Find the measure of $\angle C$.

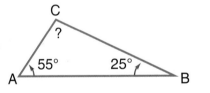

Triangle ABC or $\triangle$ABC

By mental math
$180° - 55° = 125°$
$125° - 25° = 100°$

By number sentence
$\angle A + \angle B + \angle C = 180°$
$55° + 25° + \angle C = 180°$
$80° + \angle C = 180°$ ◀ related addition
$\angle C = 180° - 80°$ ◀ and subtraction
$\angle C = 100°$ sentences

Draw an example of each type of triangle on dot paper.

1. isosceles   **2.** acute   **3.** obtuse   **4.** scalene

5. right   **6.** equilateral   **7.** isosceles obtuse   **8.** scalene acute

Classify each triangle by the measure of its sides and by the measure of its angles.

9.
17 mm, 75°, 25 mm, 65°, 40°, 27 mm

10.
10 m, 60°, 10 m, 60°, 60°, 10 m

11.
40 cm, 106°, 40 cm, 37°, 37°, 60 cm

12.
60°, 5 m, 3 m, 30°, 4 m

Find the measure of the third angle of the triangle.

13. $\angle A = 75°$, $\angle B = 35°$, $\angle C = $?

14. $\angle X = 110°$, $\angle Y = 35°$, $\angle Z = $?

15. $\angle M = 45°$, $\angle N = 45°$, $\angle O = $?

16. $\angle P = 90°$, $\angle Q = 30°$, $\angle R = $?

17. $\angle D = 43°$, $\angle T = 129°$, $\angle F = $?

18. $\angle W = 16°$, $\angle E = 5°$, $\angle U = $?

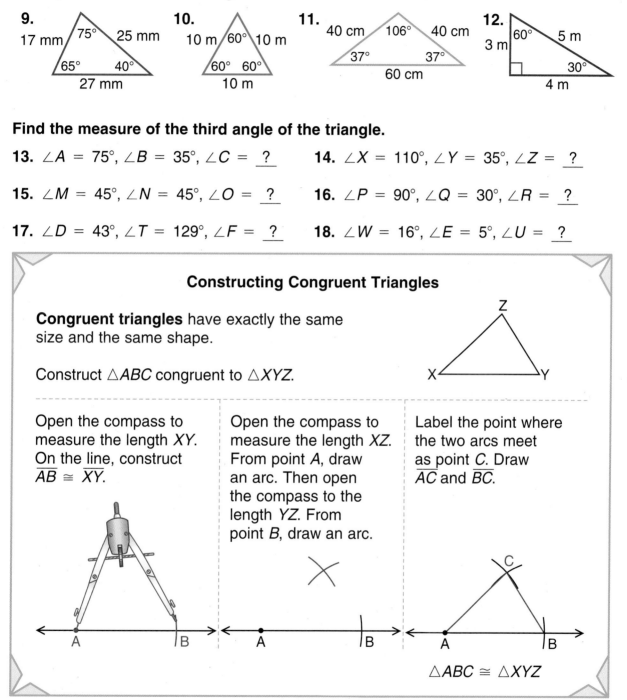

Constructing Congruent Triangles

Congruent triangles have exactly the same size and the same shape.

Construct $\triangle ABC$ congruent to $\triangle XYZ$.

Open the compass to measure the length XY. On the line, construct $\overline{AB} \cong \overline{XY}$.

Open the compass to measure the length XZ. From point A, draw an arc. Then open the compass to the length YZ. From point B, draw an arc.

Label the point where the two arcs meet as point C. Draw $\overline{AC}$ and $\overline{BC}$.

$\triangle ABC \cong \triangle XYZ$

Draw the triangle indicated. Then construct a triangle congruent to it.

19. acute $\triangle DEF$

20. obtuse $\triangle JKL$

21. right $\triangle PQR$

9-8 Classifying Quadrilaterals

A **quadrilateral** is a four-sided polygon. There are many different types of quadrilaterals.

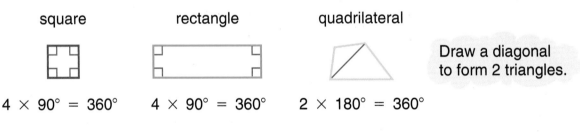

A **trapezoid** is a quadrilateral with exactly one pair of parallel sides.

A **parallelogram** is a quadrilateral with two pairs of parallel congruent sides.

A **rectangle** is a parallelogram with four right angles.

A **square** is a parallelogram with four right angles and four congruent sides.

A **rhombus** is a parallelogram with four congruent sides.

▶ Cocheta wanted to know the number of degrees in a quadrilateral. She drew these three figures.

square	rectangle	quadrilateral
$4 \times 90° = 360°$	$4 \times 90° = 360°$	$2 \times 180° = 360°$

Draw a diagonal to form 2 triangles.

Cocheta discovered: The **sum of the measures** of the angles of any quadrilateral is **360°**.

▶ Use a number sentence to find a missing angle measure in a quadrilateral, as shown below.

$$\angle A + \angle B + \angle C + \angle D = 360°$$
$$\angle A + 90° + 50° + 85° = 360°$$
$$\angle A + 225° = 360°$$
$$\angle A = 360° - 225°$$
$$\angle A = 135°$$

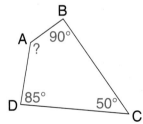

Copy and complete the table. Write *Yes* or *No* for each description.

	Description	Trapezoid	Parallelogram	Rectangle	Square	Rhombus
1.	All sides congruent	?	?	?	?	?
2.	4 right angles	?	?	?	?	?
3.	Opposite sides parallel	?	?	?	?	?

Explain the statement in your Math Journal. Draw a picture to help you.

4. A square is a rhombus.　　**5.** A trapezoid is *not* a parallelogram.

6. A square is a rectangle.　　**7.** Some quadrilaterals are parallelograms.

Find the measure of the fourth angle of the quadrilateral.

8. $\angle A = 85°$, $\angle B = 85°$, $\angle C = 65°$, $\angle D = \underline{\ ?\ }$

9. $\angle E = 65°$, $\angle F = 90°$, $\angle G = 90°$, $\angle H = \underline{\ ?\ }$

10. $\angle Q = 89°$, $\angle R = 67°$, $\angle S = 102°$, $\angle T = \underline{\ ?\ }$

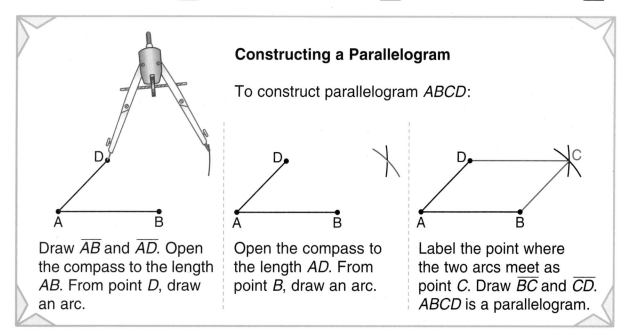

Constructing a Parallelogram

To construct parallelogram *ABCD*:

Draw $\overline{AB}$ and $\overline{AD}$. Open the compass to the length *AB*. From point *D*, draw an arc.

Open the compass to the length *AD*. From point *B*, draw an arc.

Label the point where the two arcs meet as point *C*. Draw $\overline{BC}$ and $\overline{CD}$. *ABCD* is a parallelogram.

Construct the figure described.

11. parallelogram *WXYZ*　　　　　　**12.** parallelogram *MNOP*

13. rhombus *EFGH* with each side 3 centimeters long

Circles

A **circle** is a simple closed curve. You can represent a circle on a circle geoboard or on dot paper as shown below.

Circle P

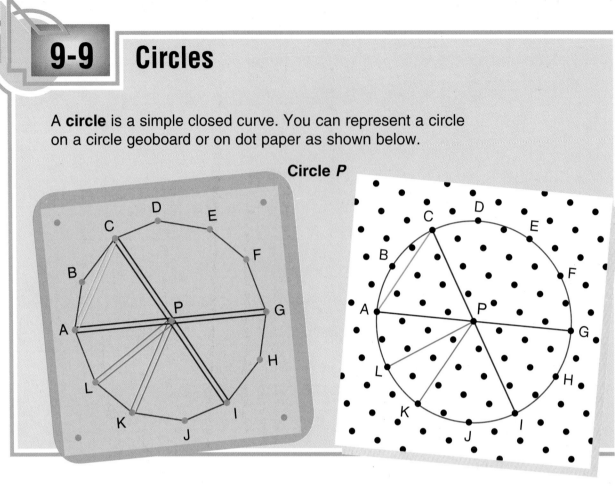

Record your answers on a sheet of paper.

Hands-On Understanding

Materials Needed: circle geoboard or dot paper, rubber bands, paper labels, cellophane tape, paper, pencil

Step 1 Make a circle on the geoboard as shown above. Use paper labels to name the twelve points *A–L*. Point *P* is the center of the circle.

Step 2 Connect the points *A* and *G*. Connect the points *C* and *I*. $\overline{AG}$ and $\overline{CI}$ are **diameters** of the circle.

Show some other diameters on your geoboard. Write their names on your paper. (*Hint:* A diameter passes through the center of the circle, with its two endpoints on the circle.)

Step 3
Connect the points L and P. Connect the points A and C. $\overline{LP}$ is a **radius** of the circle and $\overline{AC}$ is a **chord**.

One endpoint of a radius is the center of the circle and the other endpoint is on the circle. Both endpoints of a chord are on the circle.

Step 4
Connect the points K and P.
∠KPL is called a **central angle** of the circle.

Show some other central angles on your geoboard or dot paper. How many central angles can you find that have the same measure as ∠KPL? Write their names on your paper.

Communicate

1. Show diameter $\overline{DJ}$ and radius $\overline{DP}$ on your geoboard. Complete: In a given circle, the length of a diameter is __?__ the length of a radius.

2. A chord is a line segment whose endpoints are on the circle. Is a diameter of a given circle also a chord? Explain your answer.

Discuss ✓

3. An **arc** of a circle is the "curved path" between any two points of the circle. Use your finger to trace the arc on the geoboard between points C and F. How is arc CF different from chord CF?

 **FINDING TOGETHER**

4. Find the degree measure of the circle with center point A. (*Hint:* Angles EAD and DAE are straight angles.) Complete: The degree measure of a circle is __?__ degrees.

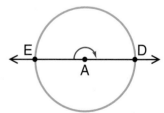

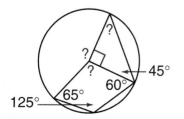

5. Use what you have learned about triangles, quadrilaterals, and circles to help you find the missing measures in the diagram at the left.

313

9-10 Classifying Space Figures

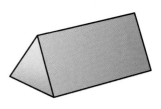

How might Juanita describe the cardboard model she made?

The cardboard model is an example of a **space figure** or **solid**. Space figures are three-dimensional. Their parts are not all contained in the same plane.

Space figures formed by *flat* plane surfaces are called **polyhedrons**.

Some Space Figures

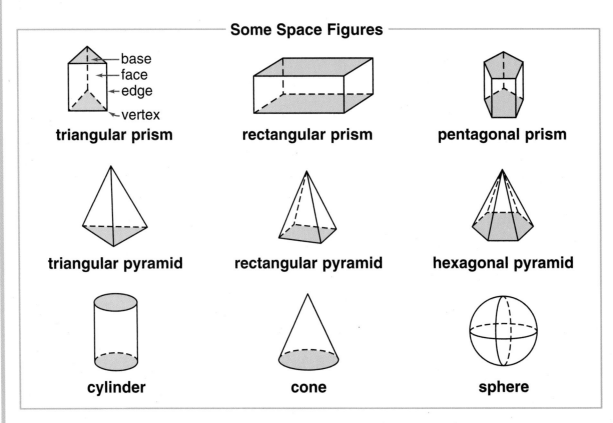

base
face
edge
vertex

triangular prism

rectangular prism

pentagonal prism

triangular pyramid

rectangular pyramid

hexagonal pyramid

cylinder

cone

sphere

Juanita opened up her cardboard model to get the **net** at the right.

She described the model as follows:

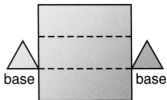

base   base

- It has 2 triangular bases.
- It has 3 rectangular faces between the bases.
- It has 6 vertices and 9 edges.
- It is a triangular prism.

314

**Is the figure a prism or a pyramid? Tell how many
faces, edges, and vertices it has.** (*Hint:* The bases are also faces.)

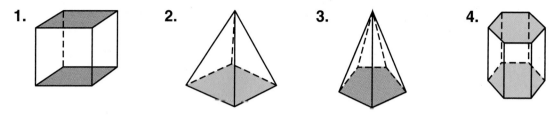

Tell which space figure(s) can have a base like the one shown.

Write *True* or *False*. If false, explain why.

Communicate ✓

9. A prism has 5 or more surfaces, or faces.

10. A pyramid has 3 or more surfaces, or faces.

11. Prisms, pyramids, and spheres are polyhedrons.

12. A cube is a special type of rectangular pyramid.

13. Eight polygons are needed to make the net for an octagonal prism.

FINDING TOGETHER

Use the space figures below for exercises 14–16.

Discuss ✓

14. Name each figure.

15. Draw the net for each figure.

16. Describe each figure completely. Discuss your descriptions with the class.

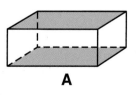
A

B

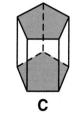
C

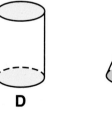
D

E

Congruent and Similar Polygons

The students in Class 6-B are matting their photos for gifts to their grandparents. There are two mat sizes: 6 x 8 and 9 x 12.

Polygons A and B are **congruent.** **Congruent polygons** have exactly the same size and shape.

Polygons A and C are **similar.** **Similar polygons** have the same shape. They may or may *not* have the same size.

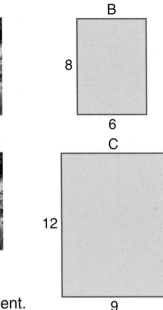

▶ The **corresponding parts** (matching sides and matching angles) of congruent polygons are congruent.

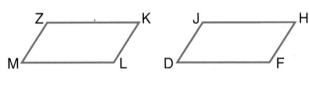

Parallelogram $MLKZ \cong$ Parallelogram $DFHJ$

$\overline{ML} \cong \overline{DF}$ $\angle M \cong \angle D$
$\overline{LK} \cong \overline{FH}$ $\angle L \cong \angle F$
$\overline{KZ} \cong \overline{HJ}$ $\angle K \cong \angle H$
$\overline{ZM} \cong \overline{JD}$ $\angle Z \cong \angle J$

▶ The **corresponding angles** (matching angles) of similar polygons are congruent.

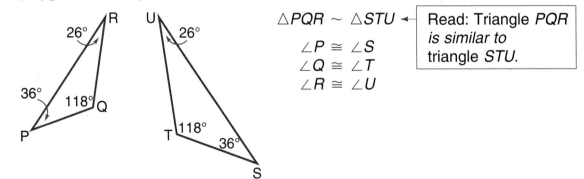

$\triangle PQR \sim \triangle STU$ ◄── Read: Triangle *PQR* is *similar to* triangle *STU*.

$\angle P \cong \angle S$
$\angle Q \cong \angle T$
$\angle R \cong \angle U$

Parallelograms *MLKZ* and *DFHJ* are congruent and similar.
Triangles *PQR* and *STU* are similar, but *not* congruent.

Are the polygons congruent? Write *Yes* or *No*.
Trace one polygon and compare.

1.

2.

3.

Are the polygons similar? Write *Yes* or *No*.

4.

5.

6.

Copy and complete.

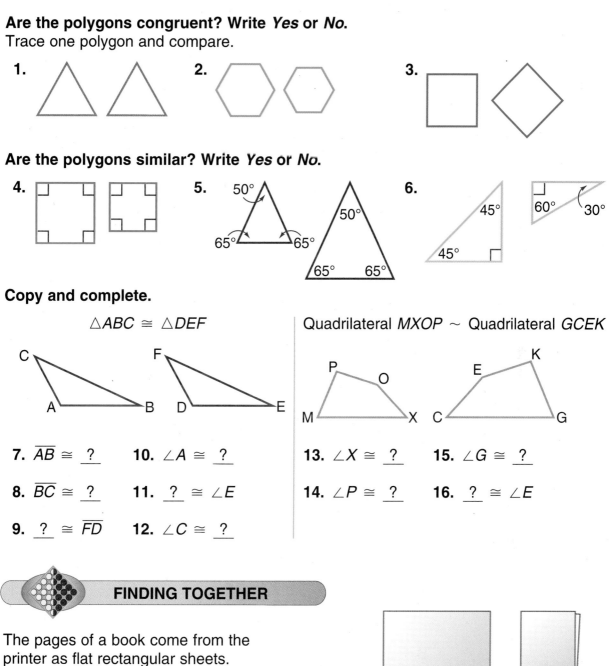

$\triangle ABC \cong \triangle DEF$

Quadrilateral *MXOP* ~ Quadrilateral *GCEK*

7. $\overline{AB} \cong$ <u> ? </u>

10. $\angle A \cong$ <u> ? </u>

13. $\angle X \cong$ <u> ? </u>

15. $\angle G \cong$ <u> ? </u>

8. $\overline{BC} \cong$ <u> ? </u>

11. <u> ? </u> $\cong \angle E$

14. $\angle P \cong$ <u> ? </u>

16. <u> ? </u> $\cong \angle E$

9. <u> ? </u> $\cong \overline{FD}$

12. $\angle C \cong$ <u> ? </u>

FINDING TOGETHER

The pages of a book come from the printer as flat rectangular sheets.

The original sheet is folded along a **line of symmetry** to get a folio. The folio is folded to get a quarto. The quarto is folded to get an octavo.

original sheet

folio

quarto

octavo

17. Fold a rectangular sheet of paper to get a folio, a quarto, and an octavo. Which pairs of figures are similar?

317

Transformations

In a plane, there are three basic types of movements of geometric figures:

- **slide** (or **translation**) — Every point of a figure moves the same distance and in the same direction.

- **flip** (or **reflection**) — A figure is flipped over a line so that its *mirror image* is formed.

- **turn** (or **rotation**) — A figure is turned around a center point.

Slide

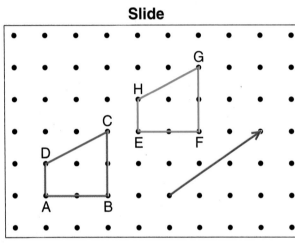

Flip

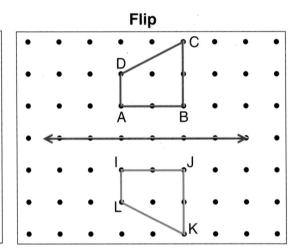

The **slide arrow** shows that *ABCD* is moved 3 units right and 2 units up.
Slide image *EFGH* ≅ *ABCD*.

ABCD is flipped over the **flip line** to form its mirror image.
Flip image *IJKL* ≅ *ABCD*.

Turn

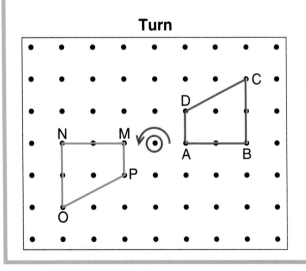

The **turn arrow** shows that *ABCD* is turned (or rotated) one half turn counterclockwise around the turn center.
Turn image *MNOP* ≅ *ABCD*.

Half turn is 180°.

**Name the corresponding congruent sides and congruent angles
in the transformations on page 318.**

1. *ABCD ≅ EFGH*   **2.** *ABCD ≅ IJKL*   **3.** *ABCD ≅ MNOP*

Tell where each slide arrow would move the figure.

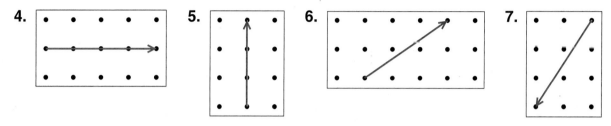

**Copy each figure on dot paper. Draw a slide, flip, and turn
image of each figure.**

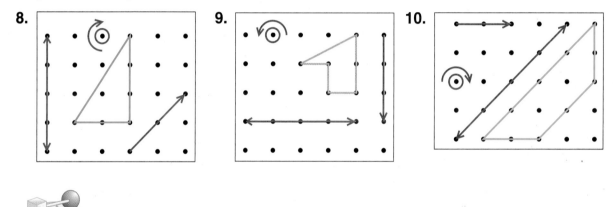

CHALLENGE

**Write *True* or *False* for each statement. If false, explain why.
Draw transformations on dot paper to help you answer.**

Communicate

11. The distance between pairs of corresponding points of the
original figure and its slide image is the same.

12. A slide, flip, or turn does not change the shape of a figure
but may change its size.

13. You can flip a figure twice to get a half turn.

14. Corresponding sides of a transformed figure (slide, flip, or turn)
and its image are parallel.

15. In a flip, corresponding points of the figure and its image
are the same distance from the flip line.

Tessellations

Interesting patterns are often used on floors, wallpaper, and fabrics. The designs are often made of polygons.

A design like the one shown at the right is called a **tessellation**.

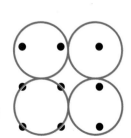

A tessellation is made from congruent figures placed so that they completely cover a surface without overlapping or leaving gaps.

> A tessellation may contain one or more different types of polygons.

Many polygons will form a tessellation. Follow these steps to form a tessellation made of squares.

- Draw square *ABCD* on dot paper.

- Turn *ABCD* one half turn clockwise around the midpoint (halfway point) of $\overline{BC}$. You get square 2.

- Turn square 2 one half turn clockwise around the midpoint of its side to get square 3.

- Turn square 3 one half turn clockwise around the midpoint of its side to get square 4.

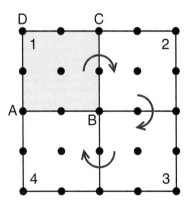

Copy the figure on dot paper and continue to draw half-turn images. You should see that a square tessellates (covers) the plane.

Study these examples.

Parallelogram tessellates.

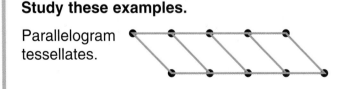

Circle does not tessellate. There are gaps.

Draw these polygons on dot paper. Try to make a tessellation using each polygon.

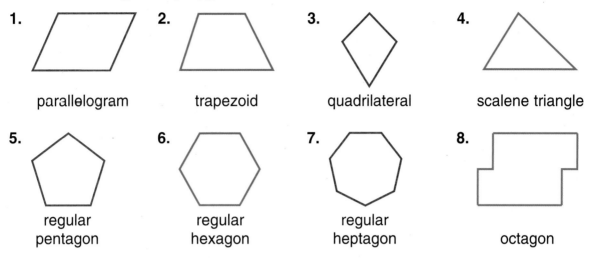

1. parallelogram
2. trapezoid
3. quadrilateral
4. scalene triangle

5. regular pentagon
6. regular hexagon
7. regular heptagon
8. octagon

9. Which of the polygons above could not be used for a tessellation?

Answer each question and then share your conclusions with the class.
Draw polygons on dot paper to help you answer.

Communicate

10. Do all triangles tessellate?

11. Do all rectangles, parallelograms, and squares tessellate?

12. How does the answer to exercise 10 suggest the answer to exercise 11?

13. Can you tessellate the plane using a combination of squares and equilateral triangles? rectangles and regular hexagons?

   **FINDING TOGETHER**

Discuss

14. Fit several cubes together to show that cubes tessellate space.

15. Use models to help you find other solids that tessellate space. Discuss your findings with the class.

9-14 Problem Solving: Logic/Analogies

Problem: Helen has to choose two different polygons to complete this analogy. Which two polygons should she choose?

2 diagonals are to a _?_ as 9 diagonals are to a _?_ .

1 IMAGINE   Picture yourself drawing all the diagonals in different polygons.

2 NAME   *Facts:*   a polygon with 2 diagonals
a polygon with 9 diagonals

Question: Which two polygons should she choose?

3 THINK   To complete the analogy, first draw some polygons and then count the number of diagonals.

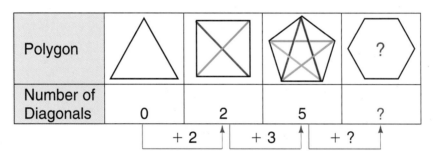

Polygon	△	⊠	⬠	?
Number of Diagonals	0	2	5	?

+ 2   + 3   + ?

4 COMPUTE   A square is a quadrilateral.
A square has 2 diagonals.
All quadrilaterals have 2 diagonals.

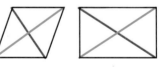

Look at the pattern in the table.

What number should be added to 5 to find the number of diagonals in a hexagon? 4

$$5 + 4 = 9 \quad \text{number of diagonals in a hexagon}$$

So 2 diagonals are to a quadrilateral as 9 diagonals are to a hexagon.

5 CHECK   Draw and count the number of diagonals in a hexagon.

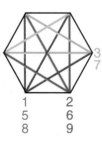

Solve each problem. Use logical reasoning to help you.

1. Ms. Geldof wants to place these five solids in a row so that no two prisms or pyramids are side by side. From the left, the fourth figure should have six congruent faces. The square pyramid should be to the right of the figure with a curved surface. What is the order of the solids?

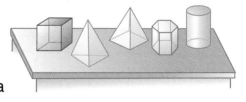

IMAGINE	Picture yourself arranging these solids.

NAME *Facts:* No two prisms or pyramids are side by side.
fourth figure—6 congruent faces
square pyramid—to the right of the figure
with a curved surface

Question: What is the order of the solids?

THINK Use the clues in the problem and logical reasoning to find the correct order.

⬜ ⬜ ⬜ ⬜ ⬜
1st 2nd 3rd 4th 5th

COMPUTE ⟶ **CHECK**

2. An acute angle is to an equilateral triangle as a right angle is to a __?__ .

3. Abby, Kara, Ed, and Ben each drew a different pattern. Match each person with the pattern each drew if Ben's pattern has exactly 6 right triangles, Abby's has 6 squares and 4 pentagons, and Ed's has 12 trapezoids and 8 right triangles. Which square did each person draw?

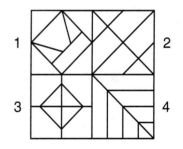

4. Find two ways Rose can complete this analogy:
180° is to a triangle as __?__° is to a __?__ .

5. Find two ways Larry can complete this analogy:
7 is to 49 as 5 is to __?__ .

  MAKE UP YOUR OWN

6. Use polygons and geometric solids to make up your own analogies.

Problem-Solving Applications

Solve each problem. Explain the method you used.

Ms. Widsky's class created a class pennant with this design:

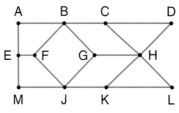

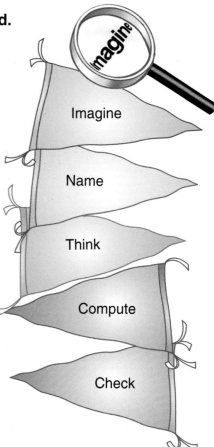

Imagine

Name

Think

Compute

Check

1. Classify quadrilateral *GHKJ*.

2. What is the sum of the measures of the angles of quadrilateral *ABFE*?

3. Identify the square in the pennant.

4. Identify one acute angle, one right angle, and one obtuse angle in the pennant.

5. Is △*HKL* scalene, isosceles, or equilateral? Is it acute, right, or obtuse?

6. Name a line segment parallel to $\overline{DK}$; perpendicular to $\overline{DK}$.

7. The class builds a sculpture out of sheet metal. The four faces of the sculpture are equilateral triangles. Its base is a rectangle. What shape is the sculpture?

8. The class creates these four floor plans. Which two plans are congruent? Which three are similar?

Use the diagram for problems 9–11.

9. Ms. Widsky's class designed a circular rug like the one at the right. Name two chords.

10. Name three radii and name a diameter.

11. Name two central angles in this pattern. Are they acute, obtuse, or right?

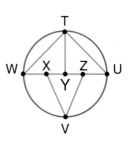

Use a strategy from the list or another strategy you know to solve each problem.

USE THESE STRATEGIES:
Logic/Analogies
Organized List
Multi-Step Problem
Use a Diagram
Write a Number Sentence
More Than One Solution

12. Mr. Gray's class creates a class logo that has six line segments. In the logo, you can find a trapezoid, two right triangles, and a square. Draw one possible design for the logo.

13. Greg is making designs with exactly three interlocking figures. He can choose from circles, squares, triangles, and trapezoids. He can use one or more than one of the figures. How many different designs can he make?

14. △ is to ⬡ as ▢ is to ? .

15. Each right triangle in the figure at the right has an area of 30 square inches. Each obtuse triangle has an area of 16.5 square inches. What is the area of the entire figure?

16. Four students each draw a rhombus, a square, a rectangle, or a trapezoid. No shape is drawn twice. Meg and Bill draw more than 2 right angles. Bill and Lyle's shapes have 4 congruent sides. What shape does Zack draw?

17. Bill measures the angles of a quadrilateral. He finds that two angles measure 55° and 87°. What is the sum of the measures of the other two angles?

Use the diagram for problems 18–21.

18. What is the measure of ∠MNQ?

19. The measures of ∠ROQ and ∠OQR are equal. What is each measure?

20. What is the measure of ∠NQO? (Notice that ∠PQR is a straight angle.)

21. Are ∠PQN and ∠QNO congruent?

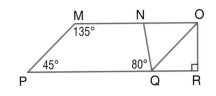

Which pairs of line segments or angles are congruent? Explain. *(See pp. 296–297.)*

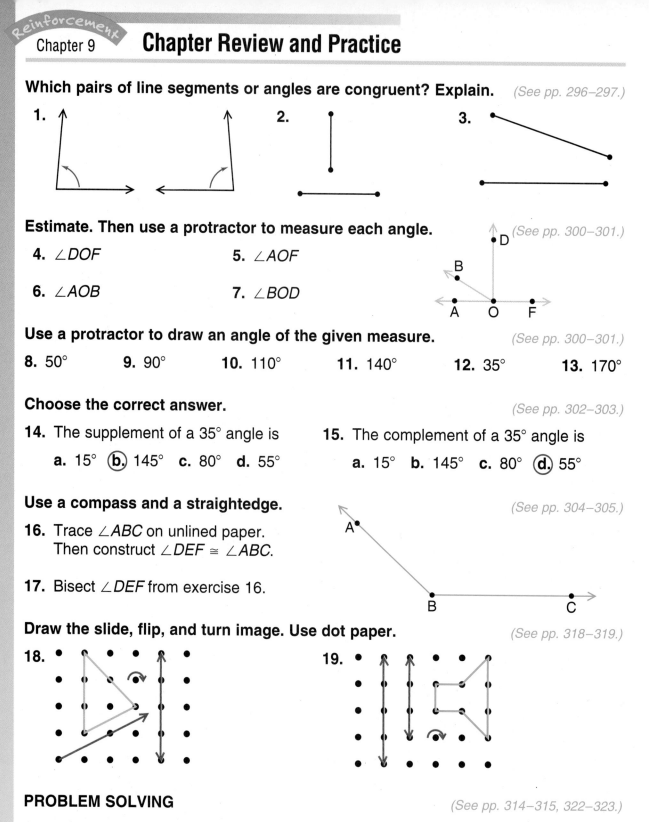

1.

2.

3.

Estimate. Then use a protractor to measure each angle. *(See pp. 300–301.)*

4. ∠*DOF*   **5.** ∠*AOF*

6. ∠*AOB*   **7.** ∠*BOD*

Use a protractor to draw an angle of the given measure. *(See pp. 300–301.)*

8. 50°   **9.** 90°   **10.** 110°   **11.** 140°   **12.** 35°   **13.** 170°

Choose the correct answer. *(See pp. 302–303.)*

14. The supplement of a 35° angle is

   a. 15°   **b.** 145°   **c.** 80°   **d.** 55°

15. The complement of a 35° angle is

   a. 15°   **b.** 145°   **c.** 80°   **d.** 55°

Use a compass and a straightedge. *(See pp. 304–305.)*

16. Trace ∠*ABC* on unlined paper.
     Then construct ∠*DEF* ≅ ∠*ABC*.

17. Bisect ∠*DEF* from exercise 16.

Draw the slide, flip, and turn image. Use dot paper. *(See pp. 318–319.)*

18.

19.

PROBLEM SOLVING *(See pp. 314–315, 322–323.)*

20. Complete the analogy and explain your answer: A rectangle is
     to a pentagonal prism as a __?__ is to a hexagonal pyramid.

(See Still More Practice, p. 514.)

SYMMETRY

▶ A figure is **symmetrical** if it can be folded so that one part matches the other part. The fold is called a **line of symmetry**. Some figures have no lines of symmetry. Some figures have more than one.

0 lines	1 line	4 lines
of symmetry	of symmetry	of symmetry
not symmetrical	symmetrical	symmetrical

▶ If a figure is rotated about a point, and it covers its original position, the figure has **rotational symmetry** about that point.

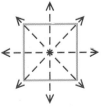

original position

What is the least rotation about point *O* that will cause the rectangle to cover its original position? In one complete turn about point *O*, how many times will the rectangle cover its original position?

$\frac{1}{4}$ turn

The rectangle will cover its original position after a half turn. So, in one complete turn, the rectangle will cover its original position 2 times. The rectangle has **rotational symmetry of order 2**.

$\frac{1}{2}$ turn

Write *symmetrical* or *not symmetrical* for each figure.
Copy the symmetrical figures and draw all lines of symmetry.

1.   2.   3.   4.

Each figure has rotational symmetry about point *O*.
Tell the order of the rotational symmetry.

5.   6.   7.   8.

327

Performance Assessment

Solve the problem. Draw each space figure and explain your reasoning.

1. Each of four students draws one of these space figures: cylinder, cone, rectangular pyramid, triangular prism. No space figure is drawn twice. Sue and Ted draw at least one circle. Ted and Ann draw at least one triangle. What space figure does Jim draw?

Draw or construct as indicated.

2. Draw $\overline{AB} \cong \overline{CD}$.   3. Draw $\angle EFG \cong \angle WXY$.   4. Construct $\overleftrightarrow{JK} \perp \overleftrightarrow{TV}$.

Identify each. Use the figures at the right.

5. regular triangle

6. hexagon

7. diagonal

8. parallelogram

9. right triangle

10. rhombus

Identify each for circle O.

11. diameter

12. 3 radii

13. 2 chords

14. 2 central angles

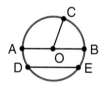

Identify each figure below.

15.   16.   17.   18.   19.

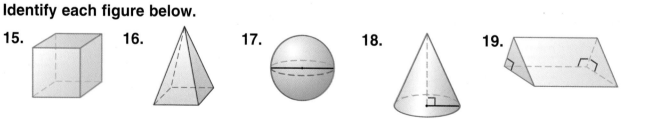

PROBLEM SOLVING   *Use a strategy or strategies you have learned.*

20. Lita says, "Similar polygons have the same size." Rafael says, "Congruent polygons are similar." Are both students correct? Explain your answer.

Choose the best answer.

1. $\frac{1}{4} + \frac{5}{8} \times 2 = \underline{?}$

   a. $\frac{7}{8}$ **b.** $1\frac{1}{2}$ **c.** 3   **d.** not given

2. $2\frac{2}{3} = j - 2\frac{1}{6}$ $j = \underline{?}$

   a. $\frac{1}{3}$ **b.** $4\frac{1}{3}$ **c.** $4\frac{5}{6}$ **d.** not given

3. 0.91×0.37

   a. 0.3367   **b.** 1.28

   c. 33.67   **d.** not given

4. $2\frac{3}{8} \div \frac{1}{4}$

   a. 2   **b.** $9\frac{1}{2}$

   c. $12\frac{3}{8}$ **d.** not given

5. What part of Ed's budget is spent on food and housing?

   a. $\frac{1}{6}$

   b. $\frac{1}{4}$

   c. $\frac{2}{3}$

   d. $\frac{3}{4}$

6. Spin each spinner one time. What is $P(\text{even}, M)$?

   a. $\frac{1}{12}$

   b. $\frac{1}{7}$

   c. $\frac{1}{4}$

   d. $\frac{7}{12}$

7. How many degrees in $\angle A$?

   a. 75°
   b. 85°
   c. 110°
   d. 175°

8. Which space figure has 5 faces, 6 vertices, and 9 edges?

   a. triangular pyramid   **b.** rectangular prism
   c. hexagonal prism   **d.** triangular prism

9. The stem-and-leaf plot shows the number of floors in several buildings. How many buildings have fewer than 40 floors?

   a. 2
   b. 6
   c. 8
   d. 12

Stem	Leaf
5	0 1
4	2 2 3 4
3	4 5 5 6
2	0 1 1 2

10. According to the graph, how many favorite sandwiches were chosen?

   a. 13
   b. 46
   c. 48
   d. 50

Favorite Sandwich

Ongoing Assessment III

For Your Portfolio

Solve each problem. Explain the steps and the strategy or strategies you used for each. Then choose one from problems 1–3 for your Portfolio.

1. Fiona's scores in six games were 27, 41, 32, 22, 36, 22. What are the range, mean, median, and mode of her scores?

2. A bowl contains 100 beads. If $P(\text{black}) = \frac{3}{5}$ and $P(\text{red}) = \frac{2}{5}$, how many black beads are in the bowl?

3. In a survey that asked if they chose soccer or tennis, 7 out of 10 people chose soccer. Based on the results of the survey, how many people would you predict to choose tennis out of a group of 1000 people? Explain.

Tell about it.

4. In problem 1, if Fiona's next score is 30, which statistic will change: range, mean, median, or mode? Explain.

Communicate ✓

5. Use this set of numbers: 22, 23, 29, 31, 34, 41. Can you include a seventh number in this set so that the range, mean, median, and mode do *not* change? Explain your answer.

For Rubric Scoring

Listen for information on how your work will be scored.

6. Visualize yourself riding a bicycle up a hill. Is your speed faster or slower than when you ride down the hill? Explain.

7. The graph shows the speed of a bicycle after 5, 10, 15, 20 minutes. What type of graph is it?

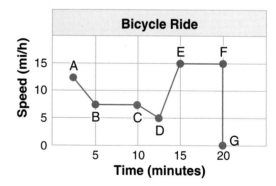

8. After riding for exactly 10 minutes, what was the speed of the bike?

9. Section B to C shows no change in speed. Which other section shows no change?

10. Which section(s) show where the bike is going uphill? downhill? Explain your answer. (*Hint:* See problem 6.)

Measurement

10

Critical Thinking/Finding Together

The 1995 Special Olympics World
Summer Games were held in New
Haven, Connecticut. Research and
report to the class the results for
three sports in these Games. Discuss
measurement used for each sport.

Joan Benoit
1984 U.S. Olympic Marathon Gold Medalist

During the third mile
not the eighteenth as expected
she surged ahead
leaving behind the press
of bodies, the breath
hot on her back
and set a pace
the experts claimed
she couldn't possibly keep
to the end.

Sure, determined,
moving to an inner rhythm
measuring herself against herself
alone in a field of fifty
she gained the twenty-six miles
of concrete, asphalt and humid weather
and burst into the roar of the crowd
to run the lap around the stadium
at the same pace
once to finish the race
and then again in victory

and she was still fresh
and not even out of breath
and standing.

Rina Ferrarelli

331

Measurement Sense: Decimal-Metric Connection

Let the flat be the standard unit of measure in the decimal system. Give it a new name—call it a **quad**.

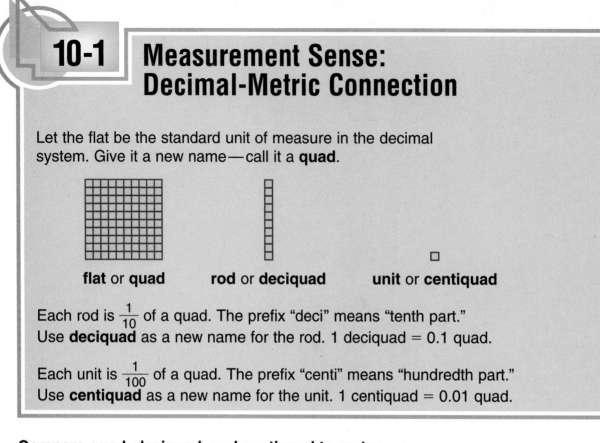

flat or **quad**   **rod** or **deciquad**   **unit** or **centiquad**

Each rod is $\frac{1}{10}$ of a quad. The prefix "deci" means "tenth part."
Use **deciquad** as a new name for the rod. 1 deciquad = 0.1 quad.

Each unit is $\frac{1}{100}$ of a quad. The prefix "centi" means "hundredth part."
Use **centiquad** as a new name for the unit. 1 centiquad = 0.01 quad.

Compare quad, deciquad, and centiquad to meter, decimeter and centimeter.

Hands-On Understanding

Materials Needed: flats, rods, units, paper, pencil, meterstick, scissors

Step 1   On a large sheet of paper (or several sheets taped together), use a meterstick to measure and mark a length of 1 meter.

Step 2   Use the length in Step 1 and make your own strip of paper 1 meter long—your own meterstick. Use scissors to cut it out.

The **meter** is the basic unit of linear measure in the metric system. You do not need to give it a new name!

Look up the word *linear* in a dictionary. How does it relate to measuring lengths?

Step 3   Divide the meter strip you made into 10 equal parts.
Use your scissors to cut the 10 parts.

What part of your original meter strip is each of the parts?
On page 332, what prefix did you use for $\frac{1}{10}$?
Copy and complete with a decimal: 1 decimeter = _?_ meter.

Step 4   Divide one of your decimeter strips from Step 3 into
10 equal parts. Use your scissors to cut the 10 parts.

What part of your original decimeter strip is each
of the parts? What part of your original meter strip
is each of the parts? (*Hint:* Find $\frac{1}{10}$ of $\frac{1}{10}$.)

Step 5   On page 332, what prefix did you use for $\frac{1}{100}$?
Copy and complete with a decimal: 1 centimeter = _?_ meter.

Step 6   Study the smallest unit of measure on the meterstick.

How many of these smallest units are on the meterstick?
The prefix "milli" means "thousandth part."
Copy and complete with a decimal: 1 millimeter = _?_ meter.

Communicate

Discuss ✓

1. How many centiquads are there in 1 quad?
   Explain using base ten blocks.

2. How many centimeters are there in 1 meter?
   Explain using your paper strips.

3. How many millimeters are there in 1 meter?
   Explain using your paper strips.

CONNECTIONS: LANGUAGE ARTS

4. Look up the meaning of each of the following
   prefixes in a dictionary: deka, hecto, kilo.

5. Explain how to make a model of each of the following:
   dekaquad, hectoquad, kiloquad.

Update your skills. See page 23.

10-2 Measuring Metric Length

Shanna uses a **meterstick** to find that the approximate height of the classroom door is 2.5 meters (m).

A meterstick is a good tool to use to measure greater lengths, widths, or heights. When the object you are measuring is somewhat shorter in length, width, or height, you may choose to use a **centimeter ruler** to complete the measurement.

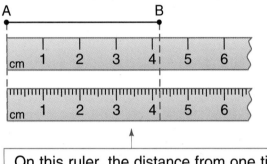

$\overline{AB}$ is about 4 centimeters (cm) long.

$\overline{AB}$ is 42 millimeters (mm) long.

On this ruler, the distance from one tick mark to the next is 1 millimeter: 1 cm = 10 mm.

For greatest lengths or distances in the metric system, the unit of measurement used is the **kilometer**. One kilometer (km) is equal to 1000 meters.

You would need 1000 metersticks laid end to end to measure a 1-kilometer bridge.

Metric units of length are related by tens.
1 m = 1000 mm   1 m = 100 cm   1 m = 10 dm   1 km = 1000 m

Measure each line segment to the nearest centimeter and to the nearest millimeter.

1. •———————•   2. •————————————•   3. •————•

Think about the actual length of each object. Estimate the metric length. Then use an appropriate measuring tool to find the exact measure.

4. length of your pencil   5. height of your chair   6. length of your shoe

7. height of the board   8. width of the board   9. thickness of a book

Draw each quadrilateral described. Then draw and measure its diagonals. Compare your measurements with those of your classmates.

Communicate ✓

10. square *ABCD* with
    AB = 5 cm

11. rhombus *EFGH* with
    EF = 25 mm

12. regular quadrilateral *MNOP*
    with *MN* = 44 mm

13. concave quadrilateral *QRST*
    with *QR* = 2.5 cm, *RS* = 3.5 cm

14. For which quadrilaterals in exercises 10–13 are the diagonals congruent? Explain your answer.

PROBLEM SOLVING

15. One piece of electrical wire is 35 mm long. A second piece is 34.9 cm long. Which piece is longer? Explain your answer.

16. The jogging track in Myers Park is 4.8 km long. Laura knows that her jogging stride is about 1 meter in length. How many of her strides would cover the distance around the track once?

SKILLS TO REMEMBER

Multiply or divide as indicated.

17. 11 × 1000   18. 7.6 × 100   19. 120 ÷ 100   20. 5.8 ÷ 100

21. 0.48 ÷ 10   22. 5.732 ÷ 100   23. 0.06 × 10   24. 15.2 × 1000

10-3 Measuring Metric Capacity and Mass

In Justin's Math Journal he wrote problems to show how capacity and mass are different.

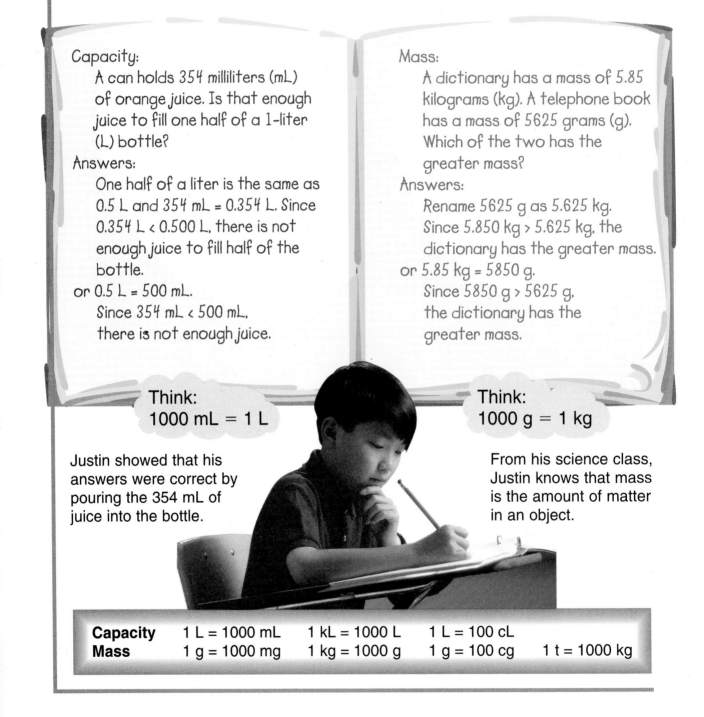

Capacity:
   A can holds 354 milliliters (mL) of orange juice. Is that enough juice to fill one half of a 1-liter (L) bottle?

Answers:
   One half of a liter is the same as 0.5 L and 354 mL = 0.354 L. Since 0.354 L < 0.500 L, there is not enough juice to fill half of the bottle.

or 0.5 L = 500 mL.
   Since 354 mL < 500 mL, there is not enough juice.

Mass:
   A dictionary has a mass of 5.85 kilograms (kg). A telephone book has a mass of 5625 grams (g). Which of the two has the greater mass?

Answers:
   Rename 5625 g as 5.625 kg. Since 5.850 kg > 5.625 kg, the dictionary has the greater mass.

or 5.85 kg = 5850 g.
   Since 5850 g > 5625 g, the dictionary has the greater mass.

Think:
1000 mL = 1 L

Think:
1000 g = 1 kg

Justin showed that his answers were correct by pouring the 354 mL of juice into the bottle.

From his science class, Justin knows that mass is the amount of matter in an object.

| **Capacity** | 1 L = 1000 mL | 1 kL = 1000 L | 1 L = 100 cL | |
| **Mass** | 1 g = 1000 mg | 1 kg = 1000 g | 1 g = 100 cg | 1 t = 1000 kg |

Compare. Use <, =, or >.

1. 24 L ? 240 mL

2. 7.3 kL ? 7300 L

3. 4000 mL ? 0.4 L

4. 24 g ? 240 mg

5. 6.6 kg ? 6600 g

6. 6550 mg ? 6.55 g

PROBLEM SOLVING

7. A glass container has a capacity of 7500 mL. Is it large enough to hold 75 L of water? Explain.

8. A can holds 354 mL of juice. How many liters of juice are there in a carton of 8 cans?

9. A **metric ton** (**t**) is equal to 1000 kg. If 1000 copies of a government report each with a mass of 5500 g is sent out, how many metric tons is that?

10. What are some measurement tools you can use to find metric length? metric capacity? metric mass?

 **FINDING TOGETHER**

11. Fill a large container with what you think is 1 L of water. Check your estimate by pouring the water into a 1-liter **graduated cylinder**. Try again if necessary.

12. Find an object or objects that you think has a mass of about 5 grams. Place the object on a double-pan balance with a 5-g unit in one pan of the scale. Keep adding or taking away objects until the scale is in balance.

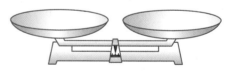

13. Repeat the activity in problem 12 using masses of 50 grams and 1 kilogram.

14. Write a summary in your Math Journal about your experiences with measuring capacity in liters and with measuring masses in grams and kilograms.

Math Journal

Renaming Metric Units

The chart shows some relationships between place value and metric units of measurement. Each unit has *ten* times the value of the next unit to its *right* and *one tenth* the value of the next unit to its *left.*

thousands	hundreds	tens	ones	tenths	hundredths	thousandths
1000	100	10	1	0.1	0.01	0.001
kilometer (km)	hectometer (hm)	dekameter (dam)	meter (m)	decimeter (dm)	centimeter (cm)	millimeter (mm)
kiloliter (kL)	hectoliter (hL)	dekaliter (daL)	liter (L)	deciliter (dL)	centiliter (cL)	milliliter (mL)
kilogram (kg)	hectogram (hg)	dekagram (dag)	gram (g)	decigram (dg)	centigram (cg)	milligram (mg)

7 km = _?_ m

7 km = (7 × 1000) m = 7000 m

240 cg = _?_ g

240 cg = (240 ÷ 100) g = 2.4 g

> Multiply to rename larger units as smaller units.

> Divide to rename smaller units as larger units.

To rename one metric unit to another (short way):

- Count the number of places from the known unit to the new unit in the chart.

- Move the decimal point the same number of places and in the same direction.

30.5 m = _?_ km

From meters to kilometers, move 3 places left.

30.5 m = 0.030.5 km

30.5 m = 0.0305 km

Study this example.

8.6 L = _?_ mL

Long way:
8.6 L = (8.6 × 1000) mL = 8600 mL

Short way:
8.6 L = 8.600. mL = 8600 mL

Copy and complete the table of metric units of measurement.

	Length	Capacity	Mass
1.	1 m = _?_ dm	1 L = _?_ dL	1 g = _?_ dg
	1 m = _?_ cm	1 L = _?_ cL	1 g = _?_ cg
	1 m = _?_ mm	1 L = _?_ mL	1 g = _?_ mg
	1 m = _?_ dam	1 L = _?_ daL	1 g = _?_ dag
	1 m = _?_ hm	1 L = _?_ hL	1 g = _?_ hg
	1 m = _?_ km	1 L = _?_ kL	1 g = _?_ kg
	1 km = _?_ m	1 kL = _?_ L	1 kg = _?_ g

Complete.

2. 6 cm = _?_ mm

3. 7 m = _?_ cm

4. 9.7 km = _?_ m

5. 11 L = _?_ mL

6. 4000 L = _?_ kL

7. 72.5 mL = _?_ L

8. 14 kg = _?_ g

9. 3000 mg = _?_ g

10. 45 000 g = _?_ kg

11. 9.1 L = _?_ kL

12. 4.025 g = _?_ mg

13. 200 dm = _?_ m

Compare. Use <, =, or >.

14. 4.5 m _?_ 45 cm

15. 6.7 kg _?_ 6700 g

16. 8575 mL _?_ 8.5 L

17. 4.8 cm _?_ 0.48 mm

18. 2000 mg _?_ 20 g

19. 257 cm _?_ 25.7 dm

PROBLEM SOLVING

Communicate

20. Last year City X reported 1.65 m of rain. City Y reported 131.5 cm of rain. Which city had more rain? Explain.

21. A water tank holds 32 000 kg of water. How many metric tons of water is that? Explain.

22. Colette's softball bat is 75 cm long. It has a mass of 11.2 g for every 1 cm of length. What is the total mass of Colette's bat? Explain the method you used to find your answer.

23. A strip of metal is 420 cm long. How many 1.4 cm strips can be cut from it? How many 14 mm strips can be cut from it? Explain how you found your answers.

339

10-5 Relating Metric Units

In the metric system, length, capacity, and mass
are related to one another.

A cube that measures 1 cm along each edge holds 1 mL of
water, which has a mass of 1 g.

A cube that measures 10 cm (or 1 decimeter) along each edge
holds 1 L of water, which has a mass of 1 kg.

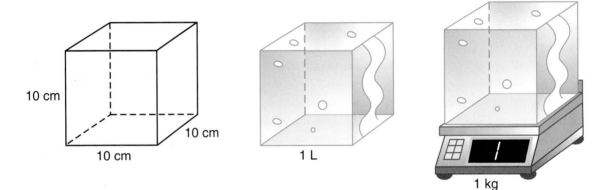

The **volume** of the 1-cm cube is:

$$1 \text{ cm} \times 1 \text{ cm} \times 1 \text{ cm} = \textbf{1 cm}^3, \text{ or } \textbf{1 cubic centimeter}.$$

The **volume** of the 10-cm cube is:

$$10 \text{ cm} \times 10 \text{ cm} \times 10 \text{ cm} = \textbf{1000 cm}^3, \text{ or } \textbf{1000 cubic centimeters}.$$

$$1 \text{ dm} \times 1 \text{ dm} \times 1 \text{ dm} = \textbf{1 dm}^3, \text{ or } \textbf{1 cubic decimeter}.$$

Study these examples.

$27 \text{ mL} = \underline{\ ?\ } \text{ cm}^3$ $9500 \text{ cm}^3 = \underline{\ ?\ } \text{ L}$ $12.5 \text{ g} = \underline{\ ?\ } \text{ mL}$

$27 \text{ mL} = 27 \text{ cm}^3$ $9500 \text{ cm}^3 = (9500 \div 1000) \text{ L} = 9.5 \text{ L}$ $12.5 \text{ g} = 12.5 \text{ mL}$

Copy and complete each chart.

	Cube	Capacity	Mass
1.	5 cm³	5 mL	5 g
2.	20 cm³	20 mL	?
3.	4000 cm³	4 L	?
4.	4 dm³	?	4 kg
5.	?	?	60 kg
6.	?	9.2 mL	?

	Cube	Capacity	Mass
7.	7.1 dm³	?	?
8.	?	?	13.9 g
9.	34.6 cm³	?	?
10.	?	8.5 L	?
11.	8500 cm³	?	?
12.	?	?	5.2 kg

Measure the edge of each cube to the nearest 0.5 cm.
Then find its capacity and its mass.

13.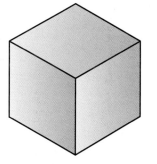

14.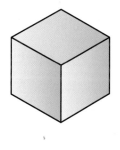

15.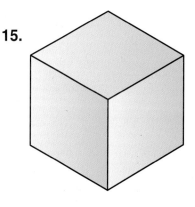

PROBLEM SOLVING

16. What is the mass in grams of the water in a cubical fish tank that measures 30 cm on each edge? in kilograms?

17. A water truck holds 27 000 kg of water. What is its capacity?

CHALLENGE

The water capacity of a cube is given.
Find the length of an edge of the cube.

18. 8 mL

19. 27 mL

20. 125 mL

21. 64 mL

22. 1 mL

23. 1000 mL

10-6 Measuring Customary Length

Use a ruler to measure the length of an object to the nearest inch or nearest part of an inch.

The length of $\overline{AB}$ is:

- 1 in. to the nearest in.

- $1\frac{1}{2}$ in. to the nearest $\frac{1}{2}$ in.

- $1\frac{1}{4}$ in. to the nearest $\frac{1}{4}$ in.

- $1\frac{3}{8}$ in. to the nearest $\frac{1}{8}$ in.

- $1\frac{5}{16}$ in. to the nearest $\frac{1}{16}$ in.

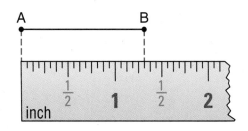

These **tools** are used to measure lengths in customary units:

ruler— measure objects that are shorter in length, width, or height

yardstick— measure greater lengths, widths, or heights, such as the width of your classroom

odometer—measure distances, such as the number of miles between two cities

▶ Multiply and divide to rename customary units of length.

$2\frac{1}{2}$ mi = $\underline{\ ?\ }$ yd   1 mi = 1760 yd

$2\frac{1}{2}$ mi = $(2\frac{1}{2} \times 1760)$ yd

$\qquad = \dfrac{5}{2} \times \dfrac{\overset{880}{\cancel{1760}}}{\underset{1}{1}}$ yd

$\qquad = 4400$ yd

102 in. = $\underline{\ ?\ }$ ft   12 in. = 1 ft

102 in. = $(102 \div 12)$ ft

$\qquad = 8$ ft 6 in.

$\qquad \qquad \qquad \boxed{\dfrac{6}{12} \text{ or } \dfrac{1}{2} \text{ ft}}$

$\qquad = 8\frac{1}{2}$ ft

Multiply to rename larger units as smaller units.

Divide to rename smaller units as larger units.

Copy and complete the statements about customary units of length.

1. **a.** 1 foot (ft) is to 12 inches (in.) as 3 feet is to ? inches.

   b. 3 feet is to ? yard (y) as 6 feet is to 2 yards.

   c. 1760 yards is to 1 mile (mi) as ? feet is to 1 mile.

Complete. Use the table on page 532 to help you.

2. 40 ft = ? yd

3. 114 in. = ? ft

4. 23,760 ft = ? mi

5. $4\frac{1}{2}$ mi = ? ft

6. $7\frac{1}{4}$ ft = ? in.

7. $6\frac{2}{3}$ yd = ? ft

8. 8 ft 9 in. = ? in. (*Hint:* 8 ft 9 in. means 8 ft + 9 in.)

9. 7 yd 2 ft = ? in.

10. 2 mi 40 ft = ? ft

11. $1\frac{1}{2}$ mi 60 yd = ? ft

Compare. Use <, =, or >.

12. 15 yd ? 50 ft

13. 18 ft ? 200 in.

14. 25,000 ft ? 5 mi

15. 96 in. ? $2\frac{2}{3}$ yd

16. $2\frac{1}{2}$ yd ? 2700 ft

17. 40 in. ? $3\frac{1}{4}$ ft

18. 49 ft ? 16 yd 2 ft

19. 294 in. ? 24 ft 6 in.

Use a ruler to measure each segment.

20. $\overline{RP}$ to the nearest 1 in., $\frac{1}{2}$ in., $\frac{1}{4}$ in., $\frac{1}{8}$ in., and $\frac{1}{16}$ in.

R ●————————————————————● P

21. $\overline{AB}$ to the nearest 1 in., $\frac{1}{2}$ in., $\frac{1}{4}$ in., $\frac{1}{8}$ in., and $\frac{1}{16}$ in.

A ●————● B

PROBLEM SOLVING

22. A rug is $7\frac{1}{2}$ ft long. How many yards long is the rug?

23. Kate's backyard is 16 yd long and $14\frac{1}{2}$ yd wide. How many feet wide is the backyard?

343

10-7 Measuring Capacity and Weight

For a 4-H project, Brittany records the amount of liquid that she drinks in a 1-week period (5 days). The result are shown in the table.

Some of Brittany's measures were exact, but other measures were estimates.

Amount of Liquid	
Milk	120 fluid ounces
Water	$12\frac{1}{2}$ cups
Other	$18\frac{1}{4}$ pints
Total	?

For her report, Brittany wants to find the total number of cups of liquid consumed. She renames the units as shown below.

120 fl oz = _?_ c

120 fl oz = (120 ÷ 16) c

$\quad = 7\frac{1}{2}$ c

16 fl oz = 1 c

$18\frac{1}{4}$ pt = _?_ c

$18\frac{1}{4}$ pt = $(18\frac{1}{4} \times 2)$ c

$\quad = \frac{73}{\overset{2}{\cancel{4}}} \times \frac{\overset{1}{\cancel{2}}}{1}$ c

$\quad = 36\frac{1}{2}$ c

1 pt = 2 c

So Brittany consumed $(7\frac{1}{2} + 12\frac{1}{2} + 36\frac{1}{2})$ c or $56\frac{1}{2}$ c of liquid in one week.

> Remember: Multiply to rename larger units as smaller units.
> Divide to rename smaller units as larger units.

Study these examples.

$5\frac{1}{2}$ gal = _?_ qt

$5\frac{1}{2}$ gal = $(5\frac{1}{2} \times 4)$ qt

$\quad = \frac{11}{\underset{1}{\cancel{2}}} \times \frac{\overset{2}{\cancel{4}}}{1}$ qt

$\quad = 22$ qt

1 gal = 4 qt

30 oz = _?_ lb

30 oz = (30 ÷ 16) lb

$\quad = 1$ lb 14 oz

$\quad = 1\frac{7}{8}$ lb

16 oz = 1 lb

14 oz = $\frac{14}{16}$ lb or $\frac{7}{8}$ lb

Copy and complete the statements.

1. **a.** 1 quart (qt) is to 2 pints (pt) as 5 quarts is to _?_ pints.

   b. 16 ounces (oz) is to _?_ pound (lb) as 64 ounces is to 4 pounds.

   c. 2000 pounds (lb) is to 1 ton (T) as 5000 pounds is to _?_ tons.

Complete. Use the table on page 532 to help you.

2. 8 gal = _?_ qt

3. 80 oz = _?_ lb

4. 50 fl oz = _?_ c

5. $2\frac{1}{2}$ lb = _?_ oz

6. $7\frac{3}{4}$ pt = _?_ fl oz

7. $10\frac{1}{8}$ lb = _?_ oz

8. 8 c 2 fl oz = _?_ fl oz

9. 16 lb 5 oz = _?_ oz

10. 15 T 920 lb = _?_ lb

11. 19 qt = _?_ gal _?_ qt

12. 17 c = _?_ pt _?_ c

13. 1 lb 48 oz = _?_ lb

PROBLEM SOLVING

14. How much more than a gallon is 7 quarts?

15. How much less than a pound is 13 ounces?

16. At $.49 a pint, what is the cost of 24 qt of milk?

17. At $.59 a quart, what is the cost of 3 gal of syrup?

18. Which weighs more: a 12-oz jar of fruit jelly or a $\frac{3}{4}$-lb jar of jam?

19. How many pint containers can be filled from 24 gal of juice?

 **PROJECT**

20. Conduct an experiment to find out how much liquid you drink in a week (5 days). Follow these steps:

    Discuss ✓

    • Estimate the number of cups of liquid you drink.

    • As closely as you can, record each day the number of cups (nearest whole number of cups) of liquid that you drink.

    • Find the total for the week and compare it with your estimate.

    • Discuss your results with the class. Then work together to find (a) the number of fluid ounces; and (b) the number of gallons for your class.

Computing Customary Units

▶ To *add* or *subtract* customary units:

- Add or subtract *like* units.
- Regroup to rename one unit to another.

$$
\begin{array}{r}
5 \text{ ft } 7 \text{ in.} \\
+\ 8 \text{ ft } 9 \text{ in.} \\
\hline
13 \text{ ft } 16 \text{ in.}
\end{array}
$$

16 in. = 12 in. + 4 in.
　　　 = 1 ft + 4 in.

13 ft 16 in. = 13 ft + 1 ft + 4 in.
　　　　　 = 14 ft 4 in.

$$
\begin{array}{r}
\overset{8}{\cancel{9}} \text{ gal } \overset{5}{\cancel{1}} \text{ qt} \\
-\ 4 \text{ gal } 3 \text{ qt} \\
\hline
4 \text{ gal } 2 \text{ qt}
\end{array}
$$

9 gal 1 qt
= 8 gal + 1 gal + 1 qt
= 8 gal + 4 qt + 1 qt
= 8 gal 5 qt

▶ To *multiply* with customary units:

- Multiply by each unit.
- Regroup to rename one unit to another.

$$
\begin{array}{r}
9 \text{ ft } 4 \text{ in.} \\
\times \qquad 3 \\
\hline
27 \text{ ft } 12 \text{ in.}
\end{array}
$$

12 in. = 1 ft

27 ft 12 in. = 27 ft + 1 ft
　　　　　 = 28 ft

or

- Rename all units as like units.

9 ft 4 in. = (9 × 12) in. + 4 in.
　　　　 = 108 in. + 4 in. = 112 in.

- Multiply.

112 in. × 3 = 336 in.

- Regroup.

= (336 ÷ 12) ft
= 28 ft

▶ To *divide* with customary units:

3 lb 8 oz ÷ 2 = _?_

1 lb = 16 oz

- Rename all units as like units.

3 lb 8 oz = (3 × 16) oz + 8 oz
　　　　 = 48 oz + 8 oz = 56 oz

- Divide.

56 oz ÷ 2 = 28 oz

- Regroup.

= (28 ÷ 16) lb
= 1 lb 12 oz

Add.

1.   8 ft   4 in.
   + 3 ft 10 in.
   _____

2.   7 yd 1 ft
   + 7 yd 2 ft
   _____

3.   10 yd 24 in.
   + 10 yd 16 in.
   _____

4.   2 gal 3 qt
   + 1 gal 3 qt
   _____

5.   6 lb 10 oz
   + 9 lb 12 oz
   _____

6.   6 pt 3 c
   + 3 pt 1 c
   _____

Subtract.

7.   10 yd 2 ft
   − 6 yd 2 ft
   _____

8.   3 ft   8 in.
   − 1 ft 10 in.
   _____

9.   4 yd 10 in.
   − 3 yd 11 in.
   _____

10.   5 gal 3 qt
   − 2 gal 1 qt
   _____

11.   8 qt 3 c
   − 4 qt 4 c
   _____

12.   10 lb 8 oz
   − 6 lb 9 oz
   _____

Multiply.

13.   2 ft 3 in.
   ×   2
   _____

14.   6 yd 2 ft
   ×   4
   _____

15.   4 yd 16 in.
   ×   5
   _____

16. 9 gal 3 qt × 7

17. 8 qt 3 pt × 6

18. 7 lb 6 oz × 6

Divide.

19. 4 yd 1 ft ÷ 3

20. 2 mi 5 yd ÷ 5

21. 3 gal 1 pt ÷ 2

22. 2 lb 1 oz ÷ 3

23. 3 qt 1 pt ÷ 7

24. 16 ft 8 in. ÷ 5

PROBLEM SOLVING

25. A 16-in. piece is cut off the end of a board 1 yd 2 in. long. How long is the board now?

26. A leaking water pipe loses $1\frac{1}{2}$ cups of water an hour. How many gallons of water does it lose in a day?

CHALLENGE

Compute.

27.   7 yd 1 ft 8 in.
   − 4 yd 2 ft 4 in.
   _____

28.   5 gal 3 qt 1 pt
   + 2 gal 4 qt 3 pt
   _____

29.   3 mi 1760 yd 100 ft
   + 4 mi 1760 yd 250 ft
   _____

30. 3 mi 6 yd 2 ft ÷ 5

31. 9 qt 2 pt 1 c × 4

32. 7 mi 140 yd 50 ft × 20

10-9 Using Perimeter

Rita wants to fence in an area of her backyard to provide a play space for her dogs. The area to be fenced in is in the shape of a polygon, as shown at the right. How many meters of fencing will Rita need?

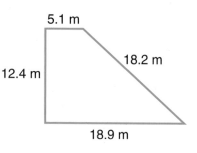

5.1 m

18.2 m

12.4 m

18.9 m

Estimate: 5 m + 18 m + 19 m + 12 m = 54 m

$$
\begin{array}{r}
5.1 \text{ m} \\
18.2 \text{ m} \\
18.9 \text{ m} \\
+ 12.4 \text{ m} \\
\hline
\end{array}
$$

Perimeter = 54.6 m

Remember: To add decimals, line up the decimal points.

Rita will need 54.6 m of fencing.

54.6 m is close to the estimate of 54 m.

To find the amount of fencing needed, Rita found the **perimeter** of the polygon by adding the lengths of all of the sides. For some ploygons, you can use formulas to find perimeter.

Perimeter of Polygons

Square or Rhombus

$P = 4 \times s$, where
s = length of a side

s

Rectangle

$P = (2 \times \ell) + (2 \times w)$ or
$P = 2 \times (\ell + w)$, where
ℓ = length, w = width

w

ℓ

Regular Triangle

$P = 3 \times s$, where
s = length of a side

s

Draw a diagram and then solve each problem. Where appropriate,
use a perimeter formula to help you.

1. What is the perimeter of a rectangular garden 20 ft long and 6 ft wide?

2. The bases on a baseball field are 90 ft apart. What is the straight line distance traveled when a player hits a home run?

3. How many meters of fringe are needed to border a triangular pennant 125 cm on a side?

4. How many yards of binding are needed to sew on the edges of a rectangular tablecloth 10 ft by 5 ft?

5. How many meters of fencing are needed for a rectangular pool area if the lengths of two sides are 25 m and 12.5 m?

6. At $2.95 per foot, what is the cost of fencing for a rectangular garden that is 20 ft wide and $30\frac{1}{2}$ ft long?

7. Find the perimeter of a triangle with sides measuring 25 mm, 2.8 cm, and 1.9 cm.

8. Find the perimeter of a quadrilateral with sides measuring $9\frac{1}{3}$ yd, 18 ft, $4\frac{1}{2}$ yd, and 45 in.

9. Find the perimeter of each regular pentagon if the length of one side (*s*) equals (a) 27 cm; (b) 3.4 ft; and (c) $5\frac{1}{5}$ in.

10. Write a formula for each: (a) perimeter of regular pentagon; and (b) perimeter of any regular polygon with *n* sides.

SHARE YOUR THINKING   *Discuss* ✓

**Measure the lengths of the sides of each polygon to the nearest
millimeter (exercises 11–12) and to the nearest $\frac{1}{8}$ in. (exercises 13–14).
Then find the perimeter of each polygon. Discuss your results.**

11.

12.

13.

14.

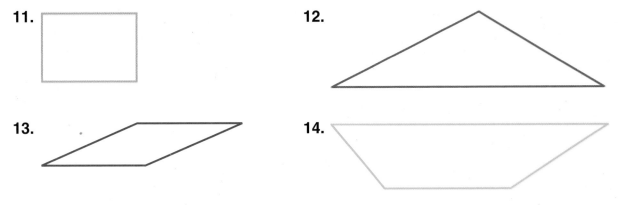

349

10-10 Area of Rectangles and Squares

The **area** of a polygon is the number of *square units* that cover its surface. The square units may be square centimeters, square meters, square inches, and so on.

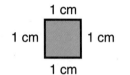

1 square centimeter (cm²)

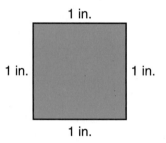

1 square inch (in.²)

What is the area of a square mirror that measures 9.5 cm on each side?

Estimate: 10 cm × 10 cm = 100 cm²

$A = s \times s$

$A = 9.5\ cm \times 9.5\ cm$

$A = 90.25\ cm^2$

What is the area of a wall that measures $16\frac{2}{3}$ ft by $10\frac{1}{2}$ ft?

Estimate: 17 ft × 11 ft = 187 ft²

$A = \ell \times w$

$A = 16\frac{2}{3}\ ft \times 10\frac{1}{2}\ ft$

$A = \dfrac{\overset{25}{\cancel{50}}}{\underset{1}{\cancel{3}}}\ ft \times \dfrac{\overset{7}{\cancel{21}}}{\underset{1}{\cancel{2}}}\ ft$

$A = 175\ ft^2$

Remember these area formulas:

Square $A = s \times s$ or
$A = s^2$, where
s = length of a side

Rectangle $A = \ell \times w,$
where ℓ = length, w = width

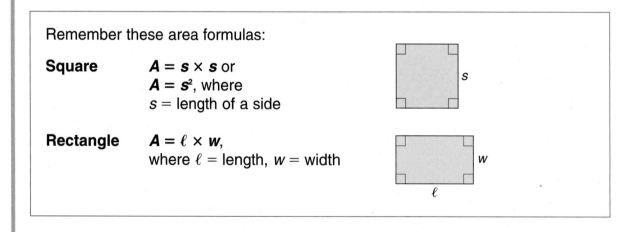

Use formulas to find the areas. Estimate to help you.

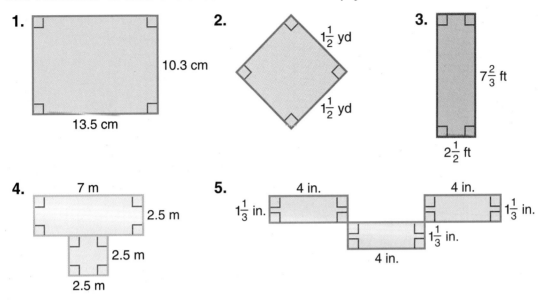

1. 10.3 cm, 13.5 cm

2. $1\frac{1}{2}$ yd, $1\frac{1}{2}$ yd

3. $7\frac{2}{3}$ ft, $2\frac{1}{2}$ ft

4. 7 m, 2.5 m, 2.5 m, 2.5 m

5. 4 in., 4 in., $1\frac{1}{3}$ in., $1\frac{1}{3}$ in., $1\frac{1}{3}$ in., 4 in.

PROBLEM SOLVING Draw a diagram and use a formula to solve each problem.

6. How many square yards of vinyl are needed to cover a square floor that measures 3 yd on each side?

7. Find the total area of 8 rectangular wooden panels if each measures 5.2 cm by 7.6 cm.

8. How many square yards of carpeting are needed to cover a floor that is 9 ft wide and 15 ft long?
   (*Hint:* Rename feet as yards or use 9 ft^2 = 1 yd^2)

CHALLENGE

Find each missing dimension. Use Guess and Test.

9. Area of square: 144 in.^{2}
   Find the length of a side.

   $s \times s = 144$ in.^{2}

   $\underline{?} \times \underline{?} = 144$ in.^{2}

   $s = 12$ in.

10. Area of square: 625 cm^2
    Find the length of a side.

11. Area of rectangle: 276 m^2
    Width = 12 m
    Find the length.

351

10-11 | Discovering Perimeter and Area

 Discover Together

Materials Needed: calculator, paper, pencil

Ms. Urbach wants to put a fence around a field she has just purchased and to plant grass seed.

(a) At $32.99 per yard, how much will it cost to completely enclose the field with a fence?

(b) Ms. Urbach must cover the entire area of the field with grass seed. Grass seed is needed for how many square yards?

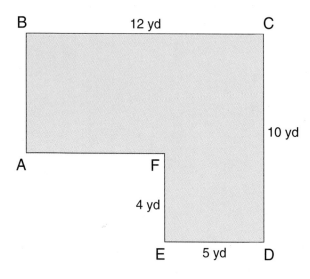

1. Copy the diagram of the field as shown above on your paper. Label the sides as shown.

The field that Ms. Urbach wants to fence in is *not* in the shape of a rectangle, square, or any other geometric figure that you are familiar with.

2. To put a fence around the field, does Ms. Urbach need to find its perimeter or its area? Explain your thinking.

Communicate ✓

352

3. You can add the lengths of all of the sides to find perimeter. Give the length of each of these sides: $\overline{BC}$, $\overline{CD}$, $\overline{DE}$, $\overline{EF}$. Which two lengths are missing?

Look carefully at your diagram. To find the length of $\overline{AB}$, subtract: 10 yd − 4 yd = 6 yd. So AB = 6 yd.

4. How can you find the length of $\overline{AF}$ by subtraction? What is this length?

Now that you know the length of each of the six sides of the field, you can find its perimeter.

5. Complete: P = 12 yd + 10 yd + 5 yd + 4 yd + __?__ + __?__ So the perimeter is __?__ yd.

6. Multiply $32.99 by the perimeter to find the total cost of the fence.

To decide how much grass seed she needs to buy, Ms. Urbach needs to know the area of the field. This is not easy since the field contains more than one geometric figure.

7. Study your diagram again. Try to draw *one line segment* that will divide the field into two rectangles. Discuss this with the class if you need help.

8. Draw a line segment to get two rectangles and then find the area of each one.

9. Add your two areas in exercise 8 to get the total number of square yards Ms. Urbach must cover with grass seed.

Communicate

10. Identify some key words in parts (a) and (b) of the problem on page 352 that tell you that part (a) involves perimeter and part (b) involves area.

Discuss

11. Is there more than one way to find the area of the field? Explain.

Area of Triangles and Parallelograms

Robin discovered the area formulas for triangles and parallelograms using the area of a rectangle.

▶ **Right Triangle**

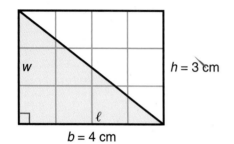

The base (*b*) and the height (*h*) of the right triangle correspond to the length and the width of the rectangle.

Think: The area of the right triangle is half the area of the rectangle.

Area of Triangle: $A = \frac{1}{2} \times b \times h$

$A = \frac{1}{2} \times b \times h$

$A = \frac{1}{\overset{1}{2}} \times \frac{\overset{2}{4}}{1} \text{ cm} \times \frac{3}{1} \text{ cm}$

$A = 6 \text{ cm}^2$

Any side of a triangle can serve as the base.
The height is the length of the perpendicular segment from the base to the opposite vertex.

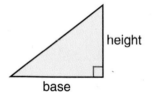

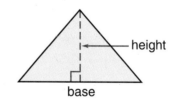

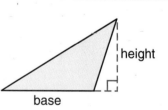

▶ **Parallelogram**

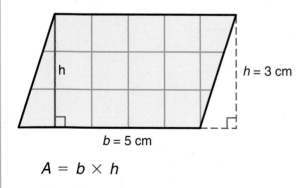

The base (*b*) and the height (*h*) of the parallelogram correspond to the length and the width of the rectangle.

Think: The area of the parallelogram is equal to the area of the rectangle.

Area of Parallelogram: $A = b \times h$

$A = b \times h$

$A = 5 \text{ cm} \times 3 \text{ cm}$

$A = 15 \text{ cm}^2$

Any side of a parallelogram can serve as the base.

Find the area of each triangle.

1.
4 cm
3 cm

2.
2 cm
6 cm

3.
2.5 cm
4 cm

4.
6 cm
8 cm

5.
8 ft
10 ft

6.
30 cm
12 cm

7.
2 yd
5 yd

Find the area of each parallelogram.

8.
3 cm
4 cm

9.
12 in.
18 in.

10.
15 mm
7.5 mm

11.
$1\frac{4}{5}$ yd
$1\frac{2}{3}$ yd

Copy and complete each table.

Area of Triangle		
Base	**Height**	**Area**
12. 10 ft	5 ft	?
13. 8.4 m	5.1 m	?
14. $5\frac{1}{2}$ yd	9 yd	?

Area of Parallelogram		
Base	**Height**	**Area**
15. 4 cm	6 cm	?
16. 8.1 m	12 m	?
17. $6\frac{1}{3}$ ft	3 ft	?

PROBLEM SOLVING

18. Find the area of each: (a) triangular traffic sign with a base of 40 cm and height of 60 cm; (b) parallelogram-shaped pennant with a base of 2 yd and a height of 15 ft.

CHALLENGE

Communicate

19. Find the length and width of the rectangle that has the least perimeter and an area of 64 ft^2. Compare your results with those of a classmate.

10-13 Surface Area

The **surface area** of a space figure is the number of square units that cover its surface.

▶ **To find the surface area of a rectangular prism:**

- Open the prism to get its net.

- Find the area of each of its 6 faces.

- Add the areas to get the total area.

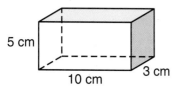

top

| | back | |

| side | bottom | side |

| | front | |

Area of bottom	= 10 cm × 3 cm =	30 cm²
Area of top	= 10 cm × 3 cm =	30 cm²
Area of side	= 5 cm × 3 cm =	15 cm²
Area of side	= 5 cm × 3 cm =	15 cm²
Area of front	= 10 cm × 5 cm =	50 cm²
Area of back	= 10 cm × 5 cm =	50 cm²
Surface Area (*S.A.*)	=	190 cm²

Each face is a rectangle.
Use $A = \ell \times w$ to find the area of each face.

▶ **To find the surface area of a cube:**

- Find the area of one face. Each face is a square. (The length of a side of the square, s, corresponds to the length of an edge of the cube, e.) Use $A = e \times e$ or $A = e^2$ to find the area of one face of the cube.

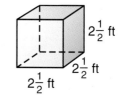

$$A = \frac{5}{2} \text{ ft} \times \frac{5}{2} \text{ ft}$$

$$A = \frac{25}{4} \text{ ft}^2$$

- Multiply the area by 6 since all 6 faces of the cube are congruent.

$$S.A. = \frac{\overset{3}{\cancel{6}}}{1} \times \frac{25}{\underset{2}{\cancel{4}}} \text{ ft}^2 = 37\frac{1}{2} \text{ ft}^2$$

Use the nets to find the surface area.

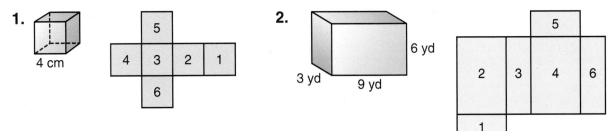

1. 4 cm

	5		
4	3	2	1
	6		

2. 6 yd 3 yd 9 yd

Find the surface area of each space figure.

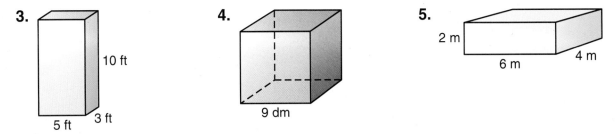

3. 10 ft 5 ft 3 ft

4. 9 dm

5. 2 m 6 m 4 m

Find the surface area of each rectangular prism.

6. $\ell = 14$ cm

$w = 9$ cm

$h = 8$ cm

7. $\ell = 5$ m

$w = 1.4$ m

$h = 5.9$ m

8. $\ell = 4\frac{1}{2}$ in.

$w = 3$ in.

$h = 6$ in.

9. $\ell = 9$ ft

$w = 2\frac{1}{3}$ ft

$h = 1\frac{1}{2}$ ft

Find the surface area of each cube.

10. $e = 15$ in.

11. $e = 8$ m

12. $e = 2.1$ cm

13. $e = 1\frac{1}{3}$ yd

PROBLEM SOLVING

Communicate ✓

14. Elisa is building a rectangular bin for firewood. The bin is 2 m long, 1.4 m wide, and 1.2 m high. What is its surface area?

15. Will 16 000 m² of wrapping paper be enough to wrap a box 50 m on each edge? Explain.

16. Pedro will paint the walls and the ceiling of a room that is 14 ft by 15 ft and 8 ft high. He will not paint the floor. What is the total surface area that Pedro will paint?

10-14 Circumference

The distance around a circle is called the **circumference (C)** of the circle.

Use a string to measure around the circle at the right. Then measure the length of the string with a metric ruler.

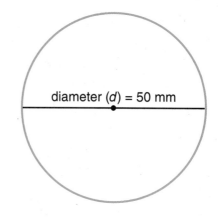
diameter (d) = 50 mm

Find the ratio of $\dfrac{\text{circumference}}{\text{diameter}}$.

$$\dfrac{\text{circumference}}{\text{diameter}} \approx \dfrac{157 \text{ mm}}{50 \text{ mm}} \approx 3.14$$

For every circle, the ratio of the circumference (C) to the length of the diameter (d) is close to 3.14. Mathematicians use the Greek letter π (**pi**) to name this ratio.

You can use formulas to find the approximate circumference of a circle.

To find circumference when the length of a diameter (d) is given:	**To find circumference when the length of a radius (r) is given:**
Use **C = π × d.**	Use **C = 2 × π × r.**

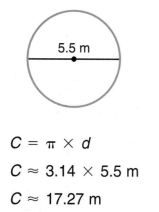
5.5 m

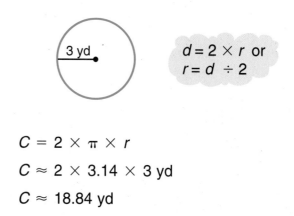
3 yd

$d = 2 \times r$ or
$r = d \div 2$

$C = \pi \times d$

$C \approx 3.14 \times 5.5 \text{ m}$

$C \approx 17.27 \text{ m}$

$C = 2 \times \pi \times r$

$C \approx 2 \times 3.14 \times 3 \text{ yd}$

$C \approx 18.84 \text{ yd}$

Find the circumference of each circle.

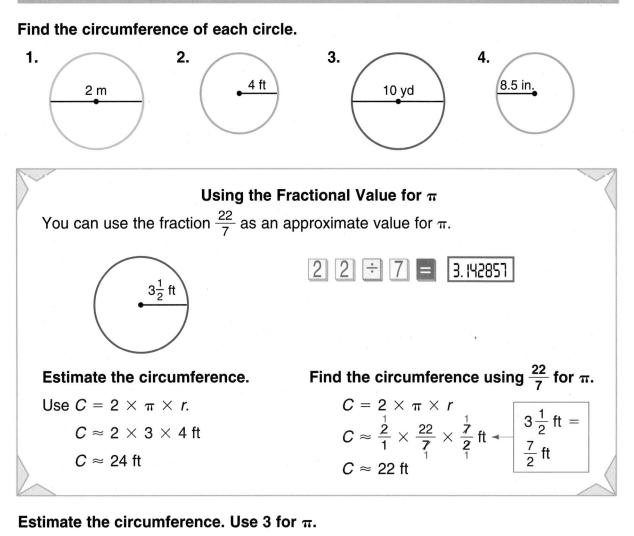

1. 2 m

2. 4 ft

3. 10 yd

4. 8.5 in.

Using the Fractional Value for π

You can use the fraction $\frac{22}{7}$ as an approximate value for π.

$3\frac{1}{2}$ ft

$2\ 2\ \div\ 7\ =\ \boxed{3.142857}$

Estimate the circumference.

Use $C = 2 \times \pi \times r$.

$C \approx 2 \times 3 \times 4$ ft

$C \approx 24$ ft

Find the circumference using $\frac{22}{7}$ for π.

$C = 2 \times \pi \times r$

$C \approx \overset{1}{\underset{1}{\frac{2}{2}}} \times \frac{22}{\underset{1}{7}} \times \overset{7}{\underset{1}{\frac{7}{2}}}$ ft ◄

$C \approx 22$ ft

$3\frac{1}{2}$ ft =
$\frac{7}{2}$ ft

Estimate the circumference. Use 3 for π.

5. $d = 3.1$ m

6. $d = 7\frac{1}{4}$ yd

7. $r = 11$ in.

8. $r = 8.7$ cm

Find the circumference. Use 3.14 or $\frac{22}{7}$ for π.

9. $d = 8$ m

10. $d = 7$ ft

11. $r = 18$ yd

12. $r = 14$ cm

13. $r = 5.2$ km

14. $r = 1.25$ m

15. $d = 21$ in.

16. $d = 35$ mm

17. $d = 1\frac{3}{4}$ ft

18. $d = 1\frac{2}{5}$ yd

19. $r = \frac{1}{2}$ yd

20. $r = 2\frac{1}{4}$ ft

PROBLEM SOLVING

21. A diameter of Earth measures about 13 000 km. What is its approximate circumference?

22. A wheel has a diameter 72 cm long. How far will a point on the wheel travel in 3 complete turns?

10-15 Area of a Circle

The **area of a circle (A)** is the number of square units that cover its surface.

You can count squares to find the area. The area of a circle can also be found by using this formula:

Area: about 16 square units

Area of circle (A) $= \pi \times r \times r$ or $A = \pi \times r^2$
where $r =$ length of a radius.

▶ Elvira wants to find the area of the circular piece of wood.

$d = 22$ ft

Since $d = 22$ ft, $r = 22$ ft $\div 2 = 11$ ft.

Estimate the area.

Use $A = \pi \times r \times r$.

$A \approx 3 \times 10$ ft $\times 10$ ft

$A \approx 300$ ft^2

Find the area using 3.14 for π.

$A = \pi \times r \times r$

$A \approx 3.14 \times 11$ ft $\times 11$ ft

$A \approx 379.94$ ft^2

The area of the wood is *about* 379.94 ft^2.

Study these examples.

$A = \underline{\ ?\ }$

4 m

Estimate: 3×4 m $\times 4$ m $= 48$ m^2

$A = \pi \times r^2$

$A \approx 3.14 \times (4$ m$)^2$

$A \approx 3.14 \times 16$ m^2

$A \approx 50.24$ m^2

$(4$ m$)^2 =$
4 m $\times 4$ m $=$
16 m^2

$A = \underline{\ ?\ }$

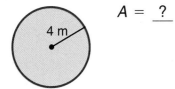

42 yd

$r = 42$ yd $\div 2$
$= 21$ yd

Estimate: 3×20 yd $\times 20$ yd $= 1200$ yd^2

$A = \pi \times r \times r$

$A \approx \dfrac{22}{7} \times \dfrac{\overset{3}{21}}{1}$ yd $\times \dfrac{21}{1}$ yd

$A \approx 1386$ yd^2

Complete.

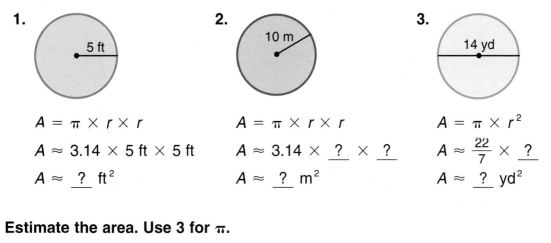

1.

5 ft

$A = \pi \times r \times r$

$A \approx 3.14 \times 5 \text{ ft} \times 5 \text{ ft}$

$A \approx \underline{\ ?\ } \text{ ft}^2$

2.

10 m

$A = \pi \times r \times r$

$A \approx 3.14 \times \underline{\ ?\ } \times \underline{\ ?\ }$

$A \approx \underline{\ ?\ } \text{ m}^2$

3.

14 yd

$A = \pi \times r^2$

$A \approx \frac{22}{7} \times \underline{\ ?\ }$

$A \approx \underline{\ ?\ } \text{ yd}^2$

Estimate the area. Use 3 for π.

4. $r = 9.7$ m   **5.** $r = 29$ ft   **6.** $d = 42$ in.   **7.** $d = 13$ yd

Find the area. Use 3.14 or $\frac{22}{7}$ for π. Estimate to help you.

8. $r = 3$ in.   **9.** $r = 10$ cm   **10.** $r = 7$ km   **11.** $r = 14$ m

12. $d = 40$ yd   **13.** $d = 35$ mm   **14.** $r = 28$ ft   **15.** $r = 36$ km

16. $r = 0.6$ m   **17.** $r = 1.5$ cm   **18.** $r = 3\frac{1}{2}$ yd   **19.** $d = 21$ ft

PROBLEM SOLVING

20. A designer has a circular piece of canvas with a radius that measures 50 cm. Find its area.

21. The length of a diameter of a metal jar lid is 4.2 cm. Find its area.

22. A circular rug is 4 m across. What is the distance around the rug?

23. A circular metal part for a machine has a radius 0.1 mm long. Find the total area of 100 of these parts.

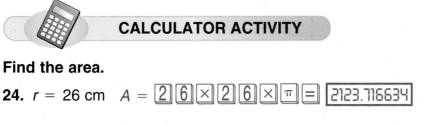

CALCULATOR ACTIVITY

Find the area.

24. $r = 26$ cm $A =$ ⬚2⬚6⬚×⬚2⬚6⬚×⬚π⬚=⬚ ⬚2123.716634⬚

25. $r = 15$ yd   **26.** $r = 1.25$ m   **27.** $d = 42$ ft   **28.** $d = 1.2$ cm

361

Volume of a Prism

The **volume** of a space figure is the number of *cubic units* that it contains.

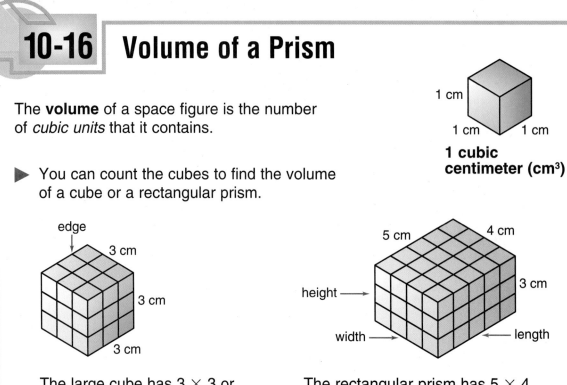

1 cm
1 cm   1 cm

1 cubic centimeter (cm³)

▶ You can count the cubes to find the volume of a cube or a rectangular prism.

edge
3 cm
3 cm
3 cm

height ——▶
width ——▶
5 cm   4 cm
3 cm
length

The large cube has 3 × 3 or 9 cubes in each layer, and there are 3 layers of cubes. The volume is 27 cm³.

The rectangular prism has 5 × 4 or 20 cubes in each layer, and there are 3 layers of cubes. The volume is 60 cm³.

▶ Formulas can be used to find the volume of cubes and rectangular solids.

Volume of cube (*V*) = *e* × *e* × *e* or **$V = e^3$**
where *e* = length of an edge.

Volume of rectangular prism (*V*) = ℓ × *w* × *h*
where ℓ = length, *w* = width, and *h* = height.

Study this example.

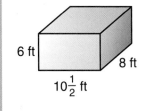

6 ft
8 ft
$10\frac{1}{2}$ ft

Estimate:
10 ft × 8 ft × 6 ft = 480 ft³

$V = \ell \times w \times h$

$V = \frac{\overset{}{21}}{\underset{1}{2}} \text{ ft} \times \frac{\overset{4}{8}}{1} \text{ ft} \times \frac{6}{1} \text{ ft}$

$V = 504 \text{ ft}^3$

Complete.

1.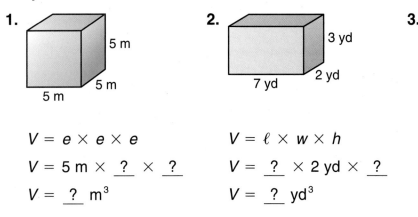

$V = e \times e \times e$

$V = 5 \text{ m} \times \underline{} \times \underline{}$

$V = \underline{} \text{ m}^3$

2.

$V = \ell \times w \times h$

$V = \underline{} \times 2 \text{ yd} \times \underline{}$

$V = \underline{} \text{ yd}^3$

3.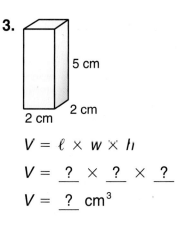

$V = \ell \times w \times h$

$V = \underline{} \times \underline{} \times \underline{}$

$V = \underline{} \text{ cm}^3$

Find the volume of each space figure.

4.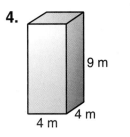

5.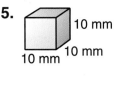

6.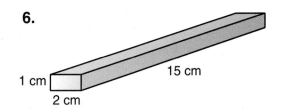

Find the volume. Estimate to help you.

7. $e = 8$ cm

8. $e = 4$ in.

9. $\ell = 3$ cm, $w = 4$ cm, $h = 5$ cm

10. $\ell = 1$ m, $w = 1$ m, $h = 5$ m

11. $e = 0.7$ m

12. $e = 1\frac{1}{2}$ ft

13. $\ell = 5$ yd, $w = \frac{1}{2}$ yd, $h = 2\frac{1}{5}$ yd

14. $\ell = 5.2$ m, $w = 5$ m, $h = 6.4$ m

PROBLEM SOLVING

15. Find the volume of a swimming pool that is 50 m long, 20 m wide, and 5 m deep.

16. A moving van is 12 m long, 3.5 m wide, and 3.8 m high. What is the volume of the van?

17. A storage room is shaped like a cube. It measures 5 m on each edge. What is the total volume of 10 storage rooms?

18. The inside of a cubical box measures 12 ft on each edge. How many cubes each measuring 1 ft on each edge will fit inside the box?

19. What happens to the volume of a cube if the length of an edge is doubled? tripled?

10-17 Computing with Time

Tim's watch reads 8:55 A.M. If he plans to go to lunch at 12:00 P.M., how much longer must he wait?

To find **elapsed time**, you may use mental math or subtract using paper and pencil.

60 seconds (s)	= 1 minute (min)
60 minutes	= 1 hour (h)
24 hours	= 1 day (d)
7 days	= 1 week (wk)
12 months (mo)	= 1 year (y)
365 days	= 1 year
100 years	= 1 century (cent.)

Mental Math

From 8:55 A.M. to 9:00 A.M. is 5 min.
From 9:00 A.M. to 12:00 P.M. is 3 h.
The elapsed time is 3 h 5 min.

Tim must wait 3 h 5 min.

Paper and Pencil

$$\begin{array}{r} 12{:}00 \\ -\ 8{:}55 \end{array} \longrightarrow \begin{array}{r} \overset{5\ \ 10}{1\ 1{:}\cancel{6}\ \cancel{0}} \\ -\ \ 8{:}5\ 5 \\ \hline 3{:}0\ 5 \end{array}$$

▶ Pam ran her laps in 3 min 20 seconds (s).
Marisa ran her laps in 2 min 35 s.
How much longer did Pam run?

Mental Math

From 2 min 35 s to 3 min is 25 s.
From 3 min to 3 min 20 s is 20 s.
The difference in the two times is 45 s.

Paper and Pencil

$$\begin{array}{r} 3 \text{ min } 20 \text{ s} \\ -2 \text{ min } 35 \text{ s} \end{array} \longrightarrow \begin{array}{r} \overset{2\ \ \ \ \ \ \ 80}{\cancel{3} \text{ min } \cancel{20} \text{ s}} \\ -2 \text{ min } 35 \text{ s} \\ \hline 0 \text{ min } 45 \text{ s} = 45 \text{ s} \end{array}$$

Pam ran 45 s longer.

Study these examples.

$$\begin{array}{r} 12 \text{ h } 15 \text{ min} \\ +\ 1 \text{ h } 55 \text{ min} \\ \hline 13 \text{ h } 70 \text{ min} = 14 \text{ h } 10 \text{ min} \end{array}$$

$$\begin{array}{r} 8 \text{ wk } 2 \text{ d} \\ \times \qquad 4 \\ \hline 32 \text{ wk } 8 \text{ d} = 33 \text{ wk } 1 \text{ d} \end{array}$$

$3 \text{ y } 8 \text{ mo} \div 2$
$44 \text{ mo} \div 2 = 22 \text{ mo}$
$22 \text{ mo} = 1 \text{ y } 10 \text{ mo}$

Find the elapsed time.

1. from 3:15 P.M. to 6:30 P.M.

2. from 5:55 A.M. to 7:30 A.M.

3. from 10:30 A.M. to 3:15 P.M.

4. from 9:20 A.M. to 1:30 P.M.

5. from 5:16 A.M. to 9:35 A.M.

6. from 6:22 A.M. to 2:10 P.M.

Complete.

7. 180 s = _?_ min

8. 28 d = _?_ wk

9. 3 y = _?_ mo

10. 18 mo = _?_ y

11. $2\frac{1}{2}$ h = _?_ min

12. 7 y = _?_ d

13. 2 y 3 mo = _?_ mo

14. 650 y = _?_ cent.

15. 3 d 2 h = _?_ h

16. 288 h = _?_ d

17. 1250 s = _?_ min _?_ s

18. 758 d = _?_ y _?_ wk

Compute.

19. 7 h 25 min
 + 3 h 15 min

20. 9 h 10 min
 + 3 h 25 min

21. 3 d 18 h
 + 1 d 15 h

22. 9 h 28 min
 − 3 h 10 min

23. 6 h 25 min
 − 2 h 40 min

24. 3 y 2 mo
 − 1 y 6 mo

25. 3 d 6 h
 × 3

26. 7 min 18 s
 × 5

27. 2 y 5 mo 12 d
 × 3

28. 12 d 18 h ÷ 3

29. 33 min 15 s ÷ 5

30. 7 cent. 250 y ÷ 2

PROBLEM SOLVING

31. Sonya set her alarm for 6:35 A.M. She woke up at 7:10 A.M. By how many minutes did Sonya oversleep?

32. Julio's clock read 9:45 P.M. when he arrived home after a 10 h 30 min trip. What time did he leave?

 **FINDING TOGETHER**

Use reference books to learn about the **24-hour clock**. Then write the 24-hour clock times.

33. 4:15 A.M.

34. 3:25 P.M.

35. 12:40 A.M.

TECHNOLOGY

Calculating Measures

You can use a calculator to help calculate perimeter, area, circumference, and volume.

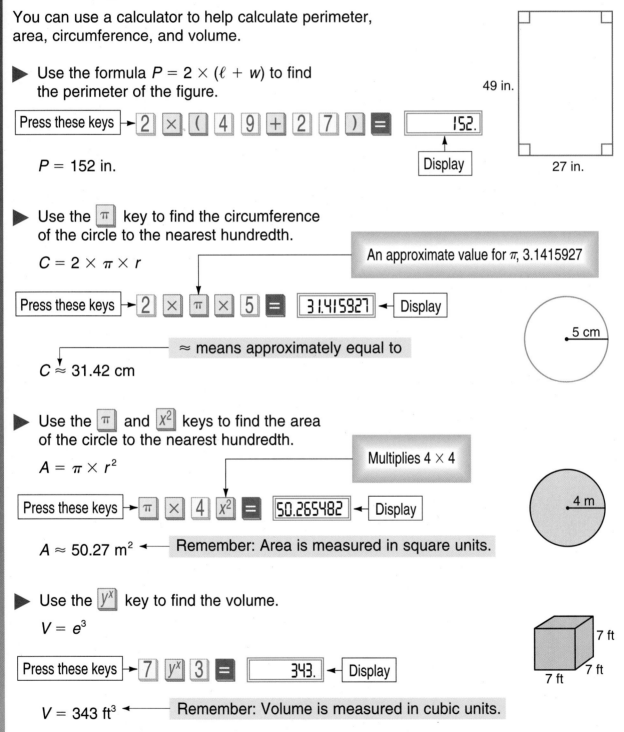

▶ Use the formula $P = 2 \times (\ell + w)$ to find the perimeter of the figure.

Press these keys → 2 $\times$ $($ 4 9 $+$ 2 7 $)$ $=$ | 152. |

$P = 152$ in.

Display

49 in.

27 in.

▶ Use the π key to find the circumference of the circle to the nearest hundredth.

$C = 2 \times \pi \times r$

An approximate value for π, 3.1415927

Press these keys → 2 $\times$ π $\times$ 5 $=$ | 31.415927 | ← Display

$\approx$ means approximately equal to

$C \approx 31.42$ cm

5 cm

▶ Use the π and x^2 keys to find the area of the circle to the nearest hundredth.

$A = \pi \times r^2$

Multiplies 4×4

Press these keys → π $\times$ 4 x^2 $=$ | 50.265482 | ← Display

$A \approx 50.27$ m² ← Remember: Area is measured in square units.

4 m

▶ Use the y^x key to find the volume.

$V = e^3$

Press these keys → 7 y^x 3 $=$ | 343. | ← Display

$V = 343$ ft³ ← Remember: Volume is measured in cubic units.

7 ft
7 ft
7 ft

366

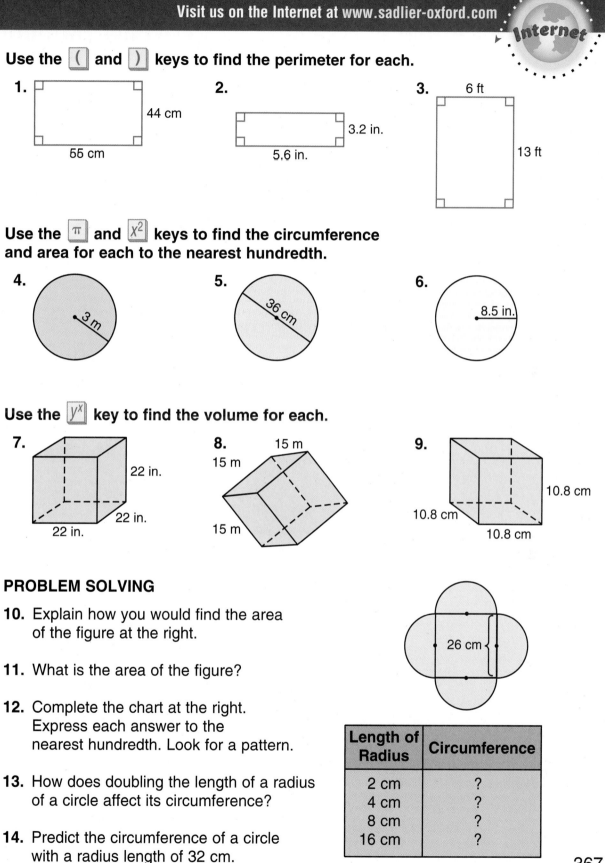

Use the (and) keys to find the perimeter for each.

1.

44 cm

55 cm

2.

3.2 in.

5.6 in.

3.

6 ft

13 ft

Use the π and x^2 keys to find the circumference and area for each to the nearest hundredth.

4.

3 m

5.

36 cm

6.

8.5 in.

Use the y^x key to find the volume for each.

7.

22 in.

22 in.

22 in.

8.

15 m

15 m

15 m

9.

10.8 cm

10.8 cm

10.8 cm

PROBLEM SOLVING

10. Explain how you would find the area of the figure at the right.

11. What is the area of the figure?

26 cm

12. Complete the chart at the right. Express each answer to the nearest hundredth. Look for a pattern.

13. How does doubling the length of a radius of a circle affect its circumference?

14. Predict the circumference of a circle with a radius length of 32 cm.

Length of Radius	Circumference
2 cm	?
4 cm	?
8 cm	?
16 cm	?

367

10-19 Problem Solving: Use Drawings/Formulas

Problem: Larry wants to fence in a rectangular garden. He plans to use 10 m of the house as one side of his garden and fencing for the other three sides. His garden is 10 m by 4 m. How much fencing does he need to buy?

4 m

10 m

4 m

1 IMAGINE Draw and label a diagram.

Remember, one side is the house and the other three sides are fencing.

2 NAME *Facts:* 10 m length
4 m width

Question: How much fencing is needed?

3 THINK Look at the diagram.

To find the amount of fencing needed to go *around* the garden, will you need to know the area or the perimeter? Perimeter

Write the formula for perimeter of a rectangle.

$P = 2 \times \ell + 2 \times w$

Do you need fencing around the *four* sides of the garden? No. One of the lengths is the house, so rewrite the formula using 1 length and 2 widths.

$P = \ell + 2 \times w$

Remember: One side of the house serves as the other length.

4 COMPUTE $P = 10 \text{ m} + 2 \times 4 \text{ m}$

$P = 10 \text{ m} + 8 \text{ m}$

$P = 18 \text{ m}$

Larry will need 18 m of fencing.

5 CHECK Is your answer reasonable?

The amount of fencing should be 10 m less than the perimeter of a 10 m by 4 m rectangle.

Solve. Draw a diagram and write a formula.

1. The new town playground has the measurements shown in the drawing. Find the total area of the playground.

IMAGINE Picture yourself measuring the playground.

NAME *Facts:* Playground measurements shown in diagram—13 m, 6 m, 9 m, 3 m, 4 m, 3 m

Question: Find the total area of the playground.

THINK Divide the playground into 2 rectangles.

Use the formula $A = \ell \times w$ to find the area of each. Then add:

Area = (4 m × 3 m) + (9 m × 6 m)

→ **COMPUTE** → **CHECK**

Diagram labels: 3 m, 4 m, 3 m, 13 m, playground, 9 m, 6 m

2. A storage box measures 1.4 m on each edge. What is the total volume of 20 storage boxes?

3. How many feet of fencing are needed to enclose a swimming pool that has a 25-ft diameter if there is a 2.5-ft deck between the edge of the pool and the fence?

4. A children's wading pool has a length and width of 5 ft and is 1.5 ft deep. How many cubic feet of water are needed to fill the pool?

5. Mr. Graycloud fenced in a part of his backyard for his dog. Find the area of the dog's yard. The dimensions are shown in the drawing.

Diagram labels: 2 m, 2 m, 3 m, 3 m, 1 m, 5 m

6. Cathy used gold trim around the edge of both the circle and the square. About how many centimeters of trim did Cathy use?

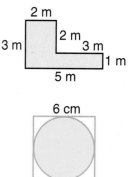

Diagram label: 6 cm

369

Problem-Solving Applications

Solve each problem and explain the method you used.

1. The Durans are renovating their apartment. Mrs. Duran measures a strip of wallpaper 2.5 m long. How many millimeters long is it?

2. The walls are 8 ft 10 in. high. Mrs. Duran cuts a strip of wallpaper that is 16 in. longer. What is the length of the strip?

3. Regina is refinishing an old wooden trunk. It measures $3\frac{1}{2}$ ft long by 2 ft wide by $1\frac{1}{2}$ ft high. What is its volume?

4. Regina will cover all surfaces of the trunk in problem 3 with translucent paper. How much paper will she need?

5. The living room is a rectangle with one side that measures 3.6 m and another side that measures 4.27 m. What is the perimeter of the living room?

6. The length of a radius of one paint can lid is 9.5 cm. What is the circumference of the lid?

7. Mr. Duran worked in the apartment for 7 h 20 min on Saturday and 5 h 47 min on Sunday. How much longer did he work on Saturday?

8. The dining alcove has this triangular shape. What is the area of the dining alcove floor?

9. Regina wants to paint the dining alcove floor with copper paint. The label says that 1 pint will cover 8000 in.^{2} Will she need more than 1 pint to cover the floor? (*Hint:* 144 in.^{2} = 1 ft^2)

10. Alvin buys a circular rug for his room. The diameter of the rug is 5 ft. What is its area?

Dining Area

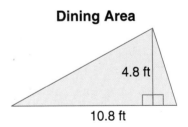

4.8 ft

10.8 ft

Use a strategy from the list or another strategy you know to solve each problem.

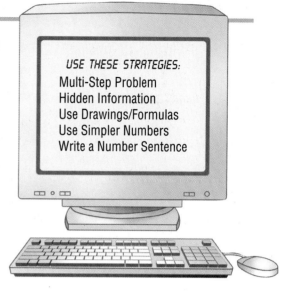

USE THESE STRATEGIES:
Multi-Step Problem
Hidden Information
Use Drawings/Formulas
Use Simpler Numbers
Write a Number Sentence

11. Ed paints Regina's room. He uses $\frac{3}{4}$ qt of lavender paint for the trim. How many fluid ounces of paint does he use?

12. Ed uses $\frac{1}{4}$ gal of white paint on the bedroom walls and $\frac{1}{3}$ gal of the same paint in the parents' room. How much paint is left in the 1-gallon container?

13. The Durans buy a circular rug for the bathroom. The rug has a circumference of about 12.56 ft. Estimate its diameter.

14. Mr. Duran has 3 boards that are each 7 ft 9 in. long. If he uses them to build 15 ft 8 in. of shelving, will he have enough left over to build a flower box that uses 6 ft 3 in. of board?

15. The hallway is a rectangle with a length of 14.5 ft and a width of 5.25 ft. Roberto estimates that they will need about 70 ft^2 of carpeting to cover the hallway floor. Is his estimate reasonable?

16. Regina buys a 25-ft roll of shelf paper for her dresser. She needs 20.5 in. for each of 5 shelves. How much shelf paper will be left after she has lined each shelf?

Use the drawings for problems 17–20.

17. What is the perimeter of Alvin's room?

18. What is the area of Regina's room?

19. Is the perimeter of Regina's room greater or less than 32 ft?

20. Is the area of Alvin's room greater or less than the area of Regina's room? by how much?

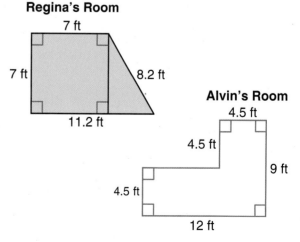

Regina's Room
7 ft
7 ft
8.2 ft
11.2 ft

Alvin's Room
4.5 ft
4.5 ft
9 ft
4.5 ft
12 ft

Chapter Review and Practice

Complete. *(See pp. 332–345.)*

1. 40 m = _?_ km

2. 1.5 cm = _?_ mm

3. 8000 mL = _?_ L

4. 4.8 kg = _?_ g

5. 32 mL = _?_ cm³

6. 1.4 kg = _?_ L

7. 30 in. = _?_ ft

8. $3\frac{1}{4}$ gal = _?_ qt

9. 64 oz = _?_ lb

(See pp. 334–335.)

Measure each line segment to the nearest centimeter and millimeter.

10. •————————————•

11. •———————————————————•

Compute. *(See pp. 346–347, 364–365.)*

12.   7 ft 5 in.
  + 6 ft 9 in.

13.   9 h 35 min
  − 2 h 40 min

14. 2 lb 5 oz ÷ 2

Find the area or surface area. *(See pp. 350–357.)*

15.
3.2 m
10 m

16.
3 ft
16 ft

17.
$2\frac{1}{2}$ in.
1 in.
4 in.

Find the circumference and the area. Use 3.14 or $\frac{22}{7}$ for π. *(See pp. 358–361.)*

18.
4 m

19.
2 yd

20.
21 ft

PROBLEM SOLVING

(See pp. 348–349, 368–371.)

21. What is the perimeter of a rectangular lawn 30 yd long and 25 yd wide?

22. What is the area of a parallelogram with b = 12 cm and h = 9.5 cm?

23. A dog is tied to a pole by a 15-m leash. What is the area in which the dog can run?

24. How many cubic centimeters are in a box 85 cm long, 25 cm wide, and 120 cm deep?

25. Find the surface area and volume of a cube with $e = 1\frac{1}{2}$ ft.

26. Find the area of a square mirror 5.2 dm on each side.

(See Still More Practice, p. 515.)

LOGIC: CONJUNCTIONS AND DISJUNCTIONS

In logic, two statements can be combined to form a compound statement using *and*, or a compound statement using *or*.

It is raining and I am leaving.   It is raining or I am leaving.

▶ A compound statement using *and* is called a **conjunction**. It is true only when *both* original statements are *true*.

A right angle measures 90°   and   a straight angle measures 180°.   True

True   True

▶ A compound statement using *or* is called a **disjunction**. It is true when *both* of the original statements are *true*, or *one* of the original statements is *true*.

A rhombus has 4 sides   or   a square has 4 sides.   True

True   True

Eleven is a prime number   or   eleven is a composite number.   True

True   False

A rectangle is a space figure   or   a prism is a plane figure.   False

False   False

Write a conjunction and a disjunction. Then tell whether each is *True* or *False*.

1. A robin is a bird.
A dime is worth exactly 5¢.

2. A boat can float.
A plane can fly.

3. Eighteen is a prime number.
One is a composite number.

4. $144 \div 12 = 12$
$56 \times 5 = 280$

5. $2400 + 80 = 2320$
$75 - 69 = 6$

6. $(17 - 3) \times 4 = 5$
$147 < 58 + 83$

Check Your Mastery

Performance Assessment

Solve each problem.

1. The perimeter of a rectangle is 60 yd. The length is twice the width. Find the length and width of the rectangle.

2. The area of a rectangle is 128 yd². The length is twice the width. Find the length and width of the rectangle.

Complete.

3. 6.5 km = __?__ m

4. 14.2 mm = __?__ cm

5. 23 L = __?__ mL

6. 58.3 L = __?__ kL

7. 5 kg = __?__ g

8. 9 t = __?__ kg

9. 40 mL = __?__ cm³

10. 2.5 g = __?__ mL

11. 3000 cm³ = __?__ L

12. 9 ft = __?__ in.

13. 48 qt = __?__ gal

14. $5\frac{1}{4}$ lb = __?__ oz

Measure each line segment to the nearest $\frac{1}{8}$ in. and $\frac{1}{16}$ in.

15.

A ———————— X

16.

B ———— Y

Find the area or surface area.

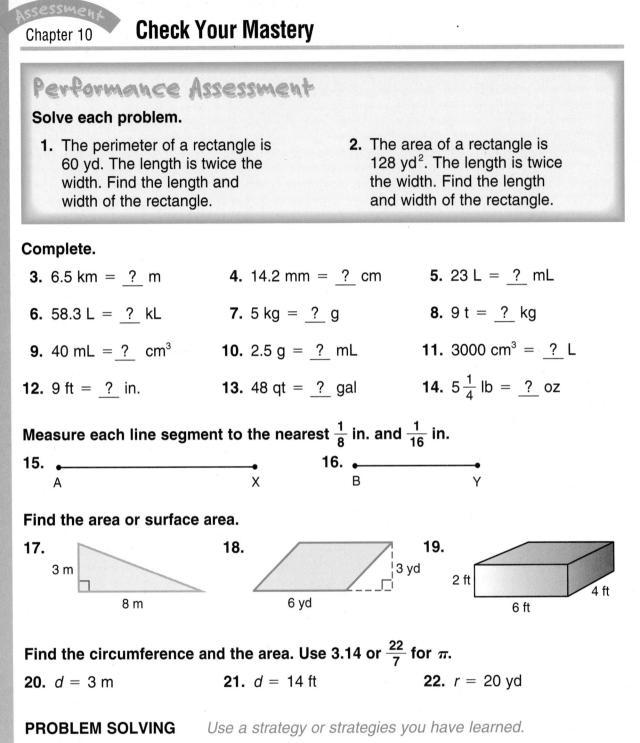

17.

3 m

8 m

18.

3 yd

6 yd

19.

2 ft

6 ft

4 ft

Find the circumference and the area. Use 3.14 or $\frac{22}{7}$ for π.

20. $d = 3$ m

21. $d = 14$ ft

22. $r = 20$ yd

PROBLEM SOLVING   *Use a strategy or strategies you have learned.*

23. Which has a greater volume: a safe in the shape of a rectangular prism that is 6 ft wide, 6 ft long, and 2 ft deep or a cubical safe that measures 4 ft on each edge? How much greater?

24. What happens to the area or volume if you double each dimension in the following formulas: $A = \ell \times w$; $A = \frac{1}{2} \times b \times h$; $V = \ell \times w \times h$? Explain.

Ratio, Proportion, and Percent

11

We've got a tree in our yard. It grows about six feet every year.

If I had grown at the same speed, I'd now be almost fifty feet tall! I wouldn't really mind, except that I'd never get clothes to fit.

From *Counting on Frank*
by Rod Clement

In this chapter you will:

Simplify ratios and rates
Write and solve proportions
Learn about scale drawings, maps, and similar figures
Relate percents, fractions, and decimals
Solve problems by combining strategies

Critical Thinking/Finding Together

Suppose that you grew at a rate of 6 feet per year. About how many inches tall would you be today?

Ratio

A **ratio** is a way of comparing two numbers or quantities by division.

The ratio of the number of fashion magazines to the number of news magazines is 3 to 5.

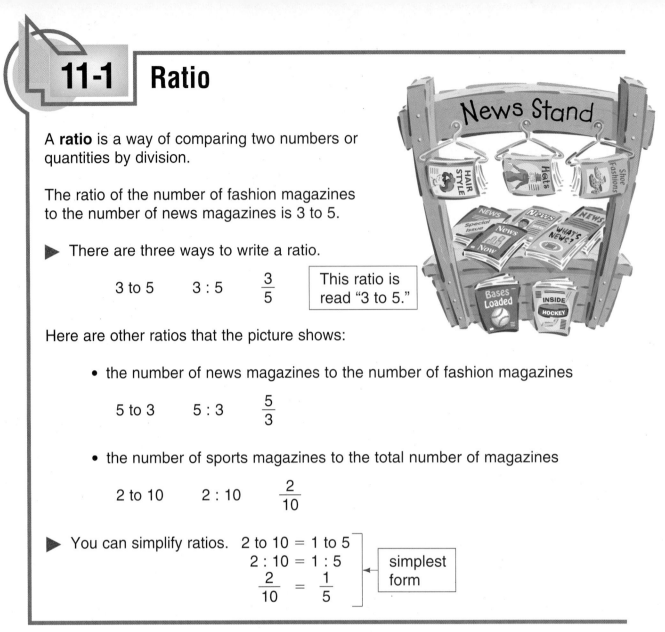

▶ There are three ways to write a ratio.

$$3 \text{ to } 5 \qquad 3 : 5 \qquad \frac{3}{5}$$

This ratio is read "3 to 5."

Here are other ratios that the picture shows:

- the number of news magazines to the number of fashion magazines

$$5 \text{ to } 3 \qquad 5 : 3 \qquad \frac{5}{3}$$

- the number of sports magazines to the total number of magazines

$$2 \text{ to } 10 \qquad 2 : 10 \qquad \frac{2}{10}$$

▶ You can simplify ratios.   2 to 10 = 1 to 5
                     2 : 10 = 1 : 5
                     $\frac{2}{10} = \frac{1}{5}$

simplest form

Use the bar graph. Write each ratio in three ways.

1. news magazines to sports magazines

2. home magazines to fashion magazines

3. news magazines to home magazines

4. sports magazines to news and home magazines

5. sports and hobby magazines to fashion and news magazines

6. news and sports magazines to fashion and home magazines

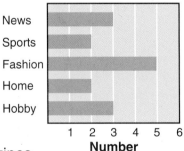

Magazines at Raoul's Newsstand

Write each ratio in simplest form.

7. 4 to 12   **8.** 5 to 10   **9.** 8 : 12   **10.** 9 : 15   **11.** 10 : 6

12. 20 : 10   **13.** 8 to 36   **14.** 9 to 45   **15.** 5 : 2   **16.** 7 : 1

17. $\dfrac{3}{15}$ **18.** $\dfrac{12}{16}$ **19.** $\dfrac{27}{36}$ **20.** $\dfrac{36}{27}$ **21.** $\dfrac{50}{30}$

The chart shows the number of coupons redeemed per household in a recent year.

Write each ratio of coupons redeemed in simplest form.

22. Spain to Italy   **23.** Italy to U.K.

24. France to U.K.   **25.** Spain to Belgium

26. U.K. to Spain   **27.** U.S.A. to Canada

Country	Coupons Redeemed
U.S.A.	81
Canada	33
Belgium	19
U.K.	16
Italy	4
Spain	2
France	1

Write a ratio in simplest form. (*Hint:* Use like units for each ratio.)

28. 1 inch to 1 foot ⟶ Think: 1 ft = 12 in. ⟶ 1 to 12

29. 1 quart to 1 gallon   **30.** 1 yard to 1 inch   **31.** 1 quart to 1 pint

32. 1 day to 1 hour   **33.** 1 half dollar to 2 quarters   **34.** 4 nickels to 1 quarter

CONNECTIONS: PHYSICAL EDUCATION

A baseball batting average is a ratio: $\dfrac{\text{Number of Hits}}{\text{Number of Times at Bat}}$

$\dfrac{16 \text{ hits}}{122 \text{ times at bat}}$ $1\ 6\ \div\ 1\ 2\ 2\ =\ \boxed{0.\ 131475}$ ⟶ 0.131

Copy and complete the table.

35.

Hits	16	43	27	38	25	29	36
Times at bat	122	201	166	154	179	111	182
Average	0.131	?	?	?	?	?	?

11-2 Equal Ratios

Equal ratios have the same value.
Equal ratios can be written as *equivalent fractions*.

The ratio of circles to squares is 5 to 15.
Write an equal ratio for 5 to 15.

To write an equal ratio:

- Write the given ratio as a fraction.

- *Multiply* or *divide* both the numerator and the denominator by the same number.

- Express the result as a fraction.

$\dfrac{5}{15}$ ← number of circles
$\phantom{\dfrac{5}{15}}$ ← number of squares

$\dfrac{5 \div 5}{15 \div 5}$ | In this case divide by 5.

$\dfrac{5 \div 5}{15 \div 5} = \dfrac{1}{3}$

The equal ratio is 1 to 3.

You can also use multiplication to find equal ratios.

$$\frac{1}{8} = \frac{1 \times 2}{8 \times 2} = \frac{2}{16} \qquad \frac{1}{8} = \frac{1 \times 3}{8 \times 3} = \frac{3}{24}$$

Write three equal ratios for each.

1. $\dfrac{5}{8}$
2. $\dfrac{3}{4}$
3. $\dfrac{8}{40}$
4. $\dfrac{4}{9}$
5. $\dfrac{32}{24}$

Write the letter of the ratio that is equal to the given ratio.

6. $\dfrac{4}{5}$ **a.** 10 to 8   **b.** 12 : 16   **c.** 12 : 14   **d.** 20 : 25

7. 7 to 10   **a.** 30 to 21   **b.** $\dfrac{14}{20}$ **c.** $\dfrac{13}{20}$ **d.** $\dfrac{10}{7}$

8. 3 : 2   **a.** $\dfrac{6}{4}$ **b.** $\dfrac{10}{12}$ **c.** $\dfrac{15}{20}$ **d.** 2 to 3

Which ratios are equal? Write = or ≠.

9. $\dfrac{6}{8}$? $\dfrac{18}{24}$
10. $\dfrac{1}{3}$? $\dfrac{3}{1}$
11. $\dfrac{15}{1}$? $\dfrac{30}{15}$
12. $\dfrac{25}{45}$? $\dfrac{5}{9}$

Complete.

13. $\dfrac{2}{8} = \dfrac{1}{?}$

14. $\dfrac{15}{18} = \dfrac{?}{6}$

15. $\dfrac{5}{6} = \dfrac{?}{12}$

16. $\dfrac{7}{4} = \dfrac{21}{?}$

17. $\dfrac{21}{24} = \dfrac{7}{?}$

18. $\dfrac{16}{20} = \dfrac{?}{5}$

19. $\dfrac{8}{1} = \dfrac{24}{?}$

20. $\dfrac{4}{7} = \dfrac{?}{28}$

21. $\dfrac{18}{6} = \dfrac{3}{?}$

22. $\dfrac{11}{?} = \dfrac{11}{1}$

23. $\dfrac{12}{36} = \dfrac{?}{3}$

24. $\dfrac{36}{9} = \dfrac{4}{?}$

25. $\dfrac{20}{16} = \dfrac{10}{?} = \dfrac{5}{4}$

26. $\dfrac{28}{24} = \dfrac{?}{12} = \dfrac{7}{?}$

27. $\dfrac{18}{12} = \dfrac{9}{?} = \dfrac{?}{?}$

28. $\dfrac{36}{24} = \dfrac{?}{4} = \dfrac{?}{?}$

Using Equal Ratios to Solve Problems

At 1 table Sue Ann can place 6 chairs.
How many chairs can she place at 3 tables?

$$\dfrac{1 \text{ table}}{6 \text{ chairs}} = \dfrac{3 \text{ tables}}{? \text{ chairs}}$$

To find the number of chairs, write a ratio equal to $\frac{1}{6}$.

$$\dfrac{1 \times 3}{6 \times 3} = \dfrac{3}{18}$$

So Sue Ann can place 18 chairs at 3 tables.

PROBLEM SOLVING Use equal ratios.

29. There are 6 boxes of fruit drink in 1 package. How many boxes are in 7 packages?

30. There are 60 pencils in 4 boxes. How many pencils are in 1 box?

31. You can fit 72 books on 3 shelves of a bookcase. How many books can you fit on 1 shelf?

32. The ratio of boys to girls in the math club is 2 to 3. There are 10 boys in the club. How many girls are in the club?

CHALLENGE

33. Iola has a collection of old magazines, comic books, and paperback novels. The ratio of magazines to comic books is 2 to 5, and the ratio of paperback novels to comic books is 1 to 3. If Iola has 6 magazines, how many paperback novels does she have? Discuss your solution with the class.

Discuss

11-3 Rates

A **rate** is a ratio that compares two different quantities. Rates are used almost every day. For example, speeds and prices are often given as rates.

▶ Moya ran 18 miles at a steady pace in 3 hours. Her rate of speed can be expressed as a ratio:

$\dfrac{18 \text{ miles}}{3 \text{ hours}}$ or 18 miles : 3 hours or 18 miles in 3 hours

To find how many miles Moya ran in 1 hour, use equal ratios.

$\dfrac{18 \text{ miles}}{3 \text{ hours}} = \dfrac{? \text{ miles}}{1 \text{ hour}}$ $\dfrac{18 \div 3}{3 \div 3} = \dfrac{6}{1}$ ◀ | 6 miles in 1 hour is called a **unit rate**.

Moya ran 6 miles in 1 hour or 6 miles per hour.

▶ Four ballpoint pens cost $2.20. This rate of cost can be given as a ratio.

$\dfrac{4 \text{ pens}}{\$2.20}$ or 4 pens : $2.20 or 4 pens for $2.20

To find the cost of one pen, use equal ratios.

$\dfrac{4 \text{ pens}}{\$2.20} = \dfrac{1 \text{ pen}}{?}$ $\dfrac{4 \div 4}{\$2.20 \div 4} = \dfrac{1}{\$.55}$ ◀ | 1 pen for $.55 is called a **unit cost**.

One pen costs $.55.

Write each rate in simplest form.

1. $\dfrac{12 \text{ books}}{2 \text{ cartons}} = \dfrac{? \text{ books}}{1 \text{ carton}}$

2. $\dfrac{20 \text{ cats}}{4 \text{ bowls}} = \dfrac{? \text{ cats}}{1 \text{ bowl}}$

3. $\dfrac{9 \text{ weeks}}{18 \text{ hours}} = \dfrac{? \text{ week}}{? \text{ hours}}$

4. $\dfrac{50 \text{ miles}}{5 \text{ hours}} = \dfrac{? \text{ miles}}{? \text{ hour}}$

5. $\dfrac{100 \text{ km}}{10 \text{ L}} = \dfrac{? \text{ km}}{? \text{ L}}$

6. $\dfrac{2 \text{ boys}}{2 \text{ wagons}} = \dfrac{? \text{ boy}}{? \text{ wagon}}$

380

Find the unit rate or unit cost. Use equal ratios.

7. 28 kilometers in 4 hours   **8.** 24 inches in 2 hours   **9.** 120 feet in 8 seconds

10. 5 envelopes for $1.50   **11.** 6 records for $30   **12.** 8 discs for $6.00

Use the unit rate or unit cost to complete.

13. 3 miles in 1 hour
  ? miles in 6 hours

14. 1 card for $.35
  ? cards for $1.05

15. 50 miles in 1 hour
  ? miles in 4 hours

16. 2 books in 1 week
  ? books in 3 weeks

17. 22 miles on 1 gallon
  ? miles on 16 gallons

18. 1 apple for $.30
  ? apples for $1.80

PROBLEM SOLVING Use equal ratios to help you.

19. Three rides on the roller coaster cost $2.25. How much does one ride cost?

20. What is the cost of one pencil if a box of 8 pencils sells for $.96?

21. A 5-lb watermelon costs $2.50. At the same rate per pound, how much would a 10-lb watermelon cost?

22. During the first hour 250 tickets to a concert were sold. At this rate, how long will it be before 1500 tickets are sold?

23. Tyrone rode his bicycle 8 miles in one hour. At the same rate, how long will it take him to ride 44 miles?

24. The school store sells paper in 175-sheet packages at $1.55 per package. How much will 3 packages of paper cost?

SKILLS TO REMEMBER

Algebra ✓

Write a related sentence to find the value of *n*.

25. $n + 16 = 26$

26. $n + 80 = 92$

27. $n + 22 = 140$

28. $n - 14 = 56$

29. $n - 16 = 68$

30. $n - 42 = 122$

31. $n \times 4 = 80$

32. $n \times 7 = 350$

33. $6 \times n = 102$

34. $n \div 5 = 20$

35. $n \div 4 = 26$

36. $n \div 16 = 30$

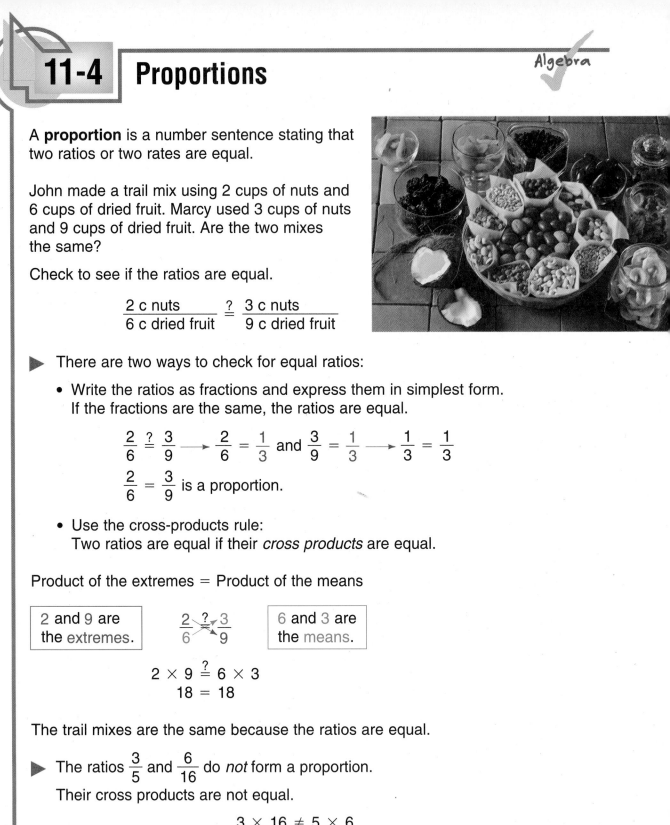

11-4 Proportions

A **proportion** is a number sentence stating that two ratios or two rates are equal.

John made a trail mix using 2 cups of nuts and 6 cups of dried fruit. Marcy used 3 cups of nuts and 9 cups of dried fruit. Are the two mixes the same?

Check to see if the ratios are equal.

$$\frac{2 \text{ c nuts}}{6 \text{ c dried fruit}} \overset{?}{=} \frac{3 \text{ c nuts}}{9 \text{ c dried fruit}}$$

▶ There are two ways to check for equal ratios:

- Write the ratios as fractions and express them in simplest form. If the fractions are the same, the ratios are equal.

$$\frac{2}{6} \overset{?}{=} \frac{3}{9} \longrightarrow \frac{2}{6} = \frac{1}{3} \text{ and } \frac{3}{9} = \frac{1}{3} \longrightarrow \frac{1}{3} = \frac{1}{3}$$

$$\frac{2}{6} = \frac{3}{9} \text{ is a proportion.}$$

- Use the cross-products rule:
  Two ratios are equal if their *cross products* are equal.

Product of the extremes = Product of the means

2 and 9 are the extremes.	$\frac{2}{6} \overset{?}{\diagdown} \frac{3}{9}$	6 and 3 are the means.

$$2 \times 9 \overset{?}{=} 6 \times 3$$
$$18 = 18$$

The trail mixes are the same because the ratios are equal.

▶ The ratios $\frac{3}{5}$ and $\frac{6}{16}$ do *not* form a proportion.
  Their cross products are not equal.

$$3 \times 16 \neq 5 \times 6$$
$$48 \neq 30$$

382

Use equivalent fractions or the cross-products rule to find out which of these are proportions.

1. $\dfrac{1}{5} \overset{?}{=} \dfrac{6}{10}$

2. $\dfrac{3}{4} \overset{?}{=} \dfrac{7}{5}$

3. $\dfrac{2}{1} \overset{?}{=} \dfrac{10}{5}$

4. $\dfrac{6}{10} \overset{?}{=} \dfrac{18}{30}$

5. $\dfrac{7}{4} \overset{?}{=} \dfrac{14}{18}$

6. $\dfrac{6}{60} \overset{?}{=} \dfrac{10}{90}$

7. $\dfrac{7}{5} \overset{?}{=} \dfrac{49}{35}$

8. $\dfrac{8}{5} \overset{?}{=} \dfrac{40}{50}$

Complete the number sentence to form a proportion.

9. $\dfrac{7}{3} = \dfrac{28}{?}$

10. $\dfrac{5}{1} = \dfrac{15}{?}$

11. $\dfrac{12}{21} = \dfrac{?}{7}$

12. $\dfrac{15}{18} = \dfrac{?}{36}$

13. $\dfrac{3}{4} = \dfrac{?}{8}$

14. $\dfrac{9}{10} = \dfrac{81}{?}$

15. $\dfrac{?}{1} = \dfrac{14}{7}$

16. $\dfrac{32}{36} = \dfrac{8}{?}$

Choose the two equal ratios. Write a proportion.

17. $\dfrac{4}{5}, \dfrac{20}{25}, \dfrac{5}{4}$

18. $\dfrac{1}{12}, \dfrac{24}{21}, \dfrac{40}{35}$

19. $\dfrac{9}{5}, \dfrac{36}{30}, \dfrac{36}{20}$

20. $\dfrac{20}{10}, \dfrac{2}{1}, \dfrac{4}{5}$

21. $\dfrac{9}{81}, \dfrac{9}{27}, \dfrac{1}{9}$

22. $\dfrac{7}{6}, \dfrac{7}{16}, \dfrac{14}{32}$

23. 6 : 2, 3 : 1, 10 : 5

24. 8 : 4, 4 : 2, 3 : 1

25. 10 : 15, 12 : 18, 14 : 16

PROBLEM SOLVING

26. Leonard's cat catches 3 mice every 2 days. Francine's cat catches 10 mice every 6 days. Do the two cats catch mice at the same rate? Explain.

27. George Ferris constructed the first Ferris wheel. It was about 250 ft high and almost 800 ft around. Name four equal ratios that compare the height of the wheel to the distance around it.

SHARE YOUR THINKING

Work with a classmate to write a proportion using each set of numbers. Share your proportions with the class.

Communicate

28. 2, 4, 6, 12

29. 3, 4, 9, 12

30. 18, 16, 90, 80

11-5 Solving Proportions

Here is a proportion with a missing term (n).

$$\frac{n}{25} = \frac{6}{5}$$

Use the cross-products rule to find the missing term.

> Remember:
> The product of the extremes equals the product of the means.

Extremes   Means

$$\frac{n}{25} \diagdown \frac{6}{5} \longrightarrow n \times 5 = 25 \times 6$$

$$n \times 5 = 150$$

$$n = 150 \div 5$$

$$n = 30$$

Think: The multiplication sentence $n \times 5 = 150$ can be written as the *related division sentence* $n = 150 \div 5$.

Check: $\frac{30}{25} \overset{?}{=} \frac{6}{5} \longrightarrow 30 \times 5 \overset{?}{=} 25 \times 6 \longrightarrow 150 = 150$

Study these examples.

Extremes   Means

$$\frac{30}{n} \diagdown \frac{6}{5} \longrightarrow 30 \times 5 = n \times 6$$

$$150 = n \times 6$$

$$150 \div 6 = n$$

$$25 = n \text{ or } n = 25$$

Check: $\frac{30}{25} \overset{?}{=} \frac{6}{5} \longrightarrow \frac{6}{5} = \frac{6}{5}$

Extremes   Means

$$\frac{3}{8} \diagdown \frac{n}{24} \longrightarrow 3 \times 24 = 8 \times n$$

$$72 = 8 \times n$$

$$72 \div 8 = n$$

$$9 = n \text{ or } n = 9$$

Check: $\frac{3}{8} \overset{?}{=} \frac{9}{24} \longrightarrow \frac{3}{8} = \frac{3}{8}$

Complete.

1. $\frac{n}{6} = \frac{5}{3} \longrightarrow n \times 3 = 6 \times 5$

$$n \times 3 = 30$$

$$n = 30 \div \underline{\ ?\ }$$

$$n = \underline{\ ?\ }$$

2. $\frac{8}{n} = \frac{32}{40} \longrightarrow 8 \times 40 = n \times 32$

$$320 = n \times 32$$

$$320 \div \underline{\ ?\ } = n$$

$$\underline{\ ?\ } = n$$

384

Find the missing term in each proportion.

3. $\dfrac{n}{12} = \dfrac{5}{20}$ 　　**4.** $\dfrac{n}{10} = \dfrac{3}{5}$ 　　**5.** $\dfrac{n}{3} = \dfrac{7}{21}$ 　　**6.** $\dfrac{6}{5} = \dfrac{24}{n}$

7. $\dfrac{4}{n} = \dfrac{16}{20}$ 　　**8.** $\dfrac{2}{n} = \dfrac{14}{28}$ 　　**9.** $\dfrac{12}{4} = \dfrac{6}{n}$ 　　**10.** $\dfrac{5}{20} = \dfrac{n}{12}$

11. $\dfrac{n}{16} = \dfrac{3}{6}$ 　　**12.** $\dfrac{n}{9} = \dfrac{5}{9}$ 　　**13.** $\dfrac{n}{7} = \dfrac{6}{2}$ 　　**14.** $\dfrac{18}{48} = \dfrac{n}{8}$

Find the value of *n*.

15. $4 : 5 = n : 10$ 　　**16.** $6 : n = 3 : 9$ 　　**17.** $n : 8 = 5 : 5$

18. $9 : 8 = 18 : n$ 　　**19.** $n : 12 = 18 : 9$ 　　**20.** $13 : 5 = n : 15$

21. $n : 6 = 0.4 : 12$ 　　**22.** $9 : 4 = 2.7 : n$ 　　**23.** $13 : 5 = n : 1.5$

Select the two ratios that form a proportion.

24. $\dfrac{1}{2}, \dfrac{1}{4}, \dfrac{2}{4}$ 　　**25.** $\dfrac{3}{5}, \dfrac{9}{15}, \dfrac{6}{9}$ 　　**26.** $\dfrac{2}{3}, \dfrac{3}{9}, \dfrac{1}{3}$

27. $\dfrac{4}{6}, \dfrac{2}{5}, \dfrac{8}{12}$ 　　**28.** $\dfrac{5}{8}, \dfrac{20}{32}, \dfrac{15}{16}$ 　　**29.** $\dfrac{3}{10}, \dfrac{9}{10}, \dfrac{9}{30}$

CRITICAL THINKING

Is the answer reasonable? Write *Yes* or *No*.
Explain your answer.

$\dfrac{3}{7} = \dfrac{n}{140}$ 　　No, 6 is not reasonable.

$n = 6?$ 　　3 is about half of 7, so
n should be about half of 140.

30. $\dfrac{33}{60} = \dfrac{n}{40}$ 　　**31.** $\dfrac{32}{160} = \dfrac{1}{n}$ 　　**32.** $\dfrac{100}{250} = \dfrac{n}{5}$ 　　**33.** $\dfrac{35}{n} = \dfrac{0.2}{0.4}$

　　$n = 30?$ 　　　　$n = 5?$ 　　　　$n = 7.5?$ 　　　　$n = 7\dfrac{1}{2}?$

11-6 Writing Proportions

Meghan's car averages 30 miles per gallon of gasoline.

To find the number of gallons of gasoline she would use on a 3600 mile trip, write a proportion.

Miles Per Gallon	
Meghan	30
Ann	35
Harold	37
Paco	38.5

Think: 30 miles per gallon means 30 miles per 1 gallon.

Write 30 miles per gallon as a ratio:

$$\frac{30 \text{ miles}}{1 \text{ gallon}}$$

Write the ratio of total miles to total gallons for the trip:

$$\frac{3600 \text{ miles}}{m \text{ gallons}}$$

Let m = the total number of gallons.

The ratios are equal since they both express miles per gallon for the same automobile. The proportion is $\frac{30}{1} = \frac{3600}{m}$.

▶ When you write a proportion, be sure that the two equal ratios compare similar things.

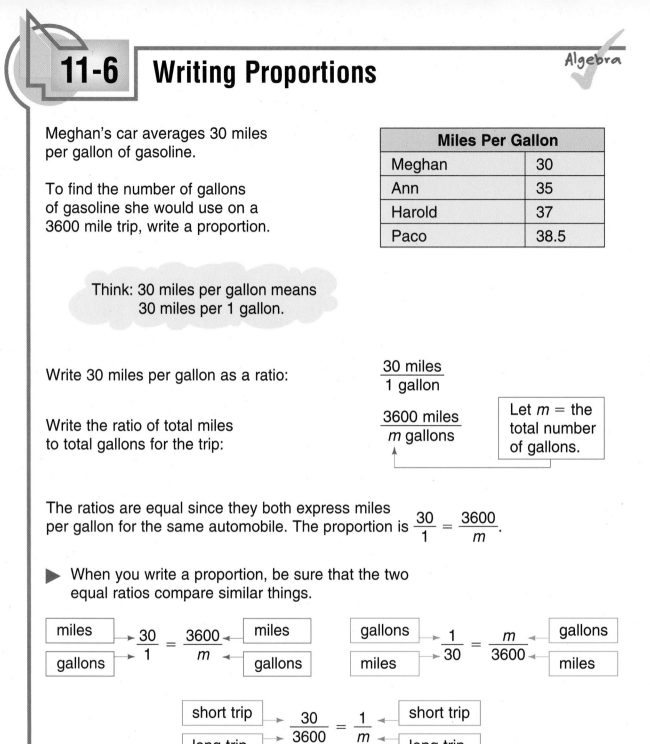

miles → $\frac{30}{1}$ = $\frac{3600}{m}$ ← miles
gallons → $\;\;\;$ ← gallons

gallons → $\frac{1}{30}$ = $\frac{m}{3600}$ ← gallons
miles → $\;\;\;$ ← miles

short trip → $\frac{30}{3600}$ = $\frac{1}{m}$ ← short trip
long trip → $\;\;\;$ ← long trip

Use the chart above to write a proportion to find the number of gallons of gasoline for each trip.

1. Harold, 185 miles   **2.** Ann, 1050 miles   **3.** Paco, 3080 miles

Write a proportion. Use a variable where necessary.

4. If apples sell at 3 for $.75, how many can be bought for $4.25?

5. If 12 calculators cost $60, what will 4 calculators cost?

6. Chin delivers 4 newspapers in 5 minutes. At that rate, how many newspapers can he deliver in 25 minutes?

7. Lorraine makes 8 out of every 10 free throws that she tries. How many can she expect to make in 45 tries?

Proportions in Similar Figures

The lengths of corresponding sides of similar figures are in proportion.

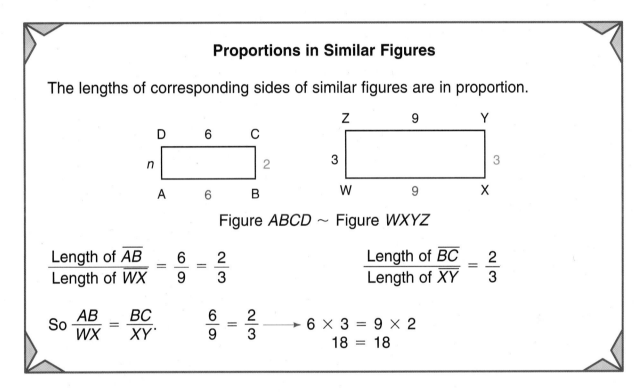

Figure *ABCD* ~ Figure *WXYZ*

$$\frac{\text{Length of } \overline{AB}}{\text{Length of } \overline{WX}} = \frac{6}{9} = \frac{2}{3} \qquad \frac{\text{Length of } \overline{BC}}{\text{Length of } \overline{XY}} = \frac{2}{3}$$

So $\dfrac{AB}{WX} = \dfrac{BC}{XY}$. $\dfrac{6}{9} = \dfrac{2}{3} \longrightarrow 6 \times 3 = 9 \times 2$
$$18 = 18$$

---

8. Complete this proportion to find *n* in the diagram above.

$$\frac{\text{Length of } \overline{AB}}{\text{Length of } \overline{WX}} = \frac{\text{Length of } \overline{DA}}{\text{Length of } \overline{ZW}} \longrightarrow \frac{6}{9} = \frac{?}{?}$$

9. The triangular sails are similar. Write a proportion to find *m*.

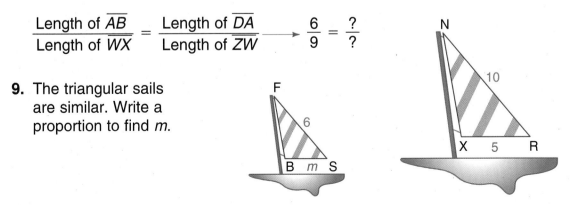

11-7 | **Using Proportions**

You can use a proportion to solve for unknown lengths
of sides of similar figures.

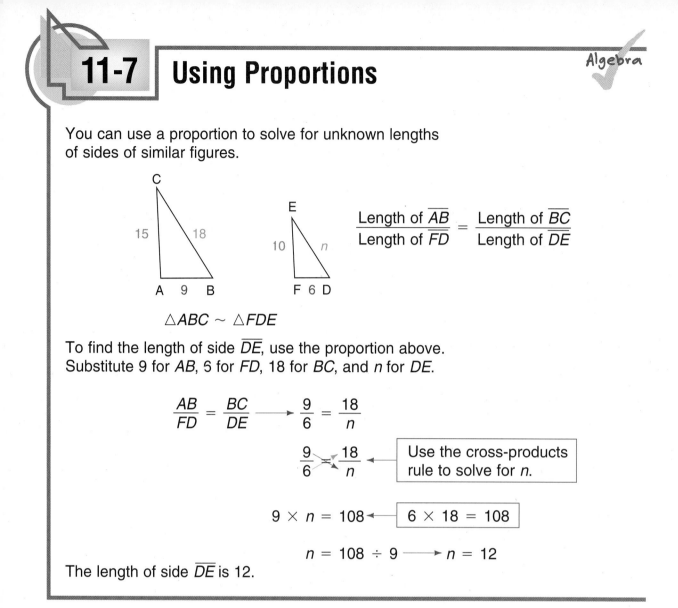

$$\dfrac{\text{Length of } \overline{AB}}{\text{Length of } \overline{FD}} = \dfrac{\text{Length of } \overline{BC}}{\text{Length of } \overline{DE}}$$

$\triangle ABC \sim \triangle FDE$

To find the length of side $\overline{DE}$, use the proportion above.
Substitute 9 for AB, 6 for FD, 18 for BC, and n for DE.

$$\dfrac{AB}{FD} = \dfrac{BC}{DE} \longrightarrow \dfrac{9}{6} = \dfrac{18}{n}$$

$$\dfrac{9}{6} \diagdown\diagup \dfrac{18}{n} \longleftarrow \boxed{\text{Use the cross-products rule to solve for } n.}$$

$$9 \times n = 108 \longleftarrow \boxed{6 \times 18 = 108}$$

$$n = 108 \div 9 \longrightarrow n = 12$$

The length of side $\overline{DE}$ is 12.

Write a proportion. Then solve.

1. Use a different proportion to find the length
   of side $\overline{DE}$ in the diagram above.

2. How far will a motorboat travel in 5 hours if it travels
   35 miles in 1 hour?

3. Luis reads 75 pages in 2 hours. At the same rate,
   how many pages will he read in 8 hours?

4. A train takes 1 hour to go 75 miles. How long will it
   take the train to go 450 miles?

Proportions and Heights

Kwam uses similar triangles and a proportion to find the height of the tree.

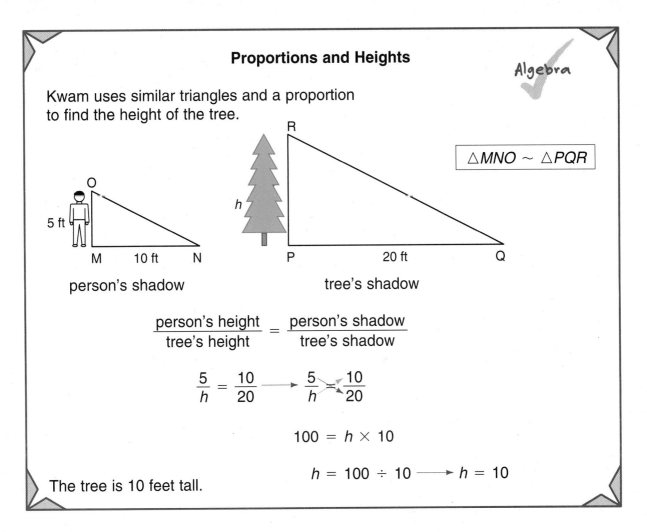

$\triangle MNO \sim \triangle PQR$

person's shadow

tree's shadow

$$\frac{\text{person's height}}{\text{tree's height}} = \frac{\text{person's shadow}}{\text{tree's shadow}}$$

$$\frac{5}{h} = \frac{10}{20} \longrightarrow \frac{5}{h} \bowtie \frac{10}{20}$$

$$100 = h \times 10$$

$$h = 100 \div 10 \longrightarrow h = 10$$

The tree is 10 feet tall.

Write a proportion. Then solve. Compare your proportions and solutions with those of your classmates.

5. An 8-foot electricity pole casts a 12-foot shadow. At the same time Ruth Ann casts a 9-foot shadow. How tall is Ruth Ann?

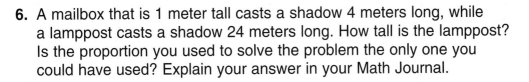

8 ft

12 ft

9 ft

6. A mailbox that is 1 meter tall casts a shadow 4 meters long, while a lamppost casts a shadow 24 meters long. How tall is the lamppost? Is the proportion you used to solve the problem the only one you could have used? Explain your answer in your Math Journal.

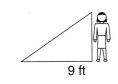

Math Journal ✓

7. On the school softball team there are 3 boys for every 4 girls. If the team has 12 girls, how many boys does it have?

8. A football team has 6 players who play offense for every 5 who play defense. There are 25 players who play defense. How many players play offense?

11-8 Scale Drawings and Maps

A **scale drawing** is an accurate picture of something, but it is different in size.

A **scale** is a ratio of the pictured measure to the actual measure.

In this scale drawing the ratio is:

$$\frac{1 \text{ in.}}{32 \text{ ft}} \xleftarrow{\hspace{1cm}} \text{measure in drawing}$$
$$\xleftarrow{\hspace{1cm}} \text{actual measure}$$

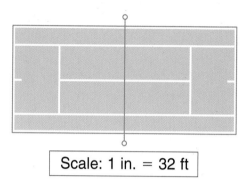

Scale: 1 in. = 32 ft

The scale length of the tennis court is $2\frac{7}{16}$ in. How long is the tennis court?

To find the actual length of the tennis court:

- Set up a proportion.

$$\frac{\text{Scale measure}}{\text{Actual measure}} = \frac{\text{Scale length}}{\text{Actual length}}$$

- Substitute.

$$\frac{1 \text{ in.}}{32 \text{ ft}} = \frac{2\frac{7}{16} \text{ in.}}{n \text{ ft}}$$

- Use the cross-products rule to solve.

$$\frac{1}{32} \bowtie \frac{2\frac{7}{16}}{n} \longrightarrow 1 \times n = 32 \times 2\frac{7}{16} \longrightarrow n = \frac{\overset{2}{\cancel{32}}}{1} \times \frac{39}{\underset{1}{\cancel{16}}} = 78$$

The actual length of the tennis court is 78 feet.

Copy and complete the table. Scale: 1 in. = 10 ft.

	Rooms	Scale Length	Scale Width	Actual Length	Actual Width
1.	Dining Room	$1\frac{1}{2}$ in.	1 in.	?	?
2.	Kitchen	$1\frac{1}{4}$ in.	$\frac{3}{4}$ in.	?	?
3.	Living Room	2 in.	$1\frac{3}{4}$ in.	?	?

**Measure the scale distance on the map
to the nearest 0.5 cm.
Use a proportion to find the actual distance.**

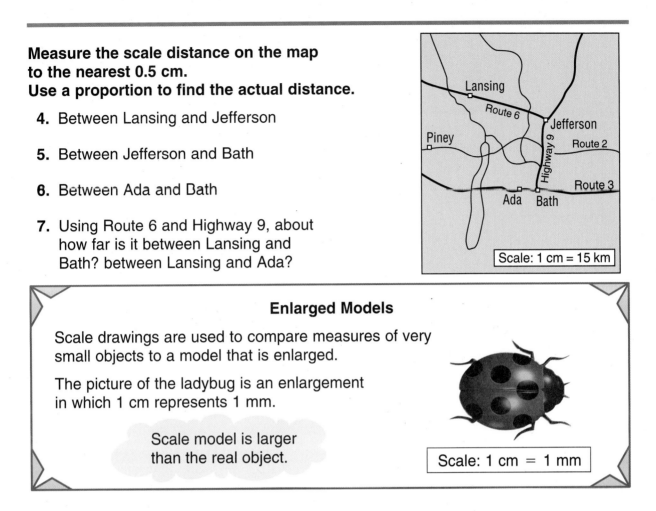

4. Between Lansing and Jefferson

5. Between Jefferson and Bath

6. Between Ada and Bath

7. Using Route 6 and Highway 9, about
   how far is it between Lansing and
   Bath? between Lansing and Ada?

Lansing

Route 6

Jefferson

Piney

Highway 9

Route 2

Route 3

Ada   Bath

Scale: 1 cm = 15 km

Enlarged Models

Scale drawings are used to compare measures of very
small objects to a model that is enlarged.

The picture of the ladybug is an enlargement
in which 1 cm represents 1 mm.

Scale model is larger
than the real object.

Scale: 1 cm = 1 mm

Use the scale drawing of the ladybug to answer each question.

8. What is the scale length of the ladybug's body?

9. What is the actual length of the ladybug's body?

10. What is the scale width of the ladybug's body? Measure to the nearest 0.5 cm.

FINDING TOGETHER

11. Use a string and a metric ruler to find these distances
    on the map at the top of the page: (a) scale distance
    between Lansing and Piney along Route 6, Highway 9, and
    Route 2; (b) "straight line" distance between Lansing and Piney.
    Compare your measurements with those of your classmates.

Communicate

12. Find the actual distances in (a) and (b) and compare them.

Percent as Ratio

A **percent** is a ratio that compares a number to 100.
Percent means "the part of each hundred."
The symbol for percent is **%**.

There are 100 squares in the grid.

20 of the 100 squares (20 : 100) are red.
20% of the grid is red.

80 of the 100 squares (80 : 100) are *not* red.
80% of the grid is *not* red.

Ratios	Percents

20 to 100 = 20 : 100 = $\dfrac{20}{100}$ = 20%

80 to 100 = 80 : 100 = $\dfrac{80}{100}$ = 80%

**Write a ratio to show the part of the grid that is shaded.
Then write the ratio as a percent.**

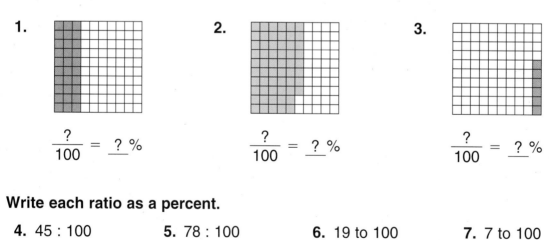

1. $\dfrac{?}{100} = \underline{\ ?\ }$ %

2. $\dfrac{?}{100} = \underline{\ ?\ }$ %

3. $\dfrac{?}{100} = \underline{\ ?\ }$ %

Write each ratio as a percent.

4. 45 : 100

5. 78 : 100

6. 19 to 100

7. 7 to 100

8. $\dfrac{36}{100}$

9. $\dfrac{88}{100}$

10. 99 to 100

11. 1 : 100

Write each percent as a ratio in three ways.

12. 70%   **13.** 46%   **14.** 11%   **15.** 5%   **16.** 8%

17. 27%   **18.** 30%   **19.** 92%   **20.** 71%   **21.** 89%

Percent Scores on 100-Item Test

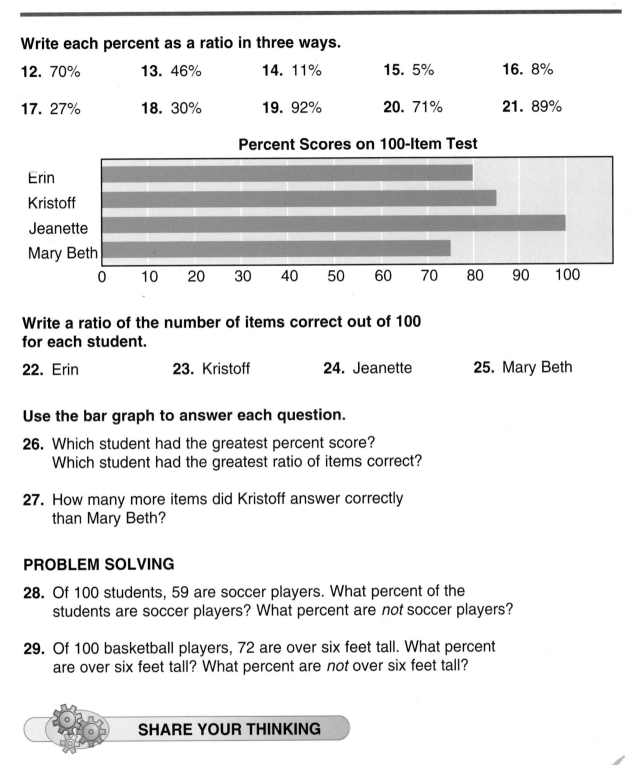

Write a ratio of the number of items correct out of 100 for each student.

22. Erin   **23.** Kristoff   **24.** Jeanette   **25.** Mary Beth

Use the bar graph to answer each question.

26. Which student had the greatest percent score?
Which student had the greatest ratio of items correct?

27. How many more items did Kristoff answer correctly
than Mary Beth?

PROBLEM SOLVING

28. Of 100 students, 59 are soccer players. What percent of the
students are soccer players? What percent are *not* soccer players?

29. Of 100 basketball players, 72 are over six feet tall. What percent
are over six feet tall? What percent are *not* over six feet tall?

SHARE YOUR THINKING

30. Using the bar graph above, explain in your Math
Journal how you know the number of test items
a person answered correctly on a 100-item test
if you know the person's percent score on the test.

Math Journal

Relating Percents to Fractions

In a recent survey, the interviewers found that television sets were turned on 25% of the time. Write this percent as a fraction.

Percent means "the part of each hundred."

▶ **To rename a percent as a fraction:**

$25\% = \underline{\ ?\ }$

- Drop the percent symbol (%).

- Write the number as the numerator and 100 as the denominator.

$\dfrac{25}{100}$

- Express the fraction in simplest form.

$\dfrac{25}{100} = \dfrac{25 \div 25}{100 \div 25} = \dfrac{1}{4}$

So $25\% = \dfrac{1}{4}$.

▶ **To rename a fraction as a percent:**

$\dfrac{4}{5} = \underline{\ ?\ }\%$

- Express the fraction as an equivalent fraction with a denominator of 100.

$\dfrac{4}{5} = \dfrac{4 \times 20}{5 \times 20} = \dfrac{80}{100}$

- Write this number of hundredths as a percent. Use the percent symbol (%).

80%

So $\dfrac{4}{5} = 80\%$.

Study these examples.

Write 32% as a fraction.

$32\% = \dfrac{32}{100} = \dfrac{32 \div 4}{100 \div 4} = \dfrac{8}{25}$

Write $\dfrac{13}{25}$ as a percent.

$\dfrac{13}{25} = \dfrac{13 \times 4}{25 \times 4} = \dfrac{52}{100} = 52\%$

Complete.

1. $\dfrac{9}{10} = \dfrac{?}{100} = \underline{\ ?\ }\%$

2. $\dfrac{3}{4} = \dfrac{?}{100} = \underline{\ ?\ }\%$

3. $\dfrac{9}{20} = \dfrac{?}{100} = \underline{\ ?\ }\%$

Write as a percent.

4. $\frac{53}{100}$ 5. $\frac{71}{100}$ 6. $\frac{6}{10}$ 7. $\frac{1}{2}$ 8. $\frac{1}{4}$

9. $\frac{1}{5}$ 10. $\frac{3}{5}$ 11. $\frac{7}{25}$ 12. $\frac{16}{40}$ 13. $\frac{42}{60}$

Write as a fraction in simplest form.

14. 75% 15. 20% 16. 60% 17. 80% 18. 13%

19. 37% 20. 64% 21. 22% 22. 5% 23. 8%

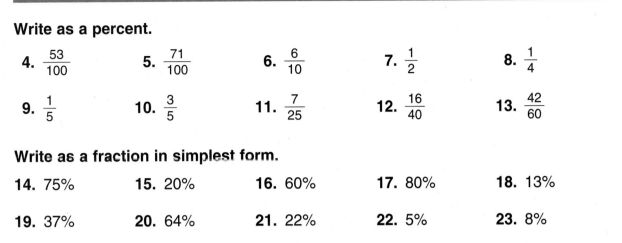

Estimating with Percents

You can use certain **benchmark** percents to help you estimate.

$50\% = \frac{50}{100} = \frac{1}{2}$ $\qquad$ $25\% = \frac{25}{100} = \frac{1}{4}$ $\qquad$ $75\% = \frac{75}{100} = \frac{3}{4}$

Some sixth grade students made designs of their initials.
Estimate what percent of each design is red.

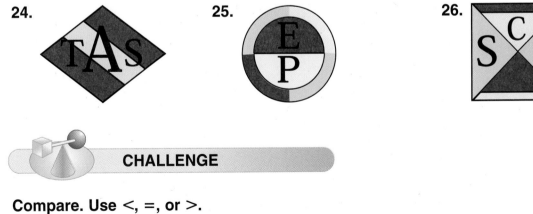

24.

25.

26.

CHALLENGE

Compare. Use <, =, or >.

27. 40% $\underline{\ ?\ }$ $\frac{2}{5}$ 28. $\frac{9}{25}$ $\underline{\ ?\ }$ 35% 29. $\frac{1}{10}$ $\underline{\ ?\ }$ 11%

Relating Percents to Decimals

Approximately 65% of the people in Virginia live in urban areas. Write this percent as a decimal.

65% means "sixty-five hundredths."

▶ **To rename a percent as a decimal:**

65% = _?_

- Drop the percent symbol (%).

- Move the decimal point *two places* to the *left.*

65% = .65. = 0.65

So 65% = 0.65. ⟵ ── sixty-five hundredths

▶ **To rename a decimal as a percent:**

0.08 = _?_ %

- Move the decimal point *two places* to the *right.*

0.08 = 0.08. = 8%

- Write the percent symbol (%).

So 0.08 = 8%.

To write percents as decimals and decimals as percents, you may need to write zero(s).

Write 2% as a decimal.

2% = .02. = 0.02

└── Write a 0.

Write 0.9 as a percent.

0.9 = 0.90. = 90%

└── Write a 0.

Write as a decimal.

1. 27%   **2.** 36%   **3.** 30%   **4.** 80%   **5.** 86%

6. 94%   **7.** 25%   **8.** 59%   **9.** 6%   **10.** 1%

Write as a percent.

11. 0.20   **12.** 0.35   **13.** 0.50   **14.** 0.70   **15.** 0.07

16. 0.7   **17.** 0.2   **18.** 0.02   **19.** 0.400   **20.** 0.100

Copy and complete each table.

	Percent	Decimal	Fraction
21.	10%	?	?
22.	50%	?	?
23.	?	0.4	?
24.	35%	?	?

	Percent	Decimal	Fraction
25.	?	0.85	?
26.	28%	?	?
27.	?	0.44	?
28.	?	?	$\frac{3}{10}$

PROBLEM SOLVING Use the double-line graph to answer each question.

29. What percent approved of each candidate in January?

30. Write a decimal for the percent approval of each candidate in April.

31. Write a fraction in simplest form for the percent approval of each candidate in May.

32. How many people out of 100 approved of Candidate A in February? Candidate B?

33. What is the difference of the approval ratings in June?

34. Assuming the trends of the graph continue, what ratings would you expect the candidates to have in July?

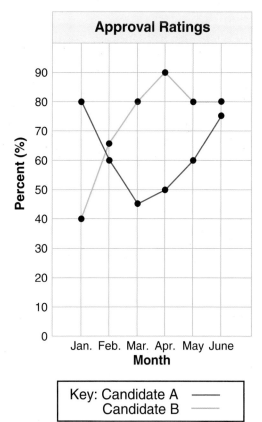

397

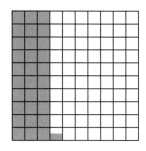

Some percents are *not* whole numbers.
In the picture 30.5% of the grid is shaded.

30.5% is greater than 30% and less
than 31%.

Decimal percents like 30.5% and 6.1%
can be renamed as decimals.

30.5% = .30.5 = 0.305   6.1% = .06.1 = 0.061

▶ Some fractions are *not* easily renamed as equivalent
fractions with a denominator of 100. So, it is more
difficult to rename such fractions as percents.

To rename a fraction like $\frac{3}{8}$ as a percent:

• Express the fraction as a decimal. (Divide the
denominator into the numerator.)

• Then express the decimal as a percent.

$$\frac{3}{8} \longrightarrow \begin{array}{r} 0.375 \\ 8\overline{)3.000} \\ \underline{2\,4} \\ 60 \\ \underline{56} \\ 40 \\ \underline{40} \end{array} \qquad 0.375 = 0.37.5 = 37.5\%$$

Study this example.

Use two ways to rename $\frac{2}{5}$ as a percent.

$$\frac{2}{5} = \frac{2 \times 20}{5 \times 20} = \frac{40}{100} = 40\% \qquad \frac{2}{5} \longrightarrow \begin{array}{r} 0.40 \\ 5\overline{)2.00} \end{array} \quad 0.40. = 40\%$$

Write as a decimal.

1. 25.3%   **2.** 64.4%   **3.** 39%   **4.** 87%   **5.** 17.2%

6. 48.5%   **7.** 53.2%   **8.** 20.1%   **9.** 8.07%   **10.** 2.16%

11. 2%   **12.** 4%   **13.** 73.02%   **14.** 84.20%   **15.** 59.99%

Write as a percent.

16. $\frac{3}{10}$ **17.** $\frac{1}{10}$ **18.** $\frac{1}{5}$ **19.** $\frac{1}{2}$ **20.** $\frac{5}{8}$

21. $\frac{7}{8}$ **22.** $\frac{9}{16}$ **23.** $\frac{7}{16}$ **24.** $\frac{24}{30}$ **25.** $\frac{42}{60}$

26. $\frac{1}{8}$ **27.** $\frac{1}{16}$ **28.** $\frac{16}{40}$ **29.** $\frac{30}{32}$ **30.** $\frac{44}{64}$

Fractional Percents

$\frac{2}{3} = \underline{\ ?\ }\%$

- Divide to the hundredths place.

- Write the remainder as a fraction.

- Write the decimal as a percent.

$$\begin{array}{r} .66 \\ 3\overline{)2.00} \\ \underline{1\ 8} \\ 20 \\ \underline{18} \\ 2 \end{array}$$

$.66 \longrightarrow 0.66\frac{2}{3} = 66\frac{2}{3}\%$

66 and $\frac{2}{3}$ hundredths equals 66 and $\frac{2}{3}$ percent.

Write as a fractional percent.

31. $\frac{5}{6}$ **32.** $\frac{1}{6}$ **33.** $\frac{1}{3}$ **34.** $\frac{1}{8}$ **35.** $\frac{2}{9}$

36. $\frac{3}{7}$ **37.** $\frac{1}{7}$ **38.** $\frac{1}{9}$ **39.** $\frac{7}{8}$ **40.** $\frac{5}{8}$

41. $\frac{3}{8}$ **42.** $\frac{1}{16}$ **43.** $\frac{1}{15}$ **44.** $\frac{1}{30}$ **45.** $\frac{11}{12}$

11-13 | Percents Greater Than 100%

A merchant sold a bicycle for 25% more than its cost to him.
The bicycle is now 125% of its original value.

100% of something
means the *whole* of it.

125% of something means
25% *more than the whole* of it.

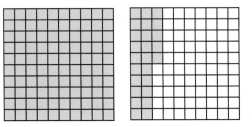

A percent *greater than 100%* can be renamed
as a decimal or as a mixed number.

▶ **To rename a percent greater than 100% as a *decimal*:**

- Drop the percent symbol (%).

- Move the decimal point *two* places to the left.

  125% = 1.25. = 1.25

▶ **To rename a percent greater than 100% as a *mixed number*:**

- Drop the percent symbol (%).

- Write the number as the numerator
  and 100 as the denominator.

- Express the fraction in simplest form.

  $125\% = \frac{125}{100} = 1\frac{25}{100} = 1\frac{1}{4}$

Write each as a decimal.

1. 175%   **2.** 220%   **3.** 440%   **4.** 350%   **5.** 101%

6. 205%   **7.** 432%   **8.** 500%   **9.** 355%   **10.** 816%

11. 200%   **12.** 550%   **13.** 625%   **14.** 130%   **15.** 760%

Write each as a mixed number in simplest form.

16. 141% **17.** 110% **18.** 350% **19.** 520% **20.** 116%

21. 212% **22.** 484% **23.** 150% **24.** 275% **25.** 680%

26. 268% **27.** 497% **28.** 720% **29.** 805% **30.** 945%

Explain the meaning of each statement.

31. The population of Maple Grove is 300% of what it was 10 years ago.

32. This year the school librarian ordered 120% of the books she ordered last year.

33. Mr. Mendoza's salary is 150% of what it was 2 years ago.

34. Kevin's sweater is 100% wool.

PROBLEM SOLVING

35. This year the cost of a bicycle is 35% higher than it was last year. What percent of last year's price is this year's price?

36. After repairs a bicycle can be sold for twice the price at which it was purchased. The bicycle is now worth what percent of its purchase price?

37. Mr. Ortega spends 13% of his budget for car repairs. What percent of his budget is used for other purposes?

38. In a public survey, 78% of the questionnaires sent out by a store were returned. What percent of the questionnaires sent out were *not* returned?

CHALLENGE

Express each percent as a fraction in simplest form.

39. $7\frac{1}{2}\% = 7.5\% = \frac{7.5}{100} = \frac{7.5 \times 10}{100 \times 10} = \frac{75}{1000} = \frac{3}{40}$

40. $8\frac{4}{5}\%$ **41.** $10\frac{3}{4}\%$ **42.** $15\frac{1}{2}\%$ **43.** $5\frac{1}{8}\%$ **44.** $16\frac{2}{5}\%$

11-14 | Problem Solving: Combining Strategies

Problem: Leon bought some stamps. He used 6 of them to mail a package. He gave 50% of what was left to Mira. Then he had 15 stamps. How many stamps had Leon bought?

1 IMAGINE Picture yourself buying stamps.

2 NAME *Facts:* Leon bought some stamps.
He used 6 stamps.
He gave 50% of the rest away.
He had 15 left.

Question: How many stamps had Leon bought?

3 THINK Some problems can be solved by combining several strategies.

To find the number of stamps Leon bought try *working backwards* and *writing equations:*

$$15 = 50\% \text{ of } \underset{\text{left}}{\underline{\quad ? \quad}} \qquad \text{and} \qquad \underset{\text{left}}{\underline{\quad ? \quad}} + 6 = \underset{\text{stamps}}{\underline{\quad ? \quad}}$$

Use the Guess and Test strategy to solve the first equation. Record your guesses in a table.

4 COMPUTE $15 = 50\%$ of $\underline{\quad ? \quad}$
"of" means $\times$. $\frac{1}{2}$ of what number equals 15?

Guess	20	40	30
Test	$\frac{1}{2} \times 20 = 10$	$\frac{1}{2} \times 40 = 20$	$\frac{1}{2} \times 30 = 15$

        too low   too high

Leon had 30 stamps left. $\underline{\quad ? \quad} + 6 = \underline{\quad ? \quad}$
                              $30 + 6 = 36$

So Leon bought 36 stamps.

5 CHECK Begin with 36 stamps. Use a calculator to check:

$$\underset{\text{bought}}{36} - \underset{\text{used}}{6} = \underset{\text{left}}{30} \qquad \text{and} \qquad 50\% \times 30 = \underset{\text{left}}{15}$$

Solve. Combine strategies to help you.

1. A business office mailed 24 packages last week. Each package weighed $1\frac{1}{2}$ lb. For every 3 packages mailed special delivery, 5 packages were bulk rate. How many packages were mailed special delivery?

IMAGINE   You are mailing the packages.

NAME   *Facts:*   24 packages mailed
$1\frac{1}{2}$ lb—weight of each package
packages—3 special delivery
                    5 bulk rate

Question: How many packages were mailed special delivery?

THINK   This is a *multi-step problem.* It contains *extra information.* First find the sum of $3 + 5$, then write and solve a proportion.

$$\frac{3}{\text{sum}} = \frac{?}{24} \begin{array}{l} \leftarrow \text{special delivery} \\ \leftarrow \text{total} \end{array}$$

COMPUTE ⟶ **CHECK**

2. Donna's birthday is a dozen days after Mark's. Mark's birthday is in May but it is after May 15th. The sum of the digits of the date is 5. His birthday falls on Sunday. What is the day and date of Donna's birthday?

3. The ratio of the length of a side of square *AXRD* to the length of a side of square *TVBN* is 2 to 7. If the perimeter of *TVBN* is 84 cm, what is the area of each square?

4. There are 630 students at South School. Three fifths of the students are girls. Four out of every 7 boys can swim. How many boys can swim?

5. Brittany mailed this puzzle to a math magazine. Solve the problems she made up.

   a. What is the ratio of the sum of the numbers outside the parallelogram to the sum of the numbers inside the circle?

   b. What is the missing number if the ratio of the missing number to the sum of the numbers inside the parallelogram equals 40%?

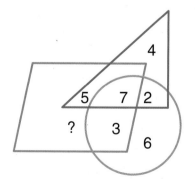

403

11-15 | Problem-Solving Applications

Solve each problem and explain the method you used.

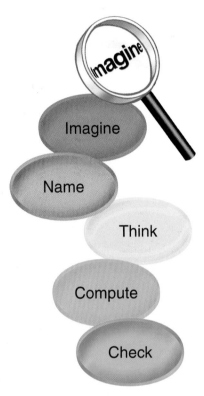

1. Mr. Barry teaches a cooking class every Saturday. The ratio of students to stoves is 4 : 1. There are 5 stoves in the classroom. How many students are in the class?

2. What is the ratio of a cup to a pint?

3. The lessons are paid for at the rate of 4 lessons for $50. How much does each lesson cost?

4. Last week, the class prepared granola. The ratio of rolled oats to raisins was 5 : 2. The class used 25 cups of rolled oats. How many cups of raisins did the class use?

5. Peg added 1 cup of almonds for every 3 cups of oats in her granola. Mark added 3 cups of almonds for every 9 cups of oats in his granola. Did Peg and Mark add the same ratio of almonds to oats?

6. If 12 ounces of almonds cost $2, how much do 18 ounces of almonds cost?

7. Out of every 100 recipes Mr. Barry teaches, 15 include peanut butter. What percent of the recipes include peanut butter?

8. Of all Mr. Barry's recipes, 90% are healthful. Write this percent as a decimal.

9. One student notices that $\frac{1}{5}$ of the recipes taught can be classified as side dishes. One half of the side dishes are pasta dishes. What percent of the recipes are for pasta dishes?

10. Of the calories in a banana bread, 75% come from carbohydrates. What part of the calories do not come from carbohydrates?

Solve. Combine strategies to help you.

USE THESE STRATEGIES:
Hidden Information
Extra Information
Use a Graph
Multi-Step Problem
Use Drawings/Formulas
Working Backwards

11. Ann and Greg each made lemonade. Ann used 10 lemons for every 1 quart of water. Greg used 5 lemons for every pint of water. Did the two friends use equal ratios of lemons to water?

12. The ratio of beans to rice in a recipe is 3 : 4. Jake cooks 2 cups of rice and 5 pounds of chicken. How many cups of beans should he use?

13. The class has 18 baking sheets. Each sheet holds 1 dozen cookies. A batch of cookie batter makes 3 dozen cookies. The class fills all of the sheets with cookies. Then they sell bags of 6 cookies each at a bake sale. How many bags of cookies did they make?

14. What is the area of the snack bar tray in the scale drawing?

15. What is the greatest number of 3 in. × 2 in. snack bar treats that Stan can cut from this tray?

16. Tara made some snack bars. She ate 8 of them. She gave Paul 50% of the bars that were left. Then she had 11 bars. How many snack bars did she make?

Snack Bar Tray

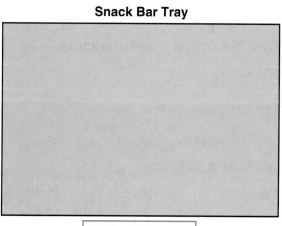

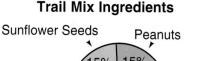

Scale: 1 in. = 4 in.

Use the circle graph for problems 17–20.

17. What part of the trail mix is raisins?

18. What is the ratio of apricots to raisins?

19. Sue makes a batch of trail mix using 1 cup of granola. Will she use more than $\frac{1}{2}$ cup apricots?

Trail Mix Ingredients

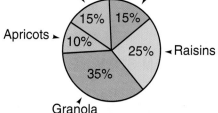

Sunflower Seeds — 15%
Peanuts — 15%
Apricots — 10%
Raisins — 25%
Granola — 35%

MAKE UP YOUR OWN

20. Use the data in the circle graph to write a problem. Then solve it.

Write each ratio in simplest form.   *(See pp. 376–377.)*

1. 8 : 12   **2.** 6 to 45   **3.** $\frac{45}{60}$ **4.** $\frac{78}{100}$

Find the missing term in each proportion.   *(See pp. 384–385.)*

5. $\frac{n}{7} = \frac{5}{21}$ **6.** $\frac{4}{n} = \frac{20}{15}$ **7.** $\frac{12}{14} = \frac{n}{7}$ **8.** $\frac{n}{1.2} = \frac{3}{2}$

Write a proportion. Then solve.   *(See pp. 386–389.)*

9. △ABC is similar to △DEF.
Find the value of n.

```
        B
       /\
  6 ft/  \6 ft
     /    \
    /_____\
   A   4 ft  C


        E
       /\
   3 ft/ \3 ft
     /___\
    D  n  F
```

Find the actual measurements.   *(See pp. 390–391.)*

10. Actual length

11. Actual width

Scale: 1 cm = 40 km

Write as a percent.   *(See pp. 392–399.)*

12. $\frac{3}{4}$ **13.** $\frac{24}{30}$ **14.** $\frac{5}{9}$ **15.** 0.72   **16.** 0.9

Write as a decimal.   *(See pp. 396–401.)*

17. 29%   **18.** 1%   **19.** 47.5%   **20.** 2.6%   **21.** 534%

Write as a fraction or mixed number in simplest form.   *(See pp. 394–395, 400–401.)*

22. 80%   **23.** 4%   **24.** 48%   **25.** 340%   **26.** 605%

PROBLEM SOLVING   *(See pp. 380–381, 390–391.)*

27. Sixteen cans of corn sell for $12.00. Find the unit cost. Explain the method you used to solve the problem.

28. A road map uses a scale of 3 in. = 9 mi. Find the distance between the cities if the map distance is 15 in.

(See *Still More Practice*, p. 516.)

PERCENTS LESS THAN 1%

Sometimes a percent is *less than 1%.*

A group of 600 students takes a test.
Only 3 students get a perfect score.
What percent of the group gets a
perfect score?

To find the percent,
divide: 3 ÷ 600 = __?__

$$\frac{3}{600} \longrightarrow 600\overline{)3.000}^{\,0.005}$$

Move the decimal point 2 places   0.005 = 0.5%
to the right. Write the % sign.

So 0.5% of the group gets a perfect score.

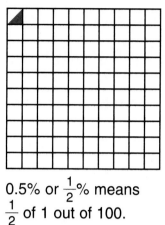

0.5% or $\frac{1}{2}$% means
$\frac{1}{2}$ of 1 out of 100.

▶ You can write a decimal or a fraction for a percent
less than 1%.

$$0.25\% = 0.25 \div 100 = 0.0025 = \frac{25}{10,000} = \frac{1}{400}$$

$$\frac{1}{4}\% = \frac{\frac{1}{4}}{100} = \frac{1}{4} \div \frac{100}{1} = \frac{1}{4} \times \frac{1}{100} = \frac{1}{400}$$

$$= 400\overline{)1.0000}^{\,0.0025}$$

Express each as a percent.

1. 0.001   2. 0.0026   3. $\frac{3}{1000}$ 4. $\frac{1}{800}$ 5. $\frac{4}{2000}$ 6. $\frac{2}{500}$

7. 0.0031   8. $\frac{1}{400}$ 9. 0.0006   10. $\frac{3}{4000}$ 11. $\frac{5}{8000}$ 12. $\frac{35}{10,000}$

Express each as a decimal.

13. 0.7%   14. 0.2%   15. 0.23%   16. $\frac{1}{5}$%   17. $\frac{3}{8}$%   18. $\frac{3}{4}$%

Express each as a fraction.

19. 0.3%   20. 0.8%   21. 0.64%   22. $\frac{1}{10}$%   23. $\frac{2}{5}$%   24. $\frac{7}{8}$%

407

Performance Assessment

Use <, =, or > to compare. Explain your answers.

1. $15 : 100$ __?__ 16%

2. 49% __?__ $\dfrac{51}{100}$

3. 5.5% __?__ $\dfrac{11}{200}$

Write each ratio in simplest form.

4. 4 to 8

5. 9 to 15

6. $20 : 30$

7. $24 : 39$

Find the missing term in each proportion.

8. $\dfrac{n}{9} = \dfrac{10}{15}$

9. $\dfrac{6}{n} = \dfrac{18}{12}$

10. $n : 2 = 0.9 : 3$

Write a proportion. Then solve.

11. A vertical meterstick casts a 6-m shadow while a telephone pole casts a 36-m shadow. How tall is the telephone pole?

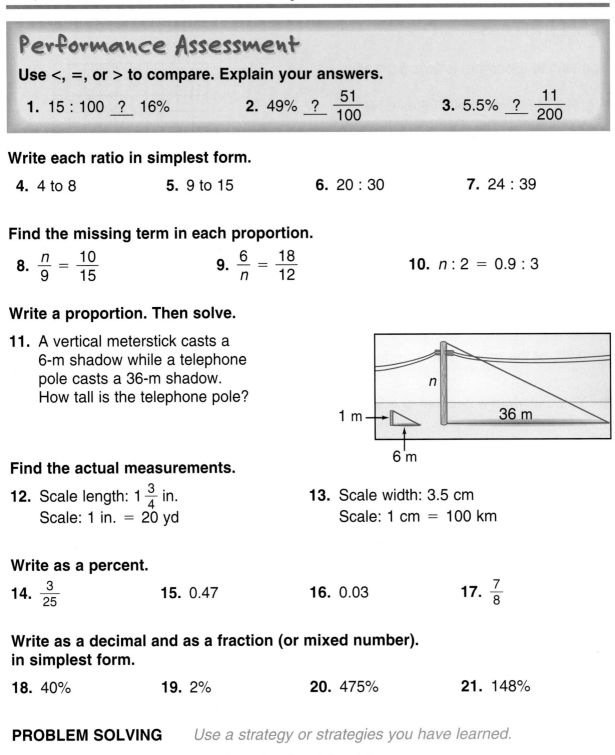

Find the actual measurements.

12. Scale length: $1\dfrac{3}{4}$ in.
Scale: 1 in. = 20 yd

13. Scale width: 3.5 cm
Scale: 1 cm = 100 km

Write as a percent.

14. $\dfrac{3}{25}$

15. 0.47

16. 0.03

17. $\dfrac{7}{8}$

Write as a decimal and as a fraction (or mixed number). in simplest form.

18. 40%

19. 2%

20. 475%

21. 148%

PROBLEM SOLVING   *Use a strategy or strategies you have learned.*

22. Out of every 9 students in the school activity club, there are 2 boys. If there are 35 girls in the activity club, how many boys are in it?

Cumulative Review IV

Choose the best answer.

1. 12.5 × 1000
- **a.** 125
- **b.** 1250
- **c.** 12,500
- **d.** not given

2. $10\frac{2}{5} \div 2\frac{1}{6}$
- **a.** $3\frac{1}{2}$
- **b.** $4\frac{4}{5}$
- **c.** $8\frac{1}{5}$
- **d.** not given

3. $0.36\overline{)0.12312}$
- **a.** 0.00342
- **b.** 0.0342
- **c.** 0.342
- **d.** not given

4. Compare.

18 yd ? 60 ft
- **a.** <
- **b.** =
- **c.** >
- **d.** cannot tell

5. Compare.

0.027 kL ? 27 L
- **a.** <
- **b.** =
- **c.** >
- **d.** cannot tell

6. $7\frac{3}{4}$ gal = ? qt
- **a.** 10
- **b.** 21
- **c.** 31
- **d.** 41

7. 6 yd 1 ft ÷ 3
- **a.** $2\frac{1}{3}$ yd
- **b.** 6 ft $\frac{1}{3}$ in.
- **c.** 6 ft 3 in.
- **d.** 6 ft 4 in.

8.   4 d 6 h
     + 1d 22 h
- **a.** 6 d   4 h
- **b.** 5 d   4 h
- **c.** 3 d 16 h
- **d.** 5 d 2 h

9. Find the area.

3.5 m · 1.5 m · 1 m · 2 m
- **a.** 5.25 m²
- **b.** 6.75 m²
- **c.** 7 m²
- **d.** 7.25 m²

10. Find the surface area.

3 cm · 6 cm · 2 cm
- **a.** 24 cm²
- **b.** 36 cm²
- **c.** 72 cm²
- **d.** 84 cm²

11. Rename as a percent.

$\frac{3}{25}$
- **a.** 3%
- **b.** 6%
- **c.** 9%
- **d.** 12%

12. Compare.

25% ? $\frac{28}{100}$
- **a.** <
- **b.** =
- **c.** >
- **d.** cannot tell

13. Rename as a percent.

0.37
- **a.** 0.37%
- **b.** 3.7%
- **c.** 37%
- **d.** 370%

14. Rename as a decimal.

12.8%
- **a.** 0.0128
- **b.** 0.128
- **c.** 1.28
- **d.** 12.8

15. Rename as a mixed number in simplest form.

275%
- **a.** $1\frac{3}{4}$
- **b.** $2\frac{2}{3}$
- **c.** $2\frac{3}{4}$
- **d.** $2\frac{75}{100}$

16. Find the value of *n*.

$\frac{n}{9} = \frac{28}{63}$
- **a.** 4
- **b.** 7
- **c.** 35
- **d.** 44

Ongoing Assessment IV

For Your Portfolio

Solve each problem. Explain the steps and the strategy or strategies you used for each. Then choose one from problems 1–4 for your Portfolio.

1. Norma reads 120 pages in 3 hours. At that rate, how many pages can she read in 9 hours?

2. On a map 1 inch = 10 miles. What is the actual distance between two towns if the map distance is 5.5 inches?

Use the circular area at the right for problems 3 and 4.

5 yd

3. Sintora has a 16-yard tape measure. Is her tape measure long enough to find the circumference of the circular area? Explain.

4. Sintora and Shyam plan to cover the circular area with crepe paper. How many square yards of paper will they need?

Tell about it.

5. **a.** Explain how you can rename the circumference in problem 3 from yards to feet.

   b. Explain how you can rename your answer for problem 4 from square yards to square feet. (*Hint:* Visualize drawing a 1-yd square on the board. How many 1-foot squares does it contain?)

Communicate

For Rubric Scoring

Listen for information on how your work will be scored.

6. You and four friends are planning a vacation. All expenses are to be shared equally.

   Housing: $1575 → $?
   Food: $250 → $ 50
   Recreation: $400 → $ 80

   $? per person

   a. You plan to take $550 in cash with you. After the above expenses, what percent of your money do you have left? (*Hint:* % = dollar amount left ÷ total cash taken.)

   b. After the above expenses, what percent of your $550 have you spent?

Percent Applications

12

SKY

Decimal point
meteors
streak
through
the night—

Fractions
of moonbeams
gleam
white-bright—

Percentages
of stars
seem
to multiply—

in the
finite
dramatic
mathematic-filled
sky.

Lee Bennett Hopkins

In this chapter you will:

Use patterns to compute mentally
Find percentage and rate
Investigate discount, sales tax,
  commission, and better buy
Make circle graphs
Learn about electronic spreadsheets
Solve problems by writing an equation

Critical Thinking/Finding Together
Which prize would you choose:
20% of (20% of $100) or 30% of
(10% of $100)? Explain.

10%
10%
10%
10%
10%
10%
10%
10%
10%
10%

20%

20%

33.3%

411

12-1 Mental Math: Percent

Here are some percents that are equivalent to common fractions.

0%	25%	50%	75%	100%
$\frac{0}{4}$	$\frac{1}{4}$	$\frac{2}{4} = \frac{1}{2}$	$\frac{3}{4}$	$\frac{4}{4} = 1$

$25\% = \frac{1}{4}$ $50\% = \frac{1}{2}$ $75\% = \frac{3}{4}$ $100\% = 1$

Study the pattern for these percents and common fractions.

$10\% = \frac{1}{10}$ $20\% = \frac{2}{10}$ or $\frac{1}{5}$

$30\% = \frac{3}{10}$ $40\% = \frac{4}{10}$ or $\frac{2}{5}$

$50\% = \frac{5}{10}$ or $\frac{1}{2}$ $60\% = \frac{6}{10}$ or $\frac{3}{5}$

$70\% = \frac{7}{10}$ $80\% = \frac{8}{10}$ or $\frac{4}{5}$

$90\% = \frac{9}{10}$ $100\% = \frac{10}{10}$ or $\frac{5}{5}$ or 1

▶ To find a percent of a number mentally, use common fractions.

50% of 20	25% of 36	40% of 35
Think: $50\% = \frac{1}{2}$	Think: $25\% = \frac{1}{4}$	Think: $40\% = \frac{2}{5}$
$\frac{1}{2}$ of 20 = 10	$\frac{1}{4}$ of 36 = 9	$\frac{1}{5}$ of 35 is 7, so $\frac{2}{5}$ of 35 is 14.

Complete.

1. $\frac{1}{5}$ of 30 is 6, so 20% of 30 is _?_

2. $\frac{2}{5}$ of 30 is 12, so 40% of 30 is _?_

> Remember: "of" means "times."

Complete. Compute mentally.

3. $\frac{3}{5}$ of 30 is 18, so 60% of 30 is _?_

4. $\frac{3}{10}$ of 30 is 9, so 30% of 30 is _?_

5. $\frac{1}{4}$ of 44 is _?_ , so 25% of 44 is _?_

6. $\frac{1}{2}$ of 44 is _?_ , so 50% of 44 is _?_

7. $\frac{3}{4}$ of 44 is 33, so 75% of 44 is _?_

8. $\frac{1}{10}$ of 30 is _?_ , so 10% of 30 is _?_

More Percents and Common Fractions

Study these percents and common fractions.

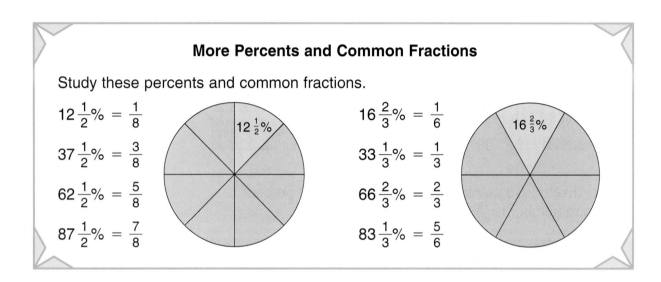

$12\frac{1}{2}\% = \frac{1}{8}$

$37\frac{1}{2}\% = \frac{3}{8}$

$62\frac{1}{2}\% = \frac{5}{8}$

$87\frac{1}{2}\% = \frac{7}{8}$

$16\frac{2}{3}\% = \frac{1}{6}$

$33\frac{1}{3}\% = \frac{1}{3}$

$66\frac{2}{3}\% = \frac{2}{3}$

$83\frac{1}{3}\% = \frac{5}{6}$

Find the percent of the number. Compute mentally.

9. $12\frac{1}{2}\%$ of 16

Think: $\frac{1}{8}$ of 16 = _?_

10. $33\frac{1}{3}\%$ of 60

Think: $\frac{1}{3}$ of 60 = _?_

11. 20% of 100

Think: $\frac{1}{5}$ of 100 = _?_

12. 10% of 50

Think: $\frac{1}{10}$ of 50 = _?_

13. 80% of 20

Think: $\frac{1}{5}$ of 20 = _?_

$\frac{4}{5}$ of 20 = _?_

14. 60% of 45

Think: $\frac{1}{5}$ of 45 = _?_

$\frac{3}{5}$ of 45 = _?_

15. 30% of 60

16. 70% of 60

17. $62\frac{1}{2}\%$ of 16

18. $83\frac{1}{3}\%$ of 24

19. 90% of 90

20. 100% of 90

12-2 | Percent Sense

Which is greater:

3% of 100 or 30% of 100?

The pictures show that

30% of 100 > 3% of 100.

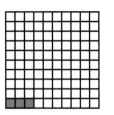

Which is less:

30% of 60 or 30% of 80?

The pictures show that

30% of 60 < 30% of 80.

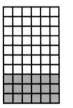

▶ You can use mental math to estimate what percent one number is of another.

Math class is 55 minutes long.
Ten minutes are used to check homework.
True or false: 10 minutes is more than 50% of the class time?

Is $\frac{10}{55} > 50\%$?

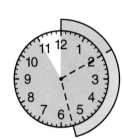

Think: 50% is $\frac{1}{2}$. $\frac{1}{2}$ of 55 is more than 10.
10 minutes is less than 50% of the class time.

The statement is false.

Compare. Use < or >.

1. 50% of 20 _?_ 50% of 40

2. 20% of 30 _?_ 20% of 10

3. 2% of 40 _?_ 2% of 80

4. 3% of 10 _?_ 3% of 4

414

Compare. Use < or >.

5. 3% of 60 _?_ 30% of 60

6. 62% of 40 _?_ 52% of 40

7. 19% of 300 _?_ 29% of 300

8. 86% of 50 _?_ 66% of 50

9. $16\frac{2}{3}$% of 12 _?_ $66\frac{2}{3}$% of 12

10. $83\frac{1}{3}$% of 24 _?_ $33\frac{1}{3}$% of 24

11. $37\frac{1}{2}$% of 16 _?_ $\frac{1}{8}$ of 16

12. $\frac{7}{8}$ of 40 _?_ $62\frac{1}{2}$% of 40

Write *True* or *False* for each situation. Explain your answer. Communicate
Draw a picture to help you.

Tina has one half hour for lunch. She finishes in 20 minutes.

13. Tina uses exactly 50% of her lunchtime to eat.

14. Tina uses less than 50% of her lunchtime to eat.

15. Tina uses more than 50% of her lunchtime to eat.

Alberto allots one hour to deliver newspapers. He finishes in 45 minutes.

16. Alberto uses more than 50% of his newspaper time.

17. Alberto uses more than 25% of his newspaper time.

18. Alberto uses less than 100% of his newspaper time.

Forty students are in the class. Fourteen receive A's.

19. 50% of the students receive A's.

20. Less than 50% receive A's.

21. Less than 25% receive A's.

Ninety animals are in the shelter. Thirty are adopted.

22. 50% of the animals are adopted.

23. More than 25% are adopted.

24. More than 50% are *not* adopted.

 **FINDING TOGETHER**

25. There are 6 ducks in the pond at the park. This is 10% of the ducks in the park. True or False: 100 ducks are in the park?

a. Shade a ten-by-ten grid to show 6 ducks and a different ten-by-ten grid to show 10% of 100 ducks. Compare the two shadings.

b. Is 10% of 100 = 6? Is the statement above true or false? How many ducks are in the park? (*Hint:* 10% of _?_ = 6?)

10% of 100 ducks

Finding a Percent of a Number

Find: 45% of 360 = __?__

To estimate a percent of a number, use benchmark percents such as 10%, 25%, and 50% and their common fractions.

45% of 360 ⟶ 50% of 360

$\frac{1}{2}$ of 360 = 180

estimate

45% is close to 50%.
50% = $\frac{1}{2}$

You may use a formula to find a percent of a number.

rate (**r**) × base (**b**) = percentage (**p**) ⟶ **r × b = p**

Rename the percent (rate) as a decimal or as a fraction.

$r \times b = p$

45% of 360 = p

0.45 × 360 = p

162 = p

$r \times b = p$

45% of 360 = p

$\frac{45}{100} \times 360 = p$

$\frac{9}{\underset{1}{20}} \times \frac{\overset{18}{360}}{1} = p$

162 = p

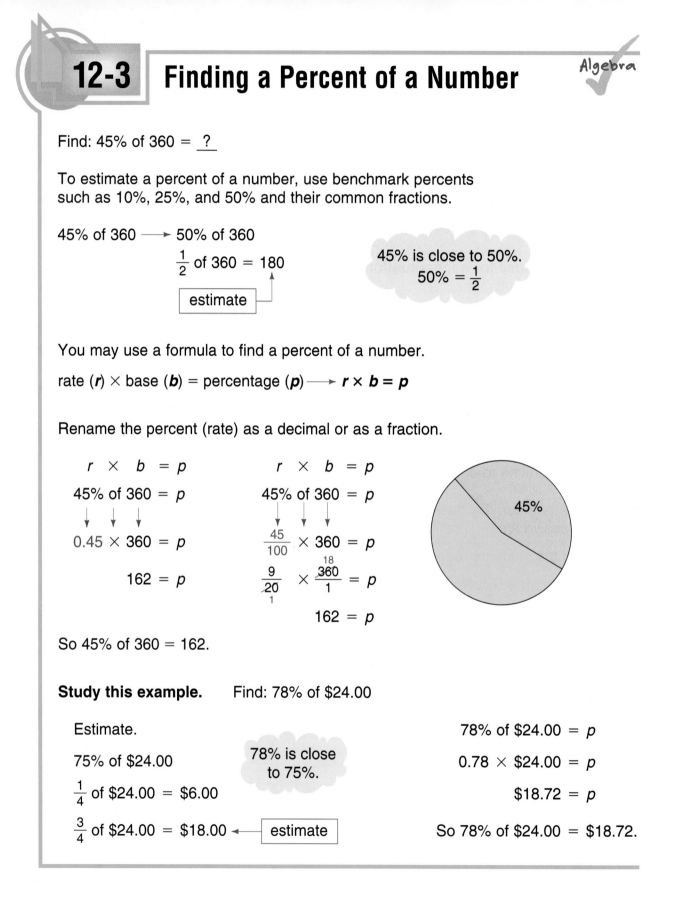

45%

So 45% of 360 = 162.

Study this example.   Find: 78% of $24.00

Estimate.

75% of $24.00

$\frac{1}{4}$ of $24.00 = $6.00

$\frac{3}{4}$ of $24.00 = $18.00 ← estimate

78% is close to 75%.

78% of $24.00 = p

0.78 × $24.00 = p

$18.72 = p

So 78% of $24.00 = $18.72.

Use the formula and decimals to find the percent of the number.

1. 30% of 60

2. 50% of 32

3. 75% of 80

4. 25% of 44

5. 10% of $18

6. 20% of $70

7. 5% of 800

8. 6% of 700

9. 35% of 120

Use the formula and fractions to find the percent of the number.

10. 40% of 20

11. 60% of 60

12. 75% of 120

13. 15% of 40

14. 2% of $8.00

15. 3% of $5.00

16. 56% of 400

17. 88% of 250

18. 95% of 240

Percent of a Number Using Proportions

In the basketball game between the sixth and seventh grades, 62.5% of 24 free throws were made. How many free throws were made?

Write a proportion and solve. Let n = the number of free throws made.

part → $\dfrac{n}{24}$ = $\dfrac{62.5}{100}$ ← part
whole →   ← whole

$\dfrac{n}{24} \diagup\!\!\!\!\diagdown \dfrac{62.5}{100}$

$n \times 100 = 24 \times 62.5$

$n = 1500 \div 100 = 15$

So 15 free throws were made.

Use a proportion to find the percent of the number.

19. 40% of 25

20. 25% of 96

21. 80% of 90

22. 55% of 200

23. 76% of 475

24. 12% of 625

25. 37.5% of 56

26. 62.5% of 320

27. 87.5% of 480

MENTAL MATH

Find the percent of the number.

28. 150% of 4   100% of 4 = 4; 50% of 4 = $\dfrac{1}{2}$ of 4 = 2; 4 + 2 = 6

29. 150% of 80

30. 110% of 200

31. 250% of 600

Missing Percent

Of 125 professional baseball players named most valuable player, 24 were also batting champions. What percent of the most valuable players were batting champions?

Estimate: 25 out of 125 $= \dfrac{25}{125} = \dfrac{1}{5}$ or 20%

Use one of these two methods to find the percent, r, that one number is of another.

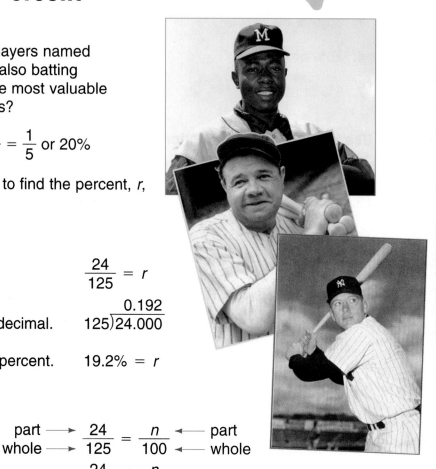

Method 1

- Use the formula $\dfrac{p}{b} = r$.

$$\dfrac{24}{125} = r$$

- Rename the fraction as a decimal.

$$125\overline{)24.000} \quad 0.192$$

- Rename the decimal as a percent.

$$19.2\% = r$$

Method 2

- Write a proportion.

part $\longrightarrow \dfrac{24}{125} = \dfrac{n}{100} \longleftarrow$ part
whole $\longrightarrow$ $\qquad\qquad\quad\longleftarrow$ whole

- Solve the proportion.

$$\dfrac{24}{125} \diagdown \diagup \dfrac{n}{100}$$

$$24 \times 100 = 125 \times n$$
$$n = 2400 \div 125 = 19.2$$

$$\dfrac{n}{100} = \dfrac{19.2}{100} = 19.2\%$$

So 19.2% were batting champions.

Study this example.

What percent of 8 is 20?

$$\dfrac{20}{8} = r \longrightarrow r = 2.5 = 250\%$$

> Think: 20 is greater than 8.
> The percent is greater
> than 100%.

Check.

Enter: 20 $\boxed{\div}$ 8 $\boxed{\%}$ $\boxed{=}$ $\boxed{250.}$

Find the percent. Estimate first.

1. What percent of 5 is 3?

2. What percent of 8 is 4?

3. What percent of 100 is 11?

4. What percent of 100 is 27?

5. 90 is what percent of 120?

6. 60 is what percent of 240?

7. What percent of 180 is 63?

8. What percent of 140 is 91?

9. What percent of 20 is 25?

10. What percent of 10 is 80?

11. 4.4 is what percent of 80?

12. 4.6 is what percent of 50?

Match each sentence on the left with a sentence on the right that has the same value for *n*.   Algebra ✓

13. __?__ $50 = n\%$ of 100

a. $20 = n\%$ of 8

14. __?__ $2 = n\%$ of 16

b. $5 = n\%$ of 40

15. __?__ $15 = n\%$ of 6

c. $20 = n\%$ of 25

16. __?__ $28 = n\%$ of 35

d. $25 = n\%$ of 50

PROBLEM SOLVING

17. At the school picnic, 30 of the 50 teachers came by car. What percent of the teachers came by car?

18. In the basketball game, 12 baskets were made in 25 attempts. What percent is this?

19. Janet earned $420 at the golf course last summer. She put 70% of her earnings in the bank. How much money did she save?

20. This year the price of a baseball glove is 105% of last year's price of $40. What is the price this year?

21. Of 125 players named as major league baseball's most valuable player, 44 were also RBI champions. What percent of the most valuable players were RBI champions?

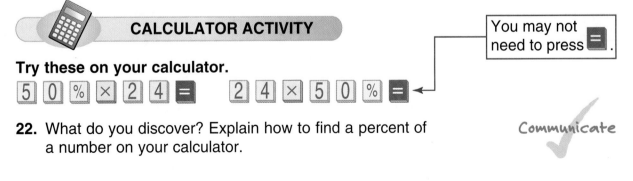

CALCULATOR ACTIVITY

You may not need to press ▤.

Try these on your calculator.

⑤ ⓪ % × ② ④ =   ② ④ × ⑤ ⓪ % ▤

22. What do you discover? Explain how to find a percent of a number on your calculator.

Communicate ✓

419

12-5 Using Percent to Solve Problems

The population of Manchester is 25,100.
Twenty-three percent of the population attends school.
How many people in Manchester attend school?

| Find a percent of a number. Use $r \times b = p$. |

Estimate: 23% of 25,100 $\longrightarrow$ 25% of 24,000

$$\frac{1}{4} \text{ of } 24,000 = 6000$$

Solve: 23% of 25,100 = p

 0.23 $\times$ 25,100 = p

Check.

 5773 = p

Enter: 25,100 ⊠ 23 % ▤ 5773.

So 5773 people in Manchester attend school.

Manchester Middle School has 960 students enrolled. Six hundred
seventy-two of these students ride the bus to school. What
percent of the students ride the bus to school?

| Find the percent one number is of another. Use $\frac{p}{b} = r$. |

Estimate: $\frac{672}{960} \longrightarrow \frac{700}{1000} = \frac{70}{100} = 70\%$

Solve: $\frac{672}{960} = r$ $960\overline{)672.00}$ 0.70

Check.

 70% = r

Enter: 672 ÷ 960 % ▤ 70.

So 70% of the students ride the bus to school.

PROBLEM SOLVING Estimate first.

1. The population of Newton is 1460. Twenty percent of the people in Newton read the local newspaper. How many people read the local newspaper?

2. There are 360 animals in the New City Zoo. Fifteen percent of the animals are monkeys. How many monkeys are in the New City Zoo?

3. Jeff used 10 oranges to make juice this morning. There were 25 oranges in the bag. What percent of the oranges in the bag did he use?

4. There are 16 teenagers in Sarah's neighborhood. Four of them are available to baby-sit. What percent of the teenagers are available to baby-sit?

5. A report states that 72% of 1250 middle school students enjoy playing team sports. How many of the middle school students enjoy playing team sports?

6. There are 960 students in sixth grade. Two hundred eighty-eight of them play basketball. What percent of the students in sixth grade play basketball?

7. The Girl Scouts want to collect 500 pounds of old newspapers. They have already collected 150 pounds. What percent of their goal have they reached?

8. On a typical day at Emerson Middle School, 6% of the 650 students are absent. How many students are absent on a typical day?

9. Nevada's population increased 49% from 1980 to 1990. The population in 1980 was 800,500. How many more people lived there in 1990?

10. The population of California in 1980 was about 24,000,000. In 1990 the population had increased by about 6,000,000. What percent of the 1980 population was the increase?

SKILLS TO REMEMBER

11. Out of each 2500 pairs of athletic shoes imported from Asia, about how many pairs come from each country?

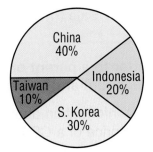

A **discount** is a reduction of the list price (regular price) of an item. The **rate of discount** is given as a percent.

The **sale price** is the difference between the list price and the discount.

Cassette tapes at Posio's Record World regularly sell for $9 each. This week they are being sold at a 15% rate of discount. What is the discount? What is the sale price?

▶ To find the discount, write a number sentence and solve for *D*, the discount.

$$D = 15\% \text{ of } \$9.00$$
$$D = 0.15 \times \$9.00$$
$$D = \$1.35$$

The discount on each cassette tape is $1.35.

▶ To find the sale price, write a number sentence and solve for *SP*, the sale price.

$$SP = \$9.00 - \$1.35$$
$$SP = \$7.65$$

The sale price of each cassette tape is $7.65.

Discount = Rate of Discount × List Price	$D = R \times LP$
Sale Price = List Price − Discount	$SP = LP - D$

Write a letter to a parent or another adult to explain how you will find the discount and sale price in exercise 1. Then complete the exercise.

1. Use the information above to find the discount and the sale price of a CD at Posio's Record World.

Communicate

Copy and complete the table.

	Item	List Price	Rate of Discount	Discount	Sale Price
2.	Radio	$130	20%	?	?
3.	Big Screen TV	$3200	10%	?	?
4.	Headset	$33	15%	?	?
5.	Turntable	$180	12%	?	?

Using Proportion to Find Discount

A VCR with a regular price of $495 is marked to sell
at 30% off. What is the discount?

Algebra ✓

"30% off" means a discount rate of 30%.

Write a proportion and solve.
Let D = the discount.

$$\text{part} \rightarrow \frac{D}{\$495} \bowtie \frac{30}{100} \leftarrow \text{part}$$
$$\text{whole} \rightarrow \hspace{1.2cm} \leftarrow \text{whole}$$

$$D \times 100 = \$495 \times 30$$
$$D = \$14{,}850 \div 100$$
$$D = \$148.50$$

The discount is $148.50.

PROBLEM SOLVING

6. The rate of discount on a $47 clock radio is 20%. Find the discount and sale price.

7. The rate of discount on a $12 CD is 15%. Find the discount and sale price.

8. How much is saved on a $625 television set at a 25%-off sale?

9. What is the sale price of a $150 tape deck marked 15% off?

10. Carlos has $25 to spend at the record store. He buys a $24 record album. The rate of discount on the album is $33\frac{1}{3}\%$. How much money does Carlos have left to spend?

11. Find the answer in problem 10 using a rate of discount of $12\frac{1}{2}\%$.

 **SHARE YOUR THINKING**

12. Solve the discount problem in the box above by writing a number sentence (as shown on page 422). Write a short note to your teacher to explain which method is easier for you.

423

Finding Sales Tax and Total Cost *Algebra* ✓

A **sales tax** is the amount of tax added to the price of an item by a state or local government.
The **rate of sales tax** is given as a percent.

The **total cost** is the sum of the marked (list) price and the tax.

A remote control toy costs $224.50, plus 5% sales tax. Find the sales tax and total cost of the toy.

▶ To find the sales tax, T, write a number sentence and solve: $T = 5\%$ of $224.50.

Estimate: 5% of $220 = $\frac{1}{20} \times$ $220 = $11

Solve: $T = 5\%$ of $224.50

$T = 0.05 \times$ $224.50

$T = 11.225 or $11.23

The sales tax is $11.23.

> Round to the nearest cent.

▶ To find the total cost, write a number sentence and solve: $TC = $224.50 + $11.23.

$TC = $224.50 + $11.23

$TC = $235.73

> Let $TC =$ the total cost.

The total cost of the toy is $235.73.

> Sales Tax = Rate of Sales Tax × Marked Price $T = R \times MP$
> Total Cost = Marked Price + Sales Tax $TC = MP + T$

Find the sales tax and total cost of each item.

1. $23.40 sneakers, 5% sales tax

2. $59.75 camera, 4% sales tax

Use the tables below for exercises 3–6. The rate of sales tax is 6%.

Item	Price
Calculator	$88.25
Video Game	$12.95
Blank Tapes	2 for $7.50

Item	Price
Color Film	1 roll for $6.50
Tape Player	$62.75
Skateboard	$49

3. Find the sales tax on 2 blank tapes.

4. Find the sales tax on a skateboard.

5. Find the total cost of a tape player.

6. Find the total cost of a calculator.

Using Proportion to Find Sales Tax

Find the sales tax on 1 roll of color film. (Use the table above.)

Let T = the sales tax.

$$\text{part} \longrightarrow \frac{T}{\$6.50} \quad \frac{6}{100} \longleftarrow \text{part}$$
$$\text{whole} \longrightarrow \qquad \qquad \longleftarrow \text{whole}$$

$$T \times 100 = \$6.50 \times 6$$
$$T = \$39.00 \div 100$$
$$T = \$.39$$

The sales tax is $ .39.

PROBLEM SOLVING Use the table above for exercises 7–9.

7. Find the sales tax on 2 rolls of color film.

8. Find the sales tax on 4 blank tapes.

9. Janell buys 3 video games. Find her change from a $50 bill.

10. A $220 bicycle is on sale at 20% off. The rate of sales tax is 5%. What is the total cost of the bicycle?

11. Find the total cost in exercise 8 if the rate of sales tax is $6\frac{1}{2}$%.

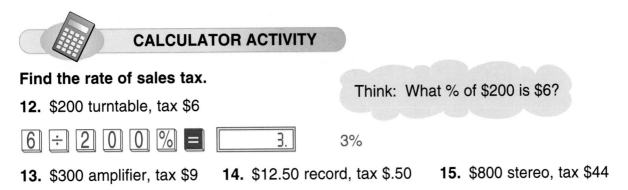

CALCULATOR ACTIVITY

Find the rate of sales tax.

12. $200 turntable, tax $6

Think: What % of $200 is $6?

$6 \div 200 \% = \boxed{\quad 3. \quad}$ 3%

13. $300 amplifier, tax $9

14. $12.50 record, tax $.50

15. $800 stereo, tax $44

425

12-8 Better Buy

The regular-size box of Le Frisky dog food contains 12 oz and sells for $1.50. The jumbo-size box contains 15 oz and sells for $1.71. Which is the better buy?

To decide which is the better buy, find the **unit cost** of each type of dog food and compare them.

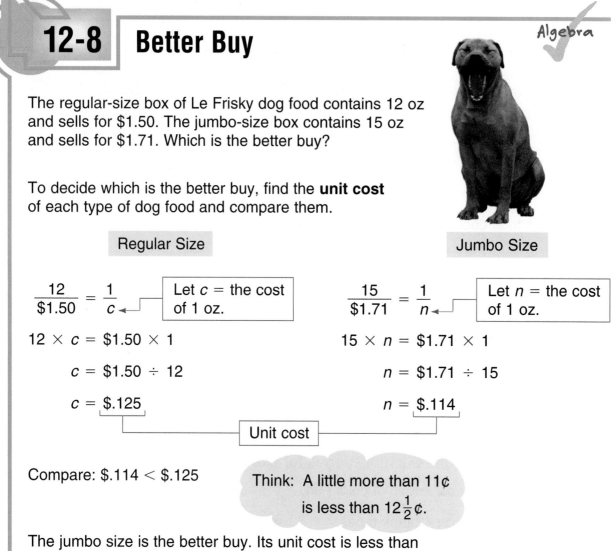

Regular Size

$$\frac{12}{\$1.50} = \frac{1}{c}$$

Let c = the cost of 1 oz.

$$12 \times c = \$1.50 \times 1$$

$$c = \$1.50 \div 12$$

$$c = \$.125$$

Jumbo Size

$$\frac{15}{\$1.71} = \frac{1}{n}$$

Let n = the cost of 1 oz.

$$15 \times n = \$1.71 \times 1$$

$$n = \$1.71 \div 15$$

$$n = \$.114$$

Unit cost

Compare: $\$.114 < \$.125$

Think: A little more than 11¢ is less than $12\frac{1}{2}$¢.

The jumbo size is the better buy. Its unit cost is less than the unit cost of the regular-size dog food.

Which is the better buy? Discuss your answers with a classmate.

1.	Drinking Glasses	4 for $1.80 6 for $2.40
2.	Corn	6 ears for 96¢ 8 ears for $1.12
3.	Detergent	2 lb box for $1.26 5 lb box for $3.05
4.	Cereal	12 oz box for $1.74 15 oz box for $2.04

Discuss

Estimation and Better Buy

Estimate to decide if it is less expensive to buy the bag of apples or to buy 6 apples individually.

$$6)\overline{\$1.29} \rightarrow 6)\overline{\$1.20}$$

Round down to get a compatible number.

Each apple in the bag costs a little more than 20¢. The bag of apples is less expensive.

6 for $1.29 or 30¢ each

Estimate to decide which is the better buy.

5. 5 cakes of soap for $1.49
1 cake of soap for 25¢

6. Bag of a dozen lemons for $2.50
Lemons: 15¢ each

7. Picture frames: 2 for $5.67
Picture frames: $2.45 each

8. 3 boxes of crackers for $4.19
Crackers: $1.50 per box

PROBLEM SOLVING

9. At Rosada's Market, a 10-oz can of mushrooms sells for 40¢ and a 6-oz can sells for 30¢. Which is the better buy?

10. A package of 6 coasters is marked $1.44 and a package of 8 of the same coasters is marked $1.76. Which is the better buy?

11. An 8-oz can of peaches costs $.68. A 14-oz can costs $1.05. Which is the better buy?

12. A 12-oz box of Crispy Cereal costs $2.10 and the 15-oz box of the same cereal costs $2.94. Which is the better buy?

Find the sale price in each store. Then decide which is the better buy. Discuss your answer with the class.

Discuss

13. Sweaters in Store A marked $18.50 at a 30%-off sale; the same sweaters in Store B marked $19.00 at a $\frac{1}{3}$-off sale

14. Skirts in Store C marked $32.00 at a 15% discount; the same skirts in Store D marked $44.00 at a $\frac{1}{4}$-off sale

12-9 Finding Commission

Commission is the amount of money that a salesperson is paid for selling a product or service. The **rate of commission** is given as a percent.

A salesperson works on **straight commission** if the commission is the only pay he or she receives.

Marvella makes a $3\frac{1}{2}$% commission on all of her clothing sales. What is her commission on sales of $6500?

Use one of the following two methods to find commission, C.

Method 1	Method 2

Method 1

Write a number sentence.

$C = 3\frac{1}{2}$% of $6500

$C = 0.035 \times \$6500$

$C = \$227.50$

Method 2

Write a proportion.

part $\longrightarrow$ $\dfrac{C}{\$6500}$ $\searrow\nearrow$ $\dfrac{3\frac{1}{2}}{100}$ $\longleftarrow$ part
whole $\longrightarrow$ $\phantom{\dfrac{C}{\$6500}}$ $\phantom{\dfrac{3}{100}}$ $\longleftarrow$ whole

$C \times 100 = \$6500 \times 3.5$

$C = \$22{,}750 \div 100 = \$227.50$

So Marvella's commission is $227.50.

Find the commission.

Algebra ✓

1. Amount sold = $550
Rate of commission = 4%

2. Amount sold = $480
Rate of commission = 2%

3. Amount sold = $1080
Rate of commission = 5%

4. Amount sold = $4800
Rate of commission = 3%

5. Amount sold = $820
Rate of commission = 3.5%

6. Amount sold = $5000
Rate of commission = 1.5%

7. Amount sold = $4250
Rate of commission = $3\frac{1}{2}$%

8. Amount sold = $6500
Rate of commission = $4\frac{1}{2}$%

PROBLEM SOLVING

9. Find the commission Ms. Levine receives for selling electronic equipment worth $13,000 if her rate of commission is 4%.

10. Ms. Velarde sold $825 worth of cosmetics last year. Her rate of commission was 8%. What was her commission?

11. Mr. Jenkins sells major appliances at an $8\frac{1}{2}$% commission rate. Last month his total sales were $9675. How much commission did he make?

12. Ms. Farber had carpet sales of $15,215 last month. Her rate of commission is $3\frac{1}{2}$%. What is her commission?

 Auto Salesperson, Hoody's Auto, Experienced Only
Salary Plus 4% Commission on Sales

13. Vicente plans to take a job at Hoody's Auto at a salary of $50 per month. If his total sales for the first month are $20,000, find his earnings for the month. (*Hint:* Earnings = Salary + Commission.)

14. Stella anticipates total monthly auto sales at Hoody's of $37,500. With a regular salary of $550 per month, how much would she make in salary plus commission?

15. Hoshi is offered a choice of jobs at Hoody's:
(a) regular monthly salary of $550 plus 4% commission; or
(b) straight commission of $7\frac{1}{2}$% on all sales.
If Hoshi expects monthly sales of $20,000, which is the better offer? Explain in your Math Journal.

Math Journal

 **CHALLENGE**

16. Aboul's boss offers him a $7\frac{1}{2}$% commission on all sales. What must Aboul's total monthly sales be in order to receive a $3000 commission?
(*Hint:* Total Sales = Commission ÷ Rate of Commission.)

12-10 Making Circle Graphs

Hakan surveyed the students in his class to find the number of television sets in each home. You can help Hakan make a circle graph of his findings.

Number of TVs Per Home	Number of Homes	Percent of Total	Angle Measure
1	6	20%	72°
2	12	?	?
3	9	?	?
4 or more	3	?	?
Totals	30	100%	360°

Make a circle graph of the data in the table.

Hands-On Understanding

Materials Needed: straightedge, paper, pencil, protractor, compass

Step 1 Copy the chart above onto your own paper.

Step 2 Six of the homes in the survey had 1 TV. Find what *percent* of the total number of homes (30) is represented by these 6 homes. Use a proportion as shown below.

part → $\dfrac{6}{30}$ ⤬ $\dfrac{n}{100}$ ← part (%)
whole → ⟍ ← whole (%)

$\dfrac{6}{30}$ ← Homes with 1 TV
← Total homes

$$6 \times 100 = 30 \times n$$
$$n = 600 \div 30 = 20$$

Complete: $\dfrac{20}{100} = \underline{\ ?\ }$ %. The correct percent should be written in your chart.

430

Step 3   Find the percent of the total homes with 2 TVs ($\frac{12}{30}$), 3 TVs ($\frac{9}{30}$), and 4 or more TVs ($\frac{3}{30}$). Write each of these percents in your chart.

Does the total percent column add up to 100%? If not, check your work.

Step 4   Find the number of degrees that 20% (homes with 1 TV) represents. (*Hint:* Number of degrees in a circle is 360°.) The correct number of degrees should be written in your chart.

20% of 360° = d

$\frac{1}{5} \times 360° = d$

$72° = d$

Think: One angle of the circle graph should be 72°.

Step 5   Find the number of degrees for each of the other percents. Write each number of degrees in your chart.

Does the Angle Measure column add up to 360°? If not, check your work.

Step 6   Use your compass to draw a large circle. Draw radius $\overline{PA}$ of the circle.

Step 7   Place the center mark of your protractor on P and draw a central angle of 72° as shown.

Step 8   Draw the other central angles. Use the degrees in Step 5. Start your next angle using radius $\overline{PB}$.

Step 9   Label each section of the graph with 1 TV, 2 TVs and so on. Give your graph a title.

Communicate

1. Change the number of homes in your table, in order, from 6, 12, 9, 3 to 9, 30, 6, 15. Discuss with your class and then draw a circle graph for the new data.

Discuss

TECHNOLOGY

Electronic Spreadsheets

An **electronic spreadsheet** is a computer-based software program that allows you to arrange data in a column-and-row format.

Each intersection of a column and row is called a **cell.** A cell can contain data, a label, or a formula. A cell is identified by the column and row in which it lies.

Column A

	A	B	C	D	E
1	Account	Balance	Interest Rate (%)	Interest Earned	New Balance
2	Checking	$112.00	3	(C2/100)*B2	
3	Savings	$430.00	3.5		
4	CD	$2500.00	4.25		
5	IRA	$8248.24	4		
6	Total		xxxxxxxxxx	xxxxxxxxxx	

Row 5   Cell A5

▶ Find the amount that will appear in cell D2.

The result of the formula (C2/100)*B2 will appear in cell D2.

(C2/100) * B2 ◀ — Data in cell C2 is *divided* by 100, then multiplied by the data in cell B2.

(3/100) * $112.00 = 0.03 * $112.00 = $3.36

So $3.36, the interest earned on checking, will appear in cell D2.

▶ What formula can be entered in cell E2?

To find the new balance, add: balance + interest earned

cell location ⟶ B2   D2

A formula consists of a combination of numbers, cell locations, or both. If a formula begins with a cell location, it must be preceded by a + sign or be enclosed within parentheses.

So the formula can be +B2 + D2 or (B2 + D2).

Use the spreadsheet on page 432 to answer each question.

1. What amount will appear in cell E2?

2. What formula can be entered in cell D3? What value will it display?

3. What formula can be entered in cell E3? What value will it display?

4. What formula can be entered in cell B6 to find the total beginning balance? What value will it display?

5. Copy and complete the spreadsheet.

6. What formula did you enter in cell E6?

7. Add a new column, F, and the label, Percent of Total, to your spreadsheet. In which cell will the label appear?

8. What formula would you enter in cell F2 to find the percent the new checking balance is of the total new balance?

9. What value will be displayed in each cell of column F?

Henry went shopping for some new clothes. Below is a stack of sales tags from the items he bought.

PROBLEM SOLVING

10. Use the information above to create a spreadsheet. Include columns for original price, amount of discount, quantity, percent discount, sales tax, sale price, and total price. The sales tax is 6% on leather items and 4% on all other items.

11. What items will appear in column A?

12. Complete the spreadsheet to find Henry's total shopping cost.

12-12 | Problem Solving: Write an Equation

Problem: A camera costs $29.95 plus 5% sales tax. Find the sales tax and total cost of the camera.

1 IMAGINE Picture yourself in the problem.

2 NAME *Facts:* camera cost—$29.95
rate of sales tax— 5%

Question: What are the sales tax and total cost of the camera?

3 THINK To find the sales tax, multiply the marked price by the rate of sales tax.

Algebra ✓

Write a percent equation.
Let *T* represent the sales tax.
   T = marked price × rate of sales tax
   T = $29.95   ×   5%

To find the total cost, add the sales tax to the marked price.
Let *TC* represent the total cost.
   TC = marked price + sales tax
   TC = $29.95   +   *T*

4 COMPUTE *T* = $29.95 × 5%

$$
\begin{array}{r}
{\scriptstyle 4\ 4\ 2} \\
\$2\,9.9\,5 \\
\times\quad 0.0\,5 \\
\hline
\$1.4\,9\,7\,5
\end{array}
$$ Round to the nearest cent. $\longrightarrow$ $1.50

The sales tax is $1.50.

TC = $29.95 + $1.50
TC = $31.45

The total cost is $31.45.

5 CHECK Is the total cost greater than the marked price? Yes.
Use a calculator to check your computation.

Solve each problem. Write an equation to help you.

1. On Henley's farm there are 90 animals. Of these animals, 60% are cows. How many of the animals are cows?

IMAGINE	Draw and label a picture.	⟵ 60% cows
NAME	*Facts:* 90 animals 60% cows	
	Question: How many of the animals are cows?	
THINK	Look at the picture. Write a percent equation. Let *n* represent the number of cows.	

$$\begin{array}{ccc}\text{Number} & \text{Percent} & \text{All} \\ \text{of cows} & = \text{of cows} \times & \text{animals} \\ n & = 60\% \times & 90\end{array}$$

COMPUTE ⟶ **CHECK**

2. The principal received 240 complimentary tickets to a concert. She gave 5% of the tickets to Ms. DeSilva's class. How many tickets did Ms. DeSilva's class receive?

3. The school received a delivery of 700 textbooks. Of these, 140 are math books. What percent of the delivery are math books?

4. There are 1250 students in the middle school. Of this number, 30% are in sixth grade. How many students are in sixth grade?

5. A swimming pool that costs $850 is on sale for $637.50. What is the rate of discount on the pool?

6. Mr. Schultz sold 3 cars for the following amounts: $11,995, $30,985, and $22,175. If his rate of commission on these 3 sales was 4%, what was his commission for all 3 cars?

7. Of the 750 children who went to the fair, 600 had yogurt. What percent of the children did *not* have yogurt?

Solve each problem and explain the method you used.

1. Of the 25 food booths at the Elm Street Fair, 20% serve vegetarian meals. How many booths serve vegetarian meals?

2. The Golden Dragon restaurant serves 375 meals at the fair. Of those 40% are lo mein. How many lo mein meals are served?

3. The Mexican Hat serves tacos at the fair. The usual price for one taco is $2.60. For the fair the store discounts the price by 35%. How much is one taco at the fair?

4. Marissa works for 5 hours at the Mexican Hat. She spends 40% of her time cooking and the rest serving. How much time does she spend serving?

5. It costs $350 to rent a booth at the fair. This year, 15% of the rental fee is donated to city charities. How much money is donated to city charities for each booth rented?

6. Of 150 booths at the fair, 45 feature games. What percent of the booths feature games?

7. The Children's Hospital sells T-shirts at the fair. Each T-shirt costs $8.00 plus 6.5% sales tax. What is the total cost of one T-shirt?

8. For each $8.00 T-shirt, the hospital earns $4.80. What percent of the selling price is profit?

9. The Potter's Place sells mugs at the fair. A mug that usually costs $15 sells at the fair for $11.25. By what percent is the usual price reduced?

10. The Potter's Place sells these percentages of goods at the fair: 30% mugs   10% wind chimes
    15% plates   15% vases
    20% bowls   10% miniature animals
    Make a circle graph to show this data.

Imagine

Name

Think

Compute

Check

**Use one or more strategies to solve
each problem.**

11. At one game booth, 10 players use air pumps to fill balloons. The first two players to pop their balloons win prizes. What percent of the players win prizes?

12. The ring-toss booth gives giant stuffed animals as prizes. At the beginning of the day the booth had 220 animals. So far, 75% of the animals have been won. Of the remaining stuffed animals, 15 are pandas. How many are not pandas?

13. Of 48 spaces on a game wheel, 10 show a fish. A player who spins a fish wins a goldfish. Are the chances of winning a goldfish better than 25%?

14. Glittering Prizes sells earrings. It sells 25% of its earrings by the end of the day. If it has 51 pairs left, how many pairs did it have to start with?

15. A pair of earrings sells for $6.75. The local sales tax is 6.5%. Necklaces have a marked price of $8.50. What is the total price for a necklace?

16. Wanda's Wickerware sells 50% of its baskets before noon. It sells 80% of the remaining baskets before 5:00. At 5:00, it has 10 baskets left. How many baskets did the store have at the beginning of the day?

Use the bar graph for problems 17–20.

17. What percent of this year's booths sell clothing?

18. Which type of booth increased by 100% between last year's fair and this year's fair?

19. By how many did the number of booths increase this year?

20. Last year, 60% of the game booths gave away stuffed animals. How many booths gave away stuffed animals last year?

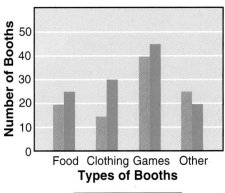

Fair Booths

Key : ■ Last Year
■ This Year

437

Find the percent of the number.   *(See pp. 412–417.)*

1. 25% of 20

2. 3% of $7.20

3. 52% of 60

4. $33\frac{1}{3}$% of 48

5. 62.5% of 800

6. 150% of 40

Find the percent.   *(See pp. 418–419.)*

7. What percent of 8 is 2?

8. What percent of 100 is 12?

9. 48 is what percent of 192?

10. 2.2 is what percent of 40?

Estimate to decide which is the better buy.   *(See pp. 426–427.)*

11. 3 pairs of socks for $2.08
  1 pair of socks for 60¢

12. Dozen oranges for $3.15
  Oranges: 25¢ each

Find the sales tax and total cost of each item.   *(See pp. 424–425.)*

13. $35.20 radio, 5% sales tax

14. $500 refrigerator, $6\frac{1}{2}$% sales tax

Draw a circle graph of the favorite pets of a Grade 6 class.   *(See pp. 430–431.)*

15. Dog   40 students   Bird   5 students
  Cat   30 students   Other   10 students
  Fish   15 students

PROBLEM SOLVING   *(See pp. 420–429, 434–437.)*

16. Seventeen percent of the 500 pages of a magazine contain photos. How many pages contain photos?

17. This year the price of a calculator is 110% of last year's price of $50. What is the price this year?

18. The rate of discount on a $560 VCR is 15%. Find the discount and the sale price.

19. Find the commission on sales of $2000 if the rate of commission is 8%.

20. Which is the better buy: 8-oz can of corn for $.56 or 12-oz can of corn for $.78?

21. Find the total cost of a $150 suitcase with a $4\frac{1}{2}$% sales tax.

22. Mr. Kirkpatrick's regular salary is $1500 per month. His rate of commission is 4%. How much does he make in a month when his total sales are $10,000?

(See Still More Practice, p. 517.)

INSTALLMENT PLAN BUYING

A camcorder has a marked price of $900. Jo buys it on an **installment plan**. She makes a down payment of 15% and agrees to make 24 monthly payments of $40 each. What is Jo's total cost on the installment plan? What is her finance charge?

To solve, follow these steps:

- **Find the amount of the down payment.**

  down payment = rate × marked price

  = 15% × $900

  = 0.15 × $900

  = $135

  Down payment is 15% of the marked price.

- **Find the total of the monthly payments.**

  total of monthly payments = monthly payment × number of months

  = $40 × 24

  = $960

- **Find the total cost on the installment plan.**

  total cost on plan = down payment + total of monthly payments

  = $135 + $960

  = $1095

- **Find the finance charge.**

  finance charge = total cost on plan − marked price

  = $1095 − $900

  = $195

Copy and complete the chart.

	Marked Price	Down Payment %	Down Payment $	Monthly Payment	Number of Months	Total of Monthly Payments	Total Cost On Plan	Finance Charge
1.	$12,500	20%	$2500	$250	48	?	?	?
2.	$3000	30%	?	$75	36	?	?	?
3.	$800	5%	?	$70	12	?	?	?
4.	$20,000	25%	?	$375	60	?	?	?

Check Your Mastery

Performance Assessment

Answer each question and explain your answer.

1. Which is the better discount: 15% off the price or $15 off every $100 you spend?

2. Is 100% off always a good buy? When is 15% off *not* a good buy?

Find the percent of the number.

3. 50% of 44

4. 4% of $6.00

5. 34% of 500

Find the percent.

6. What percent of 25 is 5?

7. 78 is what percent of 120?

Estimate to decide which is the better buy.

8. 4 loaves of bread for $4.99
1 loaf of bread for $1.50

9. 5 puzzles for $4.45
1 puzzle for 95¢

Find the commission.

10. Amount sold = $640
Rate of commission = 5%

11. Amount sold = $3500
Rate of commission = $4\frac{1}{2}$%

Draw a circle graph of Mr. Lapid's monthly budget.

12. Mortgage payment $400   Transportation $200
Food $500   Miscellaneous $800
Utilities $100

PROBLEM SOLVING   *Use a strategy or strategies you have learned.*

13. There are 520 students in the sixth grade. Forty-five percent are girls. How many are girls?

14. Fifteen out of 60 machine parts are defective. What percent are defective?

15. The rate of discount on a $14.80 CD is 15%. Find the sale price.

16. The rate of sales tax is 5%. Find the sales tax on a $40 radio.

17. Which is the better buy: $28 pants at 20% off or the same pants for $32 at a $\frac{1}{4}$-off sale?

Integers and Coordinate Graphing

13

Some Opposites

What is the opposite of *riot*?
It's *lots of people keeping quiet.*

The opposite of *doughnut*? Wait
A minute while I meditate.
This isn't easy. Ah, I've found it!
A *cookie with a hole around it.*

What is the opposite of *two*?
A *lonely me, a lonely you.*

The opposite of a *cloud* could be
A *white reflection in the sea,*
Or *a hugh blueness in the air,*
Caused by a cloud's not being there.

The opposite of *opposite*?
That's much too difficult. I quit.

Richard Wilbur

In this chapter you will:

Compare, order, and graph integers
Use models to compute with integers
Relate integers and temperature
Graph transformations
Solve problems by making a table

Critical Thinking/Finding Together

House A: 5 mi east, 3 mi south of a
point; House B: 5 mi west, 3 mi north of
the point. What is the greatest possible
distance between the houses?

441

Integers

In the United States, the town with the highest altitude is 11,560 ft above sea level. The town with the lowest altitude is 185 ft below sea level.

You can use positive and negative numbers to represent such opposite situations.

0 ft	sea level
$^+$11,560 ft	above sea level
$^-$185 ft	below sea level

The whole numbers and their opposites, including zero, make up the set of numbers called **integers**.

▶ Integers can be shown on a number line.

negative integers   **zero**   **positive integers**

$^-5$ $^-4$ $^-3$ $^-2$ $^-1$ 0 $^+1$ $^+2$ $^+3$ $^+4$ $^+5$

opposites

Integer		Opposite
$^+2$	Read: positive two	$^-2$
$^-5$	Read: negative five	$^+5$
0	Read: zero	0

0 is neither positive nor negative.

Name an integer to represent the situation. Then describe the opposite situation and give an integer to represent it.

1. gain of 8 dollars

2. loss of 15 yards

3. 22 degrees warmer

4. 5 seconds before liftoff

5. withdrawal of $50

Write the letter that matches the integer on this number line.

```
      A    B    C    D    E    F    G    H    I    J    K
◄─────┼────┼────┼────┼────┼────┼────┼────┼────┼────┼────┼─────►
     ⁻5   ⁻4   ⁻3   ⁻2   ⁻1    0   ⁺1   ⁺2   ⁺3   ⁺4   ⁺5
```

6. 0   **7.** ⁻5   **8.** ⁺4   **9.** ⁻1   **10.** ⁺3

11. ⁻2   **12.** ⁺1   **13.** ⁺5   **14.** ⁻3   **15.** ⁻4

Write the integer that matches the letter on this number line.

```
      L    M    N    O    P    Q    R    S    T    U    V
◄─────┼────┼────┼────┼────┼────┼────┼────┼────┼────┼────┼─────►
     ⁻5   ⁻4   ⁻3   ⁻2   ⁻1    0   ⁺1   ⁺2   ⁺3   ⁺4   ⁺5
```

16. *R*   **17.** *M*   **18.** *L*   **19.** *U*   **20.** *P*

21. *T*   **22.** *S*   **23.** *V*   **24.** *Q*   **25.** *N*

For each integer, name the integer that is just before and just after it on a number line.

26. ⁺7   **27.** ⁻2   **28.** ⁺1   **29.** ⁻6   **30.** ⁻1

31. ⁺16   **32.** ⁻3   **33.** 0   **34.** ⁺13   **35.** ⁻8

36. ⁻7   **37.** ⁻14   **38.** ⁺4   **39.** ⁻10   **40.** ⁺25

Name the opposite of each integer.

41. ⁻1   **42.** ⁺7   **43.** ⁺4   **44.** ⁻6   **45.** ⁻9

46. ⁺15   **47.** 0   **48.** ⁺20   **49.** ⁺16   **50.** ⁻15

51. ⁺13   **52.** ⁻10   **53.** ⁻12   **54.** ⁺8   **55.** ⁺5

CRITICAL THINKING

Name each integer on a horizontal number line.

56. five to the right of negative five

57. seven to the right of negative eight

58. four to the left of positive five

59. three to the left of positive three

60. three to the left of positive one

61. four to the left of zero

62. two to the right of zero

63. seven to the right of negative two

443

13-2 Comparing and Ordering Integers

You can use the number line to compare and to order integers.

$$\overset{\longleftarrow}{\underset{^-5 \quad ^-4 \quad ^-3 \quad ^-2 \quad ^-1 \quad 0 \quad ^+1 \quad ^+2 \quad ^+3 \quad ^+4 \quad ^+5}{\rule{9cm}{0.4pt}}}\longrightarrow$$

On a horizontal number line, any number is greater than a number to its left.

$^+5 > {^+3}$ since $^+3$ is to the left of $^+5$. $^+3 < {^+5}$

$^-1 < {^+2}$ since $^+2$ is to the right of $^-1$. $^+2 > {^-1}$

$^-2 > {^-4}$ since $^-4$ is to the left of $^-2$. $^-4 < {^-2}$

> A positive number is greater than any negative number.

On a vertical number line, any number is greater than a number below it.

$$^+3 > 0 \quad \text{or} \quad 0 < {^+3}$$

$$^+1 > {^-3} \quad \text{or} \quad ^-3 < {^+1}$$

$$^-2 < 0 \quad \text{or} \quad 0 > {^-2}$$

$$\uparrow \quad \begin{array}{l} ^+3 \\ ^+2 \\ ^+1 \\ 0 \\ ^-1 \\ ^-2 \\ ^-3 \end{array} \quad \downarrow$$

Study this example.

Order $^+2$, $^-5$, and 0 from least to greatest.

Think: $^-5$ is farthest to left and $^+2$ is farthest to right on a horizontal number line; 0 is between $^-5$ and $^+2$.

The order from least to greatest is $^-5$, 0, $^+2$.

Choose the greater integer.

1. $^+7, {^+10}$ **2.** $^-9, {^-3}$ **3.** $^+3, {^-5}$ **4.** $^-7, {^+6}$

5. $0, {^-9}$ **6.** $^+8, 0$ **7.** $^-12, {^-25}$ **8.** $^+20, {^-20}$

444

Compare. Write <, =, or >.

9. $^-10$ _?_ $^+6$ 10. $^+4$ _?_ $^+8$ 11. $^-3$ _?_ $^-6$ 12. $^-3$ _?_ $^+4$

13. $^+7$ _?_ 0 14. $^-4$ _?_ $^+4$ 15. 0 _?_ $^-3$ 16. $^-2$ _?_ $^-5$

17. $^-8$ _?_ $^+7$ 18. 0 _?_ $^-8$ 19. $^-6$ _?_ $^-6$ 20. $^-7$ _?_ $^+4$

21. $^-11$ _?_ $^-13$ 22. $^+10$ _?_ $^-20$ 23. $^-12$ _?_ $^+12$ 24. $^-13$ _?_ 0

Arrange in order from least to greatest.

25. $^+6, ^+8, ^+7$ 26. $^-10, ^-8, ^-6$ 27. $^-6, 0, ^-3$

28. $^+9, 0, ^+3$ 29. $^-5, ^-6, ^-3, ^-7$ 30. $^+4, ^-2, ^+5, ^-4$

Arrange in order from greatest to least.

31. $^-6, ^+3, ^-4$ 32. $^-2, ^-10, ^+5$ 33. $0, ^-7, ^-12$

34. $^-4, ^+5, ^-3$ 35. $^+8, ^+12, ^-15, ^-30$ 36. $^+20, 0, ^-2, ^-1$

CONNECTIONS: HISTORY

The vertical number line shows important dates in the history of Africa.

Write *True* or *False*.

37. The Sahara began to become a desert before Upper and Lower Egypt united.

38. The Kingdom of Aksum converted to Christianity after the time when the Empire of Ghana flourished.

39. Sea trade became important after the Kingdom of Kush began.

40. Most colonies in Africa had become independent by A.D. 1961.

41. The Kingdom of Kush began before Upper and Lower Egypt united.

$^+1961$ — Most colonies in Africa have become independent nations.

$^+1000$ — Empire of Ghana flourishes.

$^+300$ — Aksum converts to Christianity.

0 (A.D.)

$^-500$ — Kingdom of Kush established in the Sudan.

$^-1000$

$^-2000$ — Sea trade becomes important.

$^-3000$ — Upper and Lower Egypt unite.

$^-4000$ — The Sahara begins to become a desert.

13-3 Addition Model for Integers

Melissa uses positive, negative, and zero cards to add integers.

 positive card represents $^+1$

 negative card represents $^-1$

 zero card represents 0

Use integer cards to compute.

Hands-On Understanding

Materials Needed: integer cards, paper, pencil

Step 1

Write this example on your paper:
$^+3 + ^+2 = \underline{\ ?\ }$.

What is the sign between the two numbers in the example? What operation are you being asked to use in the example?

Step 2

On your desk or table, place integer cards in a row to represent the first number, $^+3$. Directly underneath, place integer cards to represent the second number, $^+2$.

$^+3$

$^+2$

Think: $^+3$ means positive 3, *not* add 3.

446

Step 3   Count the integer cards to find the total: positive 1, positive 2, positive 3, and so on.

Complete: $^+3 + {^+2} = \underline{\ ?\ }$. Are the signs on the numbers in the example like signs (either both positive or both negative)? Is the sign on your answer the same or different from the signs of the numbers?

Step 4   Write this example on your paper: $^-1 + {^-3} = \underline{\ ?\ }$.

Step 5   Place integer cards on your desk or table to represent each number, one set of cards underneath the other.

$^-1$ [−]

> Think: -1 means negative 1, *not* subtract 1.

$^-3$ [−] [−] [−]

Step 6   Count the integer cards to find the total: negative 1, negative 2, negative 3, and so on.

Complete: $^-1 + {^-3} = \underline{\ ?\ }$. Are the signs on the numbers in the example like signs? What sign do you have on your answer? Is it the same sign as the sign of each of the two numbers?

Communicate

1. Use positive and negative cards to find these sums:
   a. $^+6 + {^+1}$ **b.** $^-6 + {^-5}$ **c.** $^-2 + {^-2} + {^-2}$

   Discuss ✓

2. Complete: The sum of two or more positive integers is a ___?___ . The sum of two or more negative integers is a ___?___ . Discuss your answers with the class.

   **CRITICAL THINKING**

3. Use integer cards to find these sums:
   a. $^+5 + 0$ **b.** $0 + {^-6}$ **c.** $0 + 0$

> Think: Use the 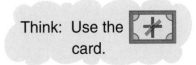 card.

13-4 | Another Addition Model

Antonio uses integer cards and the number line to model the sum of ⁻2 and ⁺6. Follow the directions below to make each model.

▶ ⁻2 + ⁺6 = ⸏?⸏

- Use integer cards to show ⁻2.
- Use integer cards to show ⁺6.
- Replace each pair of positive and negative cards with a zero card.
- Count the remaining cards to find the total.

One negative and one positive cancel each other.

⁻2 + ⁺6 = ⁺4

▶ ⁻2 + ⁺6 = ⸏?⸏

- Start at zero.
- Move 2 places to the *left.*
- Move 6 places to the *right.*

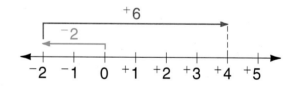

⁻2 + ⁺6 = ⁺4

Study these examples.

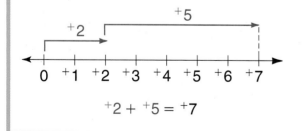

⁺2 + ⁺5 = ⁺7

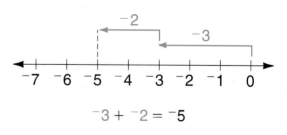

⁻3 + ⁻2 = ⁻5

Write a number sentence for each number line.

1.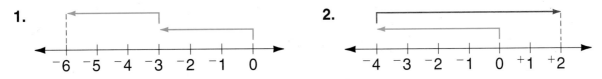

2.

Add. Use integer cards or a number line to help you.

(*Hint:* On a number line, move to the right for positive integers and move to the left for negative integers. 0 means *no move.*)

3. $^+2 + {}^+1$ **4.** $^-4 + {}^-3$ **5.** $^-1 + {}^-4$ **6.** $^+6 + {}^+1$

7. $^-6 + {}^+4$ **8.** $^+7 + {}^-5$ **9.** $^-5 + {}^-6$ **10.** $^+4 + {}^+4$

11. $^-4 + {}^+8$ **12.** $^-8 + {}^+5$ **13.** $^+6 + 0$ **14.** $0 + {}^-8$

15. $^+4 + {}^-5 + {}^-6$ **16.** $^-6 + {}^-2 + {}^+4$ **17.** $^-3 + {}^-3 + {}^-3$

Replace each letter with the correct sum.

Algebra

18. $^+7 + {}^+9 = a$ **19.** $^-5 + {}^+8 = b$ **20.** $^+11 + {}^-2 = c$

21. $^-14 + {}^+7 = d$ **22.** $^-5 + {}^-8 = e$ **23.** $^+3 + {}^-15 = f$

PROBLEM SOLVING

24. During June the water level in the reservoir rose 5 feet. During July the level fell 9 feet. Use an integer to represent the total change in water level for the two months.

25. A geologist studied rock formations at a site 5 meters below sea level. Later she moved to a site 9 meters higher. How far above or below sea level was the new site?

CALCULATOR ACTIVITY

26. Find the sum of $^-3$ and $^+3$ on a calculator. Use the change sign key, ⊬, to input $^-3$.

3 ⊬ + 3 =

27. Try $^+5 + {}^-5$ and other examples. Write a rule with examples in your Math Journal.

Math Journal

449

13-5   Adding Integers

Mr. Kickingbird, an oceanographer, uses a sonar device to map the ocean floor.

On a recent expedition, the first sonar reading was taken at a depth of 2 km below sea level. The next reading was taken at a depth that was 4 km deeper than the first reading. What was the depth below sea level of the second reading?

2 kilometers below sea level ⟶ ⁻2

4 kilometers deeper ⟶ ⁻4

To find the total depth below sea level, add: ⁻2 + ⁻4 = __?__ .

To add integers with like signs:

- Add the numbers.

- Use the sign of the addends.

$$^-2 + {}^-4 = {}^-6$$

> 2 + 4 = 6
> Use a negative sign.

So the depth below sea level was 6 kilometers.

To add integers with unlike signs:

- Find the difference. (Drop the signs and subtract the numbers.)

- Use the sign of the addend farther from zero.

$$^-8 + {}^+5 = \underline{\ ?\ }$$

unlike signs

Find the difference.

Use the sign of the addend farther from zero.

$$8 - 5 = 3$$

$$^-8 + {}^+5 = {}^-3$$

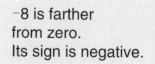

> ⁻8 is farther from zero. Its sign is negative.

Use the rules on page 450 to find each sum. Use integer cards or number lines to check your answers.

1. $^-7 + {}^+5$ 2. $^+1 + {}^-4$ 3. $^-2 + {}^-5$ 4. $^+8 + {}^+5$

5. $^+2 + {}^-6$ 6. $^-8 + {}^+6$ 7. $^+4 + {}^-2$ 8. $^-8 + {}^+4$

9. $^+8 \mid {}^-2$ 10. $^-3 + {}^+4$ 11. $^+2 + {}^-4$ 12. $^-9 + {}^+2$

13. $^-7 + {}^+1$ 14. $^+5 + {}^-3$ 15. $^-4 + {}^+9$ 16. $^-4 + {}^-9$

Choose the correct answer to complete each statement. Explain each answer.

Communicate

17. (pos.) + (pos.) = (?)   **a.** pos.   **b.** neg.   **c.** cannot tell

18. (neg.) + (neg.) = (?)   **a.** pos.   **b.** neg.   **c.** cannot tell

19. (pos.) + (neg.) = (?)   **a.** pos.   **b.** neg.   **c.** cannot tell

20. (neg.) + (zero) = (?)   **a.** pos.   **b.** neg.   **c.** cannot tell

PROBLEM SOLVING

21. On Monday Sally deposited $60 in her savings account. On Tuesday she withdrew $45. What was the net change in savings for the two days?

22. In January Raul lost 5 pounds. He gained back 3 pounds in February. What was his total weight gain or loss for the two months?

23. Rita started a checking account with $500. She later wrote a check for $50, made a deposit of $250, and wrote another check for $100. How much money was left in Rita's account?

24. An elevator starts at the 23rd floor, goes down 5 floors and then up 8 floors. At what floor is it then? Draw a vertical number line to illustrate.

 **SHARE YOUR THINKING**

25. Can you use the rules on page 450 for adding with zeros (such as $0 + {}^-7$) or with opposites (such as $^+7 + {}^-7$)? Explain to your teacher how to do such computations.

Communicate

Subtracting Integers

Catherine wants to know how to complete this subtraction:

$4 - 5 = \underline{\ ?\ }$

To study the relationship between adding and subtracting integers, she makes the table.

x	$x + {}^-5$	$x - 5$
7	$7 + {}^-5 = 2$	$7 - 5 = 2$
6	$6 + {}^-5 = 1$	$6 - 5 = 1$
5	$5 + {}^-5 = 0$	$5 - 5 = 0$
4	$4 + {}^-5 = {}^-1$	$4 - 5 = ?$

Look for a pattern in the table. The sequence of numbers in the last column is 2, 1, 0, ... Catherine determines that $4 - 5 = {}^-1$ or ${}^+4 - {}^+5 = {}^-1$. She also makes the following general conclusion.

> Subtracting an integer is the same as *adding the opposite* of that integer.

To subtract an integer:

- Add its opposite.
- Rewrite as an addition sentence.
- Then use the rules for adding integers.

Opposites

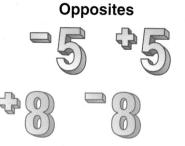

$$^+4 - {}^+5 = \underline{\ ?\ }$$ The opposite of ${}^+5$ is ${}^-5$.

$$^+4 + {}^-5 = \underline{\ ?\ }$$ Rewrite as an addition sentence.

$$^+4 + {}^-5 = {}^-1$$ Add ${}^+4$ and ${}^-5$.

$${}^-6 - {}^-8 = \underline{\ ?\ }$$

$${}^-6 + {}^+8 = \underline{\ ?\ }$$

$${}^-6 + {}^+8 = {}^+2$$

Study these examples.

$${}^-9 - {}^+6 = \underline{\ ?\ } \qquad {}^+7 - {}^-10 = \underline{\ ?\ } \qquad {}^-5 - 0 = \underline{\ ?\ }$$

$${}^-9 + {}^-6 = {}^-15 \qquad {}^+7 + {}^+10 = {}^+17 \qquad {}^-5 + 0 = {}^-5$$

0 is its own opposite.

Subtract.

1. $^+8 - {}^+4$
2. $^+5 - {}^+8$
3. $^-4 - {}^+5$
4. $^-6 - {}^+2$

5. $^-3 - {}^-7$
6. $^+9 - {}^+7$
7. $^+7 - {}^-4$
8. $^-5 - {}^+8$

9. $^+8 - {}^+10$
10. $^+3 - {}^-2$
11. $^-8 - {}^-10$
12. $^-3 - {}^+5$

13. $^-7 - {}^+2$
14. $^+9 - {}^+11$
15. $^-5 - {}^-3$
16. $^+7 - {}^-9$

17. $^+5 - {}^-10$
18. $^-6 - {}^-7$
19. $^+6 - 0$
20. $^-12 - 0$

21. $^+9 - {}^+4$
22. $^+2 - {}^-3$
23. $0 - {}^-8$
24. $0 - {}^+4$

PROBLEM SOLVING Use the table to answer the questions.

U.S. Population Changes (per hour)	
Births	$^+460$
New Immigrants	$^+100$
Deaths	$^-250$

25. What is the difference between the number of births and the number of new immigrants?

26. What is the difference between $^+460$ and $^-250$? $^+100$ and $^-250$?

27. Write an integer to show the number of people added to the U.S. population per hour from births and new immigrants.

28. Write an integer to show the net change per hour in the U.S. population.

MENTAL MATH

Look for opposites. Add.

$$\overbrace{{}^+11 + \underbrace{{}^-2 + {}^-8 + {}^+4 + {}^-11}_{} + {}^+2}^{0} = \underline{\ ?\ }$$

Think: $^-8 + {}^+4 = {}^-4$

29. $^+23 + {}^-17 + {}^-23 = \underline{\ ?\ }$

30. $^-12 + {}^+15 + {}^+12 + {}^-6 + {}^-15 + {}^-4 = \underline{\ ?\ }$

31. $^+35 + {}^+65 + {}^-65 = \underline{\ ?\ }$

32. $^+22 + {}^+14 + {}^-10 + {}^+10 + {}^-22 + {}^+14 = \underline{\ ?\ }$

453

The temperature in New York City was 15°C. After several hours it dropped 25 degrees. What was the new temperature?

A temperature of 15°C above zero (rise of 15°) is represented by $^+15$. A drop of 25° is represented by $^-25$.

To find the new temperature, add: $^+15 + {}^-25 = $ __?__

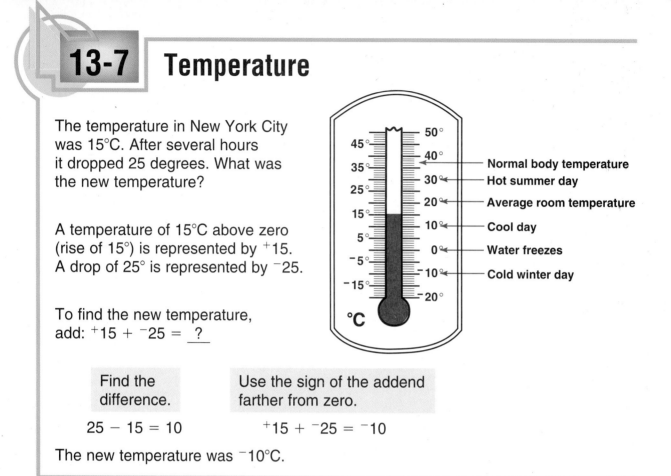

Find the difference.	Use the sign of the addend farther from zero.
$25 - 15 = 10$	$^+15 + {}^-25 = {}^-10$

The new temperature was $^-10$°C.

Choose the most reasonable temperature for each.

1. your classroom   **a.** 10°C   **b.** $^-5$°C   **c.** 22°C

2. boiling water   **a.** 50°C   **b.** 100°C   **c.** 10°C

3. normal body temperature   **a.** 37°C   **b.** 100°C   **c.** 0°C

4. snow   **a.** 32°C   **b.** 12°C   **c.** $^-2$°C

Compute the new temperature in each exercise. Explain the method you used.

Communicate

5. 35°C; rises 15°

6. 52°C; falls 15°

7. $^-10$°C; rises 20°

8. $^-5$°C; falls 10°

9. $^-20$°C; increases 6°

10. 0°C; decreases 12°

11. 36°C; drops 40°

12. $^-8$°C; climbs 4°

Temperature in Degrees Fahrenheit

The temperature in Minneapolis was 10°F. After several hours it dropped to ⁻15°F. How many degrees did the temperature drop?

To find the number of degrees, subtract: ⁺10 − ⁻15 = __?__

$$^+10 - {}^-15 = {}^+10 + {}^+15$$
$$= {}^+25 \text{ or } 25$$

The temperature dropped 25°F.

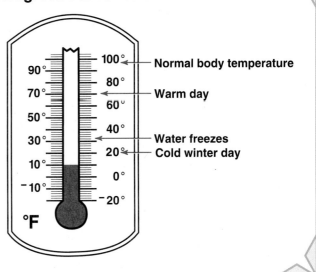

Normal body temperature

Warm day

Water freezes
Cold winter day

PROBLEM SOLVING

13. The temperature was ⁻12°C. It dropped 4 degrees. What was the new temperature?

14. The temperature was ⁻20°C. It rose 32 degrees. What was the new temperature?

15. At noon the temperature was 8°C. At midnight the temperature was ⁻10°C. How many degrees did the temperature drop?

16. What is the difference in temperature of a hot summer day of 30°C and a cold winter day of ⁻10°C?

17. On one winter morning last year, the temperature in Boston was ⁻4°F. By noon, the temperature had gone up 10 degrees. What was the temperature in Boston at noon?

18. What is the difference between normal body temperature and the freezing point of water on the Celsius scale? on the Fahrenheit scale?

 **SKILLS TO REMEMBER**

Compute. Use the correct order of operations.

19. $5 \times 6 + 4$

20. $8 \times (7 + 3)$

21. $21 \div 3 - 2$

22. $9 + 8 \times 2$

23. $(2 \times 3 - 1) + 8$

24. $2 \times (3 - 1 + 8)$

Ordered Pairs of Numbers

One of the tallest persons in medical history was Robert Wadlow, born in Alton, Illinois, in 1918. On his 13th birthday, he stood 7 ft $1\frac{3}{4}$ in. By age 17, he had reached more than 8 ft in height.

The table below shows his height in inches as a function of his age.

Age	5	8	10	20
Height in inches	64	72	77	103

The pairs of numbers in the table can be written as **ordered pairs** of (age, height).

(5, 64)   (8, 72)   (10, 77)   (20, 103)

▶ The ordered pairs above do not have an obvious rule that relates the second number in the pair to the first number. The ordered pairs below are related by a specific rule.

first number ⌐   ⌐ second number

n	$n + 3$	Ordered Pair
6	6 + 3 = 9	(6, 9)
2	2 + 3 = 5	(2, 5)
0	0 + 3 = 3	(0, 3)

Rule: Add 3 to first number to get the second number.

Copy and complete each table.

1.

n	$n + 5$	Ordered Pair
6	?	?
0	?	?
12	?	?
48	?	?
99	?	?

2.

n	$n - 4$	Ordered Pair
12	?	?
8	?	?
4	?	?
55	?	?
73	?	?

Complete each ordered pair for the given rule.

$n + 4$ **3.** (11, ?)   **4.** (0, ?)   **5.** (6, ?)   **6.** (9, ?)

$n - 8$ **7.** (8, ?)   **8.** (20, ?)   **9.** (35, ?)   **10.** (12, ?)

$n \times 3$ **11.** (3, ?)   **12.** (10, ?)   **13.** (0, ?)   **14.** (13, ?)

$n \div 2$ **15.** (14, ?)   **16.** (26, ?)   **17.** (40, ?)   **18.** (0, ?)

The rule for the ordered pairs is given by a formula. Copy and complete each table.

19. $d = 5 \times r$

r	3	5	9	1	10	12
d	$5 \times 3 = 15$	?	?	?	?	?
(r, d)	(3, 15)	?	?	?	?	?

20. $f = y \div 3$

y	6	12	21	15	36	48
f	?	?	?	?	?	?
(y, f)	?	?	?	?	?	?

The approximate age of a lobster can be found from its weight. Copy and complete the table of ordered pairs of (weight, age). Write a formula for the relationship. Begin the formula with $y =$.

21.

Weight in pounds, x	2	3	4	5	6
Age in years, y	14	21	28	?	?
(x, y)	(2, 14)	?	?	?	?

 CALCULATOR ACTIVITY

Use your calculator to complete each table.

$\boxed{2} \boxed{\times} \boxed{(} \boxed{4} \boxed{+} \boxed{5} \boxed{)} \boxed{=} \boxed{\quad 18.}$

22.

n	$2 \times (n + 5)$	Ordered Pair
4	18	(4, 18)
6	?	?
8	?	?

23.

n	$2 \times n + 7$	Ordered Pair
9	?	?
11	?	?
0	?	?

Graphing Ordered Pairs of Numbers

A horizontal and a vertical number line can be used to form a pair of **coordinate axes**. The coordinate axes form a **grid**.

> The *horizontal* number line is called the **x-axis**.

> The *vertical* number line is called the **y-axis**.

The point at which the two axes meet is called the **origin**.

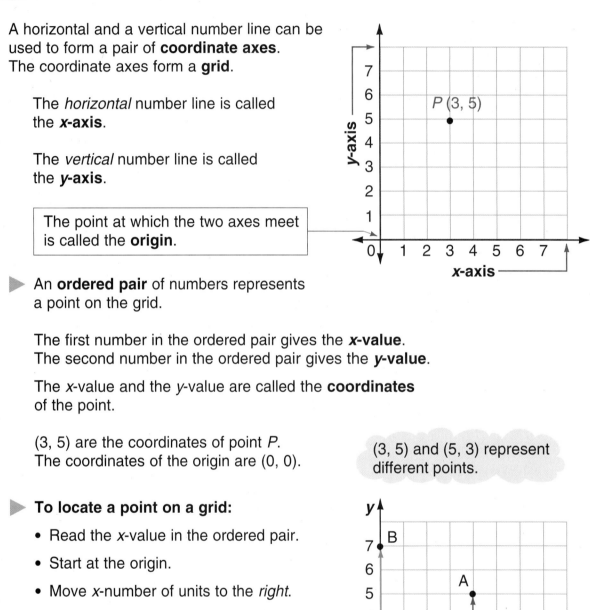

▶ An **ordered pair** of numbers represents a point on the grid.

The first number in the ordered pair gives the **x-value**.
The second number in the ordered pair gives the **y-value**.

The *x*-value and the *y*-value are called the **coordinates** of the point.

(3, 5) are the coordinates of point *P*.
The coordinates of the origin are (0, 0).

(3, 5) and (5, 3) represent different points.

▶ **To locate a point on a grid:**

• Read the *x*-value in the ordered pair.

• Start at the origin.

• Move *x*-number of units to the *right*.

• Read the *y*-value in the ordered pair. Then move *y*-number of units *up*.

Point *A*: (4, 5)   Point *B*: (0, 7)

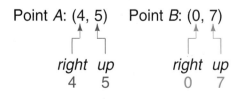

right up   right up
  4   5   0   7

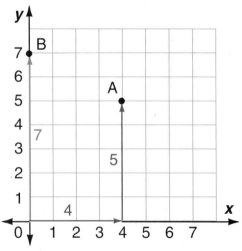

Use the grid at the right for exercises 1–25.

Write the coordinates of each point.

1. *A*   2. *C*   3. *E*

4. *G*   5. *I*   6. *K*

7. *M*   8. *O*   9. *B*

10. *D*   11. *F*   12. *H*

Name the point for each ordered pair.

13. (0, 5)   14. (2, 7)   15. (7, 4)

16. (4, 3)   17. (0, 0)   18. (6, 8)   19. (8, 0)   20. (8, 5)

21. (9, 9)   22. (6, 3)   23. (0, 8)   24. (9, 4)   25. (3, 2)

Use a grid to locate the points. Connect the points in order. Then identify the figure that is formed and find its area.

Algebra ✓

26. *A*(2, 2)   *B*(2, 4)   *C*(6, 4)   *D*(6, 2)

27. *H*(1, 1)   *I*(7, 1)   *J*(7, 7)   *K*(1, 7)

28. *R*(3, 3)   *S*(3, 6)   *T*(7, 3)

29. *M*(1, 3)   *N*(1, 8)   *O*(4, 8)   *P*(4, 3)

30. *O*(0, 0)   *X*(9, 0)   *Y*(0, 8)

CHALLENGE

Use a grid to locate the points.

31. *C*(1, 8)   *E*(1, 2)   *G*(6, 2)
Find the coordinates of point *H* if *CEGH* is a rectangle.

32. *A*(8, 1)   *B*(8, 5)
Find the coordinates of points *C* and *D* if *ABCD* is a square.

Graphing Ordered Pairs of Integers

The coordinate axes divide the coordinate grid into 4 sections.
An ordered pair of integers, (x, y), is used to locate a point on the grid.

To locate a point on a grid:

- Start at the origin and move x-number of units to the *right* or *left*.

- Then from that point move y-number of units *up* or *down*.

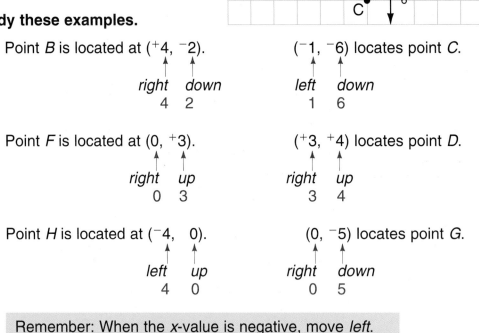

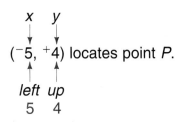

$(^-5, ^+4)$ locates point *P*.

left up
5   4

Study these examples.

Point *B* is located at $(^+4, ^-2)$.

right   down
4   2

$(^-1, ^-6)$ locates point *C*.

left   down
1   6

Point *F* is located at $(0, ^+3)$.

right   up
0   3

$(^+3, ^+4)$ locates point *D*.

right   up
3   4

Point *H* is located at $(^-4, 0)$.

left   up
4   0

$(0, ^-5)$ locates point *G*.

right   down
0   5

> **Remember:** When the *x*-value is negative, move *left*.
> When the *y*-value is negative, move *down*.

Use the grid at the right for exercises 1–22.
Name the ordered pair for each point.

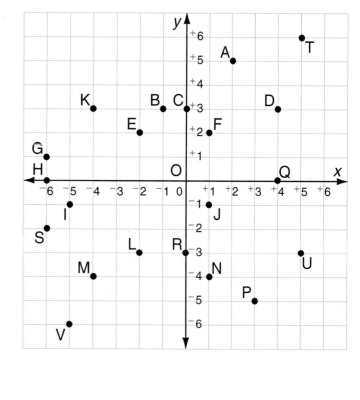

1. *D*
2. *I*
3. *B*
4. *G*
5. *M*
6. *J*
7. *O*
8. *C*
9. *A*
10. *H*
11. *L*
12. *E*

Name the point for each ordered pair.

13. ($^-$4, $^+$3)
14. ($^+$1, $^+$2)
15. ($^+$3, $^-$5)
16. (0, $^-$3)
17. ($^+$4, 0)
18. ($^+$1, $^-$4)
19. ($^-$6, $^-$2)
20. ($^+$5, $^+$6)
21. ($^+$5, $^-$3)
22. ($^-$5, $^-$6)

Use a grid to locate the points.
Then connect the points and identify the figure.

23. *A*($^-$2, $^+$1)   *B*($^+$3, $^+$1)   *C*($^+$2, $^-$2)   *D*($^-$3, $^-$2)

24. *R*($^-$8, $^+$3)   *S*($^-$3, $^+$6)   *T*($^-$3, $^+$3)

25. *M*($^+$2, 0)   *N*($^+$8, 0)   *O*($^+$6, $^+$5)   *P*($^+$4, $^+$5)

FINDING TOGETHER

Discuss ✓

Use a grid to locate the points. Then connect them.

26. ($^+$2, $^+$2), ($^+$8, $^+$2), ($^+$5, $^+$8);
    ($^+$4, $^+$3), ($^+$6, $^+$3), ($^+$5, $^+$5);
    ($^-$8, $^+$2), ($^-$2, $^+$2), ($^-$5, $^+$8)

27. ($^+$3, $^-$3), ($^+$7, $^-$3), ($^+$7, $^-$7), ($^+$3, $^-$7);
    ($^-$7, $^-$3), ($^-$3, $^-$3), ($^-$3, $^-$7), ($^-$7, $^-$7);
    ($^-$4, $^-$4), ($^-$6, $^-$4), ($^-$6, $^-$6), ($^-$4, $^-$6)

28. What relationships exist among the figures you drew
    for exercise 26? for exercise 27? Discuss with the class.

13-11 Graphing Transformations

When a geometric figure is moved without changing its size or shape, the movement is called a **transformation**. You can slide, flip, or turn a figure on a coordinate grid.

Slide

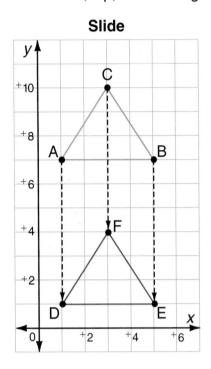

Flip

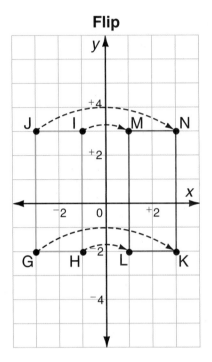

Figure *ABC* is moved down 6 units on the grid.
Figure *DEF* is its **slide image**.

Figure *GHIJ* is flipped over the *y*-axis line.
Figure *KLMN* is its **flip image**.

The coordinates of the vertices of each figure above are given. Give the coordinates of the vertices of its slide or flip image.

1. Figure *ABC*
   A(⁺1, ⁺7), B(⁺5, ⁺7), C(⁺3, ⁺10)

2. Figure *GHIJ*
   G(⁻3, ⁻2), H(⁻1, ⁻2), I(⁻1, ⁺3), J(⁻3, ⁺3)

Locate each triangle on a coordinate grid. Then draw the image and give the coordinates of its vertices.

3. N(⁻5, 0), O(⁻2, 0), P(⁻2, ⁺4)
   Move right 7 units.

4. Q(⁺2, ⁺2), R(⁺4, ⁺5), S(⁺6, ⁺2)
   Flip over *x*-axis line.

Graphing Turns

Triangle *ABC* is rotated (or turned) 180° around the origin in a *clockwise* direction.

Triangle *DBF* is the **half-turn image** of triangle *ABC*.

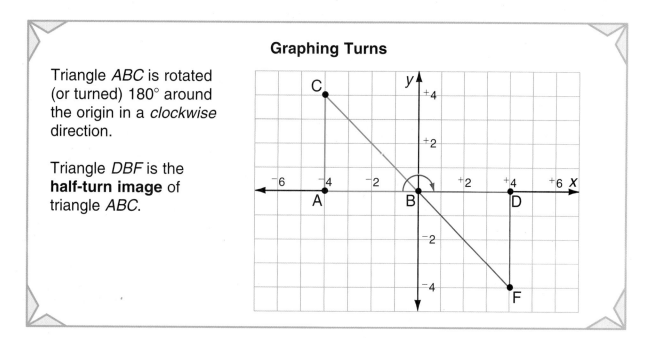

Use the graph above for exercises 5–6.

5. Give the coordinates of the vertices of triangle *ABC*.

6. Give the coordinates of the vertices of triangle *DBF*.

Use the same coordinate grid for exercises 7–8.

7. Locate triangle *X*($^-$3, $^-$5), *Y*($^-$5, $^-$1), *Z*(0, 0). Draw the half-turn image of triangle *XYZ* rotated *counterclockwise* around the origin. Give the coordinates of its vertices.

8. Draw the half-turn image of triangle *XYZ* rotated *clockwise* around the origin. What did you discover?

Locate the figures on a coordinate grid. Tell whether the result is a slide or a flip.

9. Join *G*($^-$3, $^-$5), *H*($^-$6, $^-$5), and *I*($^-$3, $^-$2). Make a new triangle (*JKL*) by using the *opposites* of the *x*-coordinates and the same *y*-coordinates. Join *J*, *K*, and *L*.

10. Join *M*($^+$3, $^-$4), *N*($^+$7, $^-$4), *O*($^+$6, $^-$2), and *P*($^+$2, $^-$2). Make a new figure (*QRST*) by adding $^-$8 to the *x*-coordinates and using the same *y*-coordinates. Join *Q*, *R*, *S*, and *T*.

MAKE UP YOUR OWN

Draw a slide, flip, and half-turn image. Share your drawings with the class.

Communicate ✓

11. Square *A*(0, 0), *B*($^+$4, 0), *C*($^+$4, $^+$4), *D*(0, $^+$4)

13-12 Problem Solving: Make a Table

Problem: The outside temperature at 11:00 P.M. is 14°F. The weather forecaster predicts that it will drop 5° each hour. Then the temperature will rise 3° each hour after 2 A.M. What will the temperature be at 6:00 A.M.?

1 IMAGINE Put yourself in the problem.

2 NAME *Facts:* At 11:00 P.M. the temperature is 14°F.

The temperature drops 5° each hour until 2:00 A.M.

The temperature rises 3° each hour after 2:00 A.M.

Question: What will the temperature be at 6:00 A.M.?

3 THINK When the temperature drops, use a negative integer.

When the temperature rises, use a positive integer.

Make a table to record the hourly temperature change.

4 COMPUTE Complete the table.

Add ⁻5° (or subtract 5°) for each hour until 2:00 A.M.
Add 3° for each hour until 6:00 A.M.

Time	11:00	12:00	1:00	2:00	3:00	4:00	5:00	6:00
Temperature	⁺14°	⁺9°	⁺4°	⁻1°	⁺2°	⁺5°	?	?

‑5 ‑5 ‑5 +3 +3 +3 +3

At 6:00 A.M. the temperature will be 11°F.

5 CHECK Use a calculator to check your computation.

Solve each problem. Make a table to help you.

1. Henry saved $130 to buy a $250 bicycle. His dad paid him for doing chores each week. Henry received $10 the first week. For each additional week Henry received $4 more than the preceding week. How many weeks must he work to have enough money to pay for the bicycle?

IMAGINE   Put yourself in the problem.

NAME   *Facts:*   amount saved—$130
cost of bicycle—$250
first week—$10
each week—$4 more than the preceding week

Question: How many weeks must he work to have enough money?

THINK   Make a table. Add to find how long it will take to have at least $250.

Week	1st	2nd	3rd	4th
Amount	$130 + $10	$140 + $14	$154 + $18	$172 + ?

COMPUTE ⟶ **CHECK**

2. A baker uses 3 c of sugar, 3 c of flour, and 2 sticks of butter for each pound cake. How much of each ingredient is needed for 8 cakes?

3. Sue played a game in which she won 8 points and lost 5 points in each round. In which round did her score reach 15 points?

4. Adam inspected apples. He found 2 out of every 15 apples to be of poor quality. How many apples of poor quality could he expect to find in a shipment of 165 apples?

5. During an experiment Dana recorded the following temperatures: 22°C, 12°C, 15°C, 5°C, 8°C. If this pattern continues, predict the tenth temperature in the series.

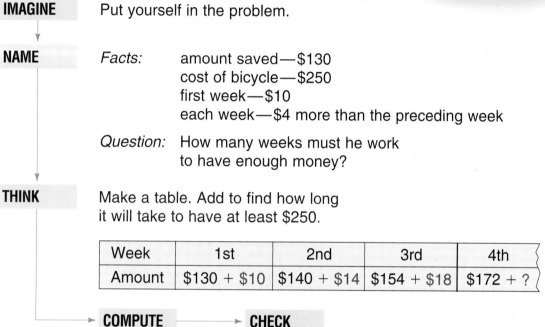

465

Solve each problem and explain the method you used.

1. Scientists built earthquake stations at different elevations. One station is 75 m above sea level, and a second is 35 m below sea level. What is the difference in height between the two stations?

2. A submarine is at a depth of ⁻300 ft. What must happen for the submarine to reach sea level?

3. How many negative integers are between ⁻4 and ⁺4?

4. How many integers greater than ⁻10 are less than ⁻3?

5. A balloon is 218 m above sea level. A submersible is 220 m below sea level. Which is closer to sea level? How much closer?

6. A diamond-mine entrance begins at 75 ft above sea level. Workers discover diamonds 48 ft below sea level. How deep is the mine at that point?

7. The temperature change from 6 A.M. to 7 A.M. was ⁺3°C. If the temperature at 6 A.M. was ⁻2°C, what was the temperature at 7 A.M.?

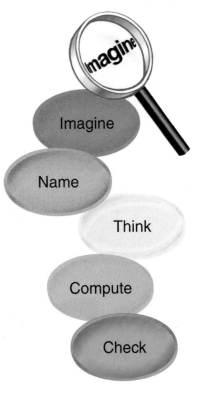

Imagine

Name

Think

Compute

Check

Identify each statement as _True_ or _False_.

8. The greatest negative integer is ⁻1.

9. On a horizontal number line, zero is always to the right of a negative integer.

10. The sum of a positive integer and a negative integer is never zero.

11. Subtracting zero is the same as adding zero.

12. The second coordinate of each point on the y-axis is always zero.

Use one or more of the strategies from the list or another strategy you know to solve each problem.

13. Leslie forgot her score from round 1. In the next three rounds she scored $^+4$, $^-11$, and $^+15$. Her total score was $^+25$. What was her first score?

14. Claire tripled a number and added $^-14$ to it. Her answer was $^+4$. What was her number?

15. $^+8 + ^+8 + ^+8$ is to $3 \times ^+8$ as $^-5 + ^-5 + ^-5$ is to __?__ .

16. A sonar device was positioned 100 m in the sea. The device was then lowered 175 m. Write an integer to indicate the final depth of the device.

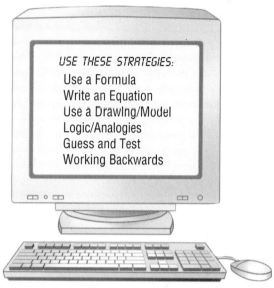

USE THESE STRATEGIES:

Use a Formula
Write an Equation
Use a Drawing/Model
Logic/Analogies
Guess and Test
Working Backwards

17. The temperature dropped 16°F to a low of 1°F. What was the original temperature?

18. Each time a hot-air balloon rose 65 ft, a downdraft pushed it back down 35 ft. Its original altitude was 185 ft. What would be its altitude after the third downdraft?

19. Laurie drew a quadrilateral by connecting the points $A(1, 2)$, $B(4, 2)$, $C(4, -1)$, and $D(1, -1)$. What are the perimeter and area of Laurie's quadrilateral?

20. Can Laurie graph a similar quadrilateral of twice the size by doubling only the first coordinate of each point? Explain.

Use the table for problems 21–23.

21. The wind was 40 mph and the wind chill was 21°F below zero. What was the air temperature?

22. How much colder does an air temperature of 30°F feel at 30 mph than at 20 mph?

23. At only 15 mph an air temperature of 15°F feels 7°F colder than when the air temperature is 20°F. What is the wind chill at 15°F?

Wind Chill Table

Miles Per Hour (mph)	35	30	25	20
5	27	21	16	12
10	16	10	3	−3
15	9	2	−5	−11
20	4	−3	−10	−17
25	1	−7	−15	−22
30	−2	−10	−18	−25
35	−4	−12	−20	−27
40	−5	−13	−21	−29

Air Temperature (°F)

Chapter Review and Practice

Express each as an integer.   *(See pp. 442–443.)*

1. an increase of 14 dollars

2. a gain of 9 meters

3. 4 hours before arrival

4. a depth of 12 meters

Name the opposite of each integer.   *(See pp. 442–443.)*

5. $^+15$ 6. $^-13$ 7. $^+7$ 8. $^+22$ 9. $^-1$

Compare. Write $<$, $=$, or $>$.   *(See pp. 444–445.)*

10. $^-9$ _?_ $^+9$ 11. $^-4$ _?_ $^-7$ 12. $^+5$ _?_ $^-14$ 13. $^-6$ _?_ $^-6$

Compute.   *(See pp. 446–453.)*

14. $^+6 + ^-3$ 15. $^-4 + ^+9$ 16. $0 + ^-5$ 17. $^-7 + ^-5$

18. $^+5 + ^+6$ 19. $^+7 + ^-10$ 20. $^+4 + ^-14$ 21. $^+6 + ^-8$

22. $^+9 - ^+5$ 23. $^-6 - ^-8$ 24. $^+7 - ^-11$ 25. $0 - ^+4$

26. $^-3 - ^-12$ 27. $^-9 - ^+13$ 28. $^+5 - ^+7$ 29. $^+12 - ^-4$

Use a grid to locate the points.   *(See pp. 458–461.)*

30. $K(^+2, ^+7)$ 31. $L(^-2, ^-4)$ 32. $M(^+6, ^-8)$ 33. $N(^-1, ^+6)$

34. $O(^-5, 0)$ 35. $P(0, ^+6)$ 36. $Q(0, 0)$ 37. $R(^-5, ^-5)$

Locate each triangle on a coordinate grid. Then draw the image   *(See pp. 462–463.)*
and give the coordinates of its vertices.

38. $S(0, ^+1)$, $T(^+4, ^+1)$, $U(^+2, ^+4)$
    Move down 5 units.

39. $V(^-4, ^+5)$, $W(^-4, ^+2)$, $X(^-1, ^+3)$
    Flip over *x*-axis line.

PROBLEM SOLVING   *(See pp. 466–467.)*

40. The temperature was $^+12°C$ at noon. By nine o'clock it was $^-2°C$. How many degrees did the temperature drop?

41. The football team gained 23 yards on 1st down and were penalized 5 yards on 2nd down. What was the net result?

(See Still More Practice, p. 518.)

ANALYZING SETS OF DATA

An **outlier** is a number that is "far away"
from the other numbers in a set of data. In the
temperature data below, ⁻20°C is an outlier.

Temperatures: ⁻20°C, 2°C, 3°C, 3°C, 4°C, 5°C

outlier

Think: The temperatures
cluster around 3°C.
⁻20°C is much less than
the other temperatures.

A given set of data may have no outliers or
it may have one or more outliers. The existence
of outliers may affect which statistic—mean, median,
or mode—is the best measure to use to describe the data.
Examine the two sets of data that follow.

Data Set A
Quick Quiz Scores: 20 85 90 90 95 95 95 100 100 mean = 85.$\overline{5}$, median = 95, mode = 95

Data Set B
Quick Quiz Scores: 75 75 80 80 80 85 85 90 90 mean = 82.$\overline{2}$, median = 80, mode = 80

Data Set A has an outlier, 20, which causes the mean to be much
less than the median or the mode. The mean is *not* the best measure
to use as the average of the data since most of the quiz scores
are greater than 85.$\overline{5}$.

Data Set B has no outlier. The quiz scores tend to cluster around
a score of 80. The mean, median, or mode could be used as the
average of the data.

**Which measure—mean, median, or mode—best describes
the data? Explain your answer.**

1.

Basketball Salaries		
$175,000	$3,000,000	$500,000
$600,000	$400,000	$60,000

2.

Week 1: Daily Shoe Sales		
8750	9020	5250
5250	8900	8950

3. Write a data set with 15 numbers for each description:
   (a) 1 outlier; (b) no outliers; and (c) two outliers. Explain
   whether mean, median, or mode best describes the data.

Performance Assessment

Complete each magic square.

Laura added the same integer to each number in the magic
square at the left to get the two new magic squares at
the right. Find the integer added and complete the squares.

+1	+2	-3
-4	0	+4
+3	-2	-1

1.

+3	+4	-1

2.

	-7	

3. Make another magic square for classmates to complete.

Arrange in order from least to greatest.

4. $^+9$, $^-9$, 0

5. $^-9$, $^+6$, $^-2$

6. $^-60$, $^+30$, 0, $^-70$

Compute.

7. $^+6 + {}^+9$

8. $^-11 + {}^-7$

9. $^-8 + {}^+4$

10. $0 + {}^+7$

11. $^+10 - {}^+8$

12. $^-2 - {}^+7$

13. $^+7 - {}^-4$

14. $0 - {}^+8$

Complete each ordered pair for the given rule.

$n \times 4$

15. (3, _?_)

16. (10, _?_)

17. (0, _?_)

18. (12, _?_)

Use a grid to locate the points.

19. $K(^-2, {}^-3)$

20. $L(^+1, {}^-3)$

21. $M(^+2, 0)$

22. $N(^-1, 0)$

23. Join points K, L, M, N in order. Move the figure right
3 units. Give the coordinates of the vertices of the slide image.

PROBLEM SOLVING   *Use a strategy or strategies you have learned.*

24. Andrew writes an integer pattern by adding $^+5$ and
subtracting $^-2$ in order. The eighth number in
the pattern is $^+20$. What number did he start with?

Cumulative Test II

Choose the best answer.

1. Choose the short word name for the number.

   537,200,000,000

   a. 537 million, 200 thousand
   b. 537 billion, 200 million
   c. 537 billion, 200 thousand
   d. 537 million, 200

2. 5,007,273
   − 846,047

   a. 416,126
   b. 4,161,226
   c. 4,161,236
   d. not given

3. Estimate.

   $529.47
   × 623

   a. $30,000
   b. $36,000
   c. $300,000
   d. $500,000

4. What is the value of m?

   $m \times \$182.75 = \$8223.75$

   a. 45
   b. 450
   c. $8041
   d. $8406.50

5. $\dfrac{2}{3}$
   $+ \dfrac{5}{9}$

   a. $\dfrac{7}{12}$
   b. $\dfrac{7}{9}$
   c. $1\dfrac{2}{11}$
   d. not given

6. What is the value of c?

   $c + 3\dfrac{5}{6} = 8\dfrac{1}{4}$

   a. $4\dfrac{5}{12}$
   b. $4\dfrac{7}{12}$
   c. $5\dfrac{7}{12}$
   d. $5\dfrac{2}{3}$

7. $\$39 \div 1\dfrac{1}{2}$

   a. $19.50
   b. $26.00
   c. $58.50
   d. not given

8. Triangle XYR

   $\angle X = 42°$, $\angle Y = 56°$,
   $\angle R = \underline{\ ?\ }$

   a. 82°
   b. 98°
   c. 102°
   d. 262°

9. Which polygon has perpendicular diagonals?

   a. rectangle
   b. rhombus
   c. trapezoid
   d. none of these

10. 8.2 g = _?_ kg

    a. 0.0082
    b. 0.082
    c. 820
    d. 8200

11. $\triangle RST \sim \triangle WXY$. Find the value of n.

    a. 10
    b. 12
    c. 15
    d. 25

12. Find the surface area.

    a. 86 m²
    b. 144 m²
    c. 172 m²
    d. 344 m²

471

Use the double bar graph for problems 13–15.

Books Read

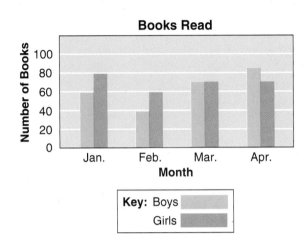

13. In which month(s) did the girls read more books than the boys?

14. Find the mean, median, and mode number of books read by the girls.

15. Estimate the total number of books read by the boys in the four-month period.

Compute or find the answer.

16. 25% of 40

17. 6% of $5.80

18. 24 is what percent of 60?

19. $^+6 + {}^-4$

20. $^-2 + {}^-8$

21. $^+6 - {}^+7$

22. $^+8 - {}^-9$

PROBLEM SOLVING

23. How much greater is the volume of a cubical storage bin $4\frac{1}{2}$ ft on each edge than a sandbox that is 6 ft long, 5 ft wide, and 2 ft deep?

24. Mariquita's weekly salary is $575 and she gets a 4% commission on all sales. What is her total income on sales of $3500?

25. From a bag containing the letters A, E, I, O, U, one letter is drawn and not replaced. Then a second letter is drawn. What is $P(A, I)$? $P(A$ or $E, I)$?

For Rubric Scoring

Listen for information on how your work will be scored.

26. Locate triangle A (0, 0), B ($^-9$, 0), C (0, $^-6$) on a coordinate grid. ABC is what type of triangle? Use $A = \frac{1}{2} \times b \times h$ to find its area.

27. On a coordinate grid, draw another triangle of a different type that has the same area as $\triangle ABC$ in problem 26.

28. Cut out $\triangle ABC$ from problem 26. Place your cutout triangle in the same position on a new coordinate grid. Slide the triangle 9 units to the right and then flip it over the x-axis. Give the coordinates of the new triangle after these two transformations. Describe one transformation of $\triangle ABC$ (slide, flip, turn) that would result in the same new triangle.

Symmetry

A pair of golden orioles sings in the green willows,
A line of white egrets flies across the blue sky.
Through my west window, snows of a thousand autumns cap the mountains.
Beyond my east door, boats from ten thousand miles away dot the river.

Du Fu

$$n + 9 = 14$$

$$3b = 27$$

$$5x - 3 = 22$$

In this chapter you will:

Learn about expressions and equations
Solve additions, subtraction,
  multiplication, and division equations
Solve problems having more than
  one solution

Critical Thinking/Finding Together

Explain with models how symmetry in
nature is similar to the balance that occurs
in an equation such as $n + 9 = 14$.

473

Moving-On: Algebra

14-1 Algebraic Expressions

Ms. Rios is a checkout clerk in a supermarket. She takes three minutes to pass an average customer's groceries over the bar code scanner. The following table shows how long it will take Ms. Rios to wait on a given number of customers.

Number of Customers	Minutes Required
1	$1 \times 3 = 3$
2	$2 \times 3 = 6$
3	$3 \times 3 = 9$
4	$4 \times 3 = 12$
5	$5 \times 3 = 15$

Each of the expressions for minutes fits the pattern $n \times 3$, where n stands for 1, 2, 3, 4, 5, . . . customers.

The letter n is called a **variable** and the numbers 1, 2, 3, 4, 5 are called **values** of the variable. The expression $n \times 3$ is called an **algebraic expression**.

Study these examples.

English Expression	**Algebraic Expression**
3 added to y	$y + 3$
8 less than a number (Let x = the number.)	$x - 8$
a number divided by 4 (Let n = the number.)	$n \div 4$ or $\dfrac{n}{4}$
7 times a number (Let g = the number.)	$7 \cdot g$ or $7g$ or $7(g)$

When a number is unknown, use a variable to represent (stand for) the number.

Translate each English expression to an algebraic expression.
Use a variable when necessary.

1. a number minus 50

2. a number plus 35

3. 12 times a number

4. 16 more than a number

5. the sum of a number and 42

6. the product of 31 and *m*

7. *n* divided by 7

8. the quotient of 45 and *p*

9. 22 less than *y*

10. 9 less than a number

More Than One Operation

An algebraic expression may contain more than one operation.

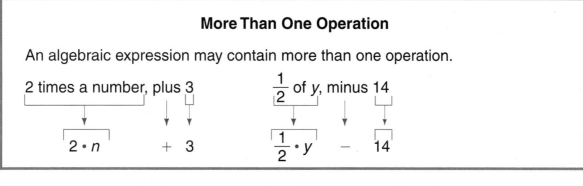

2 times a number, plus 3

$$2 \cdot n \qquad + \quad 3$$

$\frac{1}{2}$ of *y*, minus 14

$$\frac{1}{2} \cdot y \quad - \quad 14$$

Translate to an algebraic expression. Use a variable when necessary.

11. *n* doubled, minus 10

12. twice *x*, minus 34

13. $\frac{1}{2}$ of a number, plus 62

14. 17 plus twice a number

15. 2 times the width, plus 10

16. double Nancy's age, minus 22.

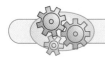 **SHARE YOUR THINKING**

The table shows the number of rows of
strawberries that Osvaldo can pick per hour.

17. Let *t* stand for the number of hours worked.
Write an expression for the number of rows picked.

18. Let *r* stand for the number of rows picked.
Write an expression for the number of hours worked.

Number of Hours	Rows Picked
1	2
2	4
3	6
4	8

475

14-2 Equations

An **equation** is a statement that two mathematical expressions are equal.

An English sentence may be translated to an equation.

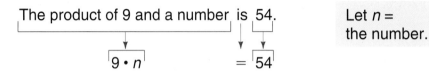

The product of 9 and a number is 54.

$9 \cdot n \qquad = 54$

Let $n =$ the number.

The equation is $9 \cdot n = 54$ or $9n = 54$.

▶ A double pan balance is a good model of an equation because both sides must be the same to keep the balance.

$$x + 2 \qquad = \qquad 9$$

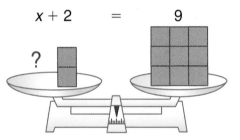

There are 9 blocks in the pan on the right side of the scale. There are 2 blocks in the pan on the left side of the scale. Seven more blocks are needed on the left side to keep the balance.

$$x + 2 = 9 \longleftarrow \boxed{\text{The value of x is 7.} \\ \text{7 is the solution of the equation.}}$$

To **solve** an equation, find the value of the variable that makes the equation true.

The value of the variable that makes an equation true is the **solution** of the equation.

Translate each of the following to an equation.
Use a variable when necessary.

1. The sum of t and 45 is 78.

2. n plus 16 is 34.

3. 8 less than a number is 4.

4. The product of 6 and c is 72.

5. A number divided by 6 equals 30.

6. 22 equals 4 more than a number.

Write an English sentence for each equation.

7. $z + 22 = 40$

8. $b - 25 = 55$

9. $18 + s = 19$

10. $9 \cdot m = 36$

11. $\dfrac{v}{6} = 45$

12. $44 = 11f$

Draw a balance-scale picture for each of these equations.
Then solve the equation.

13. $a + 4 = 12$

14. $15 = d + 9$

15. $2y = 12$

 **FINDING TOGETHER**

Dustin had to distribute pill bugs evenly among 7 science lab teams.
He was given 30 pill bugs. After giving an equal number
to each team he had 2 pill bugs left over.

$$7b + 2 \qquad = \qquad 30$$

16. To keep the balance, how many pill bugs must be in
the pan on the left side of the scale?

17. What number plus 2 equals 30? $\underline{\ ?\ } + 2 = 30$
How many pill bugs in all must be in the 7 bags?

18. How many pill bugs are in each bag? What is the value of b?

Moving-On: Algebra

14-3 Solving Equations: Guess and Test

In a math contest students were asked to complete as many problems as they could. Each correct answer was worth 3 points. Toshio had a total of 87 points. How many problems did he work correctly?

To find the number of problems that Toshio worked correctly, write an equation and solve it. Let p = the number correct.

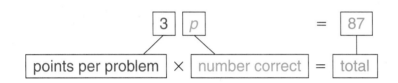

To solve the equation, use the Guess and Test strategy:

- *Guess*—Substitute numbers for the variable.

- *Test*—Check to see if you get a true statement.

Guess	Test	Conclusion
30	$3 \cdot 30 = 90$	$90 \neq 87$ Try again.
25	$3 \cdot 25 = 75$	$75 \neq 87$ Try again.
29	$3 \cdot 29 = 87$	$87 = 87$ So $p = 29$.

Think: 90 is *too much*. 75 is *not enough*.

Toshio worked 29 problems correctly.

Solve the equation. Use the Guess and Test strategy.

1. $a + 16 = 24$ **2.** $21 = x + 10$ **3.** $z - 6 = 19$

4. $b - 32 = 6$ **5.** $5n = 65$ **6.** $42 = 3y$

7. $c \div 3 = 11$ **8.** $\dfrac{m}{6} = 20$ **9.** $\dfrac{k}{9} = 12$

10. $20 = r \div 7$ **11.** $11 = \dfrac{t}{5}$ **12.** $0 = \dfrac{a}{6}$

478

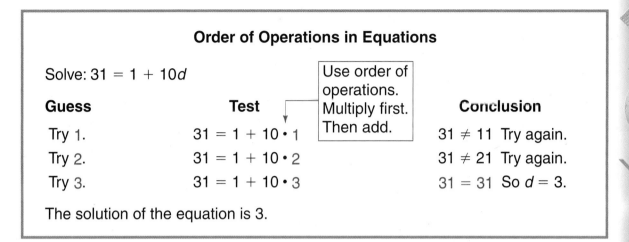

Order of Operations in Equations

Solve: $31 = 1 + 10d$

Use order of operations. Multiply first. Then add.

Guess	**Test**	**Conclusion**
Try 1.	$31 = 1 + 10 \cdot 1$	$31 \neq 11$ Try again.
Try 2.	$31 = 1 + 10 \cdot 2$	$31 \neq 21$ Try again.
Try 3.	$31 = 1 + 10 \cdot 3$	$31 = 31$ So $d = 3$.

The solution of the equation is 3.

Solve the equation. Watch the order of operations.

13. $8x - 1 = 15$

14. $2n + 1 = 15$

15. $18 + b = 39$

16. $40 = 9 + m$

17. $24 = 2 + 11g$

18. $5a - 5 = 40$

19. $10t = 100$

20. $78 = 6r$

21. $c \div 4 = 7$

22. $2 = e \div 8$

23. $60 = 2f - 20$

24. $16 + 4h = 36$

25. $\dfrac{r}{9} = 1$

26. $5 + \dfrac{v}{2} = 10$

27. $\dfrac{w}{5} - 9 = 1$

PROBLEM SOLVING

Choose the correct equation. Then solve.
Discuss your solution with the class.

Discuss

28. Paddy bought 3 sets of tomato plants. He bought a total of 12 plants. How many plants were in each set?

    a. $s + 3 = 12$ **b.** $3s = 12$ **c.** $s - 3 = 12$

29. Rodney bought a rose bush that was marked $2 off. He paid $25. What was the original price?

    a. $p + 2 = 25$ **b.** $2p = 25$ **c.** $p - 2 = 25$

30. Consuelo used 35 liters of soil to fill 10 identical flower pots with potting soil. How much soil did she put in each pot?

    a. $10f = 35$ **b.** $10 + f = 35$ **c.** $\dfrac{f}{10} = 35$

14-4 Solving Equations: Add and Subtract

Notice what happens when you add or subtract the same number on both sides of an equation.

Add 4 to both sides.

$$15 - 4 = 11$$
$$15 - 4 + 4 = 11 + 4$$
$$15 = 15 \quad \text{True}$$

Subtract 6 from both sides.

$$7 + 6 = 13$$
$$7 + 6 - 6 = 13 - 6$$
$$7 = 7 \quad \text{True}$$

When you add or subtract the same number on both sides of an equation, you get a true statement.

▶ You can solve equations with variables using the inverse operations—addition and subtraction.

Solve: $n + 6 = 13$

- See what operation is used with the variable. (Adding 6)

- What is the inverse of this operation? (Subtracting 6) Do this on both sides of the equation.

$$n + 6 = 13$$
$$n + 6 - 6 = 13 - 6$$

Subtract 6 from both sides.

$$n = 7$$

- Write the solution.

The solution is 7.

Check

- Check by replacing the variable with the solution to see if you get a true statement.

$$n + 6 = 13$$
$$7 + 6 \overset{?}{=} 13$$
$$13 = 13 \quad \text{True}$$

Study this example.

Solve: $x - 4 = 11$

$$x - 4 = 11 \quad \longleftarrow \boxed{\text{4 is subtracted from } x.}$$
$$x - 4 + 4 = 11 + 4 \quad \longleftarrow \boxed{\text{Add 4 to both sides.}}$$
$$x = 15$$

Choose the operation to solve the equation.

1. $w + 5 = 13$

   a. add 5 to both sides
   b. subtract 5 from both sides
   c. add 13 to both sides

2. $y - 7 = 22$

   a. add 22 to both sides
   b. subtract 22 from both sides
   c. add 7 to both sides

3. $n + 28 = 51$

   a. subtract 28 from both sides
   b. add 28 to both sides
   c. subtract 51 from both sides

4. $p - 18 = 40$

   a. subtract 40 from both sides
   b. add 18 to both sides
   c. subtract 18 from both sides

More Solving Equations

Solve: $24 = g - 24$ ← 24 is subtracted from g.

$24 + 24 = g - 24 + 24$ ← Add 24 to both sides.

$48 = g$ or $g = 48$

Complete.

5.
$$y - 13 = 72$$
$$y - 13 + 13 = 72 + \underline{\ ?\ }$$
$$y = \underline{\ ?\ }$$

6.
$$z + 21 = 50$$
$$z + 21 - \underline{\ ?\ } = 50 - \underline{\ ?\ }$$
$$z = \underline{\ ?\ }$$

7.
$$33 = a - 16$$
$$33 + 16 = a - 16 + \underline{\ ?\ }$$
$$\underline{\ ?\ } = a$$

8.
$$67 = h + 45$$
$$67 - \underline{\ ?\ } = h + 45 - \underline{\ ?\ }$$
$$\underline{\ ?\ } = h$$

Solve and check.

9. $x + 7 = 12$

10. $y + 8 = 20$

11. $z - 5 = 28$

12. $b - 17 = 17$

13. $12 = c - 14$

14. $30 = d + 5$

15. $9 + f = 19$

16. $0 + m = 23$

17. $p + \dfrac{1}{2} = \dfrac{1}{2}$

18. $n - \dfrac{1}{4} = 1$

19. $r - 0.2 = 0.4$

20. $s + 0.3 = 3$

Moving-On: Algebra

14-5 Solving Equations: Multiply and Divide

Notice what happens when you multiply or divide both sides of an equation by the same number.

Multiply both sides by 7.

$$28 \div 7 = 4$$
$$28 \div 7 \cdot 7 = 4 \cdot 7$$
$$28 = 28 \quad \text{True}$$

Divide both sides by 6.

$$6 \cdot 5 = 30$$
$$6 \cdot 5 \div 6 = 30 \div 6$$
$$5 = 5 \quad \text{True}$$

When you multiply or divide both sides of an equation by the same number, you get a true statement.

▶ You can solve equations with variables using the inverse operations—multiplication and division.

Solve: $6b = 30$

- See what operation is used with the variable. (Multiplying by 6)

- What is the inverse of this operation? (Dividing by 6) Do this on both sides of the equation.

$$\frac{6b}{6} = \frac{30}{6}$$

Divide both sides by 6.

$$b = 5$$

- Write the solution.

The solution is 5.

Check

- Check by replacing the variable with the solution to see if you get a true statement.

$$6b = 30$$
$$6 \cdot 5 = 30$$
$$30 = 30 \quad \text{True}$$

Study this example.

Solve: $a \div 7 = 4$

$$a \div 7 = 4$$
$$a \div 7 \cdot 7 = 4 \cdot 7$$
$$a = 28$$

← *a* is divided by 7.

← Multiply both sides by 7.

482

Choose the operation to solve the equation.

1. $2n = 12$

    a. multiply by 2 on both sides
    b. divide by 2 on both sides
    c. divide by 12 on both sides

2. $y \div 8 = 9$

    a. multiply by 9 on both sides
    b. divide by 8 on both sides
    c. multiply by 8 on both sides

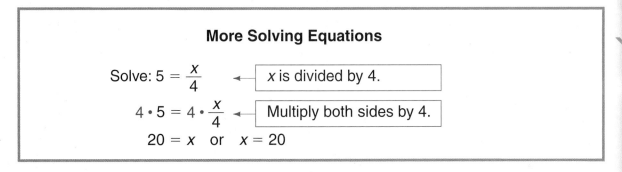

More Solving Equations

Solve: $5 = \dfrac{x}{4}$ ← x is divided by 4.

$4 \cdot 5 = 4 \cdot \dfrac{x}{4}$ ← Multiply both sides by 4.

$20 = x$ or $x = 20$

Complete.

3. $3k = 39$

$\dfrac{3k}{3} = \dfrac{39}{?}$

$k = \underline{\ ?\ }$

4. $\dfrac{z}{3} = 8$

$3 \cdot \dfrac{z}{3} = \underline{\ ?\ } \cdot 8$

$z = \underline{\ ?\ }$

5. $56 = 8m$

$\dfrac{56}{?} = \dfrac{8m}{?}$

$\underline{\ ?\ } = m$

Solve and check.

6. $7x = 14$

7. $4b = 40$

8. $c \div 7 = 3$

9. $r \div 6 = 1$

10. $32 = 2d$

11. $11 = 11f$

12. $10 = \dfrac{g}{9}$

13. $0 = h \div 6$

14. $2n = 0.6$

15. $\dfrac{p}{0.4} = 0.1$

16. $\dfrac{1}{2}y = 5$

17. $8 = \dfrac{1}{3}z$

CRITICAL THINKING

Communicate

Let $a = 8$ and $b = 9$. Write $<$, $=$, or $>$. Explain each answer.

18. $19 + a \ \square\ 28$

19. $7b \ \square\ 61$

20. $40 \div 5 \ \square\ a$

21. $\dfrac{a}{8} \ \square\ \dfrac{b}{9}$

22. $a + a \ \square\ 2b$

23. $\dfrac{a}{b} \ \square\ \dfrac{7}{8}$

Moving-On: Algebra

14-6 Evaluating Formulas

A **formula** is a rule that describes a mathematical relationship involving two or more quantities. Formulas may contain numbers, variables (letters) or a combination of numbers and variables

To **evaluate** a formula, replace all the variables except one with their number values.

Distance formula: $d = r \times t$, where d = distance, r = rate or speed, and t = time

Evaluate $d = r \times t$ for d when $r = 45$ miles per hour and $t = 3$ hours.

$$d = r \times t$$
$$d = 45 \times 3$$
$$d = 135 \text{ miles}$$

Evaluate $d = r \times t$ for t when $d = 135$ miles and $r = 45$ miles per hour.

$$d = r \times t$$
$$135 = 45 \times t \quad \leftarrow \text{Write a related division sentence.}$$
$$135 \div 45 = t$$
$$3 \text{ hours} = t$$

▶ The formula $h = m \div 60$ relates hours to minutes. Evaluate the formula for m when h = $5\frac{1}{4}$ hours.

$$h = m \div 60$$
$$5\frac{1}{4} = m \div 60 \quad \leftarrow \text{Write a related multiplication sentence.}$$
$$5\frac{1}{4} \times 60 = m$$
$$\frac{21}{\cancel{4}_{1}} \times \frac{\cancel{60}^{15}}{1} = m$$
$$315 \text{ minutes} = m$$

Replace h by $5\frac{1}{4}$ in the formula. Then solve for m.

Write the formula that you would use. Then evaluate the formula to solve the problem.

1. A train travels at a rate of 130 miles per hour. How far does it travel in 4 hours?

2. Sarah spends 300 minutes doing homework. How many hours is that?

3. A plane travels 2750 miles. If it flies at a rate of 500 miles per hour, how many hours does it fly?

4. Last weekend, Jimenez rode his bicycle for 4 h 20 min. How many minutes is that?

5. Write a formula that expresses the following relationship: Savings (s) are what is left after subtracting taxes (t) and expenses (e) from wages (w).

6. Compare your formula in exercise 5 with that of a classmate. Then find s when $w = \$950.50$, $t = \$266.14$, and $e = \$499$.

Communicate

Evaluate the formula to find the volume or height as indicated.

7. $V = e \times e \times e$ or $V = e^3$, where
$e = 10.5$ ft,
$V = \underline{\ ?\ }$ ft^3

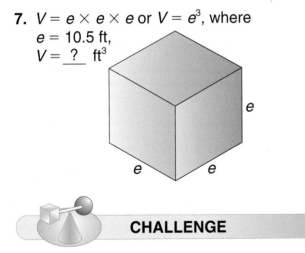

8. $V = \ell \times w \times h$, where $V = 1320$ m^3,
$\ell = 15$ m, $w = 4.4$ m,
$h = \underline{\ ?\ }$ m

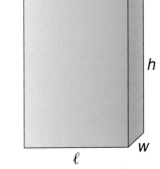

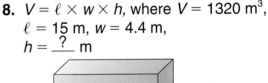

CHALLENGE

Write a new formula for finding the indicated variable.

(*Hint:* Write a related sentence.)

9. $S = C + P$, Selling Price = Cost + Profit
Formula for finding C

10. $C = \pi \times d$, Circumference = $\pi \times$ length of diameter
Formula for finding d

Moving-On: Algebra

14-7 Evaluating Volume Formulas

You can use the formula $V = B \times h$
to find the volume of a prism. B
represents the area of the base of
the prism and h represents the height
of the prism.

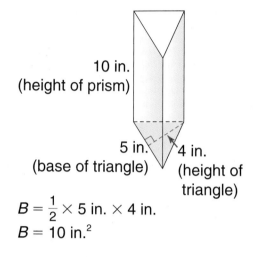

10 in.
(height of prism)

5 in.
(base of triangle)

4 in.
(height of triangle)

- Find the area of the base, B.
  The base is a triangle.

  Use $B = \frac{1}{2} \times b \times h$.

$B = \frac{1}{2} \times 5 \text{ in.} \times 4 \text{ in.}$

$B = 10 \text{ in.}^2$

- Multiply the height of the prism,
  h, by the area of the base, B.
  The product is the volume, V.

$V = B \times h$

$V = 10 \text{ in.}^2 \times 10 \text{ in.} = 100 \text{ in.}^3$

▶ To find the volume, V, of a triangular prism:

First find B, the area of
its triangular base.

Then find V.

$B = \frac{1}{2} \times b \times h$

$V = B \times h$

Study this example.

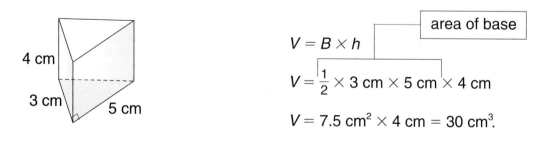

4 cm

3 cm

5 cm

area of base

$V = B \times h$

$V = \frac{1}{2} \times 3 \text{ cm} \times 5 \text{ cm} \times 4 \text{ cm}$

$V = 7.5 \text{ cm}^2 \times 4 \text{ cm} = 30 \text{ cm}^3.$

Find the volume of each triangular prism.

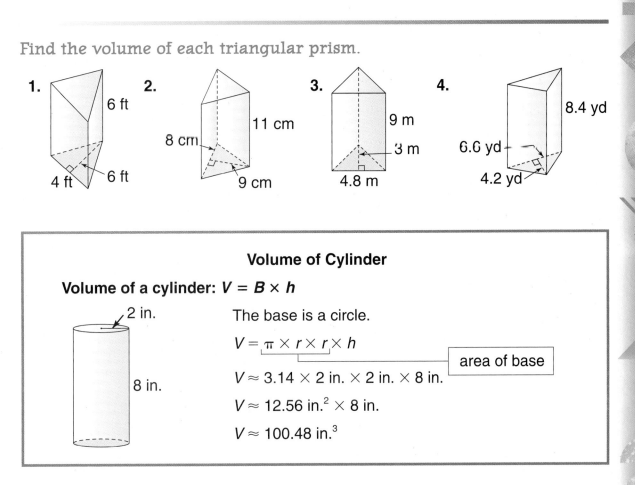

1. 6 ft, 4 ft, 6 ft

2. 11 cm, 8 cm, 9 cm

3. 9 m, 3 m, 4.8 m

4. 8.4 yd, 6.6 yd, 4.2 yd

Volume of Cylinder

Volume of a cylinder: $V = B \times h$

2 in.

8 in.

The base is a circle.

$V = \underbrace{\pi \times r \times r}_{} \times h$ → area of base

$V \approx 3.14 \times 2 \text{ in.} \times 2 \text{ in.} \times 8 \text{ in.}$

$V \approx 12.56 \text{ in.}^2 \times 8 \text{ in.}$

$V \approx 100.48 \text{ in.}^3$

Find the volume of each cylinder.

5. $r = 3$ in.
$h = 6$ in.

6. $r = 4$ ft
$h = 10$ ft

7. $r = 5$ in.
$h = 12$ in.

8. $d = 6$ yd
$h = 8$ yd

PROBLEM SOLVING

9. The height of a triangular prism is 6.5 cm and the area of its base is 24 cm². Find its volume.

10. The height of a triangular prism doubles. What happens to the volume?

SHARE YOUR THINKING

11. Work with a classmate to compare the formulas for volume of triangular prism and volume of cylinder. How are they alike? different? Discuss with the class.

Discuss ✓

14-8 Multiplying Integers

 Discover Together

Materials Needed: paper, pencil

Copy and complete the table. Look for patterns.

	1	2	3
A	$^+2 \times {}^+3 = \; ?$	$^+3 \times {}^+3 = \; ?$	$^+3 \times {}^-3 = \; ?$
B	$^+2 \times {}^+2 = {}^+4$	$^+2 \times {}^+3 = \; ?$	$^+2 \times {}^-3 = {}^-6$
C	$^+2 \times {}^+1 = {}^+2$	$^+1 \times {}^+3 = \; ?$	$^+1 \times {}^-3 = {}^-3$
D	$^+2 \times \; 0 = \; 0$	$0 \times {}^+3 = \; ?$	$0 \times {}^-3 = \; 0$
E	$^+2 \times {}^-1 = {}^-2$	$^-1 \times {}^+3 = {}^-3$	$^-1 \times {}^-3 = {}^+3$
F	$^+2 \times {}^-2 = {}^-4$	$^-2 \times {}^+3 = {}^-6$	$^-2 \times {}^-3 = {}^+6$
G	$^+2 \times {}^-3 = \; ?$	$^-3 \times {}^+3 = \; ?$	$^-3 \times {}^-3 = \; ?$

Compare the signs of the factors with the sign of the product.

4. What patterns do you see in your completed table?

5. Examine the products in row D. What is the product when one of the factors is 0?

Predict the products. Use the patterns above to help you.

6. $0 \times {}^+6$
$^+1 \times {}^+5$
$^+2 \times {}^+4$
$^+3 \times {}^+3$
$^+4 \times {}^+2$

7. $^+2 \times {}^-4$
$^+2 \times {}^-5$
$^+2 \times {}^-6$
$^+2 \times {}^-7$
$^+2 \times {}^-8$

8. $^-4 \times {}^+3$
$^-5 \times {}^+3$
$^-6 \times {}^+3$
$^-7 \times {}^+3$
$^-8 \times {}^+3$

9. $^-4 \times {}^-3$
$^-5 \times {}^-3$
$^-6 \times {}^-3$
$^-7 \times {}^-3$
$^-8 \times {}^-3$

10. Write a rule that you can use to multiply integers.

Use your rule from exercise 10 to find the product.

11. $^+9 \times {}^+8$

12. $^+7 \times {}^-16$

13. $^-8 \times {}^+14$

14. $^-24 \times {}^-6$

15. $^+12 \times {}^+11$

16. $^+6 \times {}^-13$

17. $^-5 \times {}^+21$

18. $^-15 \times {}^-12$

Use your rule to find the sign of the underlined factor
for the given product.

19. $^-5 \times \underline{9} = {}^+45$

20. $^+8 \times \underline{12} = {}^-96$

21. $^-9 \times \underline{15} = {}^-135$

22. $\underline{11} \times {}^+12 = {}^+132$

23. $\underline{7} \times {}^-26 = {}^+182$

24. $\underline{17} \times {}^-22 = {}^-374$

Communicate

25. What is the sign of the product of two integers with *like* signs?
of two integers with *unlike* signs? Discuss your answers
with the class.

Discuss ✓

26. Examine what happens to a bank account when
transactions are made as shown in the chart below. Explain
how the signs of the integers relate to the transactions.

Transaction	Representation	Result
3 deposits of $50	$^+3 \times {}^+50 = {}^+150$	increase of $150
2 withdrawals of $20	$^+2 \times {}^-20 = {}^-40$	decrease of $40
take away 4 deposits of $10	$^-4 \times {}^+10 = {}^-40$	decrease of $40
take away 5 withdrawals of $40	$^-5 \times {}^-40 = {}^+200$	increase of $200

CRITICAL THINKING

Write positive or negative to make a true statement.
Give examples to support your answer.

27. When an even number of negative
integers are multiplied, the
product is _?_ .

28. When an odd number of negative
integers are multiplied, the
product is _?_ .

Moving-On: Algebra

14-9 Dividing Integers

A diver went 15 meters below the surface of the water in 3 minutes. Use an integer to express the diver's average change in position per minute.

To find the integer, divide: $^-15 \div 3 =$ ___?___

Division and multiplication are inverse operations. You can write a related division sentence for any multiplication sentence.

Multiplication	Division
$^+5 \times {}^+3 = {}^+15$ ⟶	$^+15 \div {}^+3 = {}^+5$
$^+5 \times {}^-3 = {}^-15$ ⟶	$^-15 \div {}^-3 = {}^+5$
$^-5 \times {}^-3 = {}^+15$ ⟶	$^+15 \div {}^-3 = {}^-5$
$^-5 \times {}^+3 = {}^-15$ ⟶	$^-15 \div {}^+3 = {}^-5$

$^-15 \div 3 = {}^-5$ or $^-15 \div {}^+3 = {}^-5$

So the diver's average change in position per minute can be expressed as $^-5$ meters.

▶ The rules for determining the signs of quotients of integers is basically the same as that for determining the signs of products of integers.

> The quotient of two integers:
> • is positive if the integers have the same sign.
> • is negative if the integers have different signs.
> • is zero if the dividend is zero.

Think:
You cannot divide an integer by zero.

Study these examples.

$^+18 \div {}^+6 = {}^+3$ $^-18 \div {}^+6 = {}^-3$ $\dfrac{^+18}{^-6} = {}^-3$ $\dfrac{^-18}{^-6} = {}^+3$

$0 \div {}^+6 = 0$ $0 \div {}^-6 = 0$ $^-6 \div 0 =$ impossible

Use the rules on page 490 to find each quotient.

1. $^-9 \div ^+3$ **2.** $^+12 \div ^-6$ **3.** $^-10 \div ^-2$ **4.** $^+40 \div ^+5$

5. $^+54 \div ^-6$ **6.** $^-25 \div ^+5$ **7.** $^-80 \div ^+10$ **8.** $0 \div ^+9$

9. $^-10 \div 0$ **10.** $^+11 \div ^-1$ **11.** $^-20 \div ^+1$ **12.** $^-4 \div ^-4$

13. $\dfrac{^-36}{^-6}$ **14.** $\dfrac{^-1}{^+1}$ **5.** $\dfrac{50}{^-10}$ **16.** $\dfrac{^-80}{5}$

Complete each chart. Write the rule.

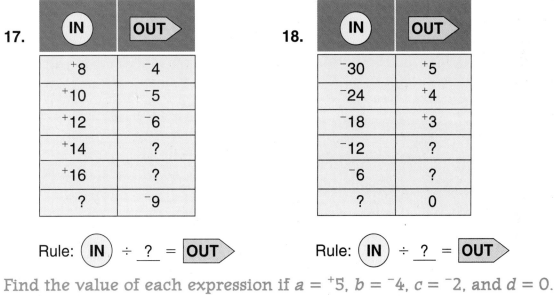

17.

IN	OUT
$^+8$	$^-4$
$^+10$	$^-5$
$^+12$	$^-6$
$^+14$	?
$^+16$	?
?	$^-9$

Rule: IN ÷ ? = OUT

18.

IN	OUT
$^-30$	$^+5$
$^-24$	$^+4$
$^-18$	$^+3$
$^-12$	?
$^-6$	?
?	0

Rule: IN ÷ ? = OUT

Find the value of each expression if $a = {}^+5$, $b = {}^-4$, $c = {}^-2$, and $d = 0$. (*Hint:* Use the rules for order of operations.)

19. $b \div c$ **20.** $a - b$ **21.** $a + b \cdot c$ **22.** $c - d$

23. $\dfrac{ab}{2c}$ **24.** $a - \left(\dfrac{b}{c}\right)$ **25.** $cd + a$ **26.** $b \div (c \cdot d)$

PROBLEM SOLVING

27. Lisa's foreign stock fund changed by $^-81¢$ during a 3-day period. If it changed at the same rate each day, what is the rate?

28. A submarine is at a depth of 510 meters. If it ascends at a rate of 15 meters per minute, can it reach the surface in a half hour?

14-10 Equations with Integers

From 9:00 P.M. to 6:00 A.M., the temperature rose 7 degrees Celsius. At 6:00 A.M., it was 2°C. What was the temperature at 9:00 P.M.?

To find the temperature at 9:00 P.M., write an equation and solve it. Let t = temperature.

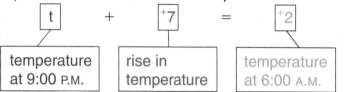

t	+7	+2
temperature at 9:00 P.M.	rise in temperature	temperature at 6:00 A.M.

To solve the equation, use the *Guess and Test* strategy.

Guess	Test	Conclusion
$^+9$	$^+9 + {}^+7 = {}^+16$	$^+16 > {}^+2$ Try again.
$^-3$	$^-3 + {}^+7 = {}^+4$	$^+4 > {}^+2$ Try again.
$^-5$	$^-5 + {}^+7 = {}^+2$	$^+2 = {}^+2$ So $t = {}^-5$.

Think: $^+9$ is too much.
$^-3$ is too much.

The temperature at 9:00 P.M. was $^-5$°C.

Study this example.

Solve $n - {}^+16 = {}^-40$. Replacement numbers for n: $^+14, {}^-56, {}^-24$

$$n - \qquad ^+16 \quad = \quad ^-40$$

subtract	positive sixteen	negative forty

Think: Subtracting $^+16$ is the same as adding its opposite, $^-16$.

Try $^+14$: $^+14 - {}^+16 \longrightarrow {}^+14 + {}^-16 = {}^-2$ No
Try $^-56$: $^-56 - {}^+16 \longrightarrow {}^-56 + {}^-16 = {}^-72$ No
Try $^-24$: $^-24 - {}^+16 \longrightarrow {}^-24 + {}^-16 = {}^-40$ Yes

So $n = {}^-24$.

Solve each equation. Use the *Guess and Test* strategy.

1. $x + {}^+8 = {}^+13$
2. $y + {}^+6 = {}^+3$
3. $z + {}^-5 = {}^-10$

4. $b + {}^-2 = {}^-8$
5. $c + {}^+1 = {}^-10$
6. $d + {}^+7 = {}^+7$

7. ${}^+9 + f = {}^+19$
8. ${}^-8 = m + {}^-16$
9. $0 + p = {}^-25$

Solve each equation. Use these replacement numbers for *n*:
${}^+5, {}^-5, 0, {}^+25, {}^-25$. (*Hint:* Watch for − and +.)

10. $n - {}^+10 = {}^-15$
11. $n - {}^+10 = {}^+15$
12. $n - {}^-10 = {}^+15$

13. $n + {}^+10 = {}^-15$
14. ${}^+25 = n + 0$
15. $n - {}^+5 = 0$

More Solving Equations

Use the *Guess and Test* strategy to solve multiplication and division equations.

Solve: ${}^-3m = {}^+48$

Try ${}^-14$: ${}^-3 \cdot {}^-14 = {}^+42$ No
Try ${}^-18$: ${}^-3 \cdot {}^-18 = {}^+54$ No
Try ${}^-16$: ${}^-3 \cdot {}^-16 = {}^+48$ Yes

So $m = {}^-16$.

Solve: $r \div {}^-3 = {}^-9$

Try ${}^+3$: ${}^+3 \div {}^-3 = {}^-1$ No
Try ${}^+9$: ${}^+9 \div {}^-3 = {}^-3$ No
Try ${}^+27$: ${}^+27 \div {}^-3 = {}^-9$ Yes

So $r = {}^+27$.

Solve each equation. Use the *Guess and Test* strategy.

16. ${}^+7n = {}^-98$
17. $p \div 6 = {}^-2$
18. ${}^-14a = {}^-140$

19. ${}^-8z = 0$
20. $x \div {}^-4 = 32$
21. $y \div {}^-1 = {}^-1$

PROBLEM SOLVING
Write an equation and solve.

22. Fifteen more than a number, *r*, equals ${}^-22$. What is the number?

23. A number *z* divided by ${}^-8$ equals ${}^-20$. What is the number?

Moving-On: Algebra

14-11 Function Tables

Sometimes one set of data depends upon another. This kind of relationship is called a **function**.

For example, suppose that the old town clock loses exactly 5 minutes every day. The table shows the minutes lost after given numbers of days.

A table like the one at the right is called a **function table**. It shows the input and output of the function.

Input, x (Days)	Output, y (Minutes Lost)
1	$^-5$
2	$^-10$
3	$^-15$
4	$^-20$
5	$^-25$

You can write an equation for the function above:
$y = {}^-5 \cdot x$ or $y = {}^-5x$, where $x = $ number of days and $y = $ minutes lost.

Study this example.

Complete the function table for $y = 3x + 4$. Write the ordered pairs of (x, y) and graph the equation on a coordinate grid.

x	$3x + 4$	y
$^-3$	$3(^-3) + 4 = {}^-9 + 4 = {}^-5$	$^-5$
$^-2$	$3(^-2) + 4 = {}^-6 + 4 = {}^-2$	$^-2$
$^-1$	$3(^-1) + 4 = {}^-3 + 4 = {}^+1$	$^+1$
0	$3(0) + 4 = \ 0 + 4 = {}^+4$	$^+4$

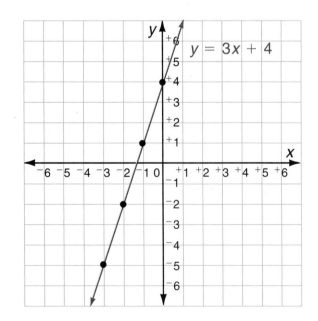

$y = 3x + 4$

Ordered pairs: $(^-3, {}^-5)$, $(^-2, {}^-2)$, $(^-1, {}^+1)$, $(0, {}^+4)$

$y = 3x + 4$ is a **linear function**.

The graphs of the ordered pairs for the function are points that form a straight line.

Copy and complete each function table. Then graph on a coordinate grid. Is the function a linear function?

1. $y = x + {}^-2$

x	x + ⁻2	y
⁺2	⁺2 + ⁻2 = 0	0
⁺1	?	?
0	?	?
⁻1	?	?

2. $y = 2x + 1$

x	2x + 1	y
⁺2	?	?
⁺1	?	?
0	?	?
⁻1	?	?

Graph each function on a coordinate grid. Use ⁺2, ⁺1, 0, ⁻1 for **x**.

3. $y = x - 4$ **4.** $y = 3x - 2$ **5.** $y = 2x$ **6.** $y = {}^-2x$

Solutions of Linear Functions

(⁻3, ⁻5) and (0, ⁺4) are **solutions** of $y = 3x + 4$. When you substitute for x and y, you get a true statement.

(⁻3, ⁻5) $y = 3x + 4$
  ↑ ↑ $^-5 = 3(^-3) + 4$
  x y $^-5 = ^-9 + 4$
        $^-5 = ^-5$ True

(0, ⁺4) $y = 3x + 4$
  ↑ ↑ $^+4 = 3(0) + 4$
  x y $^+4 = 0 + 4$
        $^+4 = ^+4$ True

---

7. Give three solutions for each equation in exercises 3-6. Can you read your solutions from the graph? Explain.

PROBLEM SOLVING

8. Using the function table at the top of page 494, how many minutes are lost in 22 days? How many days will it take to lose 200 minutes?

9. For each $9 tie shipped by Tie World, the mailing charge is $2. Use x for the number of ties shipped. Write a linear function for finding the total cost (y).

10. The number of *inches* in a length (y), is a function of the number of yards (x). Make a function table for this function using any five measures for x. Write an equation of the function.

Moving-On: Algebra

14-12 Rational Numbers: Number Line

Karin's class is learning about the stock market. Each group of students "manages" a fantasy portfolio of stocks. The table shows how the stocks in Karin's group's portfolio performed in a one-week period. It shows how much the value of a share of stock rose ($^+$) or fell ($^-$) that week.

Earl's Software	$^+1\frac{1}{2}$
Ultimate Graphics	$^-2$
Pelican Steel	$^-3\frac{1}{4}$
BAC Stores	$^+2$
Clark Electronics	$^-\frac{3}{4}$
Mike's Bikes	$^+1$
Ellen Stores	$^+\frac{1}{8}$

The table shows that the value of a share of Earl's Software rose $1\frac{1}{2}$ points. It shows that the value of a share of Ultimate Graphics fell 2 points. You can use a number line to show all of the positive and negative numbers.

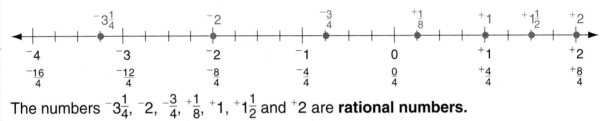

The numbers $^-3\frac{1}{4}$, $^-2$, $^-\frac{3}{4}$, $^+\frac{1}{8}$, $^+1$, $^+1\frac{1}{2}$ and $^+2$ are **rational numbers.** 0 is also a rational number.

▶ Since you can rename fractions as decimals, you can write any rational number as a decimal. On the number line below, the decimal form of the rational numbers above are shown.

Compare the two number lines. Notice that $^-3\frac{1}{4} = {}^-3.25$, $^-\frac{3}{4} = {}^-0.75$, $^+\frac{1}{8} = {}^+0.125$, and $^+1\frac{1}{2} = {}^+1.5$.

▶ Every rational number has an opposite.

The opposite of $^+2\frac{1}{2}$ is $^-2\frac{1}{2}$.   The opposite of $^-\frac{3}{4}$ is $^+\frac{3}{4}$.

The opposite of $^+3.5$ is $^-3.5$.   The opposite of $^-0.2$ is $^+0.2$.

Identify the point on the number line.

1. C   **2.** F   **3.** A   **4.** E   **5.** B   **6.** D

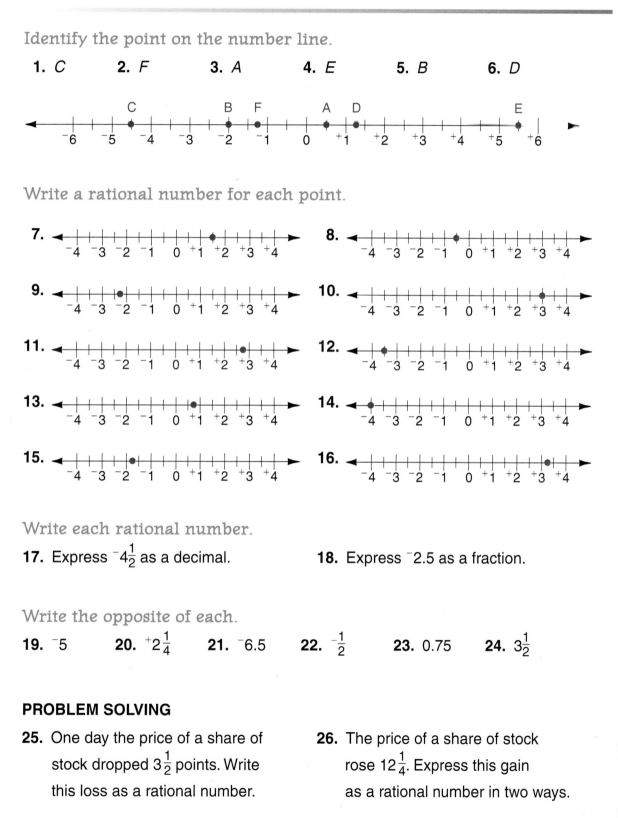

Write a rational number for each point.

7.

8.

9.

10.

11.

12.

13.

14.

15.

16.

Write each rational number.

17. Express $^-4\frac{1}{2}$ as a decimal.

18. Express $^-2.5$ as a fraction.

Write the opposite of each.

19. $^-5$ **20.** $^+2\frac{1}{4}$ **21.** $^-6.5$ **22.** $^-\frac{1}{2}$ **23.** 0.75 **24.** $3\frac{1}{2}$

PROBLEM SOLVING

25. One day the price of a share of stock dropped $3\frac{1}{2}$ points. Write this loss as a rational number.

26. The price of a share of stock rose $12\frac{1}{4}$. Express this gain as a rational number in two ways.

Moving-On: Algebra

14-13 Compare/Order Rational Numbers

One year, the average low temperature for January in Arno's town in Alaska was $^-8\frac{1}{2}°$F. In that same year, the average low temperature in the Siberian town where Yuri lives was $^-6°$F. Whose town was colder that January?

Compare: $^-6$ _?_ $^-8\frac{1}{2}$

▶ You can use a number line to compare rational numbers.

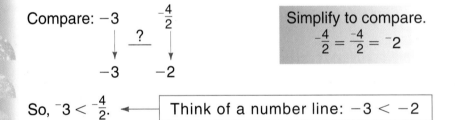

The number farther to the right is the *greater* number.
$^-6$ is farther to the right.

So, $^-6 > ^-8\frac{1}{2}$. Arno's town was colder.

Compare: -3 $\quad \dfrac{^-4}{2}$
$\qquad\quad \downarrow \;\; \dfrac{?}{\quad} \;\; \downarrow$
$\qquad\quad -3 \qquad -2$

> Simplify to compare.
> $\dfrac{^-4}{2} = \dfrac{-4}{2} = \, ^-2$

So, $^-3 < \dfrac{^-4}{2}$. ◄——— | Think of a number line: $-3 < -2$ |

Order from least to greatest: $^-2\frac{1}{2},\ ^-4,\ ^-1.5$

▶ To order rational numbers, first write them all as either decimals or fractions. Then use a number line.

- Use fractions: $^-2\frac{1}{2},\ ^-4,\ ^-1\frac{1}{2}$

  From least to greatest: $^-4,\ ^-2\frac{1}{2},\ ^-1.5$

- Use decimals: $^-2.5,\ ^-4,\ ^-1.5$

Compare. Write $<$, $=$, or $>$.

1. $\frac{-1}{2}, \frac{-3}{4}$

2. $^-0.5, {}^+0.75$

3. $^-3.5, {}^-4.25$

4. $^+3\frac{1}{4}, {}^+3\frac{1}{8}$

5. $^-4, \frac{-6}{3}$

6. $^+2.5, {}^-3\frac{1}{2}$

7. $^-5\frac{1}{8}, {}^+4$

8. $6, 5.75$

9. $0, {}^-3.25$

10. $^+\frac{3}{4}, 0$

11. $\frac{-8}{2}, {}^-4$

12. $\frac{-1}{8}, {}^-0.125$

Write in order from least to greatest. Use the number line to help you.

13. $^-3, {}^-4\frac{1}{2}, 2$

14. $0, \frac{-1}{2}, 2\frac{1}{4}$

15. $5, 0, \frac{2}{1}$

16. $^-4, 3\frac{1}{4}, {}^-1.5$

17. $^-2.25, {}^+0.25, {}^-1.5$

18. $^-2\frac{1}{2}, 2.5, {}^-1\frac{1}{4}$

19. $\frac{1}{4}, \frac{-1}{4}, 0$

20. $5\frac{1}{4}, {}^-1, {}^-2\frac{3}{4}$

21. $\frac{6}{3}, \frac{-3}{4}, {}^-4$

22. $\frac{-3}{4}, \frac{2}{1}, 1\frac{1}{4}$

23. $\frac{3}{2}, {}^-2\frac{1}{2}, 3$

24. $\frac{-4}{2}, {}^-1.5, {}^-2\frac{1}{2}$

PROBLEM SOLVING

25. Two metals were cooled to temperatures of $2\frac{1}{2}°$F and $^-3\frac{3}{4}°$F. Which of the two is the greater temperature?

26. Over a 5-day period, a share of stock showed the following changes: $^-6\frac{1}{4}, {}^-2\frac{1}{2}, {}^+1\frac{3}{4}, {}^+1,$ and $^-\frac{1}{2}$. Which was the greatest gain? greatest loss?

CRITICAL THINKING

27. Name three rational numbers equivalent to $^-5$.

28. Name three rational numbers between $^-1$ and $^-2$.

499

14-14 Problem Solving Strategy: More Than One Solution

Problem: Amy bought a sweater that was on sale. She paid $29.45 for it. She gave the clerk $30. In how many different ways might she have received her change if she received no pennies?

1 IMAGINE Put yourself in the problem.

2 NAME *Facts:* A sweater cost $29.45.
Amy gave the clerk $30.

Question: In how many different ways might the clerk have given Amy her change?

3 THINK To find some ways of making change, make a table of possible coins that could be combined to give the correct change.

4 COMPUTE

$$
\begin{array}{r}
{\scriptstyle 2\ 9\ 9\ 10} \\
\$\,\cancel{3}\,\cancel{0}.\cancel{0}\,\cancel{0} \\
-\ 2\ 9.4\ 5 \\
\hline
\$\qquad .5\ 5
\end{array}
$$ ← change

Let h represent half-dollars, q represent quarters, d represent dimes, and n represent nickels.

h	1											
q		2	1	1	1	1						
d			3	2	1		5	4	3	2	1	
n	1	1		2	4	6	1	3	5	7	9	11

There are 12 ways Amy might have received $.55 change with no pennies.

5 CHECK Use a calculator to check the solutions.

500

Solve each problem. Be sure that you find all possible solutions.

1. Cliff gave the perimeter of his rectangular garden
as 20 feet, but he could not remember the specific
dimensions. How many different whole-number
dimensions might give a perimeter of 20 feet?

IMAGINE Create a mental picture. $P = 20$ ft

↓

NAME   *Facts:* $P = 2 \times \ell + 2 \times w$
$P = 20$ ft

Question: How many different whole-number
solutions can be found?

↓

THINK   There can be several solutions to the problem.
If $2 \times \ell + 2 \times w = 20$, then $\ell + w = 10$.
How many number facts have a sum of 10?

→ **COMPUTE** → **CHECK**

2. What operation could Stan use to make the
equation $4 \bigcirc 2 = 2$ true?

3. Mr. Lentz bought a video recorder listed at $480. It
was on sale for 40% off. How much did he pay for the
recorder? He gave the clerk 3 hundred-dollar bills to
pay for his purchase. In how many different ways could
he have received his change without getting any coins?

4. Jeanine found that the difference of two single-digit
integers was $^{+}10$. What equations did Jeanine write?

5. Frank doubled the sum of two negative integers.
His answer was $^{-}12$. Find the addends.

MAKE UP YOUR OWN

6. Write a problem that has more than one
whole-number solution using this figure.

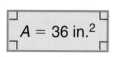

$A = 36$ in.^{2}

Moving-On: Algebra

14-15 Problem-Solving Applications

Solve each problem and explain the method you used.

Members of the Turbo-Math Club write equations for each other.

1. Alana writes this equation: $5a + 3 = 28$. What is the value of a?

2. Sherman writes this equation: $\frac{w}{4} + 22 = 36$. What is the value of w?

3. Sarah writes this equation: $3d \div d = 3$. Does this equation have one unique solution? Explain.

4. Chad is trying to solve this equation: $e + 107 = 215$. He estimates that e is between 200 and 300. Is this a reasonable estimate?

5. Oxanna notices that these two equations both include the variable f: $f + 72 = 89$, $2f = 34$. Does f have the same value in each equation?

6. Ray solves this equation: $g \div 5 = 15$. Then he finds the value of h in this equation: $h \times g = 75$. What is the value of h?

7. Find the value of j in this equation: $(j - 10) + 27 = 27$.

8. Find the value of k in this equation: $(15 \cdot 2) \cdot k = 30$.

9. Which equation has a solution greater than 55? less than or equal to 8? $16 = \frac{a}{2} + 12$; $16 = \frac{c}{2} - 12$.

10. Ms. Elliot offers this riddle: "I am a number that is exactly three times the greatest two-digit even number. What number am I?"

11. For which equation does $m = \frac{1}{2}$: $1\frac{1}{2} - m = \frac{1}{2}$ or $3 \div m = 6$?

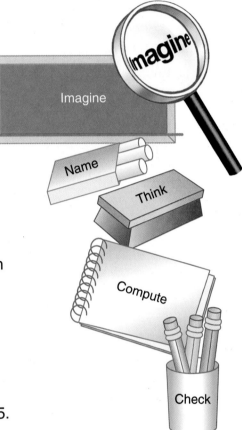

Imagine

Name

Think

Compute

Check

Use one or more strategies to solve each problem.

12. Iris wrote this riddle: "I am a number that is exactly four times the number of fluid ounces in a cup. What number am I?"

13. James thinks of a number. Sarah asks, "Are there any even digits? Is the number greater than 50? Is it divisible by 3? Is it less than 36?" James answers "no" to each of her questions. What number is he thinking of?

14. Edna has a favorite two-digit number. The sum of the digits is 9. The difference between the digits is 1. What are the two possibilities for Edna's favorite number?

15. Thomas writes this equation: $n \times q = 24$. What are some of the possible values for n and q?

16. The Math Club newsletter is twice as long as the French Club newsletter, which is one third as long as the Science Club newsletter. The Science Club newsletter is 6 pages long. How long is the Math Club newsletter?

17. What numbers can you find that, when multiplied by themselves, are greater than 240 but less than 241?

Use the magic square for problems 18–20.

18. The sum of each horizontal, vertical, and diagonal column is the same. Write an equation to find the value of w.

19. Find the values of x, y, and z.

20. Find the value of p: $(y \div x) - (w - z) = p$.

3	w	1
x	2.5	4.5
4	z	y

MAKE UP YOUR OWN

21. Use x, y, and z to represent students with black, red, or brown hair. Make up a problem modeled on problem 16. Write and solve the equation.

Translate to a mathematical expression or an equation. *(See pp. 474–477, 492–493.)*
Use a variable when necessary.

1. 6 less than a number

2. the quotient of 32 and m

3. 25 more than 8 times a number

4. the sum of z and 23 is 65.

5. A number doubled is 24.

6. Three times a number equals $^-21$.

Solve the equation. Watch the order of operations.   *(See pp. 478–483, 492–493.)*

7. $8 = z - 4$

8. $14 = n + 12$

9. $x + \dfrac{1}{2} = 2$

10. $m + 7 = 23$

11. $2b = 22$

12. $14 = x - 12$

13. $9t - 3 = 51$

14. $\dfrac{x}{4} + 3 = 27$

15. $n + {}^-16 = {}^-5$

Multiply or divide.   *(See pp. 488–491.)*

16. $^-5 \times {}^-10$

17. $^-90 \div {}^+5$

18. $0 \times {}^-6$

19. $\dfrac{^-33}{^-3}$

20. Write in order from least to greatest: $\dfrac{^-3}{2}$, $^-2.5$, $^-4$.   *(See pp. 496–499.)*

PROBLEM SOLVING
Choose the correct equation. Then solve.   *(See pp. 478–483, 486–487.)*

21. Tony bought a baseball glove that was marked $3 off.
He paid $13. What was the original price?

    a. $g - 3 = 13$
    b. $g \div 3 = 13$
    c. $g + 3 = 13$

22. Una bought 4 sets of golf balls. She bought a total of 52 balls.
How many balls were in each set?

    a. $n - 4 = 52$
    b. $52 = 4n$
    c. $4 + n = 52$

Solve.

23. Find the volume of a triangular prism whose base area   *(See pp. 486–487.)*
is 24 square meters and whose height is 8 meters.

(See Still More Practice, p. 518.)

GRAPHING EQUATIONS AND INEQUALITIES

As you have learned, an **equation** is a mathematical
sentence that contains an equals sign (**=**).
An **inequality** is a mathematical sentence that
contains an inequality sign, such as **<** or **>**.

$y + 3 = 9$ ← | equation | $x < {}^-3$ ← | inequality |

You can graph the solutions of equations and
inequalities on a number line.

▶ Graph the solution of $x + 2 = 6$.

$$x + 2 = 6$$
$$x + 2 - 2 = 6 - 2$$
$$x = 4$$

The equation has
one solution.

The solution of an equation is shown by a solid
dot on the number line. To graph the solution,
place a dot at the point that corresponds to 4, or ${}^+4$.

$$\begin{array}{ccccccccc} {}^-4 & {}^-3 & {}^-2 & {}^-1 & 0 & {}^+1 & {}^+2 & {}^+3 & {}^+4 \end{array}$$

▶ Graph the solutions of $a < 3$ if the
replacement numbers for a are integers.

Any integer less than
3 is a solution.

Draw a solid dot on the number line for each integer less than
3, or ${}^+3$. Shade the left arrow of the number line to show
that the solutions continue indefinitely.

$$\begin{array}{ccccccccc} {}^-4 & {}^-3 & {}^-2 & {}^-1 & 0 & {}^+1 & {}^+2 & {}^+3 & {}^+4 \end{array}$$

Graph each to show integer solutions.

1. $x > 5$ **2.** $k + 3 = 7$ **3.** $z < {}^-2$ **4.** $a - 4 = 9$

5. $g - 4 = 0$ **6.** $r > {}^-1$ **7.** $m < 7$ **8.** $4 = f + 3$

Performance Assessment

Match each sentence to an equation and each equation to its solution.

1. A number added to 2 is 64. $\quad\quad\quad$ $n - 2 = 64$ $\quad\quad$ 8
   A number doubled is 64. $\quad\quad\quad$ $n^2 = 64$ $\quad\quad$ 32
   A number squared is 64. $\quad\quad\quad$ $2n = 64$ $\quad\quad$ 62
   2 subtracted from a number is 64. $\quad$ $n - 64 = 0$ $\quad\quad$ 64
   64 subtracted from a number is 0. $\quad$ $2 + n = 64$ $\quad\quad$ 66

2. Write another equation-matching exercise like the one above for your classmates to complete.

3. Translate to a mathematical expression: product of 10 and b

4. Translate to an equation: 42 equals c divided by 7

Solve the equation. Watch the order of operations.

5. $26 + a = 47$

6. $2n + 17 = 21$

7. $^-15z = ^-240$

Multiply or divide.

8. $^+2 \times {}^-14$

9. $^-5 \times {}^-5$

10. $\dfrac{^-16}{^-4}$

11. $0 \div {}^-12$

Compare. Write <, =, or >.

12. $^-5 \underline{\ ?\ } {}^+2\dfrac{1}{2}$

13. $^-7 \underline{\ ?\ } {}^-6.75$

14. $^-2\dfrac{1}{8} \underline{\ ?\ } {}^+\dfrac{3}{8}$

15. Graph $y = 2x - 3$ on a coordinate grid. Use $^+2, {}^+1, 0, {}^-1$ for x.

PROBLEM SOLVING *Use a strategy or strategies you have learned.*

16. Choose the correct equation and then solve the problem: Sharon bought 4 sets of cassettes. If she bought 60 cassettes in all, how many cassettes were in each set?

    a. $s + 4 = 60$ $\quad\quad$ **b.** $4s = 60$ $\quad\quad$ **c.** $\dfrac{s}{4} = 60$

17. Which volume is greater: triangular prism with base area of 48 cm², height = 20 cm or cylinder with length of radius = 3 cm and height = 4 cm? How much greater?

Practice 1-1

Chapter 1

Write in expanded form in two ways.

1a. 83,007,100   **b.** 2,940,000,000,000

Round each number to its greatest place.

2a. 67,824   **b.** 908,543   **c.** 17,548,261

Order from least to greatest.

3. 47,396,000; 47,963,000; 47,369,000

4. 601,148; 610,148; 601,184; 610,418

Estimate the sum or difference.

5a.   816,738   **b.**   371,940   **c.** $786.75
    − 329,542   + 47,083   + 21.99

Use a related sentence to find the missing number.

6a. $n + 86 = 132$ **b.** $n - 55 = 70$

Compute. Watch for + and − signs.

7a.   9,392,738   **b.**   341,086   **c.** $970.05
    + 3,678,907   − 87,794   − 213.87

8. 3,150,167 + 94,861 + 13,848,156

PROBLEM SOLVING

9. Brazil has an area of 3,284,426 square miles. Canada has an area of 3,851,787 square miles. The United States has an area of 3,623,420 square miles. Which country has the greatest area? the least area?

10. Luis drove from New York to Chicago. The odometer read 32,949 when he started. It read 33,751 when he reached Chicago. How far did Luis travel?

11. Over a 3-day period the circus ticket office sold 12,953, 10,984, and 11,127 tickets. How many tickets were sold altogether on those 3 days?

Practice 2-1

Chapter 2

Complete. Identify the property of multiplication used.

1a. $37 \times \underline{\ ?\ } = 0$ **b.** $63 \times t = t \times \underline{\ ?\ }$

2a. $5 \times (8 + 4) = (\underline{\ ?\ } \times 8) + (5 \times 4)$
   b. $\underline{\ ?\ } \times 9 = 9$

Find the product.

3a. 40×700 **b.** 500×8000 **c.** 186×300

4a. 508×720 **b.** 709×5309 **c.** $650 \times \$38.75$

5a.   917   **b.**   8236   **c.** $795.03
    × 38   × 79   × 28

6a.   4752   **b.**   30,817   **c.** $39.87
    × 809   × 450   × 506

7a. 572×2008 **b.** $910 \times 546,019$

Estimate each product.

8a. 917×380 **b.** 817×3712 **c.** 5477×3819

Write the standard numeral.

9. $(6 \times 10^5) + (4 \times 10^3) + (2 \times 10^2) + (5 \times 1)$

10a. 5^3 **b.** 3^5 **c.** 10^7

PROBLEM SOLVING

11. A certain meteor is moving through space at 1899 miles per minute. At this rate how far will the meteor travel in 2 hours?

12. The interior temperature of the sun is about 35,000,000°F. Write this temperature in expanded form using exponents.

13. A large city has 375 office buildings. There is an average of 425 offices in each building. About how many offices are there in the city?

14. The art museum sent 160 passes to each of 176 schools. How many passes were sent?

Practice 2-2

Estimate the quotient.

1a. $31\overline{)3371}$ **b.** $297\overline{)6143}$

2a. $87\overline{)\$180{,}000}$ **b.** $117\overline{)\$54{,}000}$

Divide. Use R to write remainders.

3a. $40\overline{)1200}$ **b.** $200{,}000 \div 400$

4a. $5\overline{)7826}$ **b.** $9\overline{)3618}$ **c.** $3\overline{)\$75.21}$

5a. $29\overline{)5007}$ **b.** $82\overline{)6173}$ **c.** $12\overline{)4624}$

6a. $15\overline{)\$2208.75}$ **b.** $326\overline{)1313}$

7a. $730\overline{)25{,}550}$ **b.** $417\overline{)12{,}510}$

Compute.

8a. $3 + 7 \times 9 - 5$ **b.** $(8 \div 2) \times (7 + 9) \times 10$

9a. $9 \times 6 \div 3 + 17 - 8$ **b.** $39 - 3 \times 4 \div 3$

Tell whether each of the following numbers is divisible by 2, 3, 4, 5, 6, 9, or 10.

10a. 36,720   **b.** 3,255,075   **c.** 76,269,804

PROBLEM SOLVING

11. Every morning, 35,875 riders use public transportation to get to school or work. If a bus can hold 53 riders, estimate how many busloads of riders there are each morning.

12. Minnesota has an area of 86,943 square miles and 87 counties. What is the average number of square miles per county?

13. Kareem's Computer Store buys 19 pieces of Spelling Tutor software. The bill is $711.55. What is the average cost of each piece of software?

14. Two hundred fourteen bags of concrete mix weigh 11,984 lb. How much does one bag weigh?

Practice 3-1

> Chapter 3

Write the place of the underlined digit. Then write its value.

1a. 21.07$\underline{83}$ **b.** 6.0009$\underline{7}$ **c.** 9.10382$\underline{4}$

Write in expanded form.

2a. 0.07309   **b.** 5.008407

Round to the greatest nonzero place or to the nearest cent.

3a. 0.3512   **b.** 0.05709   **c.** 8.0957

4a. 72.807   **b.** $25.095   **c.** $18.375

Write in order from greatest to least.

5. 0.2954; 0.0298; 0.29054; 0.29504

6. 4.7; 4.77; 4.707; 4.7707; 4.077

Use a related sentence to find the missing decimal.

7a. $n + 0.8 = 0.91$ **b.** $n - 0.03 = 0.7$

Estimate the sum or difference.

8a. $\begin{array}{r} 27.14 \\ + 31.762 \\ \hline \end{array}$ **b.** $\begin{array}{r} 0.275 \\ + 3.8 \\ \hline \end{array}$ **c.** $\begin{array}{r} 43.09 \\ - 17.8 \\ \hline \end{array}$

Add or subtract.

9a. $\begin{array}{r} 73 \\ - 8.92 \\ \hline \end{array}$ **b.** $\begin{array}{r} 40.09 \\ + 7.953 \\ \hline \end{array}$ **c.** $\begin{array}{r} 7.1852 \\ - 5.706 \\ \hline \end{array}$

10a. $84.856 - 29.7$ **b.** $\$80 - \$35.97$

PROBLEM SOLVING

11. Find the sum of 8.35, 9.046, 0.7185, 30, and 6.02.

12. Jenna bought skates on sale for $37.44 instead of the full price of $50. How much did she save?

13. Ed ran 9.75 mi on Friday and 13.6 mi on Saturday. How much farther did he run on Saturday?

Multiply. Round to the nearest cent when necessary.

1a. 10×0.7 **b.** 100×1.63

2a. 1000×0.463 **b.** 45.23×100

3a.   0.09   **b.**   0.75   **c.**   5.079
      $\times$ 37 $\times$ 26 $\times$ 58

4a.   0.67   **b.**   0.734   **c.**   9.876
      $\times$ 0.9 $\times$ 0.7 $\times$ 0.05

5a. $17.05   **b.** $9.00   **c.** $1.29
      $\times$ 3.5 $\times$ 6.25 $\times$ 4.08

Estimate the product.

6a. 6.24×11.86 **b.** 0.68×9.8

7a. $\$32.09 \times 0.48$ **b.** 16.04×3.67

Write in scientific notation.

8a. 345,000   **b.** 80,000,000

Write in standard form.

9a. 7×10^5 **b.** 3.1×10^7 **c.** 9.14×10^4

Problem Solving

10. Find the cost of 6 pens if 1 pen costs $1.37.

11. Snow fell at a rate of 0.6 cm per hour. At that rate, how much snow fell in 5 hours?

12. How much greater than the product of 1.2 and 0.45 is the sum?

13. Which costs more, 9.8 gallons of gasoline at $1.17 per gallon or 10.05 gallons at $1.15 per gallon?

14. Craig needs 3.25 hours to build a bookshelf. At that rate how long does he need to build 5 bookshelves?

15. Use the properties of multiplication to compute: $(18.9 \times 4) \times 2.5$.

Practice 4-2

Divide.

1a. $36.3 \div 10$ **b.** $18.6 \div 100$

2a. $25.2 \div 1000$ **b.** $7 \div 1000$

3a. $3\overline{)0.783}$ **b.** $9\overline{)1.917}$ **c.** $4\overline{)\$32.48}$

4a. $0.3\overline{)93}$ **b.** $0.8\overline{)\$4.00}$ **c.** $0.19\overline{)38}$

5a. $0.05\overline{)2.113}$ **b.** $2.4\overline{)1.8}$

6a. $0.03\overline{)8.124}$ **b.** $0.6\overline{)1.803}$

7a. $22.6\overline{)20.34}$ **b.** $0.28\overline{)1.225}$

Estimate the quotient.

8a. $35.81 \div 5.9$ **b.** $\$394 \div 79.05$

9a. $\$57.59 \div 8.1$ **b.** $0.8 \div 0.199$

10a. $\$22.32 \div 3.1$ **b.** $\$30.15 \div 16.2$

Problem Solving

11. If 8 copies of a novel cost $38.00, find the price of one novel.

12. Golf balls are on sale for $15.69 per dozen. Determine the price of one golf ball. Round your answer to the nearest cent.

13. A section of highway 3.87 miles long is being rebuilt. If the workers can complete 0.03 miles per day, how many days will it take them to complete the job?

14. A metal worker cuts an aluminum bar into segments that measure 3.625 cm. How many segments can be cut from a bar 87 cm long?

15. What number multiplied by 0.7 will give the same product as 5.6 multiplied by 0.8?

Write a fraction for each point.

1a. R **b.** P

R   P

0   1

Complete.

2a. $\frac{5}{7} = \frac{?}{28}$ **b.** $\frac{4}{9} = \frac{24}{?}$ **c.** $\frac{18}{?} = \frac{2}{5}$

Write each fraction in simplest form.

3a. $\frac{18}{27}$ **b.** $\frac{15}{21}$ **c.** $\frac{16}{40}$

Compare. Write $<$, $=$, or $>$.

4a. $\frac{17}{23}$? $\frac{7}{23}$ **b.** $\frac{5}{6}$? $\frac{9}{10}$

5a. $\frac{7}{8}$? $\frac{49}{56}$ **b.** $\frac{1}{2}$? $\frac{3}{5}$

Write in order from least to greatest.

6a. $\frac{2}{3}, \frac{1}{5}, \frac{5}{6}$ **b.** $\frac{5}{9}, \frac{1}{4}, \frac{5}{12}$

7a. $1\frac{7}{12}, 1\frac{1}{2}, 1\frac{2}{3}$ **b.** $2\frac{2}{5}, 2\frac{2}{3}, 2\frac{2}{15}$

Tell whether each is *prime* or *composite*.

8a. 9   **b.** 19   **c.** 49

Find the prime factorization and write in exponent form.

9a. 26   **b.** 40   **c.** 56

Problem Solving

10. Which fraction is close to $\frac{1}{2}$: $\frac{6}{11}, \frac{13}{15}, \frac{1}{5}$?

11. Of 24 dogs, 9 are beagles, 5 are collies, and the rest are poodles. What fractional part are poodles?

12. Marla ate $\frac{3}{8}$ of a melon. Leah ate $\frac{2}{3}$ of a melon. Who ate more?

13. List all the prime numbers between 20 and 30.

14. Which is farthest: $7\frac{4}{5}$ mi, $7\frac{3}{4}$ mi, or $7\frac{7}{10}$ mi?

Practice 5-2

Find the GCF of each pair of numbers.

1a. 8 and 12   **b.** 15 and 24

2a. 10 and 45   **b.** 7 and 28

Find the LCM of each pair of numbers.

3a. 7 and 10   **b.** 8 and 12

4a. 6 and 15   **b.** 14 and 42

Rename as an improper fraction.

5a. $3\frac{2}{3}$ **b.** $9\frac{7}{10}$ **c.** $5\frac{1}{4}$

Rename as a fraction in simplest form.

6a. 0.54   **b.** 0.05   **c.** 0.75

Rename as a decimal.

7a. $4\frac{7}{8}$ **b.** $\frac{1}{6}$ **c.** $5\frac{2}{3}$

8a. $\frac{9}{16}$ **b.** $3\frac{4}{100}$ **c.** $6\frac{1}{8}$

Problem Solving

9. Find a pair of numbers between 12 and 24 whose GCF is 5.

10. Find a pair of numbers between 1 and 10 whose LCM is 8.

11. Rename $\frac{18}{8}$ as a mixed number in simplest form.

12. A carton holds 10 music boxes. Write a mixed number in simplest form to show how many cartons would be filled by 46 music boxes.

13. Write the quotient of 1 divided by 9 as a repeating decimal.

14. Is $\frac{54}{110}$ a little more than $\frac{1}{2}$? Write *Yes* or *No*. Explain.

15. Yolanda has $\frac{9}{10}$ of a dollar. How much money does she have?

Practice 6-1

Chapter 6

Add or subtract. Write each answer in simplest form.

1a. $\frac{5}{9} + \frac{4}{9}$ b. $\frac{7}{8} - \frac{5}{8}$ c. $\frac{11}{16} - \frac{5}{16}$

2a. $\frac{7}{8} + \frac{3}{4}$ b. $\frac{1}{3} + \frac{5}{6}$ c. $\frac{9}{10} - \frac{1}{2}$

3a. $(\frac{3}{14} + \frac{3}{14}) + \frac{6}{7}$ b. $\frac{7}{12} - \frac{1}{4}$

Estimate the sum or difference.

4a. $\frac{5}{7} + \frac{9}{10}$ b. $\frac{9}{16} - \frac{1}{7}$ c. $\frac{11}{12} - \frac{5}{9}$

5a. $9\frac{2}{3} + 3\frac{1}{8}$ b. $7\frac{1}{5} + 7\frac{5}{6}$ c. $19\frac{1}{9} + 9\frac{7}{8}$

6a. $12\frac{1}{5} - 7\frac{2}{3}$ b. $41\frac{1}{2} - 19\frac{5}{7}$ c. $10\frac{1}{3} - 3\frac{3}{4}$

Compute. Use the addition properties.

7a. $1\frac{2}{3} + (\frac{1}{6} + \frac{1}{6})$ b. $3\frac{1}{2} + 5 + 2\frac{1}{4}$

8a. $\frac{9}{10} - (\frac{1}{5} + \frac{2}{5})$ b. $2\frac{1}{4} + 2\frac{1}{3} + \frac{1}{4}$

PROBLEM SOLVING

9. How much is $\frac{1}{6}$ increased by $\frac{1}{4}$?

10. Find the sum of $\frac{1}{7}$, $\frac{4}{21}$, and $\frac{2}{3}$.

11. Anita is a runner on the school track team. Upon reaching the $\frac{7}{8}$-mi marker of the $1\frac{1}{2}$-mi track, how much farther must she run to get to the end of the track?

12. The sum of n and $1\frac{3}{4}$ is 4. Find the value of n.

13. Ethan spent $1\frac{7}{8}$ h working on a model airplane. Then he spent $2\frac{1}{6}$ h raking leaves. About how much time did Ethan spend on those two activities?

14. From the sum of $\frac{7}{10}$ and $\frac{3}{5}$, subtract $\frac{2}{3}$.

Practice 6-2

Add or subtract. Write each answer in simplest form.

1a. $2\frac{2}{3} + 3\frac{1}{5}$ b. $6\frac{5}{8} + 3\frac{1}{2}$ c. $12\frac{1}{6} + 8\frac{4}{5}$

2a. $3\frac{3}{4} - 1\frac{1}{2}$ b. $9\frac{1}{3} - 6\frac{1}{2}$ c. $11 - 8\frac{3}{7}$

3a. $7\frac{4}{7} - 5\frac{5}{6}$ b. $4 - 2\frac{2}{3}$ c. $8\frac{1}{3} - 5\frac{3}{4}$

4a. $2\frac{1}{9} + 8\frac{7}{10}$ b. $5\frac{5}{7} + 8\frac{3}{4}$ c. $7\frac{1}{5} - 2\frac{2}{3}$

5a. $3\frac{1}{7} + 5\frac{2}{3} + 4\frac{11}{21}$ b. $16 - 3\frac{5}{6}$

Use a related sentence to find the missing number.

6a. $c + \frac{2}{3} = \frac{11}{12}$ b. $t + 2\frac{1}{3} = 5\frac{1}{3}$

7a. $q - \frac{3}{5} = \frac{1}{15}$ b. $m - 3\frac{2}{5} = 7$

PROBLEM SOLVING

8. Add $2\frac{5}{6}$ to the difference between 5 and $2\frac{7}{12}$.

9. A team practiced $2\frac{1}{2}$ h before lunch and then $1\frac{3}{4}$ h after lunch. What is the total time it practiced?

10. A recipe calls for $3\frac{1}{3}$ c of white flour, $1\frac{1}{4}$ c of whole wheat flour, and $\frac{1}{2}$ c of rice flour. Find the total amount of flour in the recipe.

11. Jeannette has three jump ropes: $6\frac{1}{2}$ ft, $9\frac{2}{3}$ ft, and $7\frac{1}{8}$ ft. Estimate the total length.

12. On Monday, a certain stock opened at $67\frac{1}{8}$ points. By Friday its value was 80 points. Find its increase in value.

13. Ted weighed $145\frac{1}{2}$ lb. After 2 months of dieting, he weighed $136\frac{1}{3}$ lb. How much weight did he lose?

Multiply. Cancel whenever possible.

1a. $\frac{3}{4} \times \frac{8}{9}$ **b.** $\frac{9}{10} \times \frac{2}{3}$ **c.** $\frac{7}{8} \times \frac{12}{15}$

2a. $\frac{2}{3} \times \frac{3}{5} \times \frac{15}{16}$ **b.** $\frac{3}{4} \times \frac{5}{6} \times \frac{7}{10}$

3a. $\frac{3}{5} \times 35$ **b.** $\frac{7}{8} \times 48$ **c.** $5 \times \frac{7}{10}$

4a. $\frac{1}{3}$ of 18 **b.** $\frac{5}{6}$ of 30 **c.** $\frac{9}{10}$ of 80

Complete. Name the property of multiplication used.

5a. $\frac{3}{7} \times$? $= 0$ **b.** $\frac{7}{8} \times \frac{9}{10} =$? $\times \frac{7}{8}$

6a. $\frac{9}{11} \times 1 =$?

b. $(\frac{1}{2} \times \frac{1}{3}) \times \frac{5}{8} = \frac{1}{2} \times ($? $\times \frac{5}{8})$

Write the reciprocal of each number.

7a. $\frac{3}{4}$ **b.** $\frac{5}{10}$ **c.** 8

8a. 15 **b.** $\frac{11}{2}$ **c.** $2\frac{1}{3}$

Divide.

9a. $\frac{3}{7} \div \frac{6}{7}$ **b.** $\frac{5}{9} \div \frac{1}{3}$ **c.** $\frac{2}{5} \div \frac{8}{9}$

10a. $7 \div \frac{1}{4}$ **b.** $21 \div \frac{7}{8}$ **c.** $5 \div \frac{2}{3}$

11a. $\frac{7}{10} \div 7$ **b.** $\frac{3}{4} \div 6$ **c.** $\frac{4}{5} \div 8$

PROBLEM SOLVING

12. Jason has saved $80. He spent $\frac{3}{4}$ of it on a new camera. How much did the camera cost?

13. A baby drank $\frac{2}{3}$ of her 6-oz bottle of milk. How much milk did the baby drink?

14. A sofa was on sale for $\frac{1}{3}$ off the regular price of $360. Find the sale price of the sofa.

15. Lia bought $\frac{3}{4}$ yd of felt. She used $\frac{5}{6}$ of it to make a banner. How much of the felt did she use for the banner?

Practice 7-2

Estimate. Then multiply.

1a. $6\frac{1}{8} \times 3\frac{3}{5}$ **b.** $7\frac{1}{3} \times 3\frac{1}{7}$ **c.** $9\frac{9}{10} \times 3\frac{3}{4}$

2a. $13 \times 4\frac{1}{5}$ **b.** $3\frac{1}{6} \times 7\frac{7}{8}$ **c.** $2\frac{2}{3} \times 9$

3a. $3\frac{4}{5} \times 1\frac{7}{8} \times 2\frac{1}{2}$ **b.** $4\frac{2}{3} \times 3 \times 1\frac{1}{5}$

4a. $5 \times \frac{7}{8} \times 3\frac{1}{4}$ **b.** $7\frac{1}{2} \times 2\frac{4}{7} \times \frac{2}{3}$

Estimate. Then divide.

5a. $\frac{3}{11} \div \frac{9}{11}$ **b.** $2\frac{1}{8} \div \frac{1}{6}$ **c.** $3\frac{1}{3} \div 1\frac{7}{8}$

6a. $6\frac{1}{2} \div 2\frac{3}{4}$ **b.** $10\frac{2}{3} \div 1\frac{1}{9}$ **c.** $5\frac{2}{5} \div 2\frac{1}{4}$

7a. $8\frac{1}{3} \div 5$ **b.** $4\frac{1}{7} \div 2$ **c.** $6\frac{6}{7} \div 8$

8a. $6 \div 1\frac{1}{2}$ **b.** $\$10 \div 3\frac{1}{3}$ **c.** $\$12 \div 1\frac{1}{5}$

Compute. Use the order of operations rules.

9a. $6 \times \frac{3}{4} + \frac{1}{2}$ **b.** $\frac{2}{3} + \frac{1}{3} \times (9 + 6)$

10a. $9 \div \frac{2}{3} - \frac{7}{12}$ **b.** $\frac{3}{4} - \frac{5}{6} \div (2 + 8)$

PROBLEM SOLVING

10. A ball of yarn contains $3\frac{1}{2}$ oz. How many ounces are in 7 balls of yarn?

11. Which has the greater product: $3\frac{1}{2} \times \frac{1}{3}$ or $6\frac{1}{4} \times \frac{4}{5}$?

12. A $7\frac{1}{2}$-ft board is cut into pieces that are $\frac{5}{6}$ ft long. How many pieces can be cut?

13. The price of cashew nuts is $9.75 for $1\frac{1}{2}$ lb. What is the price of 1 lb?

14. Find the cost of $5\frac{1}{5}$ lb of grapes if each pound costs $1.25.

15. Mrs. Lopez drove 130 miles in $3\frac{1}{4}$ hours. What was her average speed in miles per hour?

16. Bill has $9\frac{1}{3}$ c of blueberries. He is using half of them to make blueberry tarts. If each tart will have $\frac{2}{3}$ c of blueberries, how many tarts can Bill make?

Practice 8-1

Chapter 8

PROBLEM SOLVING

Use the double line graph for problems 1-3.

1. Find the temperature at the summit at 7 A.M.

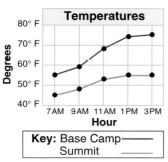

Temperatures

Degrees: 80° F, 70° F, 60° F, 50° F, 40° F

Hour: 7 AM 9 AM 11 AM 1 PM 3 PM

Key: Base Camp———
Summit ———

2. Estimate the difference in temperatures at 9 A.M.

3. Between what hours did the temperatures change least?

Use the given data for problems 4-6.

Ages of Guests at a Party

19	48	67	11	45
29	11	44	11	36
28	12	10	39	11
35	12	18	40	11

4. Organize the data in a frequency table.

5. Use the data to make:
   a. a line plot
   b. a stem-and-leaf plot

6. Find the mode, range, and median of the data in problem 4.

7. In 5 games, Jan scored 15, 18, 20, 12, and 20 points. What is her mean score?

8. Ali's scores on her first four math tests were 98, 84, 88, and 92. What score must she make on the fifth test to have the mean of the five tests equal 90?

9. The table shows how many people saw the circus. Make a double bar graph to display the data.

Attendance

Days	Matinee	Evening
Sun.	350	450
Mon.	100	150
Tue.	125	250
Wed.	300	350

10. High temperature readings during one 5-day period were 72, 63, 70, 68, and 77 degrees Fahrenheit. Give the range, median, mean, and mode for this set of data.

Practice 8-2

PROBLEM SOLVING

Use the circle graph for problems 1-4.

1. Which sport is most popular?

Sports Club Membership

Softball, Tennis, $\frac{1}{3}$, $\frac{1}{8}$, $\frac{3}{8}$, $\frac{1}{6}$, Volleyball, Soccer

2. What fractional part of the members prefer soccer or softball?

3. There are 96 members in the Sports Club. How many prefer softball?

4. Of the 96 members, which sport is preferred by exactly 12 members?

Use a coin and the spinner for problems 5-6.

5. Make a tree diagram to list all possible outcomes.

6. Find the probability.
   a. P (heads, 6)
   b. P (tails, <5)

Use a number cube labeled 1-6 to find the probability of each event.

7. a. P (1 or 3) b. P (7) c. P (1 through 6)

8. In a survey of 36 sixth graders, 16 have braces. Predict how many wear braces among the school district's 720 sixth graders.

9. A card is chosen from a bag containing cards labeled A, B, C, D, E. Then a second card is chosen. If the first card is not replaced, find P (B, D) and P (A, C or E).

Find the probability.

Experiment: A jar contains 2 red marbles, 4 green marbles, and 4 white marbles. One marble is drawn at random.

10a. P (green) b. P (red) c. P (blue)

11a. P (red or white) b. P (*not* blue)

513

Practice 9-1

Chapter 9

Use a protractor to draw an angle of the given measure.

1a. 70°   **b.** 135°   **c.** 15°

Classify each angle as *right*, *acute*, *obtuse*, or *straight*. Estimate its measure.

2a.   **b.**   **c.**

Determine whether the polygon is *concave* or *convex*. Then classify the polygon by the number of sides.

3a.   **b.**   **c.**

Complete each statement.

4. If $\angle ABC$ measures 43°, its complementary angle measures _?_ .

5. If $\angle XYZ$ measures 56°, its supplementary angle measures _?_ .

PROBLEM SOLVING

6. Draw an 80° angle. Then construct its bisector.

7. Draw hexagon *KLMNOP*. Then draw and name its diagonals.

8. Draw an isosceles right triangle. Label each angle by its measure.

9. In triangle *QRS*, $\angle Q = 39°$ and $\angle R = 76°$. Find the measure of $\angle S$.

10. Explain the difference between an equilateral triangle and a scalene triangle. Use a drawing.

11. Construct a line *CD* perpendicular to line *AB* at point *P* on $\overleftrightarrow{AB}$.

12. One of the angles of an isosceles triangle measures 68°. What are the measures of the other angles of the triangle?

Practice 9-2

Identify each quadrilateral.

1a.   **b.**   **c.**

2a.   **b.**

Identify each space figure.

3a.   **b.**   **c.**

4a.   **b.**   **c.**

Tell whether the figures are congruent. Write *Yes* or *No*.

5a.   **b.**

Which polygon is similar to *ABCD*?

6.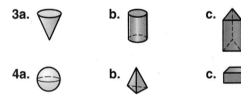

PROBLEM SOLVING

7. Draw circle *P*. Label chord $\overline{AB}$ that is a diameter. Draw central angle *APR* that is an obtuse angle.

8. Name a regular polygon that cannot be used alone in a tessellation.

9. Name the space figure that has 6 square faces, 8 vertices, and 12 edges.

10. $\triangle ABC$ is congruent to $\triangle XYZ$. Use a drawing to show this.

11. Use dot paper. Draw rhombus *MNOP*. Then draw its flip image, rhombus *ABCD*, flipped over a vertical line.

12. Which quadrilateral does *not* necessarily have opposite angles of equal measure?

13. Figure *A* is similar to figure *B*. Figure *B* is *not* similar to figure *C*. Is it possible that figure *A* is similar to figure *C*? Draw pictures to support your answer.

Practice 10-1

Complete.

1a. 7.3 m = $\dfrac{?}{}$ cm   **b.** 40 kg = $\dfrac{?}{}$ g

2a. 27.4 L = $\dfrac{?}{}$ kL   **b.** 73 dm = $\dfrac{?}{}$ m

3a. 15 ft = $\dfrac{?}{}$ yd   **b.** 14 pt = $\dfrac{?}{}$ qt

4a. 3 mi = $\dfrac{?}{}$ ft   **b.** 4T = $\dfrac{?}{}$ lb

Compute.

5a. 7 ft 11 in.
   + 4 ft 9 in.

b.   9 qt
   − 5 qt 1 pt

6a. (2 yd 5 in.) × 3   **b.** (3 h 20 min) ÷ 4

Measure each line segment to the nearest $\frac{1}{8}$ in. and $\frac{1}{16}$ in.

7.

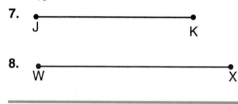

J   K

8.

W   X

Find the circumference and the area. Use 3.14 or $\frac{22}{7}$ for π.

9a. d = 5.3 m   **b.** $d = 2\frac{1}{3}$ yd   **c.** r = 9 m

PROBLEM SOLVING

10. A dump truck hauling 2 T of topsoil unloaded 1200 lb of it at a building site. How much topsoil was left in the truck?

11. Jon's science book weighs 780 g. How much do 7 such books weigh?

12. Jill is 56 in. tall and Leslie is 4 ft 10 in. tall. Who is taller? How much taller?

13. Draw line segment *FG* that is exactly $3\frac{7}{8}$ in. long.

14. A circular swimming pool has a diameter that measures $23\frac{1}{2}$ ft. Find its circumference.

15. How many cups are in $7\frac{1}{4}$ gallons?

Practice 10-2

Use formulas to find the perimeter and area.

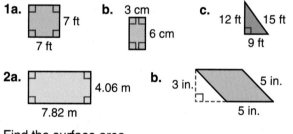

1a. 7 ft, 7 ft   **b.** 3 cm, 6 cm   **c.** 12 ft, 15 ft, 9 ft

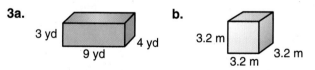

2a. 4.06 m, 7.82 m   **b.** 3 in., 5 in., 5 in.

Find the surface area.

3a. 3 yd, 9 yd, 4 yd   **b.** 3.2 m, 3.2 m, 3.2 m

PROBLEM SOLVING

4. The Pentagon building is a regular pentagon with a side of 276 m. If you walk around the Pentagon, how far will you walk in kilometers?

5. Find the volume of a cube that measures 11 cm along each edge.

6. A storage bin measures 9 ft high, 6 ft wide, and 5 ft deep. Find its volume.

7. Li is painting the walls, floor, and ceiling of her closet. The closet is shaped like a rectangular prism that measures $8\frac{1}{2}$ ft high, $5\frac{1}{2}$ ft wide, and 4 ft deep. Find the surface area.

8. Larry walked around a square field 6 times. If a side of the field measures 55 m, how far did Larry walk?

9. One can of paint will cover 144 ft². How many cans are needed to paint a wall that measures 26 ft by 15 ft?

10. The diameter of a half dollar is 30 mm. Find the area of one side of the coin in square centimeters.

Write each ratio in simplest form.

1a. 5 to 15   **b.** 4 to 24   **c.** 8 to 56

2a. 30 : 60   **b.** 27 : 42   **c.** 75 : 125

Find the missing term in each proportion.

3a. $\frac{5}{7} = \frac{25}{n}$ **b.** $\frac{3}{12} = \frac{n}{4}$ **c.** $\frac{1}{30} = \frac{1}{n}$

4a. $n : 3 = 0.5 : 5$ **b.** $2.6 : 1.7 = n : 10.2$

Write a proportion.
Then solve.

5. $\triangle ABC$ is similar to $\triangle XYZ$. Find the value of n.

Using the word EXCELLENT, write each ratio:

6a. E to all letters   **b.** consonants to vowels

7a. Ls to Es   **b.** vowels to consonants

PROBLEM SOLVING

8. Lucinda got 18 out of 20 spelling words correct on her quiz. What is the ratio of correctly spelled words to all words on the quiz?

9. The ratio of teachers to students at Hickory School is 1 : 23. There are 25 teachers at the school. How many students are at the school?

10. Four blank tapes cost $6.60. Find the cost per tape.

11. Laverne rode her bike 9 mi in 1 h. At that rate, how long will it take her to ride 30 mi?

12. If 2 dozen pencils cost $3.60, what will 3 pencils cost?

13. LeRoy makes 5 out of every 8 free throws at basketball practice. At that rate, how many free throws can he expect to make in 64 tries?

Practice 11-2

Write as a percent.

1a. 43 to 100   **b.** 7 to 100   **c.** 0.75

2a. 0.5   **b.** $\frac{1}{4}$ **c.** $\frac{7}{25}$

Write as a decimal and as a fraction or mixed number in simplest form.

3a. 30%   **b.** 5%   **c.** 81%

4a. 37.5%   **b.** 625%   **c.** 187%

5a. 150%   **b.** 1000%   **c.** 0.3%

Find the actual measurements.

6a. Scale width: $1\frac{1}{4}$ in.   **b.** Scale length: 4.5 cm
     Scale: $\frac{1}{2}$ in. = 10 mi   Scale: 1 cm = 120 km

PROBLEM SOLVING

7. In an enlarged model, 1 cm = 2 mm. A width of 5 cm is how many millimeters?

8. A road map uses a scale of 1 cm = 75 km. Find the map distance between two cities if the actual distance is 37.5 km.

9. A 9-ft telephone pole casts a 3-ft shadow. At the same time of day, Franny stands beside the pole. If she is 6 ft tall, how long is her shadow?

10. Spanish is spoken by 65 out of 100 people who work for a company. What percent of the workers speak Spanish?

11. Mrs. Gill spends 27% of her monthly income on rent and utilities. What percent of her monthly income is available for other purposes?

12. In a survey, 78% of the people said they approved of the idea of a new highway. What percent of the people did *not* approve?

13. In a ball-throwing contest, Jan scored 30 hits out of 35 tries. Al scored 0.85 of his throws and Roy's rate was 85.5%. Who had the best record?

Practice 12-1

Chapter 12

Compute mentally.

1a. 10% of 90   **b.** 50% of 60

2a. $33\frac{1}{3}$% of 75   **b.** 75% of 16

Compare. Use < or >.

3. 27% of 50 __?__ 20% of 50

4. 60% of 80 __?__ 60% of 160

Find the percent of the number.

5a. 45% of $900   **b.** 8% of $125

Find the percent.

6. What percent of 40 is 16?

7. 57.6 is what percent of 96?

PROBLEM SOLVING

8. Pam's soccer team won 15 out of 24 games. What percent of the games did Pam's team win?

9. Mel's Market sells oranges two ways: 5 for $.95 or 25¢ each. Which is the better buy?

10. In the football game, 62.5% of 24 passes were completed. How many passes were completed?

11. There are 360 members of the health club. $66\frac{2}{3}$% are adults; the rest are students. How many health club members are adults?

12. Of the 48 new library books, 12 are paperbacks. What percent are paperbacks?

13. The price of a personal stereo is 120% of last year's price of $35. Find the current price.

Practice 12-2

Compute the discount and sale price.

1a. basketball: $36   **b.** ice skates: $120
rate of discount: 20%   rate of discount: 35%

2a. swimsuit: $40   **b.** skateboard: $99
rate of discount: 25%   rate of discount: $33\frac{1}{3}$%

Find the sales tax and total cost.

3a. hat: $15   **b.** belt: $9.50
sales tax: 6%   sales tax: 5%

4a. car: $12,500   **b.** motorcycle: $4,800
sales tax: $5\frac{1}{4}$%   sales tax: $4\frac{1}{2}$%

PROBLEM SOLVING

5. Maureen earns a 4% commission on computer sales. Find her earnings on sales of $2600.

6. At a rate of commission of $6\frac{1}{2}$%, how much does Jack earn on sales of $8000?

7. A $480 VCR is on sale at 25% off. The sales tax is 3%. Find the total cost of the purchase.

8. A furniture store salesperson earns a monthly salary of $750 plus 5% commission on sales. How much does he earn in July if he sells $37,500 worth of furniture?

9. A telephone answering machine was reduced in price from $120 to $96. Find the rate of discount.

10. Draw a circle graph to show the cost of keeping a pet dog for one year.

food	$125	license	$25
vet visits	$75	dog toys	$25
boarding	$50		

11. Which is the better buy: a $60 watch at $\frac{1}{3}$ off or the same watch for $70 at 40% off?

12. The sales tax on a $48 item is $6. What percent is the sales tax?

13. Mr. Ali sold four used cars last week for $1400, $2140, $3300, and $1680. If his rate of commission was 5%, how much commission did he make on last week's sales?

Practice 13-1

Write the integer that matches each letter on the number line.

1a. J **b.** K **c.** L **d.** M

Express each as an integer.

2a. loss of 8 lb   **b.** 7 degrees warmer

3a. 50 ft below sea level   **b.** $25 raise

Name the opposite of each integer.

4a. $^-5$ **b.** $^+8$ **c.** $^-16$ **d.** $^+7$

Compare. Write $<$ or $>$.

5a. $^+6$? $^-6$ **b.** $^-3$? $^-7$

6a. 0 ? $^-2$ **b.** $^-5$? $^+1$

7a. $^-6$? $^-1$ **b.** $^+8$? $^-10$

Compute.

8a. $^+3 + {}^+8$ **b.** $^-2 + {}^+5$ **c.** $^-7 + {}^-8$

9a. $^+6 - {}^-5$ **b.** $^-8 - {}^-9$ **c.** $^-5 - {}^+3$

10a. $^+10 - {}^+4$ **b.** $^+3 - {}^-3$ **c.** $^+8 - {}^+12$

Problem Solving

11. Arrange in order from least to greatest: $^-5$; $^-8$; $^+3$; $^-4$; 0.

12. The price of a stock fell 8 points on Monday and rose 3 points on Tuesday. Find the total change over both days.

13. The temperature was $^-16°$F. It dropped 7 degrees. Find the new temperature.

14. Join the points $A\ (^+2, {}^+3)$, $B\ (^-2, {}^+3)$, $C\ (^-3, {}^-3)$, $D\ (^+1, {}^-3)$ in order on a grid. Move the figure left 2 units. Give the coordinates of the vertices of the slide image.

Practice 14-1

Translate to an expression or equation.

1. the difference between y and 16

2. c divided by 4 is 10.

Multiply or divide.

3a. $^-8 \times {}^-15$ **b.** $^-52 \div {}^+4$ **c.** $^-1 \div {}^-1$

Compare. Write $<$, $=$, or $>$.

4a. $^-12$? $^-11\frac{3}{4}$ **b.** $^+5\frac{1}{8}$? $^-20.2$

Graph on a coordinate grid. Use $^+2$, $^+1$, 0, $^-1$ for x.

5a. $y = 3x$ **b.** $y = 2x + 4$ **c.** $y = 4x - 2$

Solve.

6a. $j + 35 = 81$ **b.** $r - 8.3 = 6$

7a. $4p + 9 = 33$ **b.** $n + {}^-9 = {}^-30$

Problem Solving

Choose the correct equation. Then solve.

8. Jake bought a rake for $\frac{1}{2}$ price. He paid $7. What was the original price?

   a. $7 \times n = \frac{1}{2}$ **b.** $\frac{n}{2} = 7$ **c.** $7 \div 2 = n$

Translate into an equation. Then solve.

9. A number decreased by 7 is 30. Find the number.

10. 46 is the sum of 7 and a number multiplied by 3. Find the number.

Solve.

11. Which volume is greater: triangular prism with base area of 96 in.^{2}, height $= 40$ in. or cylinder with length of radius $= 6$ in. and height $= 8$ in. How much greater?

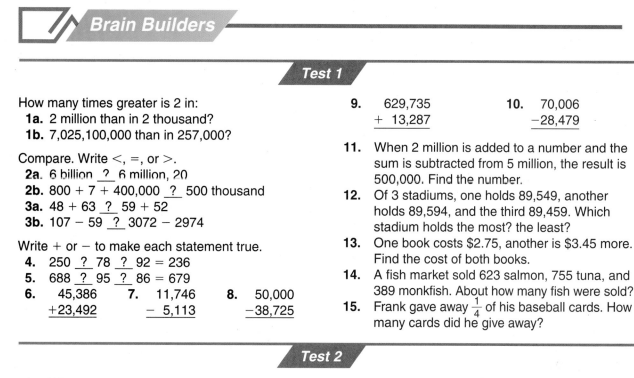
Test 1

How many times greater is 2 in:
1a. 2 million than in 2 thousand?
1b. 7,025,100,000 than in 257,000?

Compare. Write <, =, or >.
2a. 6 billion _?_ 6 million, 20
2b. 800 + 7 + 400,000 _?_ 500 thousand
3a. 48 + 63 _?_ 59 + 52
3b. 107 − 59 _?_ 3072 − 2974

Write + or − to make each statement true.
4. 250 _?_ 78 _?_ 92 = 236
5. 688 _?_ 95 _?_ 86 = 679

6. 45,386
+23,492

7. 11,746
− 5,113

8. 50,000
−38,725

9. 629,735
+ 13,287

10. 70,006
−28,479

11. When 2 million is added to a number and the sum is subtracted from 5 million, the result is 500,000. Find the number.

12. Of 3 stadiums, one holds 89,549, another holds 89,594, and the third 89,459. Which stadium holds the most? the least?

13. One book costs $2.75, another is $3.45 more. Find the cost of both books.

14. A fish market sold 623 salmon, 755 tuna, and 389 monkfish. About how many fish were sold?

15. Frank gave away $\frac{1}{4}$ of his baseball cards. How many cards did he give away?

Test 2

1. What is 250 less than 5000?
2. From 1 million take 127 thousand.
3a. 46 × 10 × 100   **b.** 832 × 10 × 100

Complete.
4a. 1700 + 30,000 = 30,000 + _?_
  b. 120 − 120 + _?_ = 40
5. 5100 + (200 + 3) = (5100 + _?_) + 3
6. 250 + (2600 + 750) = (2600 + 750) + _?_

Compare. Write <, =, or >.
7a. 382 × 36 _?_ 6 × 1735
  b. 63 × 489 _?_ 72 × 382
8a. 839 ÷ 31 _?_ 738 × 23
  b. 18,057 ÷ 221 _?_ 39,653 ÷ 481

Write × or ÷ to make each statement true.
9a. 6 _?_ 213 = 1278   **b.** 2240 _?_ 64 = 35
10a. 4218 _?_ 3 = 1406   **b.** 81 _?_ 88 = 7128

11. At $1.06 a gallon, what is the cost of 250 gallons of gasoline?

12. There are chickens and cows on the farm with a total of 36 feet. There are more cows than chickens. How many of each are on the farm?

13. There are 8125 books to be shipped. Each carton can hold 16 books. How many books will be in the carton that is not filled?

14. A train takes 102 h to travel 7140 mi. A plane takes 17 h to go the same distance. How far does the plane travel in 1 h?

15. A baker uses 5 eggs for each cake he bakes. For 25 cakes, how many dozen eggs are used?

Test 3

1. What is the smallest nonzero number divisible by 2, 3, 5, and 9?
2a. 36)824   **b.** 4986 ÷ 48
3. 6 × 2 × 3 × 2 ÷ (16 ÷ 4) = _?_

In the number 5602.347891 there are how many:
4a. millionths? **b.** hundreds? **c.** thousandths?

Round to the nearest hundredth and thousandth.
5. 13.0736   **6.** 2.1087

Compare. Write <, =, or >.
7. 6.812 _?_ 6.81   **8.** 14.006 _?_ 14.060
9. From 16.065 take 0.283.

10. Take 0.10207 from 1.

11. A package of 100 tea bags sells for $20.15. What is the approximate cost per tea bag?

12. Kim can run 6 km in 26 min. Nel can run 4 km in 15.5 min. To the nearest tenth of a minute, how much faster can Nel run 1 kilometer?

13. Lee had 380 pictures. She put a dozen pictures on each page. How many more pictures does she need to complete the last page?

14. John is 57 in. tall. Jim is 46.25 in. tall. How much taller than Jim is John?

15. What number multiplied by 0.7 gives 4.48?

1a. $8 + 0.7 - 0.53$ **b.** $2.6 - 0.3 + 1.9$
2a. 6.72×1.8 **b.** $7.856 \div 0.4$
3a. 2.06×17.04 **b.** $1.2\overline{)0.0672}$
4a. $0.4 \times (3.2 \times 1.7)$ **b.** $0.25 \div 0.005 - 50$

Order from least to greatest.
5. $627, 6.01 \times 10^3, 5.2 \times 10^2$
6. $3.9 \times 10^4, 4.1 \times 10^4, 39{,}500$

Write the multiplication rule.
7a. $0.73 \underline{\ ?\ } = 730$ **b.** $0.005 \underline{\ ?\ } = 0.5$
8a. $0.06 \underline{\ ?\ } = 0.0006$ **b.** $1.4 \underline{\ ?\ } = 0.0014$

Complete. Write $+$, $-$, $\times$, or $\div$.
9. $6.4 \underline{\ ?\ } 0.8 \underline{\ ?\ } 1.5 = 6.5$
10. $6.5 \underline{\ ?\ } 3 \underline{\ ?\ } 0.1 = 19.6$

11. Ted can swim the width of the pool in 15.24 s, while it takes Tom triple this time. How much less time does it take Ted?
12. Dana earned $6.78 on Monday. Each day after, she earned a dime more. How much did she earn in 5 days?
13. A plane averages 425.5 miles per hour. How far does it go in 6.25 hours?
14. Mr. Wilson used 5.8 gallons of gasoline to go 92.8 miles. How far did he go on 1 gallon of gasoline?
15. Earl earns $25.92 a week selling newspapers. How many weeks will it take him to earn $181.44?

Compare. Write $<$, $=$, or $>$.
1a. $2^2 \times 5 \underline{\ ?\ } 4^2$ **b.** $3^2 \cdot 2^2 \underline{\ ?\ } 2 \cdot 3^3$
2a. $\frac{2}{3} + \frac{4}{5} \underline{\ ?\ } 1\frac{1}{4}$ **b.** $11 - 8\frac{2}{3} \underline{\ ?\ } 3\frac{3}{4} - 2\frac{3}{8}$
3a. $\frac{2}{3} \underline{\ ?\ } 0.6$ **b.** $1.55 \underline{\ ?\ } \frac{13}{8}$

Complete.
4. The sum of $\frac{1}{8}$ and $\frac{3}{10}$ is close to $\underline{\ ?\ }$.
5. The sum of $\frac{7}{9}$ and $\frac{4}{15}$ is close to $\underline{\ ?\ }$.

What numbers between 20 and 28 have:
6. exactly 2 factors   **7.** exactly 4 factors
8. Take 0.172 from 2.
9. What is 243.75 more than 51.9?
10. From 5280 take 79.32.
11. What mixed numbers complete this pattern?
$2\frac{3}{4}, 4\frac{1}{8}, 3\frac{5}{8}, 5, 4\frac{1}{2}, \underline{\ ?\ }, \underline{\ ?\ }$

12. Zach's boat can travel 1 mile in 3.2 min. How many hours will it take to go 67.5 mi?
13. Ms. Sims gave out cans of juice. The 16 2nd graders each got 1. She got 5 more cans and gave 15 to the 3rd graders. After 3 girls returned theirs, she had 4 cans left. How many cans did Ms. Sims have originally?
14. A theater used 48 lb of popcorn on Thursday. On each of the next 3 days it used $1\frac{1}{4}$ times more than the day before. How much was used on Sunday?
15. Use 3 of these fractions: $\frac{1}{2}, \frac{1}{3}, \frac{1}{4}$, and $\frac{3}{4}$, to make a true number sentence.
$\underline{\ ?\ } - \underline{\ ?\ } + \underline{\ ?\ } = \frac{7}{12}$

Compare. Write $<$, $=$, or $>$.
1a. $0.1\overline{3} \underline{\ ?\ } 0.\overline{1}$ **b.** $\frac{5}{9} \underline{\ ?\ } 0.\overline{5}$
2a. $\frac{3}{4} + \frac{4}{5} \underline{\ ?\ } 1.55$ **b.** $4.5 - 3.8 \underline{\ ?\ } \frac{7}{9}$
3a. $2.0\overline{1} \underline{\ ?\ } 2\frac{1}{100}$ **b.** $\frac{1}{2}$ of $8 \underline{\ ?\ } 10 \div 2$

Complete.
4a. $2\frac{1}{6} \times \underline{\ ?\ } = 1$ **b.** $8 \times (\frac{3}{4} + \frac{1}{8}) = \underline{\ ?\ }$
5a. $\frac{1}{3}$ of $\underline{\ ?\ } = 15$ **b.** $\frac{3}{4}$ of $\underline{\ ?\ } = 15$
6. $\frac{7}{8}$ of 5.12 **7.** $\frac{4}{5}$ of 21.00

Complete. Write $+$ or $-$.
8. $\frac{4}{5} \underline{\ ?\ } \frac{1}{3} \underline{\ ?\ } \frac{1}{4} = \frac{13}{60}$ **9.** $6 \underline{\ ?\ } \frac{2}{3} \underline{\ ?\ } \frac{1}{2} = 5\frac{5}{6}$
10. $\frac{7}{8} \underline{\ ?\ } \frac{1}{5} \underline{\ ?\ } \frac{3}{20} = \frac{33}{40}$

11. From the sum of $3\frac{1}{2}$ and 0.75 subtract the sum of 3.25 and $\frac{3}{20}$.
12. Lynn bought 2 hams for a party. One ham weighed $10\frac{1}{8}$ lb; the other $12\frac{1}{4}$ lb. They ate $19\frac{1}{3}$ lb of ham. How many pounds are left?
13. Jan thought of a mixed number. She doubled it and subtracted $1\frac{3}{4}$. The result was $2\frac{11}{12}$. What was her original mixed number?
14. Of the 220 students who ate lunch, 145 ate salad, and 200 ate pizza. How many students ate both?
15. Jo, Tammy, and Drew run for president, vice president, and secretary. How many different ways might they be elected?

1a. $2\frac{5}{8} \times 4\frac{3}{7} \div 7\frac{3}{4}$ **b.** $7\frac{3}{5} + 1\frac{1}{10} - 3\frac{3}{4}$

2a. From $\frac{1}{3}$ take $\frac{5}{18}$.   **b.** Take $5\frac{5}{7}$ from $7\frac{13}{14}$.

Complete.

3a. $2\frac{5}{6} + \underline{\ ?\ } = 7\frac{1}{2}$ **b.** $\underline{\ ?\ } + 0.19 = 3$

4a. $\underline{\ ?\ } - \frac{4}{7} = 1\frac{1}{14}$ **b.** $\underline{\ ?\ } - 2.3 = 1.9$

5. Name 2 polygons that have 2 diagonals.

6. Name 2 straight angles.

7. Name 2 acute angles.

ABXD is a square.

$\overline{AB} \parallel \overline{XD}$ so:

8. $\overline{AD} \ \underline{\ ?\ } \ \overline{DX}$ **9.** $\overline{BX} \ \underline{\ ?\ } \ \overline{XD}$

10. Draw a slide, a flip, and a turn for this figure.

11. Without looking, you pick a card from cards numbered 1–9, and flip a coin. Find *P* (even, H).

12. At $9.49 each, how much will it cost to buy 2 shirts each for 3 boys?

13. Fran scored 93, 87, 95, 95, and 88 points. By how many points does the mode exceed the mean?

14. Flo had 1 quarter, 1 dime, 1 nickel, and 1 penny. She gave 2 coins away. How many different amounts might she have given away?

15. Thirty students speak at least 2 languages. Nineteen speak Spanish and English, 12 speak French and English, and 3 speak all 3 languages. How many students speak Spanish and French?

Test 8

Complete each analogy.

1. A straight angle is to a right angle as $\underline{\ ?\ }$ is to 90°.

2. $\angle ABC$ is to $\overrightarrow{BA}$ as $\angle RPT$ is to $\underline{\ ?\ }$.

3. ⬭⬭ is to 60 cans as ⬭ is to $\underline{\ ?\ }$ cans.

4. Certainty is to 1 as impossibility is to $\underline{\ ?\ }$.

5. Prism is to rectangular face as $\underline{\ ?\ }$ is to triangular face.

6.   12 yd 1 ft 3 in.
    − 8 yd 2 ft 10 in.

7.   5 h 21 min 48 s
    +3 h 39 min 15 s

8. (2 yd 2 ft 8 in.) × 2

9. (4 gal 1 qt 1 pt) ÷ 7

10a. $\frac{2}{5}$ of 1 km = $\underline{\ ?\ }$ m   **b.** $\frac{3}{4}$ of 2 ft = $\underline{\ ?\ }$ in.

11. When a store closed there were 6 newspapers left. If 42 people came in the store and every third person bought a newspaper, how many newspapers were there when the store opened?

12. On a circle graph, $\frac{1}{3}$ of Ed's day is spent sleeping, $\frac{1}{5}$ playing, $\frac{1}{10}$ eating, $\frac{1}{5}$ studying, and the rest reading. How many hours does Ed read?

13. In figure *EPRM* $\angle E = 140°$, $\angle R$ is half $\angle E$, and $\angle P$ is 20° less than $\angle R$. Find the measure of $\angle M$.

14. What is the probability of choosing a letter before N from a set of 26 alphabet cards?

15. How many ways can Leon draw 4 different quadrilaterals side by side on the board?

Test 9

1. $9\frac{1}{2} + 6\frac{3}{4} + 8\frac{1}{8}$ **2.** $6\frac{4}{7} - 2\frac{1}{3} + 1\frac{5}{7}$

Complete.

3a. $\frac{2}{3}$ of $\underline{\ ?\ }$ ft = 16 in.   **b.** $\frac{3}{4}$ of $\underline{\ ?\ }$ lb = 18 oz

4a. $0.14 \times 250 = \underline{\ ?\ }$ **b.** $0.2 \times 150 = \underline{\ ?\ }$

5. 50% is to 200% as $\frac{1}{2}$ is to $\underline{\ ?\ }$.

6. Radius is to diameter as 50% is to $\underline{\ ?\ }$.

7. 0.75 is to 75% as $\underline{\ ?\ }$ is to 4%.

8. 3 out of 5 is to 60% as 1 out of 8 is to $\underline{\ ?\ }$ %.

9. 25% is to 75% as 10% is to $\underline{\ ?\ }$ %.

10. 800% is to 5 + 3 as $\underline{\ ?\ }$ % is to $3 - 1\frac{1}{2}$.

11. Find the perimeter.

12. Find the area.

13. How much less is the volume of a rectangular prism 8 m by 14 m by 8.5 m than a cube 10.2 m on each edge?

14. A rectangular field is 178 ft long and 145 ft wide. What is its perimeter in yards?

15. You put $1000 in a bank. Each year the bank adds $\frac{1}{10}$ of your total savings to the account. How many years will it take your account to be greater than $1500?

1a. $(\frac{1}{5}$ of 30$) - (0.3$ of 20$)$ **b.** $\frac{2}{7}$ of \$1.54
2a. Take 1.046 from 3.1.   **b.** From 1.5 take $\frac{1}{4}$.
3a. 40% of \$2.00   **b.** $12\frac{1}{2}$% of \$7.20

Find each missing dimension. Use $\pi \approx 3.14$.
4. $C \approx 15.7$ in., $d \approx$ ___?___
5. $A = 16$ cm², $\ell = 6.4$ cm, $w =$ ___?___
6. $V = 128$ m³, $\ell = 8$ m, $w = 4$ m, $h =$ ___?___

Complete each analogy.
7. Circle is to πr^2 as triangle is to ___?___ .
8. Double is to 200% as triple is to ___?___ .
9. 60% of 30 is to 18 as ___?___ % of 20 is to 24.
10. 75% of 1 lb is to 12 oz as ___?___ % of 1 h is to 12 min.

11. $\frac{7}{8}$ of the distance between two towns is 147 mi. What is the total distance?
12. What is the ratio of the area of a square 6 cm on each side to the area of a rectangle that is 8 cm by 5 cm?
13. Six pounds of coffee cost \$19.74. How much will 4 lb cost?
14. A girl who weighed 97 lb lost 10% of her weight. How much did she weigh then?
15. What is the area of the shaded region of the rectangle?

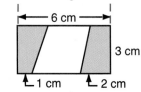

Solve for n.
1a. $3 : n :: 1.8 : 6$ **b.** $4 : 3.2 :: n : 16$
2a. $n : 1\frac{1}{2} :: 3 : \frac{1}{4}$ **b.** $2\frac{1}{4} : 9 :: \frac{1}{2} : n$
3. $^-6 + (^-3 - ^-1) = n$
4. $^-12 - (^-8 + ^+4) = n$
5a. n% of 7 is 14   **b.** n% of 1.8 is 0.9
6a. $33\frac{1}{3}$% of n is 11   **b.** 600% of $n = 42$
7a. 15% of $0.2 = n$ **b.** 1% of $50 = n$
8a. n% of $\frac{3}{4} = \frac{1}{4}$ **b.** n% of $1.2 = 3$

Find the unit cost.
9a. $1\frac{1}{2}$ lb for \$1.26   **b.** 2.5 L for \$3.60

Find the tax.
10a. 6% tax on \$17.35.
   b. 5.5% tax on \$127.40.

11. A blazer listed at \$44 was sold for 25% off. What was the selling price?
12. On a map the distance from City A to City B is 3.4 in. The scale is 1 in. = 25 mi. What is the actual distance?
13. The temperatures in 5 different cities for one day were 14°F, $^-$20°F, 31°F, 45°F, and $^-$5°F. Find the range. Find the mean.
14. The coordinates of the vertices of $\triangle FHM$ are (1,0), (2,5), and (6,0). Find the area of $\triangle FHM$.
15. A man sold a house for \$150,000. If he received \$7500 commission, what rate of commission did he receive?

Solve for n.
1a. $n - 7 = 6 \div 2$ **b.** $n + 5 = 7 \times 3$
2a. $3n = 7 + 2$ **b.** $n \div 6 = 4 + 3$
3a. $\frac{1}{2}n = 7\frac{1}{2}$ **b.** $n - 3\frac{1}{2} = 5\frac{1}{4}$
4a. $^-20 + n = ^+5$ **b.** $n - ^-6 = ^+10$
5a. $n + ^-2 = ^+8$ **b.** $^+3 - n = ^-5$

Evaluate each expression if $a = 6$, $b = 1.2$.
6. $2a + b$ **7.** $b \div a$
8. $a - 2b$ **9.** $a \times b - 7.14$

Write + or − to make each sentence true.
10a. $^-11$ _?_ $^-15$ _?_ $^+1 = ^+3$
   b. $^-1$ _?_ $^+3$ _?_ $^-7 = ^-11$

11. How many even integers are between $^-30$ and $^+30$?
12. Jim wrote down an integer. He doubled it and subtracted $^+11$. His answer was $^-5$. What integer did Jim write down?
13. If a and b represent integers between $^-2$ and $^+1$, how many equations can be written for $a - b = ^-1$?
14. Phan doubled an integer and subtracted $^-5$ from it. The result was $^-9$. Find Phan's integer.
15. In which equation does x have the greater value: $x + 2\frac{1}{2} = 5\frac{1}{4}$ or $x - 1.7 = 2.05$?

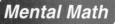
Set 1

1. Name the period. 12,<u>452</u>; <u>321</u>,589; <u>6</u>,109,372; 710,448; 626,<u>001</u>
2. Add 1 million to: 4,375,210; 508,119,042; 62,137,429; 506,317,286,902
3. Give the value of the underlined digits: 1,<u>6</u>24,<u>5</u>90,7<u>8</u>3
4. Add 2 to: 9, 6, 19, 16, 8, 28, 7, 17
5. Subtract 3 from: 10, 12, 25, 19, 32, 42
6. In the numeral 468, what is the value of 6?
7. From the sum of $9 + 6$ take $3 + 4$.
8. Mary had $3.50. She spent $1.30. How much did she have left?

9. Which is greater: 36,101 or $30,000 + 6000 + 100 + 10 + 0$?
10. What period is 25 in 25,607,384,590,012?
11. Of the numbers 36,803,251, 36,308,215, and 36,803,215, which is the greatest?
12. Name the addition property used.
    $a + (b + c) = (a + b) + c$
13. The bookstore has 48 cat calendars and 61 dog calendars. Estimate the number of calendars.
14. What is the value of 8 in 30,820?
15. On Saturday, 478 people went to the show. If 250 went to the A.M. show, about how many went to the P.M. show?

Set 2

1. Round to the nearest million. 3,733,415; 2,165,899; 7,998,115; 31,236,709
2. Round to the place of the underlined digit. 8<u>3</u>,724; 6<u>2</u>8,457; 3,<u>2</u>96,485
3. Find the missing number. $16 + \underline{\ ?\ } = 16$
    $5 + 8 = \underline{\ ?\ } + 5$ $(3 + 2) + 4 = 3 + (\underline{\ ?\ } + 4)$
4. Estimate. $18 + 19 + 17$ $32 + 29 + 25$
    $48 + 11 + 13$ $56 + 12 + 25$ $65 + 12 + 21$
5. Subtract 4 from: 9, 7, 16, 12, 23, 27
6. In the numeral 8,643,729,065 what is the value of 4?
7. Which 4 coins make $.75?
8. How much less than $9 + 8$ is $7 + 3$?

9. At a sale, the price of a rug was changed from $32 to $27. How much was it reduced?
10. Choose the operations:
    $200 \underline{\ ?\ } 75 \underline{\ ?\ } 25 = 250$
11. What is the difference in cents between 8 dimes and 8 nickels?
12. The price of eggs was $1.08 a dozen. A week later, it was $1.23. How much had the price per dozen increased?
13. If the sum of 2 numbers is 13, and one addend is 4, what is the other?
14. Round 8325 to the nearest thousand.
15. Complete.
    $10 + 4 + 3 - 6 + 9 - 11 - 3 = \underline{\ ?\ }$

Set 3

1. $1 \times \underline{\ ?\ } = 11$ $13 \times \underline{\ ?\ } = 0$
   $36 \times \underline{\ ?\ } = 36$ $42 \times \underline{\ ?\ } = 0$
   $1 \times \underline{\ ?\ } = 25$ $50 \times \underline{\ ?\ } = 0$
2. Add 4 to: 6, 7, 27, 8, 18, 9, 39, 16
3. Subtract 2 from: 8, 18, 13, 23, 17, 37
4. Name the first 4 non-zero multiples of: 6, 10, 5, 4, 7, 9, 8
5. Multiply by 2, then add 3: 4, 8, 7, 9, 3, 10, 0, 5, 6
6. How much less than 2 times 8 is 2 times 7?
7. At 2 for $.35, what will 6 marbles cost?
8. How much greater than 3×0 is 3×6?

9. At $.20 each, find the cost of 30 stamps.
10. From the difference between 16 and 7 take 2×3.
11. At 40 miles per hour, how far will a ship go in 8 hours?
12. Van's 50 cents in change contains 1 quarter, 1 nickel, and $\underline{\ ?\ }$ dimes.
13. Dan packs 77 boxes per hour. At this rate, about how many boxes will he pack in 12 hours?
14. $40 \times 30 \div 40 \times 300 = \underline{\ ?\ }$
15. The nursery plants 240 trees in each of 20 rows. How many trees are planted?

1. Multiply by 8: 6, 4, 9, 2, 7, 3, 80, 60, 40, 90, 20, 70, 30, 50
2. Add 3 to: 9, 8, 17, 27, 16, 36, 47, 19
3. Give the standard numeral. 10^3, 10^2, 10^5, 10^4, 10^6, 10^1
4. Give the exponent form: $3 \times 3 \times 3$; $5 \times 5 \times 5 \times 5$; 9×9; $7 \times 7 \times 7 \times 7 \times 7$
5. Multiply by 100: 6, 9, 10, 13, 19, 27, 32, 45, 63, 59, 83, 94, 50, 76
6. What is 10,000,000 as a power of ten?
7. Any number to the first power is __?__ .

8. If 4 bars of soap cost $1.00, how much will 12 cost?
9. Estimate the cost of 6 CDs at $9.95 each?
10. There are 30 children per class and 13 classes. How many children are there altogether?
11. At $4.05 an hour, how much will Bob earn in 9 hours?
12. What is the difference in cents between 9 dimes and 9 nickels?
13. Multiply 8 by 405.
14. At $.60 a meter, what will 8 meters of paper cost?
15. At $2 a yard, how many yards of drip-dry cotton can be bought for $18?

1. $1\overline{)7}$ $10\overline{)0}$ $3\overline{)0}$ $67\overline{)67}$ $1\overline{)18}$
   $28\overline{)0}$ $1\overline{)136}$ $258\overline{)258}$
2. $5\overline{)25}$ $5\overline{)250}$ $5\overline{)2500}$ $5\overline{)25,000}$
   $3\overline{)210}$ $4\overline{)3200}$ $6\overline{)240}$ $2\overline{)2000}$
3. Divide by 4, then add 3 to: 4, 12, 0, 20, 16, 28, 36, 32, 8, 24
4. Divide by 8: 9, 11, 13, 15, 17, 19, 21, 23
5. $2\overline{)412}$ $3\overline{)618}$ $4\overline{)328}$ $5\overline{)205}$
   $6\overline{)612}$ $7\overline{)147}$ $2\overline{)608}$ $3\overline{)312}$
6. What number divided by 2 will give 9 for the quotient and 1 for the remainder?
7. At $.92 for 4 pencils, find the cost of 2.
8. How much less than 3×1 is 2×0?

9. Rudy bought 8 meters of ribbon. If she paid $.96 for the ribbon, what was the cost of 1 meter of ribbon?
10. Divide 2432 by 4.
11. On Monday, 2076 students came to the zoo. On the average, each bus holds 49 students. About how many buses were needed?
12. At 50 mph, how long will it take to drive 300 miles?
13. Divide 3216 by 8, then subtract 2 from the quotient.
14. How much greater than 8×1 is $8 \div 1$?
15. $6 \times 2 + 8 \div 4 - 3 =$ __?__

1. Which are divisible by 3? 41, 57, 68, 363, 245, 108, 417, 239, 512, 125
2. Which are divisible by 9? 167, 2514, 3620, 428, 396, 1539, 4335, 2007
3. Divide by 7: 16, 30, 8, 37, 24, 44
4. $20\overline{)640}$ $40\overline{)600}$ $30\overline{)750}$ $50\overline{)3100}$
5. Divide by 9: 10, 19, 28, 37, 11, 20, 29
6. At $.60 for a half dozen, find the cost of 3 pencils.
7. Divide 4963 by 7.
8. Which number is divisible by 2, 3, 5, 9, and 10? 109, 364, 575, 990

9. What number divided by 3 will give 8 for a quotient and 2 for a remainder?
10. A tank containing 28,200 gallons of fuel must be emptied into smaller tanks, each holding 300 gallons. How many smaller tanks are needed?
11. From 8×7 take $108 \div 9$.
12. The Scotts pay $2832 a year for insurance. How much is that per month?
13. In 9 hours a rocket covered 7200 km. What was its average speed per hour?
14. Ashlee paid $4.75 for a hat that had been reduced by $1.25. What was the original price of the hat?
15. How many inches are in 8 feet?

1. Give the value. 0.5**6**2, 32.**4**, 1.43**7**9, 0.00**4**, 35.178**3**, 8.0267**1**, 4**9**.7
2. Read. 9.006, 21.35, 1.6285, 724.6, 3.90, 4.00763, 6.000248
3. Round to the nearest hundredth. 0.762, 2.8975, 0.261, 0.538, 16.085, 0.1992
4. Order from least to greatest. 0.4, 0.41, 4.0; 3.7, 3.3, 3.9; 52, 5., 520; 7.13, 7.31, 7.11
5. Compare. Use <, =, or >. 12.31 ? 1.23   92.3 ? 92.33   0.54 ? 0.6
6. In 3,178,242.377098, there are how many: millionths? ten thousandths?
7. From 1 take 0.7.   **8.** Add 0.3 and 0.7.

9. Estimate the total cost of a $59.95 dress and a $17.98 skirt.
10. What is the sum: 6 + 0.67 + 16.13?
11. Kimo wants to run 12 km. He has already run 7.8 km. How much farther does he have to run?
12. Place the decimal point to make the answer reasonable. Al's math score general average is 964.
13. Round 92.03729 to the nearest thousandth.
14. Which is greater: $316.25 or $361.25?
15. Fay spent $3.75 for lunch on Mon., Tues., and Wed. and $2.90 on Thurs. and Fri. Estimate how much money she had left from twenty dollars.

1. 0.6 + 0.06   7.2 + 7.02   0.3 + 1.4   0.5 + 2.1   1.30 + 0.04   0.12 + 0.07
2. 0.9 − 0.09   5.5 − 2.3   1.08 − 0.8   2.004 − 1.001   4.333 − 4.003
3. Compare. Use <, =, or >. 3.07 − 1 ? 2; 2.3 + 1.01 ? 3.4; 2.319 + 1.06 ? 3.379
4. Multiply by 10. 0.12, 0.74, 0.3, 0.11, 0.04
5. Multiply by 100. 0.2, 0.05, 0.89, 0.132
6. What number is 3.75 greater than 6.25?
7. The sum of 2.06 and another number is 9.37. Find the other number.
8. The dress factory uses 2.4 yd of fabric to make each dress. Estimate how many yards are needed to make 285 dresses.

9. The original price of a jacket was $80. It was reduced $7. For how much was it sold?
10. Find $\frac{1}{9}$ of 54, and subtract the result from 20.
11. At $1.20 a dozen, find the cost of 5 dozen eggs.
12. Find the sum of 1.8, 2, and 0.2.
13. In the numeral 6.047, what is the value of 4?
14. Complete the pattern. 0.524, 5.24, 52.4, $\underset{?}{\_\_}$
15. At $.42 each, about how many folders can be bought with $19.95?

1. Multiply by 1000. 0.1, 0.004, 0.178, 0.063, 0.5, 0.35, 0.2436, 0.789201, 0.0891
2. Multiply by 0.02. 0.3, 0.01, 0.5, 0.9, 0.08, 0.4, 0.07, 0.11, 0.06
3. Divide by 100. 300, 532, 483.1, 60.2, 8.2, 3.18, 0.06, 0.4, 0.9, 0.15
4. Divide by 10. 1.13, 24.8, 554.2, 47.6, 20, 0.3, 0.28, 0.64, 0.004
5. Divide by 1000. 6300, 700, 235.7, 4.88, 0.007, 0.08, 0.1, 0.99, 8.72
6. Van weighs $47\frac{1}{2}$ lb and Sam weighs $47\frac{7}{8}$ lb. What is the difference in their weights?

7. Express $\frac{625}{1000}$ as a decimal.
8. Each corsage uses 2.5 ft of ribbon for a bow. How many bows can be made from 62.5 ft of ribbon?
9. What number multiplied by 0.6 will give a product of 4.32?
10. A complete dictionary has a mass of 5.85 kg. A large telephone book has a mass of 5625 g. Which has greater mass?
11. At $4.50 a pound, what will 8 pounds of nuts cost?
12. Jim ran 2 km in 25 min. How far did he run in 1 minute?
13. Which is greater: 6×10^8 or 6.2×10^7?
14. Write in scientific notation: 47,000,000.
15. Compute. $0.03 \times 2 + 0.03 \times 4 = \underset{?}{\_\_}$

1. Express each as closer to 0 or to 1.
$\frac{2}{13}, \frac{15}{16}, \frac{7}{8}, \frac{3}{20}, \frac{2}{25}, \frac{5}{6}, \frac{4}{15}, \frac{14}{17}$

2. $\frac{1}{3} = \frac{?}{9} = \frac{?}{15} = \frac{?}{12} = \frac{?}{6} = \frac{?}{18} = \frac{?}{27}$

3. Identify as prime or composite. 13, 15, 21, 11, 7, 31, 18, 26, 32, 41, 54

4. Name the factors of: 6, 14, 3, 8, 12, 9, 11, 18, 10, 17

5. Find the GCF.　8 and 14　36 and 48
9 and 30　28 and 35　6 and 18

6. Of the 20 animals in the pet shop, 9 are dogs. What fractional part are dogs?

7. Take 0.4 from 2.1.

8. Complete. $\frac{34}{56} = \frac{68}{?}$

9. The prime factorization for 60 is ___?___
   a. 12×5　b. $2 \times 2 \times 3 \times 5$
   c. $2 \times 2 \times 2 \times 5$

10. Kim sleeps 8 h a day. What part of the day does Kim sleep?

11. If $\frac{3}{8}$ of a class are girls, what fractional part are boys?

12. Complete the pattern.
$\frac{1}{5}, \frac{3}{5}, \frac{2}{5}, \frac{4}{5}, \frac{3}{5}, 1, \frac{4}{5}, 1\frac{1}{5}, 1, 1\frac{2}{5}, \underline{?}$

13. From the sum of $\frac{5}{8}$ and $\frac{7}{8}$ take 1.

14. Express 0.05 as a fraction in simplest form.

15. How many dozen eggs are in 42 eggs?

1. Express in simplest form. $\frac{20}{28}, \frac{16}{32}, \frac{7}{21}, \frac{35}{45}$

2. Express as a mixed number. $\frac{9}{7}, \frac{13}{6}, \frac{19}{8}$

3. Find the LCM.　3 and 5　2 and 8
4 and 6　5 and 9　3 and 7　10 and 12

4. Give the improper fraction. $1\frac{1}{5}, 2\frac{7}{8}, 4\frac{3}{5},$
$1\frac{5}{6}, 3\frac{1}{2}, 2\frac{2}{3}, 3\frac{4}{9}, 4\frac{5}{7}$

5. Find the LCD.　$\frac{5}{6}$ and $\frac{1}{12}$　$\frac{2}{3}$ and $\frac{5}{9}$
$\frac{1}{4}$ and $\frac{9}{16}$　$\frac{11}{14}$ and $\frac{1}{7}$　$\frac{4}{5}$ and $\frac{19}{20}$

6. In the gym $\frac{1}{5}$ of the people are swimming and $\frac{3}{10}$ of the people are jogging. Which sport has more people?

7. How many thirds are there in $3\frac{1}{3}$?

8. Order from least to greatest. $\frac{1}{2}, \frac{1}{6}, \frac{2}{3}$

9. Order from greatest to least. $1\frac{1}{4}, \frac{6}{4}, 1\frac{3}{4}$

10. Give the equivalent decimal for $\frac{9}{10}$.

11. Desiree has $\frac{3}{4}$ of a dollar. How much money does she have?

12. Give the equivalent decimal for $\frac{1}{4}$.

13. How much greater than $9\overline{)0.27}$ is $2\overline{)0.08}$?

14. Express 7.5 ft as a mixed number.

15. Express $\frac{2}{3}$ as a repeating decimal.

1. Compare. Use <, =, or >. $\frac{1}{3} \; ? \; \frac{4}{12}$
$\frac{2}{5} \; ? \; \frac{7}{10}$　$\frac{1}{8} \; ? \; \frac{3}{16}$　$\frac{3}{4} \; ? \; \frac{1}{8}$

2. To 0.1 add: 0.4, 0.8, 0.03, 0.05, 0.25, 0.75

3. From 1 take: 0.6, 0.2, 0.8, 0.7, 0.1, 0.9

4. $\frac{1}{3} + \frac{2}{3}$　$\frac{3}{4} + \frac{1}{4}$　$\frac{4}{5} + \frac{1}{5}$
$\frac{3}{4} - \frac{1}{4}$　$\frac{3}{3} - \frac{2}{3}$　$\frac{5}{6} - \frac{2}{6}$

5. $\frac{3}{8} + \frac{1}{2}$　$\frac{1}{4} + \frac{1}{12}$　$\frac{1}{6} + \frac{1}{3}$
$\frac{1}{5} + \frac{1}{10}$　$\frac{1}{16} + \frac{1}{2}$　$\frac{2}{3} + \frac{1}{9}$

6. Simplify $\frac{12}{24}$ and add $1\frac{1}{4}$ to the result.

7. Julia has two ribbons. One is $1\frac{1}{9}$ yd, the other $1\frac{2}{3}$ yd. Which is closer to 1 yd?

8. Carol ran $1\frac{1}{2}$ mi, 1 mi, $2\frac{1}{2}$ mi, and $\frac{1}{2}$ mi. How many miles did she run in all?

9. To $\frac{1}{8}$ add $\frac{3}{4}$.

10. From 2 take $1\frac{3}{5}$.

11. How many $\frac{7}{8}$ are in $1\frac{3}{4}$?

12. Bruce walked $\frac{6}{7}$ mi on Monday. He walked $\frac{9}{14}$ mi less on Tuesday. How far did he walk on Tuesday?

13. One snake measures $12\frac{1}{2}$ ft. Another snake measures $3\frac{1}{3}$ ft longer. How long is the second snake?

14. Leon worked $4\frac{1}{2}$ h in the A.M. and $5\frac{1}{2}$ h in the P.M. Sam worked $6\frac{1}{4}$ h in the A.M. and $4\frac{1}{5}$ h in the P.M. Who worked longer?

15. How much greater than $\frac{33}{5}$ is $7\frac{1}{5}$?

1. $(2\frac{1}{3} + 1\frac{1}{3}) - 1\frac{2}{3}$ $3\frac{1}{5} - (1\frac{4}{5} - \frac{3}{5})$

    $(1\frac{1}{4} + 4\frac{3}{4}) + \frac{1}{4}$ $(5\frac{5}{6} - 3\frac{1}{6}) + 1\frac{1}{6}$

2. $\frac{1}{4} - \frac{1}{8}$ $\frac{1}{2} - \frac{1}{6}$ $\frac{1}{3} - \frac{1}{9}$

   $\frac{1}{2} - \frac{1}{4}$ $\frac{1}{3} - \frac{1}{27}$ $\frac{1}{2} - \frac{1}{10}$

3. $18 \times \frac{1}{3}$ $12 \times \frac{1}{4}$ $21 \times \frac{1}{7}$ $30 \times \frac{1}{5}$

   $7 \times \frac{5}{7}$ $10 \times \frac{3}{10}$ $6 \times \frac{5}{6}$ $4 \times \frac{3}{4}$

4. $\frac{2}{5} \times \frac{1}{4}$ $\frac{1}{3} \times \frac{1}{2}$ $\frac{5}{6} \times \frac{1}{4}$ $\frac{1}{6} \times \frac{1}{2}$

   $\frac{3}{5} \times \frac{1}{3}$ $\frac{3}{10} \times \frac{1}{2}$ $\frac{3}{4} \times \frac{2}{3}$ $\frac{1}{4} \times \frac{2}{3}$

5. Give the reciprocal. $\frac{1}{7}$, 16, $\frac{2}{3}$, 9, $\frac{4}{5}$, 20

6. How much greater than $\frac{1}{9}$ of 63 is $\frac{1}{7}$ of 63?

7. Nan uses $1\frac{1}{5}$ skeins of red yarn, $1\frac{3}{5}$ skeins of blue, and $1\frac{4}{5}$ skeins of white to make an afghan. How many skeins is that?

8. How many pieces $\frac{3}{4}$ m long can be cut from 6 m of string?

9. Find the value of, $\frac{3}{4} \times \frac{1}{3} \times 0$.

10. Solve. $35 \times \underline{\ ?\ } = 1$

11. On Monday, $4\frac{1}{2}$ gal of juice were served. On Tuesday, $1\frac{1}{4}$ times as much juice was served. How much juice was served on Tuesday?

12. How much less than $9\frac{1}{9}$ is $8\frac{6}{9}$?

Write what comes next in each pattern.

13. $\frac{1}{400}, \frac{1}{200}, \frac{1}{100}, \frac{1}{50}, \frac{1}{25}, \frac{2}{25}, \frac{4}{25}, \frac{8}{25}, \underline{\ ?\ }$

14. $32, 16, 8, 4, 2, \frac{1}{2}, \frac{1}{4}, \frac{1}{8}, \underline{\ ?\ }$

15. $1.5, 1.6, 1.8, 1.9, 2.1, 2.2, 2.4, \underline{\ ?\ }$

1. $7 \div \frac{1}{4}$ $3 \div \frac{1}{2}$ $6 \div \frac{1}{4}$ $3 \div \frac{3}{5}$

   $4 \div \frac{4}{7}$ $5 \div \frac{5}{9}$ $8 \div \frac{8}{15}$ $\frac{1}{2} \div \frac{1}{2}$

2. Find $\frac{1}{2}$ of: $\frac{2}{3}, \frac{2}{9}, \frac{4}{7}, \frac{4}{5}, \frac{6}{7}, \frac{6}{17}, \frac{8}{9}, \frac{8}{15}$

3. Multiply by 100. 0.25, 0.35, 0.42, 0.64

4. Divide by 4. 0.028, 0.004, 0.032, 0.020

5. $\frac{3}{4} + \frac{1}{4}$ $\frac{3}{5} + \frac{3}{5}$ $\frac{3}{5} \times \frac{3}{5}$ $\frac{3}{5} \div \frac{3}{5}$

6. $(\frac{5}{6} \div \frac{5}{6}) + 0.9 = \underline{\ ?\ }$

7. Regina has $2\frac{1}{2}$ yd of yarn. Into how many pieces $\frac{1}{2}$ yd long can the yarn be cut?

8. Don had $3\frac{1}{2}$ pizzas to share equally among 28 people. How much pizza did each person receive?

9. Zack had $7.25. He spent $\frac{1}{5}$ of his money. How much did he have left?

10. Find $\frac{1}{5}$ of 50 and subtract the result from 20.

11. From $1\frac{3}{4}$ take $1\frac{1}{2}$.

12. One twin weighed $5\frac{1}{4}$ lb at birth. The other twin weighed $1\frac{1}{3}$ times as much. How much did the second twin weigh?

13. How many sixths are there in $4\frac{5}{6}$?

14. Dorothy spent $\frac{1}{4}$ of the $33.56 she had saved. How much did she spend?

15. Add 0.6 to $\frac{1}{2}$ of 0.6.

1. Decimal points must be moved how many places?

   $0.2\overline{)4}$ $0.4\overline{)8.8}$ $0.03\overline{)6}$ $3\overline{)0.009}$

2. Multiply by 7. 0.2, 0.7, 0.9, 0.4, 0.8, 0.6, 0

3. Simplify. $\frac{5}{25}, \frac{5}{15}, \frac{5}{50}, \frac{5}{35}, \frac{5}{45}, \frac{5}{10}, \frac{5}{40}$

4. Express as a mixed number. $\frac{10}{9}, \frac{14}{9}, \frac{12}{9}$

5. Find $\frac{1}{3}$ of: 12, 21, 27, 3, 15, 24, 30, 6, 18

6. What is the difference in cents between 3 quarters and 7 nickels?

7. In the numeral 8.014 what is the value of 4?

On 5 different days, the class collected 24, 32, 28, 36, and 40 pledges.

8. Find the median.

9. Find the range.

10. Find the mode.

11. Key: Each ⬭ = 20 jars of honey. How many jars are there? ⬭⬭⬭⬭⬭⬭

12. What type of graph depends on the data adding up to 100%?

13. The probability of an event that is impossible is $\underline{\ ?\ }$.

14. The probability of an event that is certain is $\underline{\ ?\ }$.

15. In the last 3 ballgames, Emily scored 7, 11, and 15 points. What was the average number of points scored?

Set 16

1. Express as an improper fraction. $7\frac{1}{8}, 5\frac{3}{8},$ $2\frac{7}{8}, 8\frac{5}{8}, 6\frac{7}{8}, 9\frac{3}{8}, 6\frac{5}{7}$

2. $\frac{1}{2} + \frac{1}{2}$ $\frac{1}{2} - \frac{1}{2}$ $\frac{1}{2} \times \frac{1}{2}$ $\frac{1}{2} \div \frac{1}{2}$

3. $0.2 + 0.1$ $0.03 + 0.01$ $0.08 + 0.01$
   $0.3 + 0.7$ $0.03 + 0.07$ $0.4 + 0.1$

4. To $\frac{1}{2}$ add: $\frac{3}{4}, \frac{1}{6}, \frac{3}{8}, \frac{2}{5}, \frac{1}{9}, \frac{2}{9}$

5. From 1 take: $\frac{1}{8}, \frac{1}{9}, \frac{1}{4}, \frac{1}{6}, \frac{3}{5}, \frac{5}{6}, \frac{3}{7}$

6. Read. 7.72   0.772   7.072
7. Divide 714 by 7 and add 8 to the quotient.
8. $(\frac{1}{4} + \frac{1}{4}) - (6 \times 0)$

In a box there are 4 red pencils, 5 blue pencils, and 3 yellow pencils. Find:

9. P (red)
10. P (blue or yellow)
11. P (pencil)
12. P (green)
13. What fractional part of a dozen is 10?
14. The amount $.48 is equal to __?__ quarter, __?__ dimes, and 3 pennies.
15. At 3 for $.45, what will $1\frac{1}{2}$ dozen apples cost?

Set 17

1. Express as mixed numbers. $\frac{41}{8}, \frac{45}{8}, \frac{43}{8}, \frac{47}{8},$ $\frac{49}{8}, \frac{51}{8}, \frac{57}{8}, \frac{55}{8}, \frac{59}{8}, \frac{61}{8}$

2. Identify.

3. Identify.

4. Classify each angle.

5. Tell how many congruent sides are in a square, rhombus, regular pentagon, isosceles triangle, scalene triangle.

6. What type of a polygon has: exactly 6 sides? exactly 8 sides?
7. What angle is formed by the hands of a clock set at 6:00?
8. In a triangle FGH, $\angle F = 80°$ and $\angle G = 60°$. What does $\angle H$ measure in degrees?
9. In quadrilateral $ABCD$, $\angle A = 70°$, $\angle B = 95°$, $\angle C = 50°$. What does $\angle D$ measure?
10. The playground circle has a radius of 7 ft. What is its diameter?
11. Which has the greater diameter, a saucer or a dinner plate?
12. Are these polygons congruent or similar?

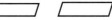

13. How many lines of symmetry are in a regular hexagon?
14. Turning a card from one side to another is a slide, flip, or turn?
15. Does the letter G have a line of symmetry?

Set 18

1. Classify each triangle as acute, obtuse, or right.

2. Classify each quadrilateral.

3. Solve. $\frac{1}{4} + \frac{1}{4}, \frac{1}{4} - \frac{1}{4}, \frac{1}{4} \times \frac{1}{4}, \frac{1}{4} \div \frac{1}{4}$

4. Find $\frac{1}{2}$ of: 8, 12, 2, 16, 14, 20, 4, 18, 6

5. $\frac{1}{8} = \frac{?}{72} = \frac{?}{48} = \frac{?}{64} = \frac{?}{32} = \frac{?}{24} = \frac{?}{56} = \frac{?}{40}$

6. In the numeral 6457.029, what is the value of 9?
7. Round 8325 to the nearest thousand.
8. Simplify $\frac{32}{56}$ and add 2 to the result.
9. Multiply 1.03 by 10.
10. Line segments or angles that have the same measure are __?__ .
11. Jessie walks to school down Street A and crosses over Street B. So Street A and Street B must __?__ .
12. A field is shaped like a rhombus. If one angle measures 150°, what are the measures of the other 3 angles?
13. The diameter of a solar disk is 11 m. What is the radius?
14. 12 is $\frac{3}{4}$ of what number?
15. How many days are 3 days less than 2 weeks?

1. Complete. 7 km = __?__ m   5 km = __?__ mm
   3 m = __?__ mm   0.12 cm = __?__ dm

2. Compare. Use <, =, or >. 60 L __?__ 6 mL
   13 L __?__ 130 mL   5.3 L __?__ 53 cL

3. Name the best unit of measure, t, kg, or g.
   a feather   a car   3 raisins   a dog
   an elephant   a baby   a penny   a pencil

4. 15 ft = __?__ yd   84 in. = __?__ ft
   27 yd = __?__ ft   2 mi = __?__ yd
   7 ft = __?__ in.   2 mi = __?__ ft

5. 2 T = __?__ lb   8 pt = __?__ qt
   8 oz = __?__ lb   2 gal = __?__ pt
   3 c = __?__ fl oz   12 c = __?__ qt

6. From 2 gal 1 qt, take 3 qt.

7. Multiply 3 ft 6 in. by 2.

8. The mass of 2 bananas is 0.34 kg. How many grams is that?

9. A ribbon 3 m long was cut into pieces each 25 cm long. How many pieces were cut?

10. The fish tank contains 26 L of water. How many mL of water does it hold?

11. 3 dm^3 = __?__ L = __?__ kg

12. Divide 0.36 by 4.

13. How many grams are there in 3 kilograms?

14. Find the perimeter of a rectanglar rug 3 m by 4 m.

15. Dividing 20 by $\frac{1}{2}$ is the same as multiplying 20 by __?__ .

1. Simplify. $\frac{2}{4}, \frac{4}{8}, \frac{8}{16}, \frac{4}{12}, \frac{4}{24}, \frac{8}{32}, \frac{4}{22}, \frac{16}{32}, \frac{8}{24}$

2. Multiply by 4: 0.03, 0.02, 0.05, 0.06, 0.08

3. 0.2 − 0.1, 3 − 0.2, 0.4 − 0.1, 5 − 0.1, 4 − 0.6, 0.6 − 0.1, 0.8 − 0.3

4. Add $\frac{5}{7}$ to: $\frac{1}{7}, \frac{3}{7}, \frac{5}{7}, \frac{6}{7}, \frac{4}{7}, \frac{2}{7}, 1\frac{2}{7}, 1\frac{1}{7}$

5. Divide by $\frac{1}{2}$: 4, 7, 9, 2, 5, 6, 1, 10, 3

6. Find the perimeter of an equilateral triangle that is 8.9 cm on each side.

7. How many square meters of tile flooring are needed to cover a floor measuring 9 m by 12 m?

8. A square trampoline measures 8 yd on each side. How many square yards of material are needed to cover it?

9. Find the area of a triangular flag that has a base of 0.5 m and a height of 2 m.

10. Find the circumference of a circular tablecloth with diameter 5 ft.

11. A circle has a diameter of 12 in. Use 3 for π and estimate the circumference.

12. Find the volume of a closet that measures 2 ft long, 4 ft wide, and 7 ft high.

13. Find the volume of a cube that measures 6 cm on each side.

14. A train due at 8:20 A.M. is 30 minutes late. At what time will it arrive?

15. It is 6:55. What time will it be $1\frac{3}{4}$ h from now?

1. Simplify. $1\frac{60}{100}, 1\frac{5}{100}, 1\frac{4}{100}, 1\frac{2}{100}, 1\frac{75}{100}$

2. Divide by $\frac{1}{3}$: 3, 5, 8, 10, 1, 4, 7, 2, 6, 9

3. $\frac{1}{50} = \frac{?}{100}$ $\frac{3}{50} = \frac{?}{100}$ $\frac{7}{50} = \frac{?}{100}$ $\frac{9}{50} = \frac{?}{100}$

4. Add 0.4 to: 1.3, 1.5, 1.2, 1.4, 1.9, 1.8, 1.0, 1.1

5. Multiply by 10; by 100: 0.34, 0.04, 0.21, 0.16, 0.32, 0.27, 0.18, 0.03, 0.07

6. Give the simplest form of the ratio:
   1 ounce to 1 pound; 1 yard to 1 inch.

7. At $1.44 a dozen, what will 4 rolls cost?

8. The ratio of baseball cards to basketball cards is 3 to 2, and the ratio of football cards to baseball cards is 1 to 2. If Ken has 36 baseball cards, how many basketball cards does he have?

9. To $3\frac{4}{7}$ add $2\frac{3}{7}$.

Complete each pattern.

10. 21, 18, 19, 16, 17, 14, 15, __?__, __?__

11. $\frac{1}{16}, \frac{1}{8}, \frac{1}{4}, \frac{1}{2}, 1, 2, \frac{?}{}, \frac{?}{}$

12. 0.2, 0.4, 0.6, 0.8, 1, __?__, __?__

13. At the rate of 200 per hour, how many envelopes can be filled in $3\frac{1}{2}$ hours?

14. A poodle eats 6 cans of food every 4 days. A collie eats 18 cans every 6 days. Do the two dogs eat food at the same rate?

15. Multiply 0.004 by 10.

1. Express each ratio in simplest form. 6 to 10
8 to 20   7 to 21   9 : 15   12 : 28

2. Complete. $\frac{15}{20} = \frac{3}{?}$ $\frac{12}{6} = \frac{?}{1}$ $\frac{14}{18} = \frac{?}{9}$
$\frac{32}{48} = \frac{?}{6}$ $\frac{27}{9} = \frac{3}{?}$ $\frac{15}{9} = \frac{?}{3}$ $\frac{8}{16} = \frac{?}{2}$

3. Are the ratios = or ≠? $\frac{6}{7}$? $\frac{18}{21}$
$\frac{5}{8}$? $\frac{25}{30}$ $\frac{1}{2}$? $\frac{7}{21}$ $\frac{4}{1}$? $\frac{16}{1}$

4. Find the value of n. 1 : 3 = n : 15
5 : n = 10 : 12 n : 1 = 4 : 4
12 : 11 = 24 : n

5. Give each as a percent. $\frac{26}{100}, \frac{7}{100}, \frac{68}{100},$
$\frac{57}{100}$, 0.41, 0.03, 0.75, 0.53, 0.39, 0.97

6. Of 100 tickets, 23 were given away free. What percent of the tickets were free?

7. On a scale drawing of a zoo, 1 in. = 12 ft. If the scale distance from the lion's den to the monkey house is 2.5 in., what is the actual distance?

8. Express 0.25 as a fraction in simplest form.

9. Write as a decimal: 75% of a class

10. Express $\frac{1}{3}$ as a percent.

11. How many seconds are there in 3 hours?

12. If golf balls sell 3 for $5.00, what is the cost of a dozen?

13. What percent expresses 10% less than 100%?

14. Two times a number is what percent of it?

15. The value of a bike is $\frac{1}{2}$ of its value when it was purchased. Express this as a percent.

1. Express as a fraction in simplest form. 25%, 50%, 75%, 20%, 40%, 60%, 10%

2. Express as a decimal. 25%, 16.2%, 3%, 82.36%, 45.9%, 6.24%, 33%, 19.8%

3. Express as a percent. 0.04, 0.02, 0.01, 0.09, 0.259, 0.17, 0.36, 0.438, 0.55, 0.623

4. Express as a percent. $\frac{1}{4}, \frac{1}{5}, \frac{1}{20}, \frac{1}{25}, \frac{1}{50}, \frac{2}{5}, \frac{1}{8},$
$\frac{4}{5}, \frac{7}{50}, \frac{5}{50}$

5. Express as a percent. 1.06, 1.08, 1.09, 1.6, 1.72, 2.5, 1.24, 2.35, 3.64

6. In a basket containing 160 apples, 20% have stems. How many have stems?

7. Express 37.5% as a fraction.

8. To 50% of 18 add 10.

9. Sally spelled 70% of 30 spelling words correctly. How many words did she spell correctly?

10. What percent is equal to $\frac{3}{50}$?

11. Write 105% as a decimal.

12. (50% of 6) + (50% of 12) = _?_

13. Marc planted 25 flower plants and 20% of them died. How many plants lived?

14. (25% of 48) ÷ 3 = _?_

15. A VCR costs $250. How much is saved if it is on sale for 20% off?

1. Express as a mixed number in simplest form. 120%, 250%, 320%, 110%, 480%

2. 5 = n% of 20   20 = n% of 80
10 = n% of 25   2 = n% of 10
40 = n% of 80   16 = n% of 100

3. Find 25% of: 24, 40, 56, 72, 48, 32, 64, 16

4. Express as a decimal. 10%, 20%, 30%, 15%, 25%, 5%, 4%, 2%, 1%, 8%

5. Find 40% of: 210, 320, 400, 300, 410, 220

6. Forty-two of 60 sixth-grade students ride the bus. What percent ride the bus?

7. A bicycle is on sale for $105. The sales tax rate is 6%. Find the sales tax.

8. Mr. Budd sold $15,000 worth of roses in one month. His rate of commission is 5%. What was his commission for the month?

9. If 10% of a number is 15, what is 30% of the number?

10. Of the 500 cars in the parking garage, 150 are on the first level. What percent of the cars are on the first level?

11. Divide 0.0081 by 0.0009.

12. Dresses were on sale for $10 off the original price of $60. What was the rate of discount?

13. Find the commission on sales of $2700 if the rate of commission is 3%.

14. Express $\frac{3}{5}$ as a decimal.   15. Find $\frac{8}{9}$ of 72.

1. Name the opposite of: $^+11$, $^-8$, $^+15$, $^-3$, $^+24$, $^-1$, $^+5$, $^+17$, $^-13$, $^-20$, $^+6$, 0
2. Compare. Use $<$, $=$, or $>$. $^-3$ _?_ $^+2$
   $^+6$ _?_ $^+11$ $^-5$ _?_ $^+1$ $^+8$ _?_ $^-8$
   $^-4$ _?_ $^-2$
3. Order from least to greatest. $^+5$, $^-4$, $^-1$; $^+10$, $^+7$, $^+4$; $^-2$, 0, $^-6$; $^+8$, $^+11$, $^-3$
4. $^+3 + {}^+6$ $^+10 + {}^+8$ $^-7 + {}^-1$
5. $^-7 + {}^+4$ $^+2 + {}^-0$ $^-12 + {}^-6$
   $^+9 + {}^-9$ $^-5 + {}^+4$
6. The temperature outside was $^-6°$. The wind made it feel $20°$ colder. What was the windchill temperature?
7. Find the sum. $^-3 + ({}^+2 + {}^+5)$

8. A store's profits for the month were: 35% for furniture sales, 20% for home appliances, 10% for clothing, and 5% for shoes. The remainder of the profits came from toys. What percent are from toys?
9. In Jan. Matt lost 6 lb. He gained 2 lb in Feb. and lost 3 lb in Mar. What was his total weight gain or loss?
10. Express 65% as a fraction.
11. Multiply 0.724 by 1000.
12. Give 1492 as a Roman numeral.
13. Divide 1020 by 5, and take 4 from the result.
14. Express 75% as a fraction.
15. The temperature went from $^+11°C$ to $^-8°C$ during the day. How many degrees did the temperature drop?

1. $^-7 - {}^-5$ $^+4 - {}^-9$ $^-6 - {}^-9$
   $^+10 - {}^+12$ $^+3 - {}^-11$ $^-13 - {}^+4$
   $^+17 - {}^+8$ $^-8 - {}^+4$
2. Find 50% of: 8, 12, 2, 10, 16, 20, 4, 14, 24
3. Express as a fraction. 20%, 25%, 50%, 75%, 80%, 15%, 10%, 5%, 60%, 35%
4. $1\frac{1}{4} + \frac{3}{4}$ $2\frac{5}{6} + \frac{1}{6}$ $2\frac{1}{2} + \frac{1}{4}$
5. Find $\frac{1}{6}$ of: 12, 30, 54, 42, 72, 48, 36, 18, 24
6. A motorboat can go 7.8 mph. How far will it go in 5 hours?
7. Write as a ratio: 3 quarters, 1 nickel to 3 dimes, 2 nickels.

8. Ella put $160 into savings. She withdrew $49. How much is left in savings?
9. In one game Ned won 9 points, lost 4 points, lost 2 points, won 7 points, and won 3 points. What was his final score?
10. 40% of 75 questions are essay. How many questions are essay?
11. 90 is $\frac{3}{4}$ of what number?
12. $(\frac{3}{5} - \frac{3}{5}) + (\frac{3}{5} \div \frac{3}{5}) = $ _?_
13. A batter has been at bat 27 times and has had 9 hits. What is his batting average?
14. Multiply 0.02 by 0.06.
15. How many pieces of wire $\frac{3}{4}$ yd long can be cut from 6 yards?

1. True or false if $x = 7$: $6 + x = 13$
   $x - 6 = 13$ $3x = 27$ $56 \div x = 8$
2. True or false if $n = 8$: $n \div 4 = 4$
   $\frac{n}{2} = 4$ $80 = 10n$ $\frac{1}{2} n = 16$
3. True or false if $n = 3$: $2n - 1 = 5$
   $\frac{n}{2} - 1 = 5$ $1 + \frac{n}{2} = 5$ $1 + 2n = 7$
4. Choose the equations. $42 \div x$ $\frac{x}{4} = 3$
   $2x - 6$ $\frac{x}{3} - 9 = 10$
5. $d = 6$ The value of $3 + 10d$ is _?_
6. 12 more than a number: $12n$ or $\frac{n}{12}$ or $n + 12$

7. Product of a number and 20: $p - 20$ or $\frac{p}{20}$ or $20p$
8. r divided by 6 is 5: $\frac{r}{6} = 5$ or $r - 6 = 5$ or $6r = 5$
9. 4 less than a number is 7: $x + 4 = 7$ or $4 - x = 7$ or $x - 4 = 7$
10. Letters a, x, n are _?_ .
11. An equation states that two expressions are _?_ .
12. Which operation solves $n + 33 = 96$?
13. Which operation solves $14n = 56$?
14. What is the value of x? $x + 17 = 39$
    $x - 22 = 50$ $9 + x = 44$
15. What is the value of n? $9n = 54$
    $\frac{n}{4} = 22$ $8 = \frac{n}{4}$

acute angle   An angle that measures less than 90°. (p. 302)

acute triangle   A triangle with three acute angles. (p. 308)

addend   Any one of a set of numbers to be added.

algebraic expression   A mathematical expression that contains one or more variables. (p. 474)

a.m.   Abbreviation that indicates time from midnight to noon.

angle   A figure formed by two rays that have a common endpoint.

area   The number of square units needed to cover a flat surface.

associative (grouping) property   Changing the grouping of the addends (or factors) does not change the sum (or product). (pp. 44, 64)

average   The quotient obtained by dividing a sum by the number of addends. Also called *mean*.

axis   The horizontal or vertical number line of a graph. (p. 458)

bar graph   A graph that uses bars to show data. The bars may be of different lengths. (p. 17)

base   One of the equal factors in a product; a selected side or face of a geometric figure. (pp. 72, 354)

BASIC   An acronym for Beginner's All-purpose Symbolic Instruction Code; a computer language used to process information.

benchmark   An object of known measure used to estimate the measure of other objects.

bisect   To divide a line segment or an angle into two congruent parts. (p. 305)

cancellation   The dividing of any numerator and denominator of a set of fractions by their greatest common factor before multiplying.

capacity   The amount, usually of liquid, a container can hold.

Celsius (°C) scale   The temperature scale in which 0°C is the freezing point of water and 100°C is the boiling point of water.

central angle   An angle whose vertex is the center of a circle. (p. 312)

chord   A line segment with both endpoints on a circle. (p. 312)

circle   A simple closed curve in which all the points are the same distance from a point called the *center.*

circle graph   A graph that uses the area of a circle to show the division of a total amount of data. (p. 276)

circumference   The distance around a circle.

commission   Money earned equal to a percent of the selling price of items sold. (p. 428)

common denominator   A number that is a multiple of the denominators of two or more fractions.

common factor   A number that is a factor of two or more numbers.

common multiple   A number that is a multiple of two or more numbers.

commutative (order) property   Changing the order of the addends (or factors) does not change the sum (or product). (pp. 44, 64)

compass   An instrument used to draw circles. (p. 313)

compatible numbers   Numbers that are easy to compute with mentally. (p. 136)

complementary angles   Two angles the sum of whose measures is 90°. (p. 303)

complex fraction   A fraction whose numerator or denominator (or both) is a fraction or a mixed number. (p. 253)

composite number   A whole number greater than 1 that has more than two factors. (p. 168)

cone   A space, or solid, figure with one circular base, one vertex, and a curved surface. (p. 314)

congruent figures   Figures that have the same size and shape. (p. 316)

conjunction   A compound statement formed by joining two statements with the connective *and*. (p. 373)

coordinate plane   A grid divided into four quadrants used to locate points by naming ordered pairs. (p. 460)

coordinates   An ordered pair of numbers used to locate a point on a grid. (p. 458)

cross products   The products obtained by multiplying the numerator of one fraction by the denominator of a second fraction and the denominator of the first fraction by the numerator of the second fraction. (p. 382)

cube   A space, or solid, figure with six congruent square faces. (p. 314)

customary system   The measurement system that uses inch, foot, yard, and mile; fluid ounce, cup, pint, quart, and gallon; ounce, pound, and ton. (See *Table of Measures*, pp. 547–548.)

cylinder   A space, or solid, figure with two parallel, congruent circular bases and a curved surface. (p. 314)

data   Facts or information.

database   A group of facts and figures that are related and can be arranged in different ways.

decagon   A polygon with ten sides. (p. 307)

decimal   A number with a decimal point separating the ones from the tenths place.

decimal point   A point used to separate ones and tenths in decimals.

degree (°)   A unit used to measure angles; a unit used to measure temperature on the Celsius (°C) or the Fahrenheit (°F) scale.

denominator   The number below the bar in a fraction.

diagonal   A line segment, other than a side, that joins two vertices of a polygon. (p. 307)

diameter   A line segment that passes through the center of a circle and has both endpoints on the circle. (p. 312)

difference   The answer in subtraction.

digit   Any one of the numerals 0, 1, 2, 3, 4, 5, 6, 7, 8, or 9.

discount   A reduction on the regular, or list, price of an item.

disjunction   A compound statement formed by joining two statements with the connective *or*. (p. 373)

distributive property   Multiplying a number by a sum is the same as multiplying the number by each addend of the sum and then adding the products. (p. 65)

dividend   The number to be divided. (p. 13)

divisible   A number is divisible by another number if the remainder is 0 when the number is divided by the other number.

divisor   The number by which the dividend is divided. (p. 13)

double bar graph   A graph that uses pairs of bars to compare two sets of data. (p. 275)

double line graph   A graph that uses pairs of line segments to compare two sets of data. (p. 274)

edge   The line segment where two faces of a space figure meet. (p. 314)

END   A BASIC command that tells the computer it has reached the end of a program.

endpoint   The point at the end of a line segment or ray.

equation   A statement that two mathematical expressions are equal. (p. 475)

equilateral triangle   A triangle with three congruent sides and three congruent angles. (p. 308)

equivalent fractions   Fractions that name the same amounts. (p. 164)

estimate   An approximate answer; to find an answer that is close to the exact answer.

evaluate   To find the number that an algebraic expression names. (p. 474)

even number   A whole number divisible by 2.

event   A set of one or more outcomes of a probability experiment.

expanded form   The written form of a number that shows the place value of each of its digits. (p. 40)

exponent   A number that tells how many times another number is to be used as a factor. (p. 72)

face   A flat surface of a space figure. (p. 314)

factor   One of two or more numbers that are multiplied to form a product.

factor tree   A diagram used to find the prime factors of a number. (p. 168)

Fahrenheit (°F) scale   The temperature scale in which 32°F is the freezing point of water and 212°F is the boiling point of water.

flip   To turn a figure to its reverse side. (p. 318)

formula   A rule that is expressed by using symbols. (p. 350)

FOR. . .NEXT loop   BASIC statements that tell the computer to repeat program lines a given number of times.

fraction   A number that names a part of a whole, a region, or a set.

frequency table   A chart that shows how often each item appears in a set of data. (p. 262)

front-end estimation   A way of estimating by using the front, or greatest, digits to find an approximate answer.

GOTO   A statement in BASIC that tells the computer to branch to a specific line.

greatest common factor (GCF)   The greatest number that is a factor of two or more numbers. (p. 170)

grid   A network of perpendicular lines used to locate points.

half turn   A turn that causes a figure to face in the opposite direction; a turn of 180°. (p. 318)

half-turn symmetry   The symmetry that occurs when a figure is turned halfway (180°) around its center point and the figure that results looks exactly the same. (p. 327)

height   The perpendicular distance between the bases of a geometric figure. In a triangle, the perpendicular distance from the opposite vertex to the line containing the base. (p. 354)

heptagon   A polygon with seven sides. (p. 307)

hexagon   A polygon with six sides. (p. 22)

hexagonal prism   A prism with two parallel hexagonal bases. (p. 315)

hexagonal pyramid   A pyramid with a hexagonal base. (p. 314)

identity property   Adding 0 to a number or multiplying a number by 1 does not change the number's value. (pp. 44, 64)

improper fraction   A fraction with its numerator equal to or greater than its denominator. (p. 174)

inequality   A statement that two mathematical expressions are not equal. It uses an inequality symbol: $<$, $>$, or $\neq$.

INPUT   A BASIC command that tells the computer to wait for a response from the user.

integers   The whole numbers and their opposites. (p. 442)

intersecting lines   Lines that meet or cross. (p. 21)

interval   The number of units between spaces on a graph's scale.

inverse operations   Mathematical operations that *undo* each other, such as addition and subtraction or multiplication and division.

isosceles triangle   A triangle with two congruent sides and two congruent angles. (p. 308)

least common denominator (LCD)   The least common multiple of the denominators of two or more fractions. (p. 180)

least common multiple (LCM)   The least number, other than 0, that is a common multiple of two or more numbers. (p. 178)

line   A set of points in order extending indefinitely in opposite directions. (p. 20)

linear measure   A measure of length.

line graph   A graph that uses points on a grid connected by line segments to show data. (p. 17)

line of symmetry   A line that divides a figure into two congruent parts. (p. 327)

line segment   A part of a line that has two endpoints. (p. 20)

lowest terms   A fraction is in lowest terms when its numerator and denominator have no common factors other than 1. (p. 172)

mass   The measure of the amount of matter an object contains.

mathematical expression A symbol or a combination of symbols that represents a number.

mean The average of a set of numbers. (p. 264)

median The middle number of a set of numbers arranged in order. If there is an even number of numbers, the median is the average of the two middle numbers. (p. 264)

metric system The measurement system based on the meter, gram, and liter. (See *Table of Measures,* pp. 547–548.)

minuend A number from which another number is subtracted. (p. 50)

mixed number A number that is made up of a whole number and a fraction. (p. 174)

mode The number that appears most frequently in a set of numbers. (p. 264)

multiple A number that is the product of a given number and any whole number. (p. 178)

negation The denial of a given statement. (p. 223)

net A flat pattern that folds into a space figure. (p. 314)

number sentence An equation or an inequality.

numeral A symbol for a number.

numerator The number above the bar in a fraction.

numerical expression A mathematical expression that contains only numbers.

obtuse angle An angle with a measure greater than 90° and less than 180°. (p. 302)

obtuse triangle A triangle with one obtuse angle. (p. 308)

octagon A polygon with eight sides. (p. 22)

odd number A whole number that is not a multiple of 2.

ordered pair A pair of numbers that is used to locate a point on a coordinate grid. (p. 456)

order of operations The order in which operations must be performed when more than one operation is involved. (p. 82)

outcome The result of a probability experiment.

parallel lines Lines in a plane that never intersect. (p. 298)

parallelogram A quadrilateral with two pairs of parallel sides. (p. 310)

pentagon A polygon with five sides. (p. 22)

pentagonal prism A prism with two parallel pentagonal bases. (p. 314)

pentagonal pyramid A pyramid with a pentagonal base. (p. 315)

percent (%) The ratio or comparison of a number to 100. (p. 392)

perimeter The distance around a figure.

period A set of three digits set off by a comma in a whole number.

perpendicular lines Lines that intersect to form right angles. (p. 298)

pi (π) The ratio of the circumference of a circle to its diameter. An approximate value of π is 3.14, or $\frac{22}{7}$. (p. 358)

pictograph A graph that uses pictures or symbols to show data. (p. 18)

place value The value of a digit depending on its position, or place, in a number.

plane A flat surface that extends indefinitely and has no thickness. (p. 20)

plane figure A two-dimensional figure that has straight or curved sides.

p.m. Abbreviation that indicates time from noon to midnight.

point An exact location, or position, usually represented by a dot.

polygon A simple closed figure with sides that are line segments. (p. 22)

polyhedron A space figure whose faces are polygons. (p. 314)

prime factorization Expressing a composite number as the product of prime numbers. (p. 168)

prime number A whole number greater than 1 that has only two factors, itself and 1. (p. 166)

PRINT A BASIC statement that tells the computer what to display on the screen.

prism A space figure with two faces called *bases* bounded by polygons that are parallel and congruent. (p. 314)

probability   A branch of mathematics that analyzes the chance that a given outcome will occur. The probability of an event is expressed as the ratio of a given outcome to the total number of outcomes possible. (p. 278)

product   The answer in multiplication.

proportion   A number sentence that shows that two ratios are equal. (p. 382)

protractor   An instrument used to measure angles. (p. 300)

pyramid   A space figure whose base is a polygon and whose faces are triangles with a common vertex. (p. 314)

quadrilateral   A polygon with four sides. (p. 22)

quotient   The answer in division.

radius (plural *radii*)   A line segment from the center of a circle to a point on the circle. (p. 312)

range   The difference between the greatest and least numbers in a set of numbers. (p. 264)

rate   A ratio that compares unlike quantities. (p. 380)

rate of commission   The percent of the total amount of goods or services sold that is earned by the seller. (p. 428)

rate of discount   The percent taken off the original, or list, price. (p. 422)

rate of sales tax   The percent of the list, or marked, price levied as tax. (p. 424)

ratio   A comparison of two numbers or quantities by division. (p. 376)

ray   A part of a line that has one endpoint and goes on forever in one direction. (p. 20)

READ   A statement in BASIC that takes entries one by one from a DATA statement and assigns each entry to a variable.

reciprocals   Two numbers whose product is 1. (p. 236)

rectangle   A parallelogram with four right angles. (p. 310)

rectangular prism   A prism with six rectangular faces. (p. 314)

rectangular pyramid   A pyramid with a rectangular base. (p. 314)

reflection   The new figure obtained by flipping a figure. (p. 318)

regular polygon   A polygon with all sides and all angles congruent.

regular price   The original, marked, or list price of an item before a discount has been given.

REM   A statement in BASIC that describes what a program or section of a program will do; a remark.

remainder   The number left over when division is completed.

repeating decimal   A decimal with digits that from some point on repeat indefinitely. (p. 190)

rhombus   A parallelogram with all sides congruent. (p. 310)

right angle   An angle that measures 90°. (p. 302)

right triangle   A triangle with one right angle. (p. 308)

Roman numerals   Symbols for numbers used by the Romans. (p. 61)

rounding   To approximate a number by replacing it with a number expressed in tens, hundreds, thousands, and so on.

RUN   A BASIC command that tells the computer to process a program.

sale price   A lowered price on an item, obtained by subtracting the discount from the regular price. (p. 422)

sales tax   The amount added to the marked price of an item and collected as tax. (p. 424)

scale   The ratio of a pictured measure to the actual measure; the tool used to measure weight; numbers along the side or bottom of a graph.

scale drawing   A drawing of something accurate but different in size.

scalene triangle   A triangle with no congruent sides. (p. 308)

sequence   A set of numbers given in a certain order. Each number is called a *term*. (p. 159)

set   A collection or group of numbers or objects.

side A line segment that forms part of a polygon; a ray that forms one part of an angle.

similar figures Figures that have the same shape. They may or may not be the same size. (p. 316)

simple closed curve A path that begins and ends at the same point and does not intersect itself.

simplest form The form of a fraction when the numerator and denominator have no common factor other than 1. (p. 172)

slide To obtain another figure by moving every point of a figure the same distance and in the same direction.

space (or solid) figure A three-dimensional figure that has volume.

sphere A curved space figure in which all points are the same distance from a point called the *center.* (p. 314)

spreadsheet A computer program that arranges data and formulas in a grid of cells.

square A rectangle with all sides congruent.

square pyramid A pyramid with a square base. (p. 314)

standard form The usual way of writing a number using digits.

statistics The study of the collection, interpretation, and display of data.

straight angle An angle that measures 180°. (p. 302)

subtrahend A number that is subtracted from another number. (p. 50)

supplementary angles Two angles the sum of whose measures is 180°. (p. 303)

surface area The sum of the areas of all the faces of a space figure.

symmetrical figure A plane figure that can be folded on a line so that the two halves are congruent. (p. 327)

terminating decimal A decimal in which digits do not show a repeating pattern. A terminating decimal results when the division of the numerator of a fraction by the denominator leaves a 0 remainder. (p. 190)

tessellation The pattern formed by fitting plane figures together without overlapping or leaving gaps. (p. 320)

translation The new figure formed by sliding a figure without flipping or turning it. (p. 318)

trapezoid A quadrilateral with only one pair of parallel sides. (p. 310)

tree diagram A diagram that shows all possible outcomes of an event or events. (p. 280)

triangle A polygon with three sides. (p. 22)

triangular prism A prism with two parallel triangular bases. (p. 314)

triangular pyramid A pyramid with a triangular base. (p. 314)

turn To obtain a new figure by rotating a figure around a point into a different position.

unit fraction A fraction with a numerator of 1. (p. 235)

unit price The cost of one item.

variable A symbol, usually a letter, used to represent a number. (p. 474)

Venn diagram A drawing that shows relationships among sets of numbers or objects. (p. 183)

vertex (plural *vertices*) The common endpoint of two rays in an angle, of two line segments in a polygon, or of three or more edges in a space figure.

volume The number of cubic units needed to fill a space figure.

weight The heaviness of an object.

whole number Any of the numbers 0, 1, 2, 3, ...

x-axis The horizontal number line on a coordinate grid.

y-axis The vertical number line on a coordinate grid.

zero property Multiplying a number by 0 always results in a product of 0. (p. 64)

537

$=$	is equal to	$\overleftrightarrow{AB}$	line AB
$\neq$	is not equal to	$\overline{AB}$	segment AB
$<$	is less than	$\overrightarrow{AB}$	ray AB
$>$	is greater than	$\angle ABC$	angle ABC
$\approx$	is approximately equal to	ABC	plane ABC
...	continues without end	$\sim$	is similar to
$+$	plus	$\cong$	is congruent to
$-$	minus	$\parallel$	is parallel to
$\times$	times	$\perp$	is perpendicular to
$\div$	divided by	π	pi
$\$$	dollars	cm^2	square centimeter
$\cancel{c}$	cents	$in.^3$	cubic inch
$\%$	percent	$\circ$	degree
$0.\overline{3}$	0.333...(repeating decimals)	$2:3$	two to three (ratio)
(3, 4)	ordered pair	$P(E)$	probablilty of an event
$\cdot$	decimal point		

Perimeter
Rectangle: $P = 2(\ell + w)$
Square: $P = 4s$

Area
Rectangle: $A = \ell w$
Square: $A = s^2$
Parallelogram: $A = bh$
Triangle: $A = \frac{1}{2}bh$
Circle: $A = \pi r^2$

Circumference of Circle
$C = \pi d = 2\pi r$

Surface Area
Rectangular Prism:
$S = 2(\ell w + \ell h + wh)$
Cube: $S = 6e^2$

Volume
Prism (general formula): $V = B \times h$
Rectangular Prism: $V = (\ell \times w) \times h$
Triangular Prism: $V = (\frac{1}{2} \times b \times h) \times h$
Cylinder: $V = (\pi \times r^2) \times h$
Cube: $V = e^3$

Distance = Rate $\times$ Time: $d = r \times t$

Discount = List Price $\times$ Rate of Discount: $D = LP \times R$ of D

Sale Price = Regular Price $-$ Discount: $SP = RP - D$

Sales Tax = Marked Price $\times$ Rate of Sales Tax: $T = MP \times R$ of T

Total Cost = Marked Price $+$ Sales Tax: $TC = MP + T$

Commission = Total Sales $\times$ Rate of Commission: $C = TS \times R$ of C

Time

60 seconds (s)	= 1 minute (min)	52 weeks	= 1 year
60 minutes	= 1 hour (h)	365 days	= 1 year
24 hours	= 1 day (d)	366 days	= 1 leap year
7 days	= 1 week (wk)	100 years	= 1 century (cent.)
12 months (mo)	= 1 year (y)		

Metric Units

Length

1000 millimeters (mm) = 1 meter (m)
100 centimeters (cm) = 1 meter
10 decimeters (dm) = 1 meter
10 meters = 1 dekameter (dam)
100 meters = 1 hectometer (hm)
1000 meters = 1 kilometer (km)

Capacity

1000 milliliters (mL) = 1 liter (L)
100 centiliters (cL) = 1 liter
10 deciliters (dL) = 1 liter
10 liters = 1 dekaliter (daL)
100 liters = 1 hectoliter (hL)
1000 liters = 1 kiloliter (kL)

Mass

1000 milligrams (mg) = 1 gram (g)
100 centigrams (cg) = 1 gram
10 decigrams (dg) = 1 gram

10 grams = 1 dekagram (dag)
100 grams = 1 hectogram (hg)
1000 grams = 1 kilogram (kg)

1000 kg = 1 metric ton (t)

Customary Units

Length

12 inches (in.) = 1 foot (ft)
3 feet = 1 yard (yd)
36 inches = 1 yard
5280 feet = 1 mile (mi)
1760 yards = 1 mile

Capacity

8 fluid ounces (fl oz) = 1 cup (c)
2 cups = 1 pint (pt)
2 pints = 1 quart (qt)
4 quarts = 1 gallon (gal)

Weight

16 ounces (oz) = 1 pound (lb)   2000 pounds = 1 ton (T)

$1\% = \frac{1}{100}$	$50\% = \frac{1}{2}$	$12\frac{1}{2}\% = \frac{1}{8}$	$87\frac{1}{2}\% = \frac{7}{8}$
$10\% = \frac{1}{10}$	$60\% = \frac{3}{5}$	$25\% = \frac{1}{4}$	$16\frac{2}{3}\% = \frac{1}{6}$
$20\% = \frac{1}{5}$	$70\% = \frac{7}{10}$	$37\frac{1}{2}\% = \frac{3}{8}$	$33\frac{1}{3}\% = \frac{1}{3}$
$30\% = \frac{3}{10}$	$80\% = \frac{4}{5}$	$62\frac{1}{2}\% = \frac{5}{8}$	$66\frac{2}{3}\% = \frac{2}{3}$
$40\% = \frac{2}{5}$	$90\% = \frac{9}{10}$	$75\% = \frac{3}{4}$	$83\frac{1}{3}\% = \frac{5}{6}$